INSTRUCTOR EDITION

WELCOME

At Course Technology, a part of Cengage Learning, our mission is to help people teach and learn about technology. This special Instructor Edition is all about helping YOU teach. We know you have a great deal to manage, and for this reason, we provide you with excellent resources to go along with every book. This Instructor Edition will save you time by helping you sort through all of this material and will enable you to choose the right combination of tools to help you teach the way you want to teach.

CONTENTS AT A GLANCE

INSTRUCTOR RESOURCES

To effectively teach your class, you need more than great textbooks. That is why we strive to provide you with outstanding instructor resources—developed by educators for educators. Our goal is to make the teaching and learning experience in your classroom the best it can be. With these resources, you can spend less time preparing and more time teaching.

INSTRUCTOR'S MANUAL

Need to compile a lecture for your class that's starting in an hour? Looking for ways to challenge your students with additional projects? Our all-new Instructor's Manual provides valuable chapter overviews; highlights key principles and critical concepts; offers sample syllabi, teaching tips, quick quizzes, learning objectives, and discussion topics; and features possible essay topics, further readings and cases, and solutions to all of the end-of-chapter questions and problems, as well as suggestions for conducting the team activities. Additional end-of-chapter questions are also included.

SAMPLE SYLLABUS

Need to keep your students on track with assignments, class policies, and due dates? A sample syllabus with sample course outlines is provided to make planning your course that much easier. The syllabi are fully customizable by you.

SOLUTIONS

Need to quickly check answers to the hands-on work at the end of each chapter? Solutions to all end-of-chapter material are provided in a separate document for your convenience, making homework corrections simpler than ever.

TEST BANK AND TEST GENERATOR

Looking for a time-efficient way to effectively test your students? ExamView® is a powerful objective-based test generator that enables instructors to create paper-, LAN-, or Web-based tests from test banks designed specifically for their Course Technology text. Instructors can utilize the ultra-efficient QuickTest Wizard to create tests in less than five minutes by taking advantage of our question banks or customizing their own exams from scratch. Page references for all questions are provided so you can cross-reference test results with the book.

POWERPOINT PRESENTATIONS

Need to deliver engaging and visually impressive lectures? A set of professionally designed Microsoft PowerPoint® slides is available for each chapter. These slides are included to serve as a teaching aid for classroom presentation, to make available to students on the network for chapter review, or to be printed for classroom distribution. You can edit the files to fit your needs. Our presentations help students focus on the main topics of each chapter, take better notes, and prepare for examinations. Instructors can also add their own slides for additional topics they introduce to the class.

FIGURE FILES

Looking for figures in the book that are not included in the PowerPoint presentation? Every figure in the textbook is available to you in electronic form. Our figure files allow you to create your own presentations using the figures taken directly from the text. If you have a personal computer and LCD device, this ancillary can be an effective tool for instruction.

BLACKBOARD AND WEBCT CONTENT

Blackboard and WebCT are the leading distance learning solutions available today. In the past few years, they've also become popular class-management platforms and include components such as Topic Reviews, Case Projects, Practice Exams, Custom Syllabi, and more. Course Technology, the premiere innovator in management information systems publishing, has partnered with Blackboard and WebCT to bring you premium online content. We are proud to present the following online courses:

- *Blackboard and WebCT Level 1 Online Content.* If you use Blackboard or WebCT, the test bank for this textbook is available at no cost in a simple, ready-to-use format. Go to *www.cengage.com/mis/stairreynolds* and search for this textbook to download the test bank.
- *Blackboard and WebCT Level 2 Online Content.* Blackboard 6.0 as well as and WebCT courses are also available for *Fundamentals of Information Systems, Fifth Edition.* Level 2 offers course management and access to a Web site that is fully populated with content for this book.

For more information on how to bring distance learning to their course, instructors should contact their Course Technology sales representative.

STUDENT COMPANION WEB SITE

We also offer an online companion for students using this book. Each student textbook comes with an individual key code that provides access to a robust Web site, located at *www.cengage.com/mis/stairreynolds*. This exciting resource includes the following features:

- **A Running Case Study:** A Running Case Study provided online includes 9 parts (one for each chapter) with 2 discussion questions and 2 critical thinking questions per part.

- **PowerPoint Slides:** Direct access is offered to the book's PowerPoint presentations, which cover the key points from each chapter. These presentations are a useful study tool.

- **Classic Cases:** A frequent request from adopters is a broader selection of cases to choose from. To meet this need, cases from the first, second, third, and fourth editions of the text are included here. These are the authors' choices of the "best cases" from these editions and span a broad range of companies and industries.

- **Links to Useful Web Sites:** Chapters in *Fundamentals of Information Systems, Fifth Edition* reference many interesting Web sites. This resource takes you to the home page links of those sites so that you can explore them. There are additional links to Web sites that the authors, Ralph Stair and George Reynolds, think you would be interested in checking out, too.

- **Hands-On Activities:** Use the Hands-On Activities to test your comprehension of IS topics and enhance your skills using Microsoft Office applications and the Internet. Using these links, you can access two critical-thinking exercises per chapter; each activity asks you to work with an Office tool or do some research on the Internet.

- **Test Yourself on IS:** This tool allows you to access 20 multiple-choice questions for each chapter, test yourself, and then submit your answers. You will immediately find out which questions you got right and which you got wrong. For each question that you answer incorrectly, you are given the correct answer and the page in your text where that information is covered. Special testing software randomly compiles 20 questions from a database of 50 questions, so you can quiz yourself multiple times on a given chapter and get some new questions each time.

- **Glossary of Key Terms:** The glossary of key terms from the text is available here.

- **Online Readings:** This feature provides you access to a computer database that contains articles relating to hot topics in information systems.

Course Technology is the world's leading IT publisher. Because we focus solely on Information Technology, we have the unique ability to address the needs of customers like you.

Find out about the latest technology trends, products, and courseware solutions on the Course Technology Web site, *www.cengage.com/mis/stairreynolds*.

Visit often to:

- Connect with your peers through our Online forums
- Learn about the latest software releases and how they will impact you in the classroom
- Browse our online catalog
- Locate and contact your local sales representative
- Register for the next Conference for Information Technology Educators and our other educational events
- Download the files that are contained on the Instructor Resources CD

Stop here before preparing for your class!

PASSWORDS AND SECURITY

INSTRUCTOR RESOURCES ON COURSE.COM

Depending on the type of Educational Institution you are affiliated with, please call one of the support services teams to obtain your password to our online Instructor Resources.

For colleges and universities in the US:
Call Course Technology at
1.800.648.7450 and select option 3 for
Support Services.

For Private Career Colleges:
Call Learning—Career at
1.800.477.3692

For High Schools:
Call Learning—School at
1.800.824.5179

**For Corporations, IT Training Centers, and
Federal Government Agencies:**
Call Course Technology at **1.800.648.7450**

Password protection is of utmost importance at Course Technology. We monitor each and every caller requesting an instructor resources password by verifying their names with the school the caller indicates as their affiliation. This ensures that students never have access to the answers, test banks, or other resources that we provide for you on course.com.

OTHER RESOURCES AND PASSWORDS

Course Technology knows that protecting these passwords from students is critical, but keeping track of all the passwords for all the tools you may adopt in a given course (one for your online instructor resources, another for SAM, another for your school's servers…) can be quite a hassle. For your convenience, we're providing space below for you to keep track of all your login information (either literally or with your own code that students won't understand!) for this course.

RESOURCE	USERNAME	PASSWORD

FEATURES

Fundamentals of Information Systems, Fifth Edition, brings a wealth of content and special features directly to your fingertips—for you and your students to explore. Here are just a few of the features that bring the exciting field of information systems to life.

Principles are highlighted at the beginning of each chapter and revisited in the summaries and self-assessment tests. These timeless, fundamental rules and guidelines reinforce concepts that all students should know and use in their careers—whatever their chosen field of study.

Detailed **Learning Objectives** break each chapter's content down into manageable learning goals so students can preview the scope of each chapter and know where they are headed.

CHAPTER · 2 ·

Hardware and Software

PRINCIPLES

- Computer hardware must be carefully selected to meet the evolving needs of the organization and its supporting information systems.

- Systems and application software are critical in helping individuals and organizations achieve their goals.

- Organizations should not develop proprietary application software unless doing so will meet a compelling business need that can provide a competitive advantage.

- Organizations should choose a programming language whose functional characteristics are appropriate for the task at hand, considering the skills and experience of the programming staff.

- The software industry continues to undergo constant change; users need to be aware of recent trends and issues to be effective in their business and personal life.

LEARNING OBJECTIVES

- Identify and discuss the role of the essential hardware components of a computer system.
- Identify the characteristics of and discuss the usage of various classes of single-user and multiuser computer systems.

- Identify and briefly describe the functions of the two basic kinds of software.
- Outline the role of the operating system and identify the features of several popular operating systems.

- Discuss how application software can support personal, workgroup, and enterprise business objectives.
- Identify three basic approaches to developing application software and discuss the pros and cons of each.

- Outline the overall evolution and importance of programming languages and clearly differentiate among the generations of programming languages.

- Identify several key software issues and trends that have an impact on organizations and individuals.

Information Systems in the Global Economy
UB Spirits, India

UB Spirits Serves up Success

The UB Group is a successful and growing international conglomerate based in Bangalore, India. The bright future of the conglomerate is illustrated by the rising steel and glass structures in the heart of Bangalore and known as UB City—a 7-acre high-rise business, shopping, and living community. The UB Group is invested in several industries including aviation, fertilizers, engineering, information technology, pharmaceuticals, and alcoholic beverages, or spirits. While this may sound like an unusual and even dangerous combination, the diverse business portfolio is making the UB Group billions of dollars.

The alcoholic beverage division of UB, called UB Spirits, is itself a successful international corporation made up of several successful distilleries: McDowell, Herbertsons, and Triumph Distillers and Vintners. Combined, these companies produce 140 brands, 15 of which are top-shelf classics. The brands are produced in 75 locations across India. UB Spirits dominates the Indian marketplace and ranks as the world's second largest distilled spirits corporation, with sales that exceed 60 million cases a year.

To comply with laws that control the distribution of alcoholic beverages in India and around the world, UB Spirits often must negotiate a selling price for their products with a national government. Because of this constraint, UB Spirits has to improve its profits by minimizing overhead and streamlining operations. One obvious waste of resources in the business was its outdated and bloated computer-based information systems. Each UB Spirits facility across India used its own information systems on its own servers—111 Microsoft Windows-based servers in all, which required constant attention. At the close of each month, the data from the disparate systems would be merged in a costly and lengthy process to produce corporate reports. It was clear to the decision makers at UB Spirits that if the company was going to grow, it would have to invest in a new system with state-of-the-art hardware.

As with many system overhauls, UB Spirits began by choosing an ERP software package before it decided on hardware. This is smart because different ERP solutions have varying hardware requirements. Working with IBM, the company selected SAP R/3, which includes information systems for every area of the business.

After selecting an ERP system, the company built the infrastructure on which the system would run. The main goal of the overhaul was to reduce the work required to maintain and administer the system. The company selected an IBM System i550 because it is powerful and simple to manage. The i550 server has multiple processors, up to 64 GB of memory, and up to 77 TB of disk storage. The i550 easily supports 400 UB Spirits users and 1.5 TB of corporate data. It can run all of the corporate systems and be accessed by 64 UB Spirits offices over a network.

By switching to a centralized server system, UB Spirits was able to replace 111 servers spread across India with one server located in its home office. The new system assisted UB Spirits in lowering manufacturing costs. It standardized business processes across the enterprise and accelerated monthly financial reporting. UB Spirits also reassigned 20 of its IT personnel from server administration to more productive projects.

UB Spirits found that the path to higher profits lay in centralizing its operations and investing in large, powerful servers. Other companies have found that they can accomplish more by distributing systems over many workstations in what is called grid computing. The hardware in which a business invests is intimately linked to the type of information

Also included in the Fifth Edition are **Why Learn About —?** boxed features. These features give students quick insight into the importance of each of the chapters and emphasize where and how students will encounter IS concepts in their everyday lives.

---

system it implements. Creating an underlying infrastructure from hardware should support the evolving needs of the business and its information systems.

As you read this chapter, consider the following:

- How does the type of hardware and software a company acquires affect the way the company operates?
- Businesses are in constant fluctuation—growing, diversifying, acquiring, and always working to reduce costs. How do these conditions and requirements affect the purchase of the firm's hardware and software?

Why Learn About Hardware and Software?

Organizations invest in computer hardware and software to improve worker productivity, increase revenue, reduce costs, and provide better customer service. Those that don't may be stuck with outdated hardware and software that is unreliable and cannot take advantage of the latest advances. As a result, obsolete hardware and software can place an organization at a competitive disadvantage. Managers, no matter what their career field and educational background, are expected to know enough about their business needs to be able to ask tough questions of those recommending the hardware and software to meet those needs. Cooperation and sharing of information between business managers and IT managers is needed to make wise IT investments that yield real business results. Managers in marketing, sales, and human resources often help IS specialists assess opportunities to apply hardware and software and evaluate the various options and features. Managers in finance and accounting especially must also keep an eye on the bottom line, guarding against overspending, yet be willing to invest in computer hardware and software when and where business conditions warrant it.

Today's use of technology is practical—it's intended to yield real business benefits, as demonstrated by UB Spirits. Employing information technology and providing additional processing capabilities can increase employee productivity, expand business opportunities, and allow for more flexibility. This chapter discusses the hardware and software components of a computer-based information system (CBIS), beginning with a definition of hardware.

hardware
Any machinery (most of which uses digital circuits) that assists in the input, processing, storage, and output activities of an information system.

Hardware consists of any machinery (most of which uses digital circuits) that assists in the input, processing, storage, and output activities of an information system. When making hardware decisions, the overriding consideration of a business should be how hardware can support the objectives of the information system and the goals of the organization.

COMPUTER SYSTEMS: INTEGRATING THE POWER OF TECHNOLOGY

To assemble an effective and efficient system, you should select and organize components while understanding the trade-offs between overall system performance and cost, control, and complexity. For instance, in building a car, manufacturers try to match the intended use of the vehicle to its components. Racing cars, for example, require special types of engines, transmissions, and tires. Selecting a transmission for a racing car requires balancing how much engine power can be delivered to the wheels (efficiency and effectiveness) with how expensive the transmission is (cost), how reliable it is (control), and how many gears it has (complexity). Similarly, organizations assemble computer systems so that they are effective, efficient, and well suited to the tasks that need to be performed.

Because the business needs and their importance vary at different companies, the IS solutions they choose can be quite different.

Marginal glossary definitions of key terms that are discussed in the text provide quick reference to essential terms and concepts. An end-of-book glossary compiles and defines all key terms in the text.

store critical data. The Bombay Company designs and markets home furnishings and decorative accessories via 422 retail outlets, catalogs, and the Internet. The firm implemented 1.5 TB of SAN data storage to effectively hold its inventory data.[11]

Figure 2.3

Storage Area Network

SAN provides high-speed connections between data storage devices and computers over a network.

Optical Discs

A common optical disc is the **compact disc read-only memory (CD-ROM)** with a storage capacity of 740 MB of data. After data is recorded on a CD-ROM, it cannot be modified—the disc is "read-only." A CD burner, the informal name for a CD recorder, is a device that can record data to a compact disc. *CD-recordable (CD-R)* and *CD-rewritable (CD-RW)* are the two most common types of drives that can write CDs, either once (in the case of CD-R) or repeatedly (in the case of CD-RW). CD-rewritable (CD-RW) technology allows PC users to back up data on CDs.

compact disc read-only memory (CD-ROM)
A common form of optical disc on which data, once it has been recorded, cannot be modified.

Digital Video Disc

A **digital video disc (DVD)** is a five-inch diameter CD-ROM look-alike with the ability to store about 135 minutes of digital video or several gigabytes of data (see Figure 2.4). Software programs, video games, and movies are common uses for this storage medium.

digital video disc (DVD)
A storage medium used to store digital video or computer data.

Figure 2.4

Digital Video Disc and Player

DVDs look like CDs but have a greater storage capacity and can transfer data at a faster rate.

(Source: Courtesy of Toshiba America Information Systems.)

DVDs have replaced recordable and rewritable CD discs (CD-R and CD-RW) as the preferred format for sharing movies and photos. Whereas a CD can hold about 740 MB of data, a single-sided DVD can hold 4.7 GB, with double-sided DVDs having a capacity of 9.4 GB. Unfortunately, DVD manufacturers haven't agreed on a recording standard, so several types of recorders and discs are currently in use. Recordings can be made on record-once discs (DVD-R and DVD+R) or on rewritable discs (DVD-RW, DVD+RW, and DVD-RAM). Not all types of rewritable DVDs are compatible with other types.

The Blu-ray high-definition video-disc format based on blue-laser technology stores at least three times as much data as a DVD now holds. The primary use for this new format is in home entertainment equipment to store high definition video, though this format can also store computer data.

A tried-and-true favorite, the **Ethical and Societal Issues** boxes in each chapter focus on a compelling issue pulled from today's headlines. Updated for this edition are **What Would You Do?** scenarios that immerse students in issues that they will read about and encounter as they progress in their studies and careers.

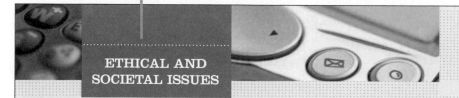

ETHICAL AND SOCIETAL ISSUES

Collecting Accurate and Verifiable Data Where It Counts

Imagine having to design or choose an input device that will satisfy every person's needs: the young, elderly, intelligent, illiterate, sighted, or blind. Then imagine that this device has to provide a 100 percent guarantee that it is easy to use for all and collects accurate data—exactly what the user wanted to enter. Sound challenging? That's the struggle that countries around the world are facing as they continue to create the perfect voting machine.

As the technology revolution races ahead, those responsible for voting systems are trying to harness technology to streamline the voting process. Submitting paper ballots now seems prehistoric in this day of movie downloads and cell phone text messaging. It was only natural that the touch screen would make its way into the voting process—with disastrous results. Touch screen machines, also called Direct Recording Electronic (DRE) units, allow the voter to press the name of the person for whom they want to vote. Each vote is either stored in the machine's storage device to be collected later and batch processed, or sent directly to a central database over a private network.

The use of touch screen machines has led to numerous questionable elections and accusations of scandal. The most prominent are the 2000 and 2004 United States presidential elections, where the close results were questioned due to voting irregularities caused by electronic voting machines.

To overcome the problems with touch screen voting machines, many experts feel that a paper backup of a citizen's vote should be generated along with the electronic vote. By providing a "paper trail" of votes, questionable elections can be easily checked. At the time of this writing, 12 U.S. states still have no paper record requirements.

Many voting administrators have given up on touch screen voting systems altogether. In the 2008 primary presidential elections, the state of New Hampshire relied on optical mark recognition (OMR) technology for their voting. Voters fill in the circle next to a candidate's name on a card. The voter's card is scanned to record the votes, and is filed away as a backup in case a recount is needed. Some precincts in New Hampshire provide voters with simple paper ballots that are counted by hand at the polling place. Visually impaired and disabled can use a touch-tone phone to place their votes. In this way, New Hampshire uses several methods to collect votes.

Other states are experimenting with other systems. Oregon holds its votes by mail. Citizens do not have to travel to a precinct center to cast their votes; instead, they simply mark their ballots, stamp them, and put them in the mailbox. The state claims record voting turnouts and little strife.

Some states seem committed to touch screen systems. Despite a report describing several methods of compromising the vote records of its voting machines, the Crawford County, Ohio, county commissioner tells the citizens that there is nothing to worry about. Since only officials from the county are provided with access codes to the inner workings of the machines, the system should be secure.

The voting machine debate extends beyond the United States to every other voting country. In Germany, a group of computer experts collected signatures to request that a court grant an injunction stopping the use of electronic voting machines. They wanted the system switched to a paper ballot system. They argue that the system had security flaws that allowed a hacker to manipulate voting outcomes. The group contended that the government didn't have the technical understanding to ensure an accurate vote count using the electronic system.

As the search for the perfect voting machine continues, one thing is clear: Collecting data into a system that is verifiably accurate, using easy and fast methods, can be a challenge. Security experts put forth three requirements for touch screen machines: They should produce a voter-verifiable paper trail, use software that is open to examination by the public, and provide verifiable ballots to safeguard against machine failure.

Discussion Questions

1. Do you think that one method of collecting data into a voting system can satisfy all the different types of voters? Or are multiple methods required?
2. What would be your concerns about elections by mail, such as the system used in Oregon?

Critical Thinking Questions

1. Of the systems described in this feature, which would you most like to use? In other words, describe your ideal voting method.
2. What are the security risks of your ideal voting method?

SOURCES: Weiss, Tod, "As primary season ramps up, an e-voting snapshot," *Computerworld*, January 8, 2008, *www.computerworld.com/action/article.do? command=viewArticleBasic&articleId=9056 098.* Smith, Jane, "Officials confident voting machines pose no problems," The Meadville Tribune, January 12, 2008, *www.meadvilletribune.com/local/local_story_009222956.html.* Kirk, Jeremy, "German activists move to block e-voting," *NetworkWorld*, January 8, 2008, *www.networkworld.com/news/2008/010808-german-activists-move-to-block.html?fsrc=rss-security.* Kim, Myung, "Most clerks pushing for mail ballots," Rocky Mountain News, January 12, 2008, *www.rockymountainnews.com/news/2008/jan/12/most-clerks-pushing-for-mail-ballots.*

73

Information Systems @ Work special-interest boxes
show how information systems or technologies are employed
in different business career fields. Students answer both
Discussion and Critical Thinking Questions for each box.

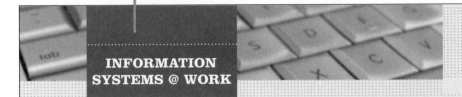

INFORMATION SYSTEMS @ WORK

Software Helps Target Radiation Treatment for Cancer

Doctors have been using radiation therapy as a treatment for cancer since the 1940s. The treatment has saved countless lives, yet has been somewhat imprecise until recently. The original method of treating a tumor with radiation used a linear accelerator that delivered radiation in rectangular beams. Doctors used lead blocks to prevent the beams from harming healthy tissue. The process was cumbersome and only partially effective. Surrounding tissue was often destroyed along with the tumor.

In the 1980s, a machine called an MLC, for multileaf collimator, was invented. The MLC had motorized leaves to disrupt the beam of radiation and focus it more closely on where it was needed. Still, the treatment was imprecise, lacking real-time control of the radiation intensity and direction.

Until the mid-1990s, most of the development of radiation treatment technologies focused on hardware. Varian Medical Systems decided that devising a more effective system would require a heavy investment in software development. Computing processors and hardware were advanced enough to precisely control beams of radiation, but the software to empower the hardware had yet to be developed. Varian transformed itself from a hardware company to a software company to get the job done.

Varian hired experts in programming embedded controls, user interfaces, treatment planning, and databases. It proceeded incrementally over many years to develop a trustworthy and powerful system called the SmartBeam IMRT (for Intensity Modulated Radiation Therapy), which is now in use at thousands of medical facilities around the world.

The SmartBeam IMRT combines an x-ray and radiation technology into one device that rotates around the patient delivering radiation at precise intensities from any angle. The machine is the first that allows physicians to examine and treat a tumor at the same time. The on-board imager produces "high-resolution images of tumors and tracks changes in a tumor's shape, size, and position... that when coupled with SmartBeam IMRT, allows clinicians to be even more precise when targeting tumors," according to *Computerworld*. The magazine awarded Varian the top prize for information systems in manufacturing in its 2007 Computerworld Honors Program.

Discussion Questions

1. What role does software play in the SmartBeam IMRT medical system?
2. Why couldn't Varian produce the SmartBeam IMRT before it did?

Critical Thinking Questions

1. What additional safeguards must be programmed into the software that runs the SmartBeam IMRT that aren't necessary in typical PC software?
2. How do you think the development of the SmartBeam IMRT launched Varian to the top of the market in cancer treatment systems?

SOURCES: Pratt, Mary K., "Software Helps Target Radiation Treatment for Cancer," *Computerworld*, December 3, 2007, *www.computerworld.com/action/article.do?command=viewArticleBasic&articleId=3048 65& pageNumber=1*. Varian Medical Systems Web site, *www.varian.com*, accessed February 3, 2007.

91

An end-of-chapter **Summary** recaps critical chapter concepts, grouped by the chapter's Principles.

SUMMARY

Principle

Computer hardware must be carefully selected to meet the evolving needs of the organization and its supporting information systems.

Hardware devices work together to perform input, processing, data storage, and output. Processing is performed by an interplay between the central processing unit (CPU) and memory. Primary sto...
age for program inst...
provides them to the...
cess data and execut...

Processing that u...
multiprocessing. A m...
more independent pr...
they can share the w...
Parallel processing i...
work together to solv...
the use of a collectio...
individuals or organiz...
to solve a common p...

Computer system...
instructions in secon...
has greater capacity...
either sequential acc...
secondary storage in...
optical disc storage, ...
independent/inexpen...
data that allows the s...
event of a hardware f...
computer servers, dis...
to provide fast and ef...

Input and output...
instructions to the co...
sequent storage and...
interface through wh...
tems. Input and outp...
common characteris...

A keyboard and c...
devices used for entr...
enables a computer t...
native means of provi...
eras record and store...
Handwriting recognit...
the screen into text. ...
technology employs a...
which is read by an R...
include facts such as...
information, or other...

Output devices pr...
from hard copy to so...
are standard output d...

size, number of colors that can be displayed, and resolution. Other output devices include printers and plotters.

The main computer system types are handheld computers, ultra portable computers, portable computers, thin clients, desktop computers, workstations, servers, mainframe computers, and supercomputers. Personal computers (PCs) are small, inexpensive computer systems. Handheld (palmtop) computers are increasingly popular for portable computing and communications needs. Portable computers...

All software programs are written in coding schemes called programming languages, which provide instructions to a computer to perform some processing activity. The several classes of programming languages include machine, assembly, high-level, query and database, and natural and intelligent languages.

Programming languages have changed since their initial development in the early 1950s. In the first generation, computers were programmed in machine language, and the second generation of languages used assembly languages. The third generation consists of many high-level programming languages that use English-like statements and commands. Fourth-generation languages include database and query languages such as SQL.

Users frequently use fourth generation and higher level programming languages to develop their own simple programs.

Principle

The software industry continues to undergo constant change; users need to be aware of recent trends and issues to be effective in their business and personal life.

Software bugs, software licensing and copyrighting, software upgrades, and global software support are all important software issues and trends.

A software bug is a defect in a computer program that keeps it from performing in the manner intended. Software bugs are common, even in key pieces of business software.

Software upgrades are an important source of increased revenue for software manufacturers and can provide useful new functionality and improved quality for software users.

Global software support is an important consideration for large, global companies putting together standardized, company-wide systems. A common solution is outsourcing global support to one or more third-party software distributors.

CHAPTER 2: SELF-ASSESSMENT TEST

Computer hardware must be carefully selected to meet the evolving needs of the organization and its supporting information systems.

1. Organizations typically make a one-time investment in the computer hardware necessary to meet their needs with little need for future changes and upgrades. True or False?

2. Executing an instruction by the CPU involves two phases: the instruction phase and the _____ phase.

3. Which of the following components performs mathematical calculations and makes logical comparisons?
   a. control unit
   b. register
   c. ALU
   d. main memory

Systems and application software are critical in helping individuals and organizations achieve their goals.

4. Application software like Microsoft Excel manipulates the computer hardware directly. True or False?

5. Today's operating systems support _____, the ability to run multiple processes seemingly simultaneously.

Organizations should not develop proprietary application software unless doing so will meet a compelling business need that can provide a competitive advantage.

6. Software that enables users to improve their personal effectiveness, increasing the amount of work they can do and its quality is called _____.
   a. personal productivity software
   b. operating system software
   c. utility software
   d. graphics software

7. Optimization can be found in which type of application software?
   a. spreadsheets
   b. word processing programs
   c. database programs
   d. graphics software

8. A program to detect and eliminate viruses is an example of what type of software?
   a. personal productivity software
   b. operating system software
   c. utility software
   d. application software

Organizations should choose a programming language whose functional characteristics are appropriate to the task at hand, taking into consideration the skills and experience of the programming staff.

9. A class of application software that helps groups work together and collaborate is called _____ .

Back by popular demand, each chapter's **Self-Assessment Test** is linked to the chapter principles and lets students quiz themselves on critical chapter concepts. Answers are provided, so students know immediately whether they have mastered concepts

INSTRUCTOR EDITION

10. More lines of code for current applications are written in the _____ programming language than in any other language.

The software industry continues to undergo constant change; users need to be aware of recent trends and issues to be effective in their business and personal life.

11. _____ are an important source of increased revenue for software manufacturers and can provide useful new functionality and improved quality for software users.

CHAPTER 2: SELF-ASSESSMENT TEST ANSWERS

(1) False (2) execution (3) c. (4) False (5) multitasking (6) a (7) a (8) c (9) workgroup application software (10) COBOL (11) software upgrades

REVIEW QUESTIONS

1. When determining the appropriate hardware components of a new information system, what role must the end user of the system play?
2. Identify two basic characteristics of RAM and ROM.
3. What is RFID technology? How does it work?
4. Identify the three components of the CPU and explain the role of each.
5. What is RAID storage technology? What advantages does it offer?
6. Identify and briefly describe the various classes of computers.
7. Give four examples of personal productivity software.
8. What is software as a service? What advantages does it provide for meeting an organization's software needs?
9. What are the two basic types of software? Briefly describe the role of each.
10. What is multiprocessing? What is multitasking?
11. What is an application service provider? What issues arise in considering the use of one?
12. Define the term utility software and give two examples.
13. What does the acronym API stand for? What is the role of an API?
14. Describe the term *enterprise resource planning* (*ERP*) system. What functions does such a system perform?

DISCUSSION QUESTIONS

1. Briefly discuss the advantages and disadvantages of frequent software upgrades from the perspective of the user of that software. How about from the perspective of the software manufacturer?
2. What would be the advantages for a university computer lab to install thin clients rather than standard desktop personal computers? Can you identify any disadvantages?
3. Which would you rather have—a hand held computer or a tablet computer? Why?
4. If cost were not an issue, describe the characteristics of your ideal computer. What would you use it for? Would you choose a handheld, portable, desktop, or workstation computer? Why? Which operating system would you want it to run?
5. Identify the three spheres of influence and briefly discuss the software needs of each.
6. Identify the two fundamental sources for obtaining application software. Discuss the advantages and disadvantages of each source.
7. Define the software as a service. Discuss some of the pros and cons of using software as a service. What precautions might you take to minimize the risk of using one?
8. In what ways is an operating system for a mainframe computer different from the operating system for a laptop computer? In what ways are they similar?
9. How can application software improve the effectiveness of a large enterprise? What are some of the benefits associated with implementation of an enterprise resource planning system? What are some issues that could keep the use of an enterprise resource planning system from being successful?

Thoroughly revised for each chapter, **Review Questions and Discussion Questions** test students' comprehension of topics from the chapter.

A variety of **exercises and activities** help students apply what they have learned in the chapters. These exercises can be used for both in-class activities and independent study.

PROBLEM-SOLVING EXERCISES

1. Develop a spreadsheet that compares the features, initial purchase price, and ongoing operating costs for three laser printers. Now do the same for three inkjet printers. Write a brief memo on which printer you would choose and why. Cut and paste the spreadsheet into a document.
2. Use word processing software to document what your needs are as a computer user and your justification for selecting either a desktop or laptop computer. Find a Web site that allows you to order and customize a computer and select those options that meet your needs in a cost-effective manner. Assume that you have a budget of $1500. Enter the computer specifications into an Excel spreadsheet that you cut and paste into the document defining your needs. E-mail the document to your instructor.
3. Use a database program to enter five software products you are likely to use at work. List the name, vendor or manufacturer, cost, and features in the columns of a database table. Use a word processor to write a report on the software. Copy the database table into the word processing program.

TEAM ACTIVITIES

1. With two or three of your classmates, visit a computer retail store and identify the most popular ultra laptop computers. Interview members of the sales staff to find out why they think this particular laptop is popular.
2. Form a group of three or four classmates. Identify and contact an employee of a local firm. Interview the individual and describe the application software the company uses and the importance of the software to the organization. Write a brief report summarizing your findings.
3. Divide your team into two groups. The first group should prepare a report using a word processing program. Make sure to include a large number of spelling, grammatical, and similar errors in the document. The second group should use the word processing program's features to locate and eliminate the errors. The entire team should write a report on the advantages and limitations of the spelling and grammar checking features of the word processing program you used. What additional features would you like to see in future word processing programs?

WEB EXERCISES

1. Use the Web to research four productivity software suites from various vendors (see *http://en.wikipedia.org/wiki/Office_Suite*). Create a table in a word processing document to show what applications are provided by the competing suites. Write a few paragraphs on which suite you think best matches your needs and why.
2. Do research on the Web and develop a two-page report summarizing the latest consumer appliance OSs. Which one seems to be gaining the most widespread usage? Why do you think this is the case?
3. Do research on the Web about application software that is used in an industry and is of interest to you. Write a brief report describing how the application software can be used to increase profits or reduce costs.

CAREER EXERCISES

1. Imagine that you are going to buy a single handheld device to improve your communication and organizational abilities. What tasks do you need it to perform? What features would you look for in this device? Visit a computer store or a consumer electronics store and see whether you can purchase such a device for under $400.

Career Exercises challenge students to take the plunge in their intended field of study and gain practical knowledge and experience.

Two all new end-of-chapter **Case Studies** per chapter provide an in-depth look at how today's organizations are capitalizing on the benefits of technology.

2. Think of your ideal job. Describe five application software packages that could help you advance in your career. If the software package doesn't exist, describe the kinds of software packages that could help you in your career.

CASE STUDIES

Case One

Advance America Implements Grid Computing

Chances are you have seen places that offer payday loans in your town. Payday loans are short-term loans designed for people that run out of money before payday, but can repay the loan when their paycheck arrives. Advance America is the leading payday loan company in the United States. It includes 3,000 centers in 37 states, and employs nearly 7,000 people, according to its Web site. Advance America is big, and growing bigger every day. Its growth in recent years is straining the capabilities of its client-server information system infrastructure and holding the company back from further growth.

Advance America used a system in which each center was equipped with an independent hardware and software environment. Installation and maintenance costs were high, and compiling data for all centers was time consuming and difficult. Each night the thousands of centers would upload their data to the main server for consolidation. With the growing number of centers, there wasn't enough time in the night to process all of the incoming data. Advance America's system had run up against a wall. It was time for a change.

Advance America decided to invest in a new system based on a grid computing architecture. They installed thin client machines to run in each center, connecting via the Web to a fault-tolerant server cluster running Oracle database software. The server cluster consists of a four-node cluster of IBM P5 series servers, which include four processors per node for a total of 16 processors. The servers in the cluster work as a grid by sharing the work load of the entire organization equally among them. A pair of Cisco load balancers make sure that processing is distributed evenly among the servers for maximum performance. The new system includes a 2 TB storage area network (SAN) that uses an IBM disk array controlled by the Oracle Automatic Storage Management (ASM) software.

System IT managers at headquarters use a central grid-management console to oversee the entire nationwide network. Problems are easily identified and fixed through the centrally managed system. So far, the system has provided 100 percent uptime at the cash-advance centers.

Advance America took a chance with its $3.8 million investment in this new technology, but it has paid off. Center managers can now tap into "a continuously updated central database and generate reports in near real time." The new system has decreased the time it takes to open a new Advance America center. Managers are getting information much more quickly, making it easier for them to analyze business performance and customer trends. The new system is also easy to expand as the business grows. It is estimated that the new system will provide total net benefits of almost $3 million over five years for an ROI of 131 percent.

Discussion Questions

1. How does grid computing provide Advance America managers with faster access to data?
2. How did grid computing assist Advance America in breaking through the wall that held it back from growth?

Critical Thinking Questions

1. Why is the new grid computing system at Advance America much easier to install, manage, and maintain than its old system?
2. How might Advance America expand its system as the company outgrows it?

SOURCES: Staff, "Advance America Grows with Oracle Enterprise Grid," *Computerworld Honors Program*, 2007, *www.cwhonors.org/viewCaseStudy.asp?NominationID=104*. Advance America Web Site, *www.advanceamerica.net/values.php*, accessed January 12, 2008. Staff, "Integrated Data Infrastructure Pays Off for Advance America, Cash Advance Centers, Inc," Oracle Customer Snapshot, *www.oracle.com/customers/snapshots/advance_america.pdf*, accessed January 12, 2008.

Case Two

Systems Management Software Helps Fight Crime

The York regional police protect 1,800 square kilometers north of Toronto, Canada. Until recently, the force has been challenged trying to keep the rugged Panasonic laptops in its 200 cruisers, boats, and helicopters up-to-date, secure, and synchronized. Although the notebooks were wirelessly networked, the data, software, and systems were not necessarily well synchronized.

To apply system updates and patches every few months, officers were required to check in at the main station where they had to wait for a few hours while the notebook was updated. With over 200 laptops on the force, this cost the department hundreds of working hours every few months.

Not only were human resources wasted, but the internal IT department was pushed past its limit. Staff members sometimes unknowingly worked to solve the same problems. They spent too much time coordinating applications, running backups, and trying to keep up with new law enforcement applications.

**FIFTH
EDITION**

Fundamentals of Information Systems
Fifth Edition

Ralph M. Stair
Professor Emeritus, Florida State University

George W. Reynolds

COURSE TECHNOLOGY
CENGAGE Learning

Australia · Canada · Mexico · Singapore · Spain · United Kingdom · United States

**Fundamentals of Information Systems,
Fifth Edition by
Ralph M. Stair and George W. Reynolds**

VP/Editorial Director: Jack Calhoun

Senior Acquisitions Editor:
    Charles McCormick, Jr.

Product Manager: Kate Hennessy

Development Editor: Lisa Ruffolo,
    The Software Resource

Editorial Assistant: Bryn Lathrop

Content Product Manager: Erin Dowler,
    Jennifer Goguen McGrail

Manufacturing Coordinator: Denise Powers

Marketing Manager: Bryant Chrzan

Marketing Coordinator: Suellen Ruttkay

Art Director: Stacy Shirley

Cover Image: Getty Images/Photonica

Compositor: Value Chain International

Proofreader: Harry Johnson

Indexer: Rich Carlson

For product information and contact us at
Cengage Learning Customer & Sales Support, 1–800–354–9706

For permission to use material from this text or product,
submit all requests online at **cengage.com/permissions**
Further permissions questions can be emailed to
permissionrequest@cengage.com

Microsoft, Windows 95, Windows 98, Windows 2000, Windows XP, and Windows Vista are registered trademarks of Microsoft® Corporation. Some of the product names and company names used in this book have been used for identification purposes only and may be trademarks or registered trademarks of their manufacturers and sellers. SAP, R/3, and other SAP product/services referenced herein are trademarks of SAP Aktiengesellschaft, Systems, Applications and Products in Data Processing, Neurottstasse 16, 69190 Walldorf, Germany. The publisher gratefully acknowledges SAP's kind permission to use these trademarks in this publication. SAP AG is not the publisher of this book and is not responsible for it under any aspect of press law.

ISBN-13:978–14239–2581–1
ISNN-10:1–4239–2581–5

Instructor Edition:
ISBN-13:978-1-4239-2583-5
ISBN-10:1-4239-2583-1

Course Technology
25 Thomson Place
Boston, MA 02210
USA

Cengage Learning is a leading provider of customized learning solutions with office locations around the globe, including Singapore, the United Kingdom, Australia, Mexico, Brazil, and Japan. Locate your local office at: **international.cengage.com/region**

Cengage Learning products are represented in Canada by Nelson Education, Ltd.

To learn more about Course Technology, visit www.cengage.com/coursetechnology
To learn more about Cengage Learning, visit www.cengage.com

Printed in the United States of America
1 2 3 4 5 6 7 13 12 11 10 09

BRIEF CONTENTS

CONTENTS

We are proud to publish the fifth edition of *Fundamentals of Information Systems.* This new edition builds on the success of the previous editions in meeting the need for a concise introductory information systems text. We have listened to feedback from the previous editions' adopters and manuscript reviewers and incorporated many suggestions to refine this new edition. We hope you are pleased with the results.

Like the previous editions, the overall goal of the fifth edition is to develop an outstanding text that follows the pedagogy and approach of our flagship text, *Principles of Information Systems,* with less detail and content. The approach in developing *Fundamentals of Information Systems* is to take the best material from *Principles of Information Systems* and condense it into a text containing nine chapters. So, our most recent edition of *Principles of Information Systems* is the foundation from which we built this new edition of *Fundamentals of Information Systems.*

We have always advocated that education in information systems is critical for employment in almost any field. Today, information systems are used for business processes from communications to order processing to number crunching and in business functions ranging from marketing to human resources to accounting and finance. Chances are, regardless of your future occupation, you will need to understand what information systems can and cannot do and be able to use them to help you accomplish your work. You will be expected to suggest new uses of information systems and participate in the design of solutions to business problems employing information systems. You will be challenged to identify and evaluate IS options. To be successful, you must be able to view information systems from the perspective of business and organizational needs. For your solutions to be accepted, you must identify and address their impact on fellow workers. For these reasons, a course in information systems is essential for students in today's high-tech world.

Fundamentals of Information Systems, Fifth Edition, continues the tradition and approach of the previous editions of this text and *Principles of Information Systems.* Our primary objective is to develop the best IS text and accompanying materials for the first information technology course required of all business students. Using surveys, questionnaires, focus groups, and feedback that we have received from adopters and others who teach in the field, we have been able to develop the highest-quality teaching materials available.

Fundamentals of Information Systems stands proudly at the beginning of the IS curriculum, offering the basic IS concepts that every business student must learn to be successful. This text has been written specifically for the first course in the IS curriculum, and it discusses computer and IS concepts in a business context with a strong managerial emphasis.

APPROACH OF THE TEXT

Fundamentals of Information Systems, Fifth Edition, offers the traditional coverage of computer concepts, but it places the material within the context of meeting business and organizational needs. Placing IS concepts in this context and taking a general management perspective has always set the text apart from general computer books thus making it appealing not only to MIS majors but also to students from other fields of study. The text isn't overly technical, but rather deals with the role that information systems play in an organization and the key principles a manager needs to grasp to be successful. These principles of IS are brought together and presented in a way that is both understandable and relevant. In addition, this book offers an overview of the entire IS discipline, while giving students a solid foundation for further study in advanced IS courses such as programming, the Internet, project management, database management, data communications, systems development, electronic commerce and mobile commerce applications, decision support, and knowledge management. As such, it serves the needs of both general business students and those who will become IS professionals.

The overall vision, framework, and pedagogy that make *Principles of Information Systems* so popular are retained in this text. In particular, this book offers the traditional coverage of computer concepts, but it places the material within the context of business and information systems. Placing IS concepts in a business context has always set the text apart from general computer books and makes it appealing not only to MIS majors but also to students from other courses of study. It approaches information systems from a general management perspective. Without being overly technical, the text deals with the role that information systems play in an organization and the general concepts a manager needs to be aware of to be successful. The text stresses IS principles, which are brought together and presented so that they are understandable and relevant. In addition, this book offers an overview of the entire IS discipline, as well as solid preparation for further study in advanced IS courses.

IS Principles First, Where They Belong

Exposing students to fundamental IS principles provides a service to students who do not later return to the discipline for advanced courses. Because most functional areas in business rely on information systems, an understanding of IS principles helps students in other course work. In addition, introducing students to the principles of information systems helps future functional area managers avoid mishaps that often result in unfortunate and sometimes costly consequences. Furthermore, presenting IS principles at the introductory level creates interest among general business students who will later choose information systems as a field of concentration.

Author Team

Ralph Stair and George Reynolds have teamed up again for the fifth edition. Together, they have more than sixty years of academic and industrial experience. Ralph Stair brings years of writing, teaching, and academic experience to this text. He has written numerous books and a large number of articles while at Florida State University. George Reynolds brings a wealth of computer and industrial experience to the project, with more than 30 years of experience working in government, institutional, and commercial IS organizations. He has written numerous texts and has taught the introductory IS course at the University of Cincinnati and College of Mount St. Joseph. The Stair and Reynolds team brings a solid conceptual foundation and practical IS experience to students.

GOALS OF THIS TEXT

Fundamentals of Information Systems has four main goals:

1. To present a core of IS principles with which every business student should be familiar
2. To offer a survey of the IS discipline that will enable all business students to understand the relationship of advanced courses to the curriculum as a whole
3. To present the changing role of the IS professional
4. To show the value of the discipline as an attractive field of specialization

Because *Fundamentals of Information Systems* is written for all business majors, we believe it is important not only to present a realistic perspective of information systems in business but also to provide students with the skills they can use to be effective leaders in their companies.

IS Principles

Fundamentals of Information Systems, Fifth Edition, although comprehensive, cannot cover every aspect of the rapidly changing IS discipline. The authors, having recognized this, provide students an essential core of guiding IS principles to use as they face the career challenges ahead. Think of principles as basic truths, rules, or assumptions that remain constant regardless of the situation. As such, they provide strong guidance in the face of tough decisions. A set of IS principles is highlighted at the beginning of each chapter. The application of these principles to solve real-world problems is driven home from the opening vignettes to the end-of-chapter material. The ultimate goal of *Fundamentals of Information Systems* is to develop effective, thinking, action-oriented employees by instilling them with principles to help guide their decision making and actions.

Survey of the IS Discipline

This text not only offers the traditional coverage of computer concepts, but stresses the broad framework to provide students with solid grounding in business uses of technology. In addition to serving general business students, this book offers an overview of the entire IS discipline and solidly prepares future IS professionals for advanced IS courses and their careers in the rapidly changing IS discipline.

Changing Role of the IS Professional

As business and the IS discipline have changed, so too has the role of the IS professional. Once considered a technical specialist, today the IS professional operates as an internal consultant to all functional areas of the organization, being knowledgeable about their needs and competent in bringing the power of information systems to bear throughout the organization. The IS professional views issues through a global perspective that encompasses the entire organization and the broader industry and business environment in which it operates, including the entire interconnected network of suppliers, customers, competitors, regulatory agencies, and other entities—no matter where they are located.

The scope of responsibilities of an IS professional today is not confined to just his/her employer but encompasses the entire interconnected network of employees, suppliers, customers, competitors, regulatory agencies, and other entities, no matter where they are located. This broad scope of responsibilities creates a new challenge: how to help an organization survive in a highly interconnected, highly competitive global environment. In accepting that challenge, the IS professional plays a pivotal role in shaping the business itself and ensuring its success. To survive, businesses must now strive for the highest level of customer satisfaction and loyalty through competitive prices and ever-improving product and service quality. The IS professional assumes the critical responsibility of determining the organization's approach to both overall cost and quality performance and therefore plays an important role in the ongoing survival of the organization. This new duality in the role of the IS employee—a professional who exercises a specialist's skills with a generalist's perspective—is reflected throughout the book.

IS as a Field for Further Study

Despite the downturn in the economy at the start of the twenty-first century, especially in technology-related sectors, the outlook for computer and IS managers is bright. In fact, employment of computer and IS managers is expected to grow much faster than the average occupation today and in the future. Technological advancements are boosting the employment of computer-related workers; in turn, this increase in hiring will create demand for managers to direct these workers. In addition, job openings will result from the need to replace managers who retire or move into other occupations.

A career in information systems can be exciting, challenging, and rewarding! This text shows the value of the discipline as an appealing field of study and the IS graduate as an integral part of today's organizations.

CHANGES IN THE FIFTH EDITION

We have implemented a number of exciting changes to the text based on user feedback on how the text can be aligned even more closely with how the IS principles and concepts course is now being taught. The following list summarizes these changes:

- **All new opening vignettes.** All of the chapter-opening vignettes are new, and continue to raise actual issues from foreign-based or multinational companies.

- **All new Information Systems @ Work special interest boxes.** Highlighting current topics and trends in today's headlines, these boxes show how information systems are used in a variety of business career areas.

- **All new Ethical and Societal Issues special interest boxes.** Focusing on ethical issues today's professionals face, these boxes illustrate how information systems professionals confront and react to ethical dilemmas.

- **All new case studies.** Two new cases at the end of every chapter provide a wealth of practical information for students and instructors. Each case explores a chapter concept or problem that a real-world company or organization has faced. The cases can be assigned as homework exercises or serve as a basis for class discussion.

Each chapter has been completely updated with the latest topics and examples. The following is a summary of the changes.

Chapter 1, An Introduction to Information Systems in Organizations

The topics and sections in Chapter 1 create a framework for the entire book. Major sections in this chapter become entire parts. Any chapter titles and material that have been changed in this edition are reflected in Chapter 1. As with all chapters, the opening material at the beginning of the chapter and all end-of-chapter material have been updated to reflect the changes in Chapter 1.

This chapter continues to emphasize the benefits of an information system, including speed, accuracy, reduced costs, and increased functionality. The Why Learn About Information Systems section has been updated to include a new example about financial advisor's use of information systems. In the section on data, information, and knowledge, we have included a new definition of a knowledge management system (KMS). New examples have been introduced in the section on computer-based information systems to give students a better understanding of these important components. The material on the Internet, for example, has been completely updated with the latest information, including Web 2.0 technologies. References of corporate IS usage have been stressed in the business information systems section. The latest material on hardware, software, databases, telecommunications, and the Internet have been included. This material contains fresh, new examples of how organizations use computer-based information systems to their benefit. The best corporate users of IS, as reported by popular computer and business journals, has been explored. We continue stress that ERP systems can replace many applications with one unified set of programs. The material on virtual reality includes new information on the use of this technology to design and manufacture Boeing's new Dreamliner 787 aircraft. The material on systems development has been updated with new examples of successes and failures.

The Ethical and Social Issues section includes the latest threats and what is being done to prevent them. This material contains new examples on the dangers of identity theft, computer mistakes, power consumption, and computer waste. Some experts believe that computers waste up to half of the energy they consume and account for about two percent of worldwide energy usage. We also discuss legal actions in this section. For example, lawsuits have been filed against various Internet sites to protect important copyrighted material from being posted and distributed.

Chapter 1 also gives an overview of business organizations and presents a foundation for the effective and efficient use of IS in a business environment. Since its inception, the primary goal of this text has been to present a core of IS principles and concepts that every business student should know. Chapter 1 stresses the importance and usage of IS within the business organization. Chapter 1 gives an overview of business organizations and presents a foundation for the effective and efficient use of IS in a business environment. We have stressed that the traditional mission of the IS organization "to deliver the right information to the right person at the right time" has broadened to include how this information is used to improve overall performance and help people and organizations achieve their goals.

New photographs throughout the chapter show how organizations use information systems to their benefit. New examples of how companies use information systems have been introduced throughout the chapter. We have stressed that a competitive advantage can result in higher quality products, better customer service, and lower costs. New examples of how companies use information systems have been included, such as new examples on the value chain, supply chain management, customer relationship management, and more. We also present new information and quotes about supply chain management and customer relationship management. The section on performance-based information systems has been updated with fresh and recent examples of how organizations have reduced costs and improved performance. The section on careers has been updated with a new list of top U.S. employers, what makes a satisfied IS worker on the job, and new information on various visas and their impact on the workforce. We have updated the information on the U.S. H-1B and L-1 visa programs. In the first few days that applications were available for the H-1B program in 2007, over 130,000 applications were filed for 65,000 positions. As in the past, some fear that the H-1B program is being abused to replace high paid U.S. workers with less expensive foreign workers. In 2007, two U.S. senators on the Senate Judiciary Subcommittee on Immigration sent letters of concern to a number of Indian firms that were using the H-1B program to staff their U.S. operations with IS personnel from foreign countries. We have also included information and a table about the best places to work in the information systems field. The changing role of the CIO has been highlighted with many new examples and quotes from CIOs at companies of all sizes. The changing roles include the involvement with strategic decisions and more involvement with customers. At the end of the section on IS Careers, we have added a new section on finding a job in IS. This new section describes the many ways that students find jobs. We have included information on IS certification and how jobs can be located using the Internet through sites such as Monster.com and Hotjobs.com. As with previous editions, this chapter continues to stress performance-based management with new examples of how companies can use information systems to improve productivity and increase return on investment.

We have trimmed material throughout the chapter to keep its length reasonable and consistent with previous editions. Table 1.3 on uses of the Internet, for example, has been deleted. These applications are covered in detail in Chapter 4 on the Internet.

Chapter 2, Hardware and Software

As with all chapters, the opening material at the beginning of the chapter including the vignette, the Information Systems @ Work and Ethical and Societal Issues special interest boxes, the end-of-chapter cases, and all end-of-chapter material have been updated to reflect the changes in Chapter 2. The chapter has also been updated to include description, screen shots, and examples of the latest and greatest hardware and software shaping the way people live and work.

In the "Computer Systems: Integrating the Power of Technology" section, Bosch Security Products and the Iowa Health System are provided as new examples of organizations applying business knowledge to reach critical hardware decisions.

In the "Processing and Memory Devices: Power, Speed, and Capacity, Hardware Components" section, the new quad core processors from Intel and AMD are covered and Viiv is identified as an exciting new quad core application. Also Folding@home (a project to research protein folding and misfolding and gain an understanding of how this protein behavior is related to diseases such as Alzheimer's, Parkinson's, and many forms of cancer) is provided as a new example of grid computing. The concept of cloud computing is introduced and the efforts of IBM, Google, and Amazon.com to offer exciting new services based on cloud computing are discussed. The potential is raised that some organizations may consider replacing part of their IT infrastructure with cloud computing.

The Secondary Storage and Input and Output Devices section is updated to identify the newest devices and the latest speeds and capacities. Dial Directions is offered as an example of an organization using speech recognition technology to provide customer service. The Large Synoptic Survey Telescope used for space research is offered as a new example of the use of digital cameras. There is also a new example of the use of OCR technology to improve the payroll function of Con-way Inc. There is a new example of the use of RFID technology to track inventory for Boekhandels, a major book retailer in the Netherlands. Updated information is provided on the use of holographic disks. A new figure of HP's new high-speed, high-volume CM8060 ink jet printer is given. There is a new figure of the MacBook Air ultra thin laptop computer and a discussion and photo of Apple iPod Touch device.

In the "Computer System Types" section, Table 2.3 depicting Types of Computer Systems has been updated. There is discussion and a photo of a pocket computer. An updated list of ultra laptop computer manufacturers and their costs is provided. A new class of computer—ultra small desktop computer is discussed. A new example is offered of the use of the Pocket PC (a handheld computer that runs the Microsoft Windows Mobile operating system) by the Coca-Cola field sales force to automate the collection of information about sales calls, customers, and prospects. Another new example is CSX Transportation, one of the nation's largest railroads, which uses DT Research's WebDT 360 to enable train conductors to monitor systems while onboard and communicate with stations for real-time updates. How mainframe computers are used at the top 25 banks and retailers is briefly discussed. Current information is provided on the IBM Blue Gene supercomputer. A table depicting the processing speed of supercomputers has been added.

In the software overview, a new section has been added covering installing and removing software. The operating system section has been updated to include detailed coverage of Windows Vista and equally thorough coverage of Apple OS X Leopard, and Linux (the various distributions). These three operating systems are compared and contrasted. The section on workgroup and enterprise operating systems has been updated to include the latest systems including Windows Server 2008, Mac OS X Leopard Server, z/OS, and HP-UX. Smartphone operating systems are presented. Embedded operating systems are also discussed including Microsoft's Sync, a popular feature on GM vehicles.

In the application software section, Software as a Service (SaaS) is introduced and discussed and examples are provided. Microsoft Office 2007 is covered thoroughly. Alternatives to Office 2007 are also provided. Online productivity software is introduced and discussed, including Google Docs, Zoho applications, Thinkfree, and Microsoft Office Live. A variety of software licensing options is presented. Open source is provided as an alternative to popular software and many examples are cited.

Throughout the chapter, many new examples are cited including: The National Aquarium's use of Windows Vista, a lawyer's use of Mac OS X, radio station KRUU's use of Ubuntu Linux, Ebay's use of Solaris, the benefits of the software provided on RIM's Blackberries to workers at the U.S. Department of Agriculture, Blue Cross Blue Shield's use of proprietary software for claims management, use of SaaS by "The Improv" for managing marketing and ticket sales, and the use of Personal Information Management software by Greenfield online, a Web survey company.

Chapter 3, Database Systems and Business Intelligence

The database chapter has been updated to include descriptions, screen shots, and examples of the latest database technologies shaping the way people live and work. A new and interesting vignette has been created to engage the student's attention from the first page of the chapter. New, thought-provoking boxes and cases have been created to help bring the chapter material to life for the reader. The growth of digital information and the importance of managing overwhelming amounts of data are discussed. The market leaders in database systems are introduced along with their market share information. After discussing traditional database technologies, Database as a Service (DaaS) is introduced as a new form of database management. New examples are provided for databases accessible on the Internet. The semantic Web is discussed and its relationship to databases is explored. All of the high-quality, still-relevant material from the eighth edition is maintained.

We have included many new examples throughout the chapter. These include a Hollywood talent agency's use of databases to store client information, and the city of Albuquerque's use of a database to provide citizens with information on water bill's and usage. We have also included an example of a database security breach at an Ivy League college, the FBI's huge database of biometric data, and Wal-Mart's medical database for use at its health clinics. We mention Microsoft's use of OneNote for presenting management training classes. Other database examples include a New Delhi's lighting manufacturer's use of a DBMS, Morphbank—a special-purpose database for scientific data and photographs, the phenomenal growth of data centers around the world, and databases used for medical records; We also explore 1–800-Flowers use of a data warehouse for customer data, the Defense Acquisition University's use of a data warehouse for student records, and the use of data mining to forecast terrorist behavior. We cover the use of data mining by MySpace for targeted marketing, the use of predictive analysis by police to forecast crimes, the use of business intelligence in the health industry, and the use of object-oriented databases by King County Metro Transit system for routing buses.

Chapter 4 Telecommunications, the Internet, Intranets, and Extranets

As with all other chapters, the opening vignette, Ethical and Societal Issues, Information Systems at Work, and chapter-ending cases are all new. New, real-world examples are sprinkled throughout the chapter to maintain the reader's interest and to demonstrate the actual application of the topics being discussed. New end-of-chapter questions and exercises are included.

The new opening vignette discusses how automobile manufacturer Lamborghini discovered the benefits of employing the latest communications and collaboration technologies to shrink the distances between its more than 100 dealerships scattered around the world.

Additional examples are offered of organizations using telecommunications and networks to move ahead including Wal-Mart, Procter & Gamble, Aflac, BT, McKesson, SAP, Verizon, and Boeing.

In the "An Overview of Telecommunications" section, near-field communication (NFC), a very short-range wireless connectivity technology designed for cell phones and credit cards, is discussed. Several pilot projects testing this technology with consumers are discussed. A brief description of how Wi-Fi operates is given along with several examples of where Wi-Fi is being used. The latest information is offered on 3G, 4G, and WiMAX communications including mention of the joint effort between Sprint Nextel and Clearwire to build the first nationwide WiMAX network. The coverage of VoIP is expanded and some of the potential VoIP advantages as well as disadvantages are covered. Merrill Lynch's implementation of VoIP to support its brokers around the world is discussed.

In the section on networks and distributed processing, Disneyland's House of the Future in Tomorrowland is offered as an example of an advanced home local area network. The Miami-Dade Police Department consists of 3,000 officers and 1,500 civilians who serve and protect more than two million citizens over a 2,100 square mile area. Their metropolitan area network is discussed to illustrate the benefits of such technology. Jenny Craig's (the weight management company) use of a Wide Area Network to capture and send daily the private information about its customers from some 500 locations in North America to its headquarters in Carlsbad, CA is discussed. A new Information Systems @ Work special feature discusses the use of telepresence to improve the operations at Dream Works. The Ticketmaster reservation service is offered as an example of a centralized system. 7-Eleven is offered as an example of a decentralized system. The concept of communications protocol is discussed and several examples including ATM, Frame Relay, IEEE 802.3 (Ethernet), and T-carrier system are identified.

T-Mobile of Austria is provided as an example of an organization that operates a collection of systems, hardware, and applications and that performs constant monitoring to detect potential device failures or system bottlenecks before they can generate customer complaints or service failures.

In the "Use and Functioning of the Internet" section, the latest information is presented about the number and geographic dispersion of the users of the Internet. Also the current status of the Next Generation Internet is discussed. Cell phone carriers who provide Internet access for handsets or notebooks equipped with connect cards are discussed. Apple's popular iPhone, a combination mobile phone, widescreen iPod, and Internet access device to support e-mail and Web browsing is covered. Table 4.4 has been added which shows the time required to perform basic tasks using various Internet connections. A new Ethical and Societal Issues special feature discusses Comcast's controversial use of packet shaping to provide even levels of service to all users.

In the World Wide Web section, Web 2.0 and the social Web are covered. Many of the social networking sites such as YouTube, Flickr, Facebook, MySpace, Twitter, Jaiku, Digg, del.icio.us, and Epinions are covered. There is a discussion of how powerful Web-delivered applications such as Google Docs, Adobe Photoshop Express, Xcerion Web-based OS, and Microsoft Maps have elevated the Web from an online library to a platform for computing. There is a discussion of rich Internet applications, software that has the functionality and complexity of traditional application software, but runs in a Web browser and does not require local installation. There are new examples of companies using Web services. Amazon has developed Amazon Web Services (AWS) to make the contents of its huge online catalog available to other Web sites or software applications. Mitsubishi Motors of North America uses Web services to link about 700 automotive dealers on the Internet.

The Internet and Web applications section provides expanded coverage of how search engines work to determine which of the hundreds or thousands of Web pages associated with a particular keyword are most useful. Search engine optimization is also covered. Video chat that enables users to speak to each other face-to-face is also discussed. New information is provided on career information and job searching. Video log and podcasting are covered. Also discussed is Hewlett-Packard's Halo videoconferencing system that makes it appear as though you are speaking with a number of people across a table, though those people may actually be located around the world.

Chapter 5, Electronic and Mobile Commerce and Enterprise Systems

As with all other chapters, the opening vignette, Ethical and Societal Issues, Information Systems at Work, and chapter-ending cases are all new. New, real-world examples are sprinkled throughout the chapter to maintain the reader's interest and to demonstrate the actual application of the topics being discussed. New end-of-chapter questions and exercises are included.

In the introduction to electronic commerce, the latest statistics are provided about the growth and size of B2B, B2C and C2C electronic commerce. New examples of companies using various forms of e-commerce are presented including Covisint, Dell, and American Eagle Outfitters. Various forms of e-Government are covered including G2C, G2B and G2G. New data is also provided about the size and growth of M-commerce.

The section on electronic and mobile commerce applications is fully updated with new material and examples of companies using electronic and mobile commerce to increase revenue and reduce costs. Many new and innovative applications of m-commerce are discussed including mobile banking, mobile price comparison, mobile advertising, and mobile coupons. The technology infrastructure required section has been expanded to include a discussion of the latest technologies used for both e-commerce and m-commerce. The Dun & Bradstreet address verification service, the Oanda.com currency exchange rate service, and the UPS shipment tracking service are discussed as examples of Web services. The limitations of wireless devices and the use of Wireless Application Protocol and Wireless Markup Language are discussed.

The section on electronic payment systems has been updated with new solutions as well as solution providers. A table has been added to compare the various payment systems. Making payments using cell phones like a credit card is also discussed.

The chapter has also been greatly revised to put a much greater emphasis on ERP systems and less on traditional transaction processing systems. Much of the discussion of traditional transaction processing systems has been reduced and/or placed in simple lists.

A section has been added that discusses the use of transaction processing systems by SMBs and identifies a number of software packages available that provide integrated transaction processing system solutions for SMBs. The use of a software package to provide an integrated set of transaction processing systems by the city of Lexington, Kentucky is mentioned.

The material on ERP and CRM has been updated with new material and examples. For example, the implementation of an ERP system by Gujarat Reclaim and Rubber Products, Amgen, BNSF Railway Company and the resulting benefits are discussed.

A new section has been added that covers the use of ERP by SMBs and highlights the special needs of those firms. Several open source ERP solutions, frequently preferred by SMBs, are identified. The adoption of ERP systems at SMBs Cedarlane, Vertex Distribution, Prevention Partners, Inc., and Galenicum is discussed. The 15 top-rated ERP and CRM packages for both large organizations and SMBs are presented.

Chapter 6, Information and Decision Support Systems

All boxes, cases, and the opening vignette are new to this edition. We have added a new learning objective about the importance of special-purpose systems. This chapter includes many new examples of how managers and decision makers can use information and decision support systems to achieve personal and corporate goals. The section on problem solving and decision making has new examples. We emphasize that many of the information and decision support topics discussed in the chapter can be built into some ERP systems, discussed in Chapter 5.

We have added a new section on the benefits of information and decision support systems at the end of the material on decision making and problem solving. This new section investigates the performance and costs of these systems and how they can benefit individuals, groups, and organizations by helping them make better decisions, solve and implement problems, and achieve their goals. We have also included new examples, including how British Airways used the problem-solving approach to address flight delays.

The section on MIS has new examples and references throughout, including many in 2008. We have included how companies and local governments have used software tools to help them in developing effective MIS reporting systems. In the section on demand reports, there are new examples and references about how people can get medical records from the Internet. The section on exception reports has a new example of the use of mashups to get reports from different data sources. This section also includes a new example of the use of texting as a way to deliver exception reports. The Financial MIS section has new examples and material. We have shown how the Internet has been used to make small, micro loans using social-networking sites, such as Facebook. We have also investigated how financial companies have used corporate news to help them make trading decisions. The section on manufacturing MIS has also been updated. We have included examples that show how companies, such as Toyota, continue to use JIT inventory control techniques. The material on marketing MIS has been updated with new material on the use of video advertising and social-networking sites to promote new products and services. This section also includes new material on how the Internet is being used to auction radio ads. The material on the human resource (HR) MIS has also been updated with new material on just-in-time talent and the use of supply chain techniques in the HR area.

The use of digital dashboards and business activity monitoring continues to be stressed in the section on decision support systems, along with new examples and references. The section on optimization includes new examples, showing the huge cost savings of the technique. The use of data warehousing, data marts, and data mining, first introduced in Chapter 3, has been stressed in this chapter. New corporate examples of the use of data warehousing, data marts, and data mining that can be used to provide information and decision support have been placed throughout the section. The use of mashups to integrate data from different sources into a data-driven DSS has been included.

This chapter continues to stress group support systems. We discuss some additional features of groupware, including group monitoring, idea collection, and idea organizing and voting features. New examples and approaches have been explored, including the use of the Web to deliver group support. We have modified the section on parallel communication to include the use of unified communication in group decision making.

To keep the chapter length reasonable, we have trimmed a number of the sections. We have deleted the section on developing effective reports and moved some of the material into other sections. The material on the characteristics of a management information system has been summarized in a table. The characteristics of a decision support system and those of an executive support system have also been put in new tables.

Chapter 7, Knowledge Management and Specialized Information Systems

We have updated all material in this chapter and added many new and exciting examples and references. The overall purpose of this chapter is to cover knowledge management and specialized business information systems, including artificial intelligence, expert systems, and many other specialized systems. These systems are substantially different from more traditional information and decision support systems, and while they are not used to the same extent as more traditional information systems, they still have an important place in business.

As with the last edition, the section on knowledge management continues to be a natural extension of the material in Chapter 6 on information and decision support systems and leads to a discussion of some of the special-purpose systems discussed in the chapter, including expert systems and knowledge bases. This chapter contains many new examples and references. We have new material on the importance of knowledge management for Tata, a large Indian company that uses it to retain and use knowledge from retiring employees. The Aerospace and Defense (A&D) organization also uses knowledge management to keep knowledge in its organization. This chapter shows how Pratt & Whitney uses knowledge management systems to help it deliver information and knowledge about its jet engine parts to the company and airlines, including Delta and United. We highlight the importance of the knowledge manager and the chief knowledge officer with new references and a new quote. The material on knowledge management includes new examples of communities of practice (COP). A group of people from the International Conference on Knowledge Management in Nuclear Facilities has a COP to investigate the use of knowledge management systems in the development and control of nuclear facilities. We also have numerous new examples on how organizations can obtain, store, share, and use knowledge. The University of South Carolina, for example, has joined with Collexis to develop and deliver new knowledge management software, based on Collexis's Knowledge Discovery Platform. We have referenced a survey that estimates that American companies have paid about $70 billion on knowledge management technology in 2007. New material and examples have been included in the section on the technology to support knowledge management. This section contains information on research into the use and effectiveness of knowledge management.

The other topics discussed in this chapter also contain numerous new examples and references in robotics, vision systems, expert systems, virtual reality, and a variety of other special-purpose systems. New robotics examples include research done at the Robot Learning Laboratory at Carnegie Mellon University, robots by iRobot, medical uses of robots, and robots in the military. New examples of voice recognition have been included. New commercial and military examples of expert systems have also been included. We have updated the information on expert systems tools and products. New material on computer vision has been included. One expert believes that in 10 years computer vision systems may be able to recognize certain levels of emotions, expressions, gestures, and behaviors, all through vision. The section on virtual reality has many new examples and references. We have added a new section on business applications to reflect the increased use of virtual reality by businesses of all types and sizes. Kimberly-Clark Corporation has developed a virtual reality system to view store aisles carrying its products. Boeing uses virtual reality to help it design and manufacture airplane parts and new planes, including the 787 Dreamliner. Clothing and fashion companies, such as Neiman Marcus and Saks Fifth Avenue, are using virtual reality on the Internet to display and promote new products and fashions.

The section on other specialized systems also has new material, examples, and references. The U.S. Defense Advanced Research Projects Agency (DARPA) and other organizations are exploring mechanical computers that are energy efficient and can stand up to harsh environments. We have included information on Segway and 3VR Security that performs a video-face recognition test to identify people from pictures or images. Ford Motor Company and Microsoft have developed a voice-activated system called *Sync* that can play music, make phones calls, and more from voice commands. The Advanced Warning System by Mobileye warns drivers to keep a safe distance from other vehicles and drivers. The Surface from Microsoft is a touch-screen computer that uses a glass-top display. Microsoft's Smart Personal Objects Technology (SPOT) allows small devices to transmit data and messages over the air. We also discuss that manufacturing is being done with inkjet printers to allow computers to "print" 3-D parts. More information has been included on RFID technology. Wearable computers used to monitor inventory levels and perform other functions have been introduced. We have also included new references in the material on game theory and informatics.

We have reduced the chapter length in a number of areas. Table 7.1 on additional knowledge management resources has been deleted. Some of the introductory AI material has been shortened. The material on expert systems development has been deleted to trim chapter length. Much of this material is covered in Chapter 8 on systems development.

Chapter 8, Systems Development

This chapter has new material, examples, and references. In addition, the end-of-chapter material has been completely updated with new questions and exercises. As with other editions, all boxes and cases are new to this edition.

This chapter has a new emphasis on entrepreneurs and small business owners. The systems development skills and techniques discussed in this chapter can help people launch their own businesses. When Marc Mallow couldn't find off-the-shelf software to schedule workers, he took a few years to develop his own software. The software he created became the core of a company he founded, located in New York. We also have new material and examples on the use of systems development for nonprofit organizations.

We have placed increased emphasis on long-range planning in systems development projects. It can result in getting the most from a systems development effort. It can also help make sure that IS goals are aligned with corporate goals. Hess Corporation, a large energy company with over 1,000 retail gasoline stations, used long-range planning to determine what computer equipment it needed and the IS personnel needed to run it.

There are new examples in the section on IS planning and aligning corporate and IS goals. Procter & Gamble (P&G), for example, uses ROI to measure the success of its projects and systems development efforts. Providing outstanding service is another important corporate goal. Coca-Cola Enterprises, which is Coca Cola's largest bottler and distributor, decided to use online services from Microsoft and SharePoint to speed its systems development process.

New examples of outsourcing are discussed. IBM, for example, has consultants located in offices around the world. In India, IBM has increased its employees from less than 10,000 people to more than 30,000. We have also included information on the scope of outsourcing, including hardware maintenance and management, software development, database systems, networks and telecommunications, Internet and Intranet operations, hiring and staffing, and the development of procedures and rules regarding information systems. Small and medium-sized firms are also using outsourcing to cut costs and get needed technical expertise that would be difficult to afford with in-house personnel. Millennium Partners Sports Club Management, for example, used Center Beam to outsource many of its IS functions, including its helpdesk operations. The Boston-based company plans to spend about $30,000 a month on outsourcing services, which it estimates to be less than it would have to pay in salaries for additional employees. The market for outsourcing services for small and medium-sized firms is expected to increase by 15 percent annually through 2010 and beyond.

The section discussing on-demand computing has new material and examples. Amazon, the large online retailer of books and other products, will offer on-demand computing to individuals and other companies of all sizes, allowing them to use Amazon's computer expertise and database capacity. Individuals and companies will only pay for the computer services they use.

We have stressed the need for cooperation and collaboration with the systems investigation team. The systems investigation team can be diverse, with members located around the world. Cooperation and collaboration are keys to successful investigation teams.

The section on design includes new material and examples. In the introductory material, we stress that system development should take advantage of the latest developments in technology. Many companies, for example, are looking into cloud computing, where applications are run on the Internet instead of being developed and run within the company or organization. Cloud computing is allowing individuals, like racecar driving instructor Tom Dyer, to do work while traveling or in a car. We also discuss that developing Web applications will become more important in the future. Microsoft's Live Mesh, for example, allows systems developers to seamlessly coordinate data among different devices and provide data backup on the Internet. Designing new systems to reduce total costs is also discussed.

We have a new section on environmental design, also called green design, that involves systems development efforts that slash power consumption, take less physical space, and result in systems that can be disposed in a way that doesn't negatively impact the environment. A *Computerworld* survey revealed that over 80 percent of IS managers considered energy efficiency when selecting new computer equipment. The Environmental Protection Agency (EPA) estimates that a 10 percent cut in data center electricity usage would be enough to power about a million U.S. homes every year. This new section discusses the companies that are developing products and services to help save energy. UPS developed its own software to reduce the miles its 90,000 trucks and other vehicles drive by routing them more efficiently. The new software helped UPS cut 30 million miles per year, slash fuel costs, and reduce carbon emissions by over 30,000 metric tons. This section also shows how companies are developing systems to dispose of old equipment. Hewlett-Packard and Dell Computer have developed procedures and machines to dispose of old computers and computer equipment in environmentally friendly ways. Old computers and computer equipment are fed into machines that disintegrate them into small pieces and sort them into materials that can be reused. The process is often called *green death.* One study estimates that more than 130,000 PCs in the U.S. are thrown out every day. The U.S. government is also involved in environmental design. It has a plan to require federal agencies to purchase energy-efficient computer systems and equipment. The plan would require federal agencies to use the *Electronic Product Environmental Assessment Tool (EPEAT)* to analyze the energy usage of new systems. The U.S. Department of Energy rates products with the *Energy Star* designation to help people select products that save energy.

The material on systems implementation has been updated to include new references, material, and examples. The section on hardware acquisition has new material and examples, including material on hardware virtualization. Tellabs, for example, acquired virtualized Dell PowerEdge Servers for its operations to consolidate its hardware, increase utilization, and save space. The software acquisition section has new material on software as a service (SaaS). The Humane Society of the United States used SaaS to obtain a secure system to receive credit-card contributions from donors. We also stress the importance of software acquisition. Companies like Google are delivering word processing, spreadsheet programs, and other office suite packages over the Internet. Allstate Insurance decided to develop a new software program, called Next Gen, to speed claims processing and reinforce its "You're in good hands" slogan. The company is expected to spend over $100 million on the new software. The material on database acquisition includes the use of open-source databases. In addition, we have discussed that virtual databases and database as a service (DaaS) are also popular ways to acquire database capabilities. XM Radio, Bank of America, and Southwest Airlines, for example, use the DaaS approach to manage many of their database operations from the Internet. We've also updated the material on user preparation. When a new operating system or application software package is implemented, user training is essential. In some cases, companies decide not to install the latest software because the amount of time and money needed to train employees is too much. In one survey, over 70 percent of the respondents indicated that they were in no hurry to install a new operating system. Additional user training was a factor in delaying the installation of the new operating system. In the section on site preparation, we stress that developing IS sites today requires energy efficiency. We begin the section on testing by describing what can go wrong without adequate testing. A $13 million systems development effort to build a vehicle title and registration system, for example, had to be shut down because inaccurate data led to vehicles being pulled over or stopped by mistake.

The section on maintenance and review contains updated material on a variety of topics. In the systems review section, we discuss that a review can result in halting a new system while it is being built because of problems. The section on systems review also stresses how problems with an existing system can trigger new systems development. The Section on systems maintenance also contains new examples.

Chapter 9, The Personal and Social Effects of Computers

As with all other chapters, the opening vignette, Ethical and Societal Issues, Information Systems at Work, and chapter ending Cases are all new. New, real-world examples are sprinkled throughout the chapter to maintain the reader's interest and to demonstrate the actual application of the topics being discussed. New end-of-chapter questions and exercises are included.

The new opening vignette discusses how managing the largest online banking service and marketplace makes eBay a huge target for online hackers and fraudsters. The vignette goes on to outline many of the security measures and tools eBay uses to combat these threats.

Additional material is added on spam filters and a list of the most highly rated filters is given. In addition, it is mentioned that there is a potential problem using spam filters in that some require that first time e-mailers be verified else their e-mails will be rejected. Image-based spam is also covered.

In the section on computer-related mistakes, several new examples of waste are provided including Moody's Corporation, NASA, Nippon Airlines, United Airlines, and Wells Fargo. The mistake by Moody's was especially damaging and resulted in the firm's stock price dropping over 20 percent.

In the section on preventing computer waste and mistakes, the need for proper user training is illustrated with an example of the Maryland Department of Transportation preparing its users to use new business intelligence software. The Société Générale scandal in France is provided as a classic example of an individual employee circumventing internal policies and procedures. A low-level trader on the arbitrage desk at the French bank created a series of fraudulent and unauthorized investment transactions that built a $72 billion position in European stock index futures. Tokyo Electron, a global supplier of semiconductor production equipment, is offered as an excellent example of a firm thoroughly reviewing its policies and procedures.

In the section on computer crime, the point is made that even good IS policies might not be able to predict or prevent computer crime. Five new examples of hackers causing problems at Citibank, the Chilean government, MS Health, a Pennsylvania school district, and on-line brokers E-Trade and Schwab. Results of the 2007 FBI Internet Crime Report are summarized showing that 206,844 complaints of crime were perpetrated over the Internet during 2007 with a dollar value of $240 million in losses. Results of the 2007 Computer Crime and Survey are also highlighted.

In the "Computer as a Tool to Commit Crime" section, the growing problem of cyberterrorism is discussed. The International Multilateral Partnership Against Cyber Terrorism (IMPACT) is identified as a global public and privately supported initiative to counteract this threat. Examples of cyberterrorism against the small Baltic nation of Estonia and against the CNN news network are given. A new section on Internet gambling has been added. The point is made that although Internet gambling is legal in more than 70 countries, the legality of these online activities is far from clear in the U.S. The Internet Wire Act of 1961 and the Unlawful Internet Gambling Enforcement Act of 2006 along with various laws passed by the individual states make this a murky area indeed. CBSSports.com and Facebook are given as examples of orgaizations that were investigated briefly by the FBI for collaborating to make it easier for Facebook users to fill out brackets for the NCAA 2008 Basketball Tournament.

The "Computer as the Object of Crime" section discusses how some criminals have started phony VoIP phone companies and sold subscriptions for services to unsuspecting customers. Instead of establishing their own network, the criminals hack into the computers that route calls over the networks of legitimate VoIP providers and use their networks to carry its customers' calls. The latest data is provided about the growth and spread of malware and several of the most current viruses are identified. New examples are provided of malware causing harm and destruction. Information has been added about the use of a rootkit, a set of programs that enable its user to gain administrator level access to a computer or network. Once installed, the attacker can gain full control of the system and even obscure the presence of the rootkit from legitimate system administrators. The highest rated antivirus software for 2007 is identified. It is mentioned that tests have shown that antivirus scans run significantly faster on computers with regularly defragmented files and free space reducing the time to do a complete scan by 18 to 58 minutes. A new section has been added that covers spyware—what it is, how you get it on your computer and precautions to take to avoid spyware. Many new examples are provided of organizations that have lost valuable data by careless handling of laptop computers. Suggestions are offered on how to reduce the number of these incidences. The safe disposal of computers, which can be a risk because of toxic chemicals released at landfills and the possibility that sensitive information that may be obtained by trash divers, is discussed. The use of disk-wiping software utilities that overwrite all sectors of your disk drive making all data unrecoverable is recommended. A definition has been added of copyright and patent to introduce the material on software piracy. New examples are offered of companies getting into trouble over software piracy. Also discussed is Operation Copycat, an ongoing undercover investigation into warez groups, which are online organizations engaged in the illegal uploading, copying, and distribution of copyrighted works such as music, movies, games, and software, often even before they are released to the public. Examples of patent infringement involving Acer, Apple, Dell, Hewlett-Packard, Fujitsu, Firestar Software, and DataTern are identified. A new Ethical and Societal Issues special feature discusses international cyber espionage. A number of victims are identified as well as the tools most frequently used by spies.

The "Preventing Computer Related Crime" section discusses the Terrorist Finance Tracking Program, which relies on data in international money transfers from the Society for Worldwide Interbank Financial Telecommunications. The goal of the program was to track and combat terrorist financing. The program was credited with helping to capture at least two terrorists; however, revelation of the secret program's existence stirred up controversy and rendered the program ineffective. Mechanisms to reduce computer crime are discussed such as fingerprint authentication devices that provide security in the PC environment by using fingerprint recognition instead of passwords. Also the JetFlash 210 Fingerprint USB Flash Drive requires users to swipe their fingerprints and match them to one of up to 10 trusted users to access the data. The data on the flash drive can also be encrypted for further protection. New material is presented on the use of security dashboard software to provide a comprehensive display on a single computer screen of all the vital data related to an organization's security defenses including threats, exposures, policy compliance and incident alerts. Associated Newspapers is provided as an example of an organization that has implemented a successful security dashboard. Several attempts by the U.S. Congress to limit children's exposure to online pornography including the Communications Decency Act, the Child Online Protection Act, and the Children's Internet Protection Act are discussed.

The Privacy Issues section has been updated and includes the latest information on computer monitoring of employees and others. Issues associated with Instant Messaging and privacy, personal sensing devices such as RFID chips, and social networking are discussed. A new Information Systems @ Work special feature discusses the issues associated with controlling the privacy of Finland's largest information system.

WHAT WE HAVE RETAINED FROM THE FOURTH EDITION

The fifth edition builds on what has worked well in the past; it retains the focus on IS principles and strives to be the most current text on the market.

- **Overall principle.** This book continues to stress a single,-all-encompassing theme: The right information, if it is delivered to the right person, in the right fashion, and at the right time, can improve and ensure organizational effectiveness and efficiency.

- **Information systems principles.** Information system principles summarize key concepts that every student should know. This important feature is a convenient summary of key ideas presented at the start of each chapter.

- **Global perspective.** The global aspects of information systems is a major theme.

- **Learning objectives linked to principles.** Carefully crafted learning objectives are included with every chapter. The learning objectives are linked to the Information Systems principles and reflect what a student should be able to accomplish after completing a chapter.

- **Opening vignettes emphasize international aspects.** All of the chapter-opening vignettes raise actual issues from foreign-based or multinational companies.

- **Why Learn About features.** Each chapter has a "Why Learn About" section at the beginning of the chapter to pique student interest. The section sets the stage for students by briefly describing the importance of the chapter's material to the students-whatever their chosen field.

- **Information Systems @ Work special interest boxes.** Each chapter has an entirely new "Information Systems @ Work" box that shows how information systems are used in a variety of business career areas.

- **Ethical and Societal Issues special interest boxes.** Each chapter includes a new "Ethical and Societal Issues" box that presents a timely look at the ethical challenges and the societal impact of information systems

- **Current examples, boxes, cases, and references.** As we have in each edition, we take great pride in presenting the most recent examples, boxes, cases, and references throughout the text. Some of these were developed at the last possible moment, literally weeks before the book went into publication. Information on new hardware and software, the latest operating systems, mobile commerce, the Internet, electronic commerce, ethical and societal issues, and many other current developments can be found throughout the text. Our adopters have come to expect the best and most recent material. We have done everything we can to meet or exceed these expectations.

- **Summary linked to principles.** Each chapter includes a detailed summary with each section of the summary tied to an associated information system principle.

- **Self-assessment tests.** This popular feature helps students review and test their understanding of key chapter concepts.

- **Career exercises.** End-of-chapter career exercises ask students to research how a topic discussed in the chapter relates to a business area of their choice. Students are encouraged to use the Internet, the college library, or interviews to collect information about business careers.

- **End-of-chapter cases.** Two new end-of-chapter cases provide a wealth of practical information for students and instructors. Each case explores a chapter concept of problem that a real-world company or organization has faced. The cases can be assigned as individual homework exercises or serve as a basis for class discussion.

- **Integrated, comprehensive, Web case.** At the end of every chapter, we have included an integrated and comprehensive case that runs throughout the text. The case follows the activities of two employees of the fictitious Whitmann Price Consulting firm as they are challenged to complete various IS-related projects. Each case has activities related to the material in the chapter in which it is presented.

STUDENT RESOURCES

Student Online Companion Web Site

We have created an exciting online companion, password-protected for students to use as they work through the Fourth Edition. At the beginning of this text is a key code that provides full access to a robust Web site, located at *www.cengage.com/mis/stairreynolds*. This Web resource includes the following features:

- **PowerPoint slides**
  Direct access is offered to the book's PowerPoint presentations that cover the key points from each chapter. These presentations are a useful study tool.

- **Classic cases**
  A frequent request from adopters is that they'd like a broader selection of cases to choose from. To meet this need, a set of over 85 cases from the fifth, sixth, and seventh editions of the text are included here. These are the authors' choices of the "best cases" from these editions and span a broad range of companies and industries.

- **Links to useful Web sites**
  Chapters in *Fundamentals of Information Systems, Fifth Edition* reference many interesting Web sites. This resource takes you to links you can follow directly to the home pages of those sites so that you can explore them. There are additional links to Web sites that the authors, Ralph Stair and George Reynolds, think you would be interested in checking out.

- **Hands-on activities**
  Use these hands-on activities to test your comprehension of IS topics and enhance your skills using Microsoft® Office applications and the Internet. Using these links, you can access three critical-thinking exercises per chapter; each activity asks you to work with an Office tool or do some research on the Internet.

- **Test yourself on IS**
  This tool allows you to access 20 multiple-choice questions for each chapter; test yourself and then submit your answers. You will immediately find out what questions you got right and what you got wrong. For each question that you answer incorrectly, you are given the correct answer and the page in your text where that information is covered. Special testing software randomly compiles 20 questions from a database of 50 questions, so you can quiz yourself multiple times on a given chapter and get some new questions each time.

- **Glossary of key terms**
  The glossary of key terms from the text is available to search.

- **Online readings**
  This feature provides you access to a computer database, which contains articles relating to hot topics in Information Systems.

INSTRUCTOR RESOURCES

The teaching tools that accompany this text offer many options for enhancing a course. As always, we are committed to providing one of the best teaching resource packages available in this market.

Instructor's Manual

An all-new *Instructor's Manual* provides valuable chapter overviews; highlights key principles and critical concepts; offers sample syllabi, learning objectives, and discussion topics; and features possible essay topics, further readings and cases, and solutions to all of the end-of-chapter questions and problems, as well as suggestions for conducting the team activities. Additional end-of-chapter questions are also included. As always, we are committed to providing the best teaching resource packages available in this market.

Sample Syllabus

A sample syllabus with sample course outlines is provided to make planning your course that much easier.

Solutions

Solutions to all end-of-chapter material are provided in a separate document for your convenience.

Test Bank and Test Generator

ExamView® is a powerful objective-based test generator that enables instructors to create paper-, LAN- or Web-based tests from test banks designed specifically for their Course Technology text. Instructors can utilize the ultra-efficient QuickTest Wizard to create tests in less than five minutes by taking advantage of Course Technology's question banks or customizing their own exams from scratch. Page references for all questions are provided so you can cross- reference test results with the book.

PowerPoint Presentations

A set of impressive Microsoft PowerPoint slides is available for each chapter. These slides are included to serve as a teaching aid for classroom presentation, to make available to students on the network for chapter review, or to be printed for classroom distribution. Our presentations help students focus on the main topics of each chapter, take better notes, and prepare for examinations. Instructors can also add their own slides for additional topics they introduce to the class.

Figure Files

Figure files allow instructors to create their own presentations using figures taken directly from the text.

DISTANCE LEARNING

Course Technology, the premiere innovator in management information systems publishing, is proud to present online courses in WebCT and Blackboard.

- **Blackboard and WebCT Level 1 Online Content.** If you use Blackboard or WebCT, the test bank for this textbook is available at no cost in a simple, ready-to-use format. Go to *www.cengage.com/mis/stairreynolds* and search for this textbook to download the test bank.

- **Blackboard and WebCT Level 2 Online Content.** Blackboard 5.0 and 6.0 as well as Level 2 and WebCT Level 2 courses are also available for *Fundamental of Information Systems, Fourth Edition.* Level 2 offers course management and access to a Web site that is fully populated with content for this book.

For more information on how to bring distance learning to your course, instructors should contact their Course Technology sales representative.

ACKNOWLEDGMENTS

A book of this size and undertaking requires a strong team effort. We would like to thank all of our fellow teammates at Course Technology and the Software Resource for their dedication and hard work. We would like to thank Dave Boelio for helping us transition from Thomson Learning to Cengage Learning. Special thanks to Kate Hennessy, our Product Manager. Our appreciation goes out to all the many people who worked behind the scenes to bring this effort to fruition including Abigail Reip, our photo researcher, and Charles McCormick, our Senior Acquisitions Editor. We would like to acknowledge and thank Lisa Ruffolo, our development editor. She deserves special recognition for her tireless effort and help in all stages of this project. Erin Dowler and Jennifer Goguen McGrail, our Content Product Managers, shepherded the book through the production process

We are grateful to the sales force at Course Technology whose efforts make this all possible. You helped to get valuable feedback from current and future adopters. As Course Technology product users, we know how important you are.

We would especially like to thank Ken Baldauf for his excellent help in writing the boxes and cases and revising several chapters for this edition. Ken also provided invaluable feedback on other aspects of this project.

Ralph Stair would like to thank the chairman and faculty members of the Department of Management Information Systems, College of Business Administration, at Florida State University. He would also like to thank his family, Lila and Leslie, for their support.

George Reynolds would like to thank his family, Ginnie, Tammy, Kim, Kelly, and Kristy, for their patience and support in this major project.

TO OUR PREVIOUS ADOPTERS AND POTENTIAL NEW USERS

We sincerely appreciate our loyal adopters of the previous editions and welcome new users of *Fundamentals of Information Systems, Fifth Edition.* As in the past, we truly value your needs and feedback. We can only hope this new edition continues to meet your high expectations.

In addition, Ralph Stair would like to thank Mike Jordan, Senior Lecturer, Division of Business Information Management Glasgow Caledonia University for his hospitality and help at the UKAIS conference.

We are indebted to the following reviewers, both past and present, for their perceptive feedback on early drafts of this text:

Jill Adams, *Navarro College*

Robert Aden, *Middle Tennessee State University*

A.K. Aggarwal, *University of Baltimore*

Sarah Alexander, *Western Illinois University*

Beverly Amer, *University of Florida*

Noushin Asharfi, *University of Massachusetts*

Kirk Atkinson, Western Kentucky University

Yair Babad, *University of Illinois—Chicago*

Cynthia C. Barnes, *Lamar University*

Charles Bilbrey, *James Madison University*

Thomas Blaskovics, *West Virginia University*

John Bloom, *Miami University of Ohio*

Warren Boe, *University of Iowa*

Glen Boyer, *Brigham Young University*

Mary Brabston, *University of Tennessee*

Jerry Braun, *Xavier University*

Thomas A. Browdy, *Washington University*

Lisa Campbell, *Gulf Coast Community College*

Andy Chen, *Northeastern Illinois University*

David Cheslow, *University of Michigan—Flint*

Robert Chi, *California State University—Long Beach*

Carol Chrisman, *Illinois State University*

Phillip D. Coleman, *Western Kentucky University*

Miro Costa, *California State University—Chico*

Caroline Curtis, *Lorain County Community College*

Roy Dejoie, *USWeb Corporation*

Sasa Dekleva, *DePaul University*

Pi-Sheng Deng, *California State University—Stanislaus*

Roger Deveau, *University of Massachusetts—Dartmouth*

John Eatman, *University of North Carolina*

Gordon Everest, *University of Minnesota*

Juan Esteva, *Eastern Michigan University*

Badie Farah, *Eastern Michigan University*

Karen Forcht, *James Madison University*

Carroll Frenzel, *University of Colorado—Boulder*

John Gessford, *California State University—Long Beach*

Terry Beth Gordon, *University of Toledo*

Kevin Gorman, *University of North Carolina—Charlotte*

Costanza Hagmann, *Kansas State University*

Bill C. Hardgrave, *University of Arkansas*

Al Harris, *Appalachian State University*

William L. Harrison, *Oregon State University*

Dwight Haworth, *University of Nebraska—Omaha*

Jeff Hedrington, *University of Wisconsin—Eau Claire*

Donna Hilgenbrink, *Illinois State University*

Jack Hogue, *University of North Carolina*

Joan Hoopes, *Marist College*

Donald Huffman, *Lorain County Community College*

Patrick Jaska, *University of Texas at Arlington*

G. Vaughn Johnson, *University of Nebraska—Omaha*

Tom Johnston, *University of Delaware*

Grover S. Kearns, *Morehead State University*

Robert Keim, *Arizona State University*

Karen Ketler, *Eastern Illinois University*

Mo Khan, *California State University—Long Beach*

Chang E. Koh, *University of North Texas*

Michael Lahey, *Kent State University*

Jan de Lassen, *Brigham Young University*

Robert E. Lee, *New Mexico State University—Carlstadt*

Joyce Little, *Towson State University*

Herbert Ludwig, *North Dakota State University*

Jane Mackay, *Texas Christian University*

Al Maimon, *University of Washington*

Efrem Mallach, University of Massachusetts Dartmouth

James R. Marsden, *University of Connecticut*

Roger W. McHaney, *Kansas State University*

Lynn J. McKell, *Brigham Young University*

John Melrose, *University of Wisconsin—Eau Claire*

Michael Michaelson, *Palomar College*

Ellen Monk, *University of Delaware*

Bertrad P. Mouqin, *University of Mary Hardin-Baylor*

Bijayananda Naik, *University of South Dakota*

Zibusiso Ncube, *Concordia College*

Pamela Neely, *Marist College*

Leah R. Pietron, *University of Nebraska—Omaha*

John Powell, *University of South Dakota*

Maryann Pringle, *University of Houston*

John Quigley, *East Tennessee State University*

Mahesh S. Raisinghani, *University of Dallas*

Mary Rasley, *Lehigh-Carbon Community College*

Earl Robinson, *St. Joseph's University*

Scott Rupple, *Marquette University*

Dave Scanlon, *California State University—Sacramento*

Werner Schenk, *University of Rochester*
Larry Scheuermann, *University of Southwest Louisiana*
James Scott, *Central Michigan University*
Vikram Sethi, *Southwest Missouri State University*
Laurette Simmons, *Loyola College*
Janice Sipior, *Villanova University*
Anne Marie Smith, *LaSalle University*
Harold Smith, *Brigham Young University*
Patricia A. Smith, *Temple College*
Herb Snyder, *Fort Lewis College*
Alan Spira, *University of Arizona*
Tony Stylianou, *University of North Carolina*
Bruce Sun, *California State University—Long Beach*
Howard Sundwall, *West Chester University*
Hung-Lian Tang, *Bowling Green State University*

William Tastle, *Ithaca College*
Gerald Tillman, *Appalachian State University*
Duane Truex, *Georgia State University*
Jean Upson, *Lorain County Community College*
Misty Vermaat, *Purdue University—Calumet*
David Wallace, *Illinois State University*
Michael E. Whitman, *University of Nevada—Las Vegas*
David C. Whitney, *San Francisco State University*
Goodwin Wong, *University of California—Berkeley*
Amy Woszczynski, *Kennesaw State University*
Judy Wynekoop, *Florida Gulf Coast University*
Myung Yoon, *Northeastern Illinois University*

OUR COMMITMENT

We are committed to listening to our adopters and readers and to developing creative solutions to meet their needs. The field of Information Systems continually evolves, and we strongly encourage your participation in helping us provide the freshest, most relevant information possible.

We welcome your input and feedback. If you have any questions or comments regarding *Fundamentals of Information Systems Fifth Edition,* please contact us through Course Technology or your local representative, or via the Internet at *www.cengage.com/mis/stairreynolds.*

Ralph Stair
George Reynolds

PART
· 1 ·

Information
Systems in
Perspective

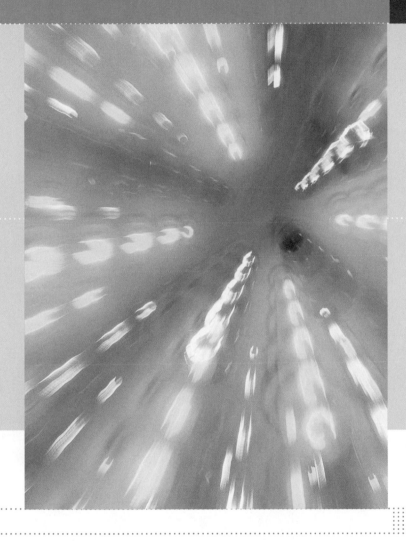

Chapter 1   An Introduction to Information Systems in Organizations

CHAPTER
· 1 ·

An Introduction to Information Systems in Organizations

PRINCIPLES	LEARNING OBJECTIVES
▪ The value of information is directly linked to how it helps decision makers achieve the organization's goals.	▪ Distinguish data from information and describe the characteristics used to evaluate the value of data.
▪ Knowing the potential impact of information systems and having the ability to put this knowledge to work can result in a successful personal career and organizations that reach their goals.	▪ Identify the basic types of business information systems and discuss who uses them, how they are used, and what kinds of benefits they deliver.
▪ System users, business managers, and information systems professionals must work together to build a successful information system.	▪ Identify the major steps of the systems development process and state the goal of each.
▪ The use of information systems to add value to the organization can also give an organization a competitive advantage.	▪ Identify the value-added processes in the supply chain and describe the role of information systems within them.
	▪ Identify some of the strategies employed to lower costs or improve service.
	▪ Define the term *competitive advantage* and discuss how organizations are using information systems to gain such an advantage.
▪ Cooperation between business managers and IS personnel is the key to unlocking the potential of any new or modified system.	▪ Define the types of roles, functions, and careers available in information systems.

▶ Information Systems in the Global Economy ▶
Fossil, United States

Computer-based Information Systems Support Best Business Practices

High-quality, up-to-date, well-maintained computer-based information systems are at the heart of today's most successful global corporations. For a business to succeed globally, it must be able to provide the right information to the right people in the organization at the right time, even if those people are located around the world. Increasingly, this means that decision-makers can view the state of every aspect of the business in real time. For example, an executive in Paris can use an information system to see that a company product was purchased from a retailer in San Francisco three minutes ago. If a company's information system is not efficient and effective, the company will lose market share to a competitor with a better information system. For a deeper understanding of how information systems are used in business, consider Fossil.

You are probably familiar with the Fossil brand. Fossil is well known for its watches, handbags, jewelry, and fashion accessories that are sold in numerous retail and department stores around the world. Fossil was founded in 1984 when it set up wholesale distribution of its products to department stores in North America, Asia, and Europe. The company quickly grew and began manufacturing products for other brands such as Burberry, Diesel, DKNY, and Emporio Armani. As Fossil grew, the information it managed expanded until it threatened to be unmanageable, so Fossil invested in a corporate-wide information system developed by SAP Corporation and designed for apparel wholesale companies. The SAP information system efficiently stored all of Fossil's business information, and organized it so that it assisted Fossil management with the tough business decisions they needed to make.

An information system's ability to organize information so that it provides fuel for smart business decisions is the real value of computer-based information systems. SAP, IBM, Oracle, and other computer-based information systems developers do much more than provide hardware systems and databases. The systems they install are governed by software that implements best business practices. These systems assist managers in designing the best business solutions, which is why selecting the right computer-based information system is crucial to a company's success.

Using the SAP information system to manage its business, Fossil continued to prosper. Fossil linked its information system to those of its customers such as Wal-Mart and Macy's to automate the task of fulfilling orders. Fossil was one of the first companies to launch an online store on the Web, and managed its evolution from a wholesale business to a retail business. Another information system was developed for Web sales that worked with its core SAP corporate-wide information system.

More recently, Fossil began experimenting by opening its own retail stores, which have now blossomed into hundreds of Fossil stores across the United States and in fifteen other countries. However, because managing a retail store is different from managing a wholesale company, Fossil turned again to SAP and IBM to design additional information systems that would service its retail needs. Since Fossil's retail and wholesale operations share production warehousing and shipping, the retail information system is designed to be integrated with its wholesale information system.

Fossil's information systems are all integrated, connecting to one central database. Using these information systems, Fossil can quickly react to market demands. For example, if Fossil sees that a particular style of watch is selling well at its retail store in

London, it can quickly ship more of that style to department stores operating in the same area. Fossil credits its information systems for simplifying its business infrastructure and supporting consistent best practices across its expanding global business.

As you read this chapter, consider the following:

- How might the information systems such as those used at Fossil make use of the various components of a computer-based information system: hardware, software, databases, telecommunications, people, and procedures?
- How do computer-based information systems like Fossil's help businesses implement best practices?

Why Learn About Information Systems in Organizations?

Information systems are used in almost every imaginable profession. Sales representatives use information systems to advertise products, communicate with customers, and analyze sales trends. Managers use them to make multimillion-dollar decisions, such as whether to build a manufacturing plant or research a cancer drug. Financial planners use information systems to advise their clients to help them save for their children's education and retirement. From a small music store to huge multinational companies, businesses of all sizes could not survive without information systems to perform accounting and finance operations. Regardless of your college major or chosen career, information systems are indispensable tools to help you achieve your career goals. Learning about information systems can help you land your first job, earn promotions, and advance your career.

Why learn about information systems in organizations? What is in it for you? Learning about information systems will help you achieve your goals. Let's get started by exploring the basics of information systems.

information system (IS)
A set of interrelated components that collect, manipulate, store, and disseminate data and information and provide a feedback mechanism to meet an objective.

People and organizations use information everyday. Many retail chains, for example, collect data from their stores to help them stock what customers want and to reduce costs. The components that are used are often called an information system. An **information system (IS)** is a set of interrelated components that collect, manipulate, store, and disseminate data and information and provide a feedback mechanism to meet an objective. It is the feedback mechanism that helps organizations achieve their goals, such as increasing profits or improving customer service. Businesses can use information systems to increase revenues and reduce costs. This book emphasizes the benefits of an information system, including speed, accuracy, and reduced costs.

Information systems are everywhere. An air traveler checks in for a flight using a kiosk, which sends the check-in information to a network to verify the traveler's reservation and flight information. The terminal processes the information and prints a boarding pass, speeding airport check-in times.

(Source: Courtesy of Joshua Lott/ Bloomberg News/Landov.)

We interact with information systems every day, both personally and professionally. We use automated teller machines at banks, access information over the Internet, select information from kiosks with touch screens, and let checkout clerks scan our purchases using bar codes and scanners. Major *Fortune* 500 companies can spend more than $1 billion per year on information systems. Knowing the potential of information systems and putting this knowledge to work can result in a successful career and organizations that reach their goals.

INFORMATION CONCEPTS

Information is a central concept of this book. The term is used in the title of the book, in this section, and in almost every chapter. To be an effective manager in any area of business, you need to understand that information is one of an organization's most valuable resources. This term, however, is often confused with *data*.

Data, Information, and Knowledge

Data consists of raw facts, such as an employee number, number of hours worked in a week, inventory part numbers, or sales orders.[1] As shown in Table 1.1, several types of data can represent these facts. When facts are arranged in a meaningful manner, they become information. **Information** is a collection of facts organized so that they have additional value beyond the value of the facts themselves.[2] For example, sales managers might find that knowing the total monthly sales suits their purpose more (i.e., is more valuable) than knowing the number of sales for each sales representative. Providing information to customers can also help companies increase revenues and profits. According to Frederick Smith, chairman and president of FedEx, "Information about the package is as important as the package itself… We care a lot about what's inside the box, but the ability to track and trace shipments and therefore manage inventory in motion revolutionized logistics."[3] FedEx is a worldwide leader in shipping packages and products around the world. Increasingly, information generated by FedEx and other organizations is being placed on the Internet. In addition, many universities are now placing course information and content on the Internet.[4] Using the Open Course Ware program, the Massachusetts Institute of Technology (MIT) places class notes and contents on the Internet for more than 1,500 of its courses.

data
Raw facts, such as an employee number, number of hours worked in a week, inventory part numbers, or sales orders.

information
A collection of facts organized in such a way that they have additional value beyond the value of the facts themselves.

Data	Represented by
Alphanumeric data	Numbers, letters, and other characters
Image data	Graphic images and pictures
Audio data	Sound, noise, or tones
Video data	Moving images or pictures

Table 1.1

Types of Data

Data represents real-world things. Hospitals and healthcare organizations, for example, maintain patient medical data, which represents actual patients with specific health situations. In many cases, hospitals and healthcare organizations are converting data to electronic form. Some have developed *electronic records management (ERM)* systems to store, organize, and control important data. However, data—simply raw facts—has little value beyond its existence. For example, consider data as pieces of railroad track in a model railroad kit. Each piece of track has little value beyond its inherent value as a single object. However, if you define a relationship among the pieces of the track, they will gain value. By arranging the pieces in a certain way, a railroad layout begins to emerge (see Figure 1.1a, top). Data and information work the same way. Rules and relationships can be set up to organize data into useful, valuable information.

The type of information created depends on the relationships defined among existing data. For example, you could rearrange the pieces of track to form different layouts. Adding new or different data means you can redefine relationships and create new information. For instance, adding new pieces to the track can greatly increase the value—in this case, variety and fun—of the final product. You can now create a more elaborate railroad layout (see Figure 1.1b, bottom). Likewise, a sales manager could add specific product data to his sales

Defining and Organizing
Relationships Among Data
Creates Information

(a)

(b)

process
A set of logically related tasks
performed to achieve a defined
outcome.

knowledge
The awareness and understanding
of a set of information and ways that
information can be made useful to
support a specific task or reach a
decision.

data to create monthly sales information organized by product line. The manager could use this information to determine which product lines are the most popular and profitable.

Turning data into information is a **process**, or a set of logically related tasks performed to achieve a defined outcome. The process of defining relationships among data to create useful information requires knowledge. **Knowledge** is the awareness and understanding of a set of information and the ways that information can be made useful to support a specific task or reach a decision. Having knowledge means understanding relationships in information. Part of the knowledge you need to build a railroad layout, for instance, is the understanding of how much space you have for the layout, how many trains will run on the track, and how fast they will travel. Selecting or rejecting facts according to their relevancy to particular tasks is based on the knowledge used in the process of converting data into information. Therefore, you can also think of information as data made more useful through the application of knowledge. *Knowledge workers (KWs)* are people who create, use, and disseminate knowledge, and are usually professionals in science, engineering, business, and other areas. A *knowledge management system (KMS)* is an organized collection of people, procedures, software, databases, and devices used to create, store, and use the organization's knowledge and experience.

In some cases, people organize or process data mentally or manually. In other cases, they use a computer. In the earlier example, the manager could have manually calculated the sum of the sales of each representative, or a computer could calculate this sum. Where the data comes from or how it is processed is less important than whether the data is transformed into results that are useful and valuable. This transformation process is shown in Figure 1.2.

The Process of Transforming
Data into Information

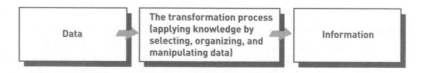

The Characteristics of Valuable Information

To be valuable to managers and decision makers, information should have the characteristics described in Table 1.2. These characteristics make the information more valuable to an organization. Many shipping companies, for example, can determine the exact location of inventory items and packages in their systems, and this information makes them responsive to their customers. In contrast, if an organization's information is not accurate or complete, people can make poor decisions, costing thousands, or even millions, of dollars. If an inaccurate forecast of future demand indicates that sales will be very high when the opposite is true, an organization can invest millions of dollars in a new plant that is not needed. Furthermore, if information is not relevant, not delivered to decision makers in a timely fashion, or too complex to understand, it can be of little value to the organization.

Characteristics	Definitions
Accessible	Information should be easily accessible by authorized users so they can obtain it in the right format and at the right time to meet their needs.
Accurate	Accurate information is error free. In some cases, inaccurate information is generated because inaccurate data is fed into the transformation process. (This is commonly called garbage in, garbage out [GIGO].)
Complete	Complete information contains all the important facts. For example, an investment report that does not include all important costs is not complete.
Economical	Information should also be relatively economical to produce. Decision makers must always balance the value of information with the cost of producing it.
Flexible	Flexible information can be used for a variety of purposes. For example, information on how much inventory is on hand for a particular part can be used by a sales representative in closing a sale, by a production manager to determine whether more inventory is needed, and by a financial executive to determine the total value the company has invested in inventory.
Relevant	Relevant information is important to the decision maker. Information showing that lumber prices might drop might not be relevant to a computer chip manufacturer.
Reliable	Reliable information can be depended on. In many cases, the reliability of the information depends on the reliability of the data-collection method. In other instances, reliability depends on the source of the information. A rumor from an unknown source that oil prices might go up might not be reliable.
Secure	Information should be secure from access by unauthorized users.
Simple	Information should be simple, not overly complex. Sophisticated and detailed information might not be needed. In fact, too much information can cause information overload, whereby a decision maker has too much information and is unable to determine what is really important.
Timely	Timely information is delivered when it is needed. Knowing last week's weather conditions will not help when trying to decide what coat to wear today.
Verifiable	Information should be verifiable. This means that you can check it to make sure it is correct, perhaps by checking many sources for the same information.

Table 1.2

Characteristics of Valuable Information

Depending on the type of data you need, some characteristics become more valuable than others. For example, with market-intelligence data, some inaccuracy and incompleteness is acceptable, but timeliness is essential. Sutter Health, for example, developed a real-time system for its intensive care units (ICUs) that can detect and prevent deadly infections, saving over 400 lives a year and millions of dollars in additional healthcare costs.[5]

The Value of Information

The value of information is directly linked to how it helps decision makers achieve their organization's goals.[6] Valuable information can help people and their organizations perform tasks more efficiently and effectively.[7] Consider a market forecast that predicts a high demand for a new product. If you use this information to develop the new product and your company makes an additional profit of $10,000, the value of this information to the company is $10,000 minus the cost of the information. Valuable information can also help managers decide whether to invest in additional information systems and technology. A new computerized ordering system might cost $30,000, but generate an additional $50,000 in sales. The *value added* by the new system is the additional revenue from the increased sales of $20,000. Most corporations have cost reduction as a primary goal. Using information systems, some manufacturing companies have slashed inventory costs by millions of dollars. Other companies have increased inventory levels to increase profits. Wal-Mart, for example, uses information about certain regions of the country and specific situations to increase needed inventory levels of certain products and improve overall profitability.[8] The giant retail store used valuable information about the needs of people in the path of Hurricane Ivan when it hit Florida. The store stocked strawberry Pop-Tarts and other food items that didn't need refrigeration or food preparation to serve people in the area and to increase its profits.

WHAT IS AN INFORMATION SYSTEM?

As mentioned previously, an information system (IS) is a set of interrelated elements or components that collect (input), manipulate (process), store, and disseminate (output) data and information and provide a corrective reaction (feedback mechanism) to meet an objective (see Figure 1.3). The feedback mechanism is the component that helps organizations achieve their goals, such as increasing profits or improving customer service.

Input, Processing, Output, Feedback

Input

input
The activity of gathering and capturing raw data.

In information systems, **input** is the activity of gathering and capturing raw data. In producing paychecks, for example, the number of hours every employee works must be collected before paychecks can be calculated or printed. In a university grading system, instructors must submit student grades before a summary of grades for the semester or quarter can be compiled and sent to the students.

Processing

processing
Converting or transforming data into useful outputs.

In information systems, **processing** means converting or transforming data into useful outputs. Processing can involve making calculations, comparing data and taking alternative actions, and storing data for future use. Processing data into useful information is critical in business settings.

Processing can be done manually or with computer assistance. In a payroll application, the number of hours each employee worked must be converted into net, or take-home, pay. Other inputs often include employee ID number and department. The required processing can first involve multiplying the number of hours worked by the employee's hourly pay rate to get gross pay. If weekly hours worked exceed 40 hours, overtime pay might also be included. Then deductions—for example, federal and state taxes, contributions to health and life insurance or savings plans—are subtracted from gross pay to get net pay.

After these calculations and comparisons are performed, the results are typically stored. *Storage* involves keeping data and information available for future use, including output, discussed next.

Output

output
Production of useful information, usually in the form of documents and reports.

In information systems, **output** involves producing useful information, usually in the form of documents and reports. Outputs can include paychecks for employees, reports for managers, and information supplied to stockholders, banks, government agencies, and other groups. In some cases, output from one system can become input for another. For example, output from a system that processes sales orders can be used as input to a customer billing system.

Feedback

In information systems, **feedback** is information from the system that is used to make changes to input or processing activities. For example, errors or problems might make it necessary to correct input data or change a process. Consider a payroll example. Perhaps the number of hours an employee worked was entered as 400 instead of 40 hours. Fortunately, most information systems check to make sure that data falls within certain ranges. For number of hours worked, the range might be from 0 to 100 hours because it is unlikely that an employee would work more than 100 hours in a week. The information system would determine that 400 hours is out of range and provide feedback. The feedback is used to check and correct the input on the number of hours worked to 40. If undetected, this error would result in a very high net pay on the printed paycheck!

Feedback is also important for managers and decision makers. For example, a furniture maker could use a computerized feedback system to link its suppliers and plants. The output from an information system might indicate that inventory levels for mahogany and oak are getting low—a potential problem. A manager could use this feedback to decide to order more wood from a supplier. These new inventory orders then become input to the system. In addition to this reactive approach, a computer system can also be proactive—predicting future events to avoid problems. This concept, often called **forecasting**, can be used to estimate future sales and order more inventory before a shortage occurs. Forecasting is also used to predict the strength of hurricanes and possible landing sites, future stock-market values, and who will win a political election.

Manual and Computerized Information Systems

As discussed earlier, an information system can be manual or computerized. For example, some investment analysts manually draw charts and trend lines to assist them in making investment decisions. Tracking data on stock prices (input) over the last few months or years, these analysts develop patterns on graph paper (processing) that help them determine what stock prices are likely to do in the next few days or weeks (output). Some investors have made millions of dollars using manual stock analysis information systems. Of course, today many excellent computerized information systems follow stock indexes and markets and suggest when large blocks of stocks should be purchased or sold (called *program trading*) to take advantage of market discrepancies.

feedback
Output that is used to make changes to input or processing activities.

forecasting
Predicting future events to avoid problems.

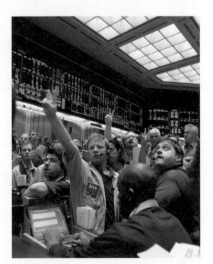

Program trading systems allow traders to keep up with swift changes in stock prices and make better decisions for their investors.

(Source: Courtesy of REUTERS/ Allen Frederickson/Landov.)

Computer-Based Information Systems

A **computer-based information system (CBIS)** is a single set of hardware, software, databases, telecommunications, people, and procedures that are configured to collect, manipulate, store, and process data into information. A company's payroll, order entry, or inventory-control system is an example of a CBIS. Lloyd's Insurance in London is starting to use a CBIS to reduce paper transactions and convert to an electronic insurance system.[9] The new CBIS allows Lloyd's to insure people and property more efficiently and effectively. Lloyd's is often used to insure the unusual, including actress Betty Grable's legs, Rolling Stone Keith Richard's hands, and a possible appearance of the Loch Ness Monster (Nessie) in Scotland, which would result in a large payment for the person first seeing the monster. CBISs can also be embedded into products. Some new cars and home appliances include computer hardware, software, databases, and even telecommunications to control their operations and make them more useful. This is often called embedded, pervasive, or ubiquitous computing.

The components of a CBIS are illustrated in Figure 1.4. *Information technology (IT)* refers to hardware, software, databases, and telecommunications. A business's **technology infrastructure** includes all the hardware, software, databases, telecommunications, people, and procedures that are configured to collect, manipulate, store, and process data into information. The technology infrastructure is a set of shared IS resources that form the foundation of each computer-based information system.

computer-based information system (CBIS)

A single set of hardware, software, databases, telecommunications, people, and procedures that are configured to collect, manipulate, store, and process data into information.

technology infrastructure

All the hardware, software, databases, telecommunications, people, and procedures that are configured to collect, manipulate, store, and process data into information.

Figure 1.4

The Components of a Computer-Based Information System

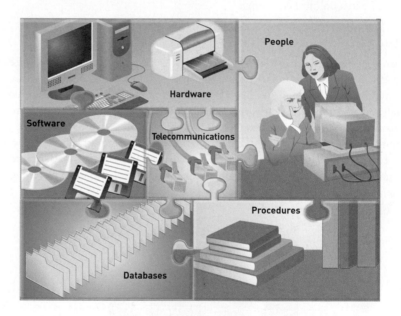

Hardware

Hardware consists of computer equipment used to perform input, processing, and output activities.[10] Input devices include keyboards, mice and other pointing devices, automatic scanning devices, and equipment that can read magnetic ink characters.[11] Investment firms often use voice-response technology to allow customers to access their balances and other information with spoken commands. Processing devices include computer chips that contain the central processing unit and main memory.[12] Advances in chip design allow faster speeds, less power consumption, and larger storage capacity.[13] Processor speed is also important. Today's more advanced processor chips have the power of supercomputers that occupied a room measuring 10 feet by 40 feet ten years ago.[14] A large IBM computer used by U.S. Livermore National Laboratories to analyze nuclear explosions might be the fastest in the world (up to 596 teraflops—596 trillion operations per second). The super fast computer, called Blue Gene, costs about $40 million. Mental Images of Germany and Pixar of the United States also need high processor speeds to produce award-winning images. Image technology is used to help design cars, such as the sleek shapes of Mercedes-Benz vehicles. Small, inexpensive computers are also becoming popular. The One Laptop Per Child

hardware

Computer equipment used to perform input, processing, and output activities.

(OLPC) computer, for example, costs under $200. The Classmate PC by Intel will cost about $300 and includes some educational software.[15] Both computers are intended for regions of the world that can't afford traditional personal computers.

The many types of output devices include printers and computer screens. Another type of output device is a printer kiosk, which are located in some shopping malls. After inserting a disc or a memory card from a computer or a camera, you can print photos and some documents. There are also many special-purpose hardware devices. Computerized event data recorders (EDRs) are now being placed into vehicles. Like an airplane's black box, EDRs record a vehicle's speed, possible engine problems, a driver's performance, and more. The technology is being used to document and monitor vehicle operation, determine the cause of accidents, and investigate whether truck drivers are taking required breaks. In one case, an EDR was used to help convict a driver of vehicular homicide.

Hardware consists of computer equipment used to perform input, processing, and output activities. The trend in the computer industry is to produce smaller, faster, and more mobile hardware.

(Source: © Alberto Pomares/ iStockphoto.com.)

Software

Software consists of the computer programs that govern the operation of the computer. These programs allow a computer to process payroll, send bills to customers, and provide managers with information to increase profits, reduce costs, and provide better customer service. With software, people can work anytime at any place. Software along with manufacturing tools, for example, can be used to fabricate parts almost anywhere in the world. Software called Fab Lab controls tools, such as cutters, milling machines, and other devices.[16] One Fab Lab system, which costs about $20,000, has been used to make radio frequency tags to track animals in Norway, engine parts to allow tractors to run on processed castor beans in India, and many other fabrication applications.

The two types of software are *system software*, such as Microsoft Windows Vista, which controls basic computer operations, including start-up and printing, and *applications software*, such as Microsoft Office 2007, which allows you to accomplish specific tasks, including word processing or creating spreadsheets.[17] Software is needed for computers of all sizes, from small handheld computers to large supercomputers.[18] Although most software can be installed from CDs, many of today's software packages can be downloaded through the Internet.[19]

Sophisticated application software, such as Adobe Creative Suite 3, can be used to design, develop, print, and place professional-quality advertising, brochures, posters, prints, and videos on the Internet.[20]

software
The computer programs that govern the operation of the computer.

Sophisticated application software, such as Adobe Creative Suite, can be used to design, develop, print, and place professional-quality advertising, brochures, posters, prints, and videos on the Internet.

(Source: Courtesy of Adobe Systems Corporation.)

database
An organized collection of facts and information.

Databases

A **database** is an organized collection of facts and information, typically consisting of two or more related data files. An organization's database can contain facts and information on customers, employees, inventory, competitors' sales, online purchases, and much more. Most managers and executives consider a database to be one of the most valuable parts of a computer-based information system.[21] A number of health insurance companies are now making their databases available to their customers through the Internet.[22] Aetna, for example, provides important health data to millions of its customers. Aetna customers can also place their own health information, such as blood pressure measurements taken at home, on the comprehensive database. However, making databases accessible can pose risks. The Department of Education decided to limit access to its database of college student loan information to banks and financial institutions.[23] The database contains over 50 million records on student loans that could be inappropriately used to market financial products to students and their families.

telecommunications
The electronic transmission of signals for communications, which enables organizations to carry out their processes and tasks through effective computer networks.

Telecommunications, Networks, and the Internet

Telecommunications is the electronic transmission of signals for communications, which enables organizations to carry out their processes and tasks through effective computer networks. The Associated Press was one of the first users of telecommunications in the 1920s, sending news over 103,000 miles of wire in the United States and almost 10,000 miles of cable across the ocean.[24] Today, telecommunications is used by organizations of all sizes and individuals around the world. The U.S. government is expected to spend almost $50 billion on upgraded telecommunications systems and equipment in the next several years.[25] With telecommunications, people can work at home or while traveling.[26] This approach to work, often called *telecommuting,* allows a telecommuter living in England to send his or her work to the United States, China, or any location with telecommunications capabilities. Today, China is the largest provider of mobile phone and telecommunications services, with over 300 million subscribers.[27]

networks
Computers and equipment that are connected in a building, around the country, or around the world to enable electronic communications.

Networks connect computers and equipment in a building, around the country, or around the world to enable electronic communication. Investment firms can use wireless networks to connect thousands of investors with brokers or traders. Many hotels use wireless telecommunications to allow guests to connect to the Internet, retrieve voice messages, and exchange e-mail without plugging their computers or mobile devices into an Internet connector. Wireless transmission also allows drones, such as Boeing's Scan Eagle, to fly using a remote control system and monitor buildings and other commercial establishments. The drones are smaller and less-expensive versions of the Predator and Global Hawk drones that the U.S. military used in the Afghanistan and Iraq conflicts.

Internet
The world's largest computer network, actually consisting of thousands of interconnected networks, all freely exchanging information.

The **Internet** is the world's largest computer network, consisting of thousands of interconnected networks, all freely exchanging information. Research firms, colleges, universities, high schools, and businesses are just a few examples of organizations using the Internet.

People use the Internet wherever they are to research information, buy and sell products and services, make travel arrangements, conduct banking, download music and videos, and listen to radio programs.

(Source: © Bob Daemmrich / Photo Edit.)

People use the Internet to research information, buy and sell products and services, make travel arrangements, conduct banking, download music and videos, and listen to radio programs, among other activities.[28] Increasingly, the Internet is used for communications, collaboration, and information sharing.[29] Internet sites like MySpace (*www.myspace.com*) and FaceBook (*www.facebook.com*) have become popular places to connect with friends and colleagues.[30] Some people, however, fear that this increased usage can lead to problems, including criminals hacking into the Internet and gaining access to sensitive personal information.[31] Large computers, personal computers, and today's cell phones, such as Apple's iPhone, can access the Internet.[32] This not only speeds communications, but also allows you to conduct business electronically. Some airline companies are providing Internet service on their flights so that travelers can send and receive e-mail, check investments, and browse the Internet.[33] Internet users can create *Web logs (blogs)* to store and share their thoughts and ideas with others around the world. Using *podcasting*, you can download audio programs or music from the Internet to play on computers or music players. One of the authors uses podcasts to obtain information on information systems and technology used throughout this book. You can also record and store TV programs on computers or special viewing devices and watch them later.[34] Often called *place shifting*, this technology allows you to record TV programs at home and watch them at a different place when it's convenient. The *World Wide Web (WWW)*, or the *Web*, is a network of links on the Internet to documents containing text, graphics, video, and sound. Information about the documents and access to them are controlled and provided by tens of thousands of special computers called Web servers. The Web is one of many services available over the Internet and provides access to millions of documents. New Internet technologies and increased Internet communications are collectively called *Web 2.0*.

The technology used to create the Internet is also being applied within companies and organizations to create **intranets**, which allow people in an organization to exchange information and work on projects. Companies, for example, often use intranets to connect its employees around the globe. An **extranet** is a network based on Web technologies that allows selected outsiders, such as business partners and customers, to access authorized resources of a company's intranet. Companies can move all or most of their business activities to an extranet site for corporate customers. Many people use extranets every day without realizing it—to track shipped goods, order products from their suppliers, or access customer assistance from other companies. If you log on to the FedEx site (*www.fedex.com*) to check the status of a package, for example, you are using an extranet.

People

People can be the most important element in most computer-based information systems. People make the difference between success and failure for most organizations. Information systems personnel include all the people who manage, run, program, and maintain the system. Large banks can hire IS personnel to speed the development of computer-related projects. Users are people who work with information systems to get results. Users include financial executives, marketing representatives, manufacturing operators, and many others. Certain computer users are also IS personnel.

Procedures

Procedures include the strategies, policies, methods, and rules for using the CBIS, including the operation, maintenance, and security of the computer. For example, some procedures describe when each program should be run. Others describe who can access facts in the database, or what to do if a disaster, such as a fire, earthquake, or hurricane, renders the CBIS unusable. Good procedures can help companies take advantage of new opportunities and avoid potential disasters. Poorly developed and inadequately implemented procedures, however, can cause people to waste their time on useless rules or result in inadequate responses to disasters, such as hurricanes or tornadoes.

Now that we have looked at computer-based information systems in general, we will briefly examine the most common types used in business today. These IS types are covered in more detail in Part 3.

intranet
An internal network based on Web technologies that allows people within an organization to exchange information and work on projects.

extranet
A network based on Web technologies that allows selected outsiders, such as business partners and customers, to access authorized resources of a company's intranet.

procedures
The strategies, policies, methods, and rules for using a CBIS.

BUSINESS INFORMATION SYSTEMS

The most common types of information systems used in business organizations are those designed for electronic and mobile commerce, transaction processing, management information, and decision support. In addition, some organizations employ special-purpose systems, such as virtual reality, that not every organization uses. Together, these systems help employees in organizations accomplish routine and special tasks—from recording sales, processing payrolls, and supporting decisions in various departments, to providing alternatives for large-scale projects and opportunities. Although these systems are discussed in separate sections in this chapter and explained in more detail later, they are often integrated in one product and delivered by the same software package. See Figure 1.5. For example, some enterprise resource planning packages process transactions, deliver information, and support decisions.

Figure 1.5

Business Information Systems

Business information systems are often integrated in one product and can be delivered by the same software package.

Figure 1.6 shows a simple overview of the development of important business information systems discussed in this section.

Figure 1.6

The Development of Important Business Information Systems

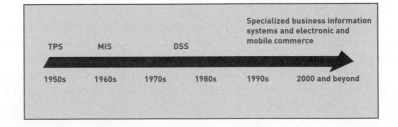

Electronic and Mobile Commerce

e-commerce

Any business transaction executed electronically between companies (business-to-business), companies and consumers (business-to-consumer), consumers and other consumers (consumer-to-consumer), business and the public sector, and consumers and the public sector.

E-commerce involves any business transaction executed electronically between companies (business-to-business, B2B), companies and consumers (business-to-consumer, B2C), consumers and other consumers (consumer-to-consumer, C2C), business and the public sector, and consumers and the public sector. You might assume that e-commerce is reserved mainly for consumers visiting Web sites for online shopping. But Web shopping is only a small part of the e-commerce picture; the major volume of e-commerce—and its fastest-growing segment—is business-to-business (B2B) transactions that make purchasing easier for corporations.[35] This growth is being stimulated by increased Internet access, growing user confidence, rapidly improving Internet and Web security, and better payment systems. PayPal, an e-commerce payment system, for example, processes about $1.5 billion in e-commerce transactions annually.[36] E-commerce also offers opportunities for small businesses to market and sell at a low cost worldwide, allowing them to enter the global market. **Mobile commerce (m-commerce)** refers to transactions conducted anywhere, anytime. M-commerce relies on wireless communications that managers and corporations use to place orders and conduct business with handheld computers, portable phones, laptop computers connected to a network, and other mobile devices. Today, mobile commerce can use cell phones to pay for goods and services.[37] After an account is set up, text messages can be sent and received using a cell phone to authorize purchases. In South Korea, cell phones are used 70 percent of the time to pay for digital content, such as digital music.

mobile commerce (m-commerce)
Transactions conducted anywhere, anytime.

With mobile commerce (m-commerce), people can use cell phones to pay for goods and services anywhere, anytime.

(Source: Courtesy of AP Photo/Itsuo Inouye.)

E-commerce offers many advantages for streamlining work activities. Figure 1.7 provides a brief example of how e-commerce can simplify the process of purchasing new office furniture from an office-supply company. In the manual system, a corporate office worker must get approval for a purchase that exceeds a certain amount. That request goes to the purchasing department, which generates a formal purchase order to procure the goods from the approved vendor. Business-to-business e-commerce automates the entire process. Employees go directly to the supplier's Web site, find the item in a catalog, and order what they need at a price set by their company. If approval is required, the approver is notified automatically. As the use of e-commerce systems grows, companies are phasing out their traditional systems. The resulting growth of e-commerce is creating many new business opportunities.

In addition to e-commerce, business information systems use telecommunications and the Internet to perform many related tasks. *Electronic procurement (e-procurement)*, for example, involves using information systems and the Internet to acquire parts and supplies. **Electronic business (e-business)** goes beyond e-commerce and e-procurement by using information systems and the Internet to perform all business-related tasks and functions, such as accounting, finance, marketing, manufacturing, and human resource activities. E-business also includes working with customers, suppliers, strategic partners, and stakeholders. Compared to traditional business strategy, e-business strategy is flexible and adaptable. See Figure 1.8.

electronic business (e-business)
Using information systems and the Internet to perform all business-related tasks and functions.

Figure 1.7

E-Commerce Greatly Simplifies Purchasing

Traditional process for placing a purchase order

E-commerce process for placing a purchase order

Figure 1.8

Electronic Business

E-business goes beyond e-commerce to include using information systems and the Internet to perform all business-related tasks and functions, such as accounting, finance, marketing, manufacturing, and human resources activities.

transaction

Any business-related exchange, such as payments to employees, sales to customers, and payments to suppliers.

transaction processing system (TPS)

An organized collection of people, procedures, software, databases, and devices used to record completed business transactions.

Enterprise Systems: Transaction Processing Systems and Enterprise Resource Planning

Transaction Processing Systems

Since the 1950s, computers have been used to perform common business applications. Many of these early systems were designed to reduce costs by automating routine, labor-intensive business transactions. A **transaction** is any business-related exchange such as payments to employees, sales to customers, or payments to suppliers. Thus, processing business transactions was the first computer application developed for most organizations. A **transaction processing system (TPS)** is an organized collection of people, procedures, software, databases, and devices used to record completed business transactions. If you understand a transaction processing system, you understand basic business operations and functions.

One of the first business systems to be computerized was the payroll system (see Figure 1.9). The primary inputs for a payroll TPS are the number of employee hours worked during the week and the pay rate. The primary output consists of paychecks. Early payroll systems produced employee paychecks and related reports required by state and federal agencies, such as the Internal Revenue Service. Other routine applications include sales ordering, customer billing and customer relationship management, and inventory control. Some automobile companies, for example, use their TPSs to buy billions of dollars of needed parts

Figure 1.9

A Payroll Transaction Processing System

In a payroll TPS, the inputs (numbers of employee hours worked and pay rates) go through a transformation process to produce outputs (paychecks).

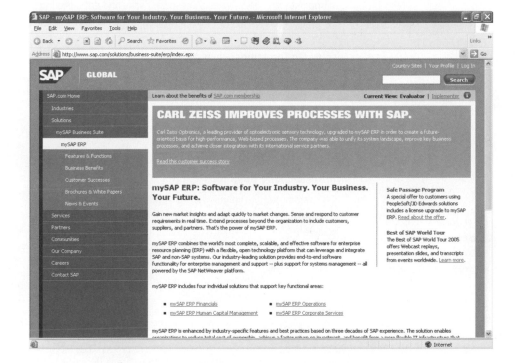

SAP AG, a German software company, is one of the leading suppliers of ERP software. The company employs more than 34,000 people in more than 50 countries.

each year through Internet sites. Because these systems handle and process daily business exchanges, or transactions, they are all classified as TPSs.

Enterprise systems help organizations perform and integrate important tasks, such as paying employees and suppliers, controlling inventory, sending invoices, and ordering supplies. In the past, companies accomplished these tasks using traditional transaction processing systems. Today, they are increasingly being performed by enterprise resource planning systems.

Enterprise Resource Planning

An **enterprise resource planning (ERP) system** is a set of integrated programs that manages the vital business operations for an entire multisite, global organization. An ERP system can replace many applications with one unified set of programs, making the system easier to use and more effective. SAP AG, a German software company, is one of the leading suppliers of ERP software. The company employs more than 34,000 people in more than 50 countries.

enterprise resource planning (ERP) system
A set of integrated programs capable of managing a company's vital business operations for an entire multisite, global organization.

Information and Decision Support Systems

The benefits provided by an effective TPS are tangible and justify their associated costs in computing equipment, computer programs, and specialized personnel and supplies. A TPS can speed business activities and reduce clerical costs. Although early accounting and financial TPSs were already valuable, companies soon realized that they could use the data stored in these systems to help managers make better decisions, whether in human resource management, marketing, or administration. Satisfying the needs of managers and decision makers continues to be a major factor in developing information systems.

Green Data Centers

Mid-sized to large businesses maintain powerful computers called servers that store data and run software to provide information system services to users on the corporate network and Internet. Large corporations might maintain hundreds or even thousands of servers in large facilities called data centers. For example, Microsoft is building a 400,000 square foot data center in San Antonio at a cost of $550 million. Google is investing $750 million in a data center facility in Goose Creek, South Carolina, and a $600 million in a facility in Lenoir, North Carolina.

Because businesses rely on information and its management, the demand for powerful data centers is rapidly growing. Unfortunately, data centers require a lot of power to run and to cool. It is estimated that the money required to cool a data center is equal to the cost of the servers themselves. With the increased awareness of global warming and the contributions of coal-burning power plants to this problem, data centers are drawing the attention of environmentalists and others who want to save energy. The electricity required to run data centers worldwide doubled between 2000 and 2005. This trend is expected to continue; one report estimates that by 2010, the world will require at least 10 new 1,000 megawatt power plants to support the increased demands of data centers.

Governments and corporations, wanting to do what they can to minimize the impact of data centers on the environment and gain some good publicity in the process, are taking action. At the end of 2006, President Bush signed a law authorizing the Environmental Protection Agency (EPA) to analyze the effect of data centers on the environment. The United States federal government has big plans to consolidate its own many data centers into smaller, more efficient facilities. The United Kingdom is evaluating its data centers and moving to greener technologies and techniques to comply with new environmental policies and laws passed in the United Kingdom and the European Union.

The report from the EPA projects that data center power consumption could be cut by as much as 20 percent if data center managers take simple steps such as using power management systems, turning off unused servers, and consolidating resources.

Manufacturers are working on new technologies to minimize power consumption in servers. One company is even experimenting with building a data center in an abandoned coal mine underground, where cooling requirements will be minimal. It is estimated that the subterranean data center will save $9 million per year.

Clearly the current power requirements of information systems and the concern over global warming are at odds. Technology companies are well aware of these concerns and are directing the power of technology at finding solutions to the problem.

Discussion Questions

1. In what ways do information systems negatively affect the environment? Are there positive effects as well? If so, what are they?
2. What can be done to minimize the effect of data centers on the environment?

Critical Thinking Questions

1. Consider our rapidly growing dependence on data centers. What is the risk of this dependence on our society?
2. How might the geography of our planet change if the growth of data centers continues increasing? Will there come a time when the growth levels out?

SOURCES: Bushell, Sue, "British Government Turns Green," *CIO*, December 20, 2007, *www.cio.com.au/index.php/id;1300344377*. Levine, Barry, "Data Center Study Looks at Global Trends," *Top Tech News*, December 14, 2007, *www.toptechnews.com/story.xhtml?story_id=57242*. Brodkin, Jon, "Server electricity use doubled from 2000 to 2005," *itWorld Canada*, December 10, 2007, *www.itworldcanada.com/a/Green-IT/149cf7ef-2d04-41b5-a23d-0732d23c5e40.html*. Mullins, Robert, "Bush signs law to study data center energy usage," *Computerworld*, December 22, 2006, *www.computerworld.com/action/article.do?command=viewArticleBasic&articleId=90066 98*. Gittlen, Sandra, "Data center land grab: How to get ready for the rush," *Computerworld*, March 12, 2007, *www.computerworld.com/action/article.do?command=viewArticleBasic&articleId=90129 63*. Mellor, Chris, "Sun to set up underground data center to save on power," Computerworld, November 16, 2007, *www.computerworld.com/action/article.do?command=viewArticleBasic&articleId=90474 78*.

Management Information Systems

A **management information system** (MIS) is an organized collection of people, procedures, software, databases, and devices that provides routine information to managers and decision makers. An MIS focuses on operational efficiency. Marketing, production, finance, and other functional areas are supported by MISs and linked through a common database. MISs typically provide standard reports generated with data and information from the TPS (see Figure 1.10). Producing a report that describes inventory that should be ordered is an example of an MIS.

management information system (MIS)
An organized collection of people, procedures, software, databases, and devices that provides routine information to managers and decision makers.

Figure 1.10

Management Information System

Functional management information systems draw data from the organization's transaction processing system.

MISs were first developed in the 1960s and typically use information systems to produce managerial reports. In many cases, these early reports were produced periodically—daily, weekly, monthly, or yearly. Because of their value to managers, MISs have proliferated throughout the management ranks. For instance, the total payroll summary report produced initially for an accounting manager might also be useful to a production manager to help monitor and control labor and job costs.

Decision Support Systems

By the 1980s, dramatic improvements in technology resulted in information systems that were less expensive but more powerful than earlier systems. People at all levels of organizations began using personal computers to do a variety of tasks; they were no longer solely dependent on the IS department for all their information needs. People quickly recognized that computer systems could support additional decision-making activities. A **decision support system** (DSS) is an organized collection of people, procedures, software, databases, and devices that support problem-specific decision making. The focus of a DSS is on making effective decisions. Whereas an MIS helps an organization "do things right," a DSS helps a manager "do the right thing."

Decision support systems are used when the problem is complex and the information needed to make the best decision is difficult to obtain and use. So, a DSS also involves managerial judgment and perspective. Managers often play an active role in developing and implementing the DSS. A DSS recognizes that different managerial styles and decision types require different systems. For example, two production managers in the same position trying to solve the same problem might require different information and support. The overall emphasis is to support, rather than replace, managerial decision making.

A DSS can include a collection of models used to support a decision maker or user (model base), a collection of facts and information to assist in decision making (database), and systems and procedures (user interface or dialogue manager) that help decision makers and other users interact with the DSS (see Figure 1.11). Software is often used to manage the database—the database management system (DBMS)—and the model base—the model management system (MMS). Not all DSSs have all of these components.

decision support system (DSS)
An organized collection of people, procedures, software, databases, and devices used to support problem-specific decision making.

Decisioneering provides decision support software called Crystal Ball, which helps business people of all types assess risks and make forecasts. Shown here is the Standard Edition being used for oil field development.

(Source: Crystal Ball screenshot courtesy of Decisioneering, Inc.)

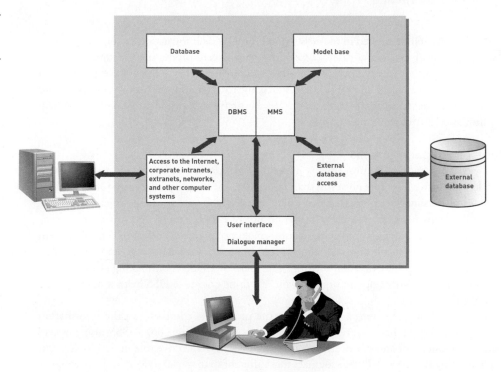

Figure 1.11

Essential DSS Elements

In addition to DSSs for managers, group support systems and executive support systems use the same approach to support groups and executives. A *group support system*, includes the DSS elements just described and software, called *groupware*, to help groups make effective decisions. An executive support system, also called an *executive information system*, helps top-level managers, including a firm's president, vice presidents, and members of the board of directors, make better decisions. An executive support system can assist with strategic planning, top-level organizing and staffing, strategic control, and crisis management.

Specialized Business Information Systems: Knowledge Management, Artificial Intelligence, Expert Systems, and Virtual Reality

In addition to TPSs, MISs, and DSSs, organizations often rely on specialized systems. Many use *knowledge management systems (KMSs)*, an organized collection of people, procedures, software, databases, and devices to create, store, share, and use the organization's knowledge and experience.[38] A shipping company, for example, can use a KMS to streamline its transportation and logistics business.

In addition to knowledge management, companies use other types of specialized systems. Experimental specialized systems in cars can help prevent accidents.[39] These new systems allow cars to communicate with each other using radio chips installed in their trunks. When two or more cars move too close together, the specialized systems sound alarms and brake in some cases. Some specialized systems are based on the notion of **artificial intelligence** (**AI**), in which the computer system takes on the characteristics of human intelligence. The field of artificial intelligence includes several subfields (see Figure 1.12). Some people predict that in the future, we will have nanobots, small molecular-sized robots, traveling throughout our bodies and in our bloodstream, keeping us healthy. Other nanobots will be embedded in products and services, making our lives easier and creating new business opportunities.

artificial intelligence (AI)
A field in which the computer system takes on the characteristics of human intelligence.

A Nissan Motor Company car swerves back into its lane on its own shortly after it ran off the track during a test of the Lane Departure Prevention feature, which also sounds a warning when the car veers out of its lane.

(Source: Courtesy of AP Photo/ Katsumi Kasahara.)

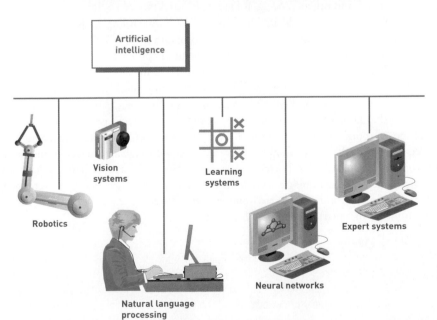

Figure 1.12

The Major Elements of Artificial Intelligence

Artificial Intelligence

Robotics is an area of artificial intelligence in which machines take over complex, dangerous, routine, or boring tasks, such as welding car frames or assembling computer systems and components. Vision systems allow robots and other devices to "see," store, and process visual images. Natural language processing involves computers understanding and acting on verbal or written commands in English, Spanish, or other human languages. Learning systems allow computers to learn from past mistakes or experiences, such as playing games or making business decisions, and neural networks is a branch of artificial intelligence that allows computers to recognize and act on patterns or trends. Some successful stock, options, and futures traders use neural networks to spot trends and make them more profitable with their investments.

Expert Systems

expert system
A system that gives a computer the ability to make suggestions and act like an expert in a particular field.

knowledge base
The collection of data, rules, procedures, and relationships that must be followed to achieve value or the proper outcome.

Expert systems give the computer the ability to make suggestions and act like an expert in a particular field. It can help the novice user perform at the level of an expert. The unique value of expert systems is that they allow organizations to capture and use the wisdom of experts and specialists. Therefore, years of experience and specific skills are not completely lost when a human expert dies, retires, or leaves for another job. Expert systems can be applied to almost any field or discipline. They have been used to monitor nuclear reactors, perform medical diagnoses, locate possible repair problems, design and configure IS components, perform credit evaluations, and develop marketing plans for a new product or new investment strategy. The collection of data, rules, procedures, and relationships that must be followed to achieve value or the proper outcome is contained in the expert system's **knowledge base**.

Virtual Reality

virtual reality
The simulation of a real or imagined environment that can be experienced visually in three dimensions.

Virtual reality is the simulation of a real or imagined environment that can be experienced visually in three dimensions. Cigna Healthcare, for example, is experimenting with a virtual reality game designed to help treat cancer in young adults and children.[40] Developed by HopeLab (*www.hopelab.org*), the virtual reality game called Re-Mission shows young adults and children how to combat cancer.

Originally, virtual reality referred to immersive virtual reality, which means the user becomes fully immersed in an artificial, computer-generated 3D world. The virtual world is presented in full scale and relates properly to the human size. Virtual reality can also refer to applications that are not fully immersive, such as mouse-controlled navigation through a 3D environment on a graphics monitor, stereo viewing from the monitor via stereo glasses, stereo projection systems, and others. Boeing, for example, used virtual reality and computer simulation to help design and build its Dreamliner 787.[41] According to Kevin Fowler, vice president of process integration, "A breakthrough program like the 787 Dreamliner needed to lead the way in performance, quality, cost and schedule supported by efficient and effective production planning." Boeing used 3D models from Dassault Systems to design and manufacture the new aircraft. Retail stores like Saks Fifth Avenue and Neiman-Marcus are using virtual reality to help advertise high-end products on the Internet.[42] In one virtual ad campaign, about $500,000 of orders from over 20 countries were received in less than a week.

A variety of input devices, such as head-mounted displays (see Figure 1.13), data gloves, joysticks, and handheld wands, allow the user to navigate through a virtual environment and to interact with virtual objects. Directional sound, tactile and force feedback devices, voice recognition, and other technologies enrich the immersive experience. Because several people can share and interact in the same environment, virtual reality can be a powerful medium for communication, entertainment, and learning.

Boeing used virtual reality and computer simulation when designing and building its Dreamliner 787 aircraft.

(Source: Courtesy of Frank Brandmaier/dpa/Landov.)

Figure 1.13

A Head-Mounted Display

The head-mounted display (HMD) was the first device to provide the wearer with an immersive experience. A typical HMD houses two miniature display screens and an optical system that channels the images from the screens to the eyes, thereby presenting a stereo view of a virtual world. A motion tracker continuously measures the position and orientation of the user's head and allows the image-generating computer to adjust the scene representation to the current view. As a result, the viewer can look around and walk through the surrounding virtual environment.

(Source: Courtesy of 5DT, Inc. *www.5dt.com.*)

It is difficult to predict where information systems and technology will be in 10 to 20 years. It seems, however, that we are just beginning to discover the full range of their usefulness. Technology has been improving and expanding at an increasing rate; dramatic growth and change are expected for years to come. Without question, a knowledge of the effective use of information systems will be critical for managers both now and in the long term. But how are these information systems created?

SYSTEMS DEVELOPMENT

Systems development is the activity of creating or modifying business systems. Systems development projects can range from small to very large in fields as diverse as stock analysis and video game development. Some systems development efforts are a huge success. Wachovia Corporation and Investment Bank, for example, used systems development to create a new computer-trading platform that increased processing capacity by a factor of three, while dramatically reducing costs.[43] According to Tony Bishop, senior vice president of the firm, "We looked at the current system and said, 'Where can we build standardized frameworks, components, and services…?' We now do pricing in milliseconds, not seconds, for either revenue protection or revenue gain." Other systems development efforts fail to meet their cost or schedule goals. A large federal database designed to track hundreds of millions of dollars of money transfers in an effort to curb terrorism was delayed by several years.[44] Scheduled to be implemented by 2007, the system may not be ready until 2010. Some also question whether the $32 million budget for the new system will be met. In another case, Colorado Governor Bill Ritter ordered that all centralized systems development efforts be placed under the state's chief information officer to curb expensive computer systems that

systems development
The activity of creating or modifying business systems.

don't work correctly.[45] In previous years, some have claimed that over $300 million was spent on systems that couldn't pay welfare benefits on time, issue overtime payments to road crews, issue license plates correctly, or accurately track unemployment benefits. A new Colorado vehicle registration system may also have to be abandoned because of faulty systems development procedures, costing Colorado taxpayers over $10 million.[46] Systems development failures can be a result of poor planning and scheduling, insufficient management of risk, poor requirements determination, and lack of user involvement.[47] Training people to use a new or modified system can be critical to the successful implementation of these systems and can help avoid systems development failures.[48]

People inside a company can develop systems, or companies can use *outsourcing*, hiring an outside company to perform some or all of a systems development project.[49] Outsourcing allows a company to focus on what it does best and delegate other functions to companies with expertise in systems development. Outsourcing, however, is not the best alternative for all companies.

Developing information systems to meet business needs is highly complex and difficult—so much so that it is common for IS projects to overrun budgets and exceed scheduled completion dates. One strategy for improving the results of a systems development project is to divide it into several steps, each with a well-defined goal and set of tasks to accomplish (see Figure 1.14). These steps are summarized next.

Figure 1.14

An Overview of Systems Development

Systems Investigation and Analysis

The first two steps of systems development are systems investigation and analysis. The goal of the *systems investigation* is to gain a clear understanding of the problem to be solved or opportunity to be addressed. After an organization understands the problem, the next question is, "Is the problem worth solving?" Given that organizations have limited resources—people and money—this question deserves careful consideration. If the decision is to continue with the solution, the next step, *systems analysis*, defines the problems and opportunities of the existing system. During systems investigation and analysis, as well as design maintenance and review, discussed next, the project must have the complete support of top-level managers and focus on developing systems that achieve business goals.

Systems Design, Implementation, and Maintenance and Review

Systems design determines how the new system will work to meet the business needs defined during systems analysis. *Systems implementation* involves creating or acquiring the various system components (hardware, software, databases, etc.) defined in the design step, assembling them, and putting the new system into operation. The purpose of *systems maintenance and review* is to check and modify the system so that it continues to meet changing business needs. Increasingly, companies are hiring outside companies to do their design, implementation, maintenance, and review functions.

ORGANIZATIONS AND INFORMATION SYSTEMS

An **organization** is a formal collection of people and other resources established to accomplish a set of goals. The primary goal of a for-profit organization is to maximize shareholder value, often measured by the price of the company stock. Nonprofit organizations include social groups, religious groups, universities, and other organizations that do not have profit as their goal.

An organization is a system, which means that it has inputs, processing mechanisms, outputs, and feedback. An organization constantly uses money, people, materials, machines and other equipment, data, information, and decisions. As shown in Figure 1.15, resources such as materials, people, and money serve as inputs to the organizational system from the environment, go through a transformation mechanism, and then are produced as outputs to the environment. The outputs from the transformation mechanism are usually goods or services, which are of higher relative value than the inputs alone. Through adding value or worth, organizations attempt to achieve their goals.

organization
A formal collection of people and other resources established to accomplish a set of goals.

Figure 1.15

A General Model of an Organization

Information systems support and work within all parts of an organizational process. Although not shown in this simple model, input to the process subsystem can come from internal and external sources. Just prior to entering the subsystem, data is external. After it enters the subsystem, it becomes internal. Likewise, goods and services can be output to either internal or external systems.

Providing value to a stakeholder—customer, supplier, manager, shareholder, or employee—is the primary goal of any organization. The value chain, first described by Michael Porter in a 1985 *Harvard Business Review* article, reveals how organizations can add

value chain
A series (chain) of activities that includes inbound logistics, warehouse and storage, production, finished product storage, outbound logistics, marketing and sales, and customer service.

value to their products and services. The **value chain** is a series (chain) of activities that includes inbound logistics, warehouse and storage, production, finished product storage, outbound logistics, marketing and sales, and customer service (see Figure 1.16). You investigate each activity in the chain to determine how to increase the value perceived by a customer. Depending on the customer, value might mean lower price, better service, higher quality, or uniqueness of product. The value comes from the skill, knowledge, time, and energy that the company invests in the product or activity. The value chain is just as important to companies that don't manufacture products, but provide services, such as tax preparers and legal firms. By adding a significant amount of value to their products and services, companies ensure success. Combining a value chain with just-in-time (JIT) inventory means companies can deliver materials or parts when they are needed. Ball Aerospace, for example, uses JIT to help reduce inventory costs and enhance customer satisfaction.[50]

Combining a value chain with just-in-time (JIT) inventory means companies can deliver materials or parts when they are needed. Ball Aerospace uses JIT to help reduce inventory costs and enhance customer satisfaction.

(Source: AP Photo/Denver Post, R. J. Sangosti.)

Figure 1.16

The Value Chain of a Manufacturing Company

Managing raw materials, inbound logistics, and warehouse and storage facilities is called *upstream management,* and managing finished product storage, outbound logistics, marketing and sales, and customer service is called *downstream management.*

Wal-Mart's use of information systems is an integral part of its operation. It gives suppliers access to its inventory system, so the suppliers can monitor the database and automatically send another shipment when stocks are low, eliminating the need for purchase orders, which speeds delivery time, lowers Wal-Mart's inventory carrying costs, and reduces stockout costs.

(Source: *www.walmart.com.*)

Managing the supply chain and customer relationships are two key parts of managing the value chain. *Supply chain management (SCM)* helps determine what supplies are required for the value chain, what quantities are needed to meet customer demand, how the supplies should be processed (manufactured) into finished goods and services, and how the shipment of supplies and products to customers should be scheduled, monitored, and controlled.[51] For example, in an automotive company, SCM can identify key supplies and parts, negotiate with vendors for the best prices and support, make sure that all supplies and parts are available to manufacture cars and trucks, and send finished products to dealerships around the country when they are needed. Increasingly, SCM is accomplished using the Internet and electronic marketplaces (e-marketplaces). When an organization has many suppliers, it can use Internet exchanges to negotiate good prices and service. SCM is becoming a global issue as companies have parts and products made around the world.[52] According to Jean Philipe Thenoz, vice president of CMA-CGM, a worldwide shipper, "The client wants to know where the blue socks in size medium are that he ordered two weeks ago from China."

Customer relationship management (CRM) programs help a company manage all aspects of customer encounters, including marketing and advertising, sales, customer service after the sale, and programs to retain loyal customers. CRM can help a company collect customer data, contact customers, educate them about new products, and actively sell products to existing and new customers. Often, CRM software uses a variety of information sources, including sales from retail stores, surveys, e-mail, and Internet browsing habits, to compile comprehensive customer profiles. CRM systems can also get customer feedback to help design new products and services. Tesco, Britain's largest retail operation, uses a CRM Clubcard program to provide outstanding customer service and deliver loyalty rewards and perks to valued customers. See Figure 1.17. Customers can earn services such as meals out, travel, dry cleaning, and car maintenance. The Clubcard loyalty program also extends to Tesco's business partners, introducing Tesco customers to other businesses. To be of most benefit, CRM programs must be tailored for each company or organization.[53] According to Amanda Zuniga, senior research analyst at pharmaceutical intelligence firm Cutting Edge Information, "Customer Relationship Management programs must be individually tailored to meet the programs' specific objectives."

Organizational Culture and Change

culture
A set of major understandings and assumptions shared by a group.

organizational culture
The major understandings and assumptions for a business, corporation, or other organization.

Culture is a set of major understandings and assumptions shared by a group, such as within an ethnic group or a country. **Organizational culture** consists of the major understandings and assumptions for a business, corporation, or other organization. The understandings, which can include common beliefs, values, and approaches to decision making, are often not stated or documented as goals or formal policies. For example, Procter & Gamble has an organizational culture that places an extremely high value on understanding its customers and their needs. For marketing recommendations to be accepted, they must be based on facts known about customers. As another example, employees might be expected to be clean-cut, wear conservative outfits, and be courteous in dealing with all customers. Sometimes organizational culture is formed over years. In other cases, top-level managers can form it rapidly by starting a "casual Friday" dress policy. Organizational culture can also have a positive effect on the successful development of new information systems that support the organization's culture.[54]

organizational change
The responses that are necessary so that for-profit and nonprofit organizations can plan for, implement, and handle change.

Organizational change deals with how for-profit and nonprofit organizations plan for, implement, and handle change. Change can be caused by internal factors, such as those initiated by employees at all levels, or external factors, such as activities wrought by competitors, stockholders, federal and state laws, community regulations, natural occurrences (such as hurricanes), and general economic conditions. Many European countries, for example, adopted the euro, a single European currency, which changed how financial companies do business and use their information systems. Organizational change also occurs when two or more organizations merge. When organizations merge, however, integrating their information systems can be critical to future success.[55] Unfortunately, many organizations only consider the integration of their different information systems late in the merger process.

Change can be sustaining or disruptive.[56] *Sustaining change* can help an organization improve the supply of raw materials, the production process, and the products and services it offers. New manufacturing equipment to make disk drives is an example of a sustaining change for a computer manufacturer. The new equipment might reduce the costs of producing the disk drives and improve overall performance. *Disruptive change,* on the other hand, often harms an organization's performance or even puts it out of business. In general, disruptive technologies might not originally have good performance, low cost, or even strong demand. Over time, however, they often replace existing technologies. They can cause good, stable companies to fail when they don't change or adopt the new technology.

User Satisfaction and Technology Acceptance

To be effective, reengineering and continuous improvement efforts must result in satisfied users and be accepted and used throughout the organization. Over the years, IS researchers have studied user satisfaction and technology acceptance as they relate to IS attitudes and usage. Although user satisfaction and technology acceptance started as two separate theories, some believe that they can be integrated into one.[57]

User satisfaction with a computer system and the information it generates often depends on the quality of the system and the information.[58] A quality information system is usually flexible, efficient, accessible, and timely. Recall that quality information is accurate, reliable, current, complete, and delivered in the proper format.[59]

The **technology acceptance model (TAM)** specifies the factors that can lead to better attitudes about the information system, along with higher acceptance and usage of the system in an organization.[60] These factors include the perceived usefulness of the technology, the ease of its use, the quality of the information system, and the degree to which the organization supports its use.[61]

You can determine the actual usage of an information system by the amount of technology diffusion and infusion.[62] **Technology diffusion** is a measure of how widely technology is spread throughout an organization. An organization in which computers and information systems are located in most departments and areas has a high level of technology diffusion.[63] Some online merchants, such as Amazon.com, have a high diffusion and use computer systems to perform most of their business functions, including marketing, purchasing, and billing. **Technology infusion**, on the other hand, is the extent to which technology permeates an area or department. In other words, it is a measure of how deeply embedded technology is in an area of the organization. Some architectural firms, for example, use computers in all aspects of designing a building from drafting to final blueprints. The design area, thus, has a high level of infusion. Of course, a firm can have a high level of infusion in one part of its operations and a low level of diffusion overall. The architectural firm might use computers in all aspects of design (high infusion in the design area), but not to perform other business functions, including billing, purchasing, and marketing (low diffusion). Diffusion and infusion often depend on the technology available now and in the future, the size and type of the organization, and the environmental factors that include the competition, government regulations, suppliers, and so on. This is often called the technology, organization, and environment (TOE) framework.[64]

technology acceptance model (TAM)
A model that describes the factors that lead to higher levels of acceptance and usage of technology.

technology diffusion
A measure of how widely technology is spread throughout the organization.

technology infusion
The extent to which technology is deeply integrated into an area or department.

Although an organization might have a high level of diffusion and infusion, with computers throughout the organization, this does not necessarily mean that information systems are being used to their full potential. In fact, the assimilation and use of expensive computer technology throughout organizations varies greatly.[65] Companies hope that a high level of diffusion, infusion, satisfaction, and acceptance will lead to greater performance and profitability.[66]

COMPETITIVE ADVANTAGE

competitive advantage
A significant and (ideally) long-term benefit to a company over its competition.

A **competitive advantage** is a significant and (ideally) long-term benefit to a company over its competition, and can result in higher-quality products, better customer service, and lower costs. Establishing and maintaining a competitive advantage is complex, but a company's survival and prosperity depend on its success in doing so. An organization often uses its information system to help achieve a competitive advantage. According to Meg McCarthy, "At Aetna, the IT organization is critical to enabling the implementation of our business strategy. I report to the chairman of our company and I am a member of the executive committee. In that capacity, I participate in all of the key business conversations/decisions that impact the company strategy and the technology strategy."[67] In his book *Good to Great,* Jim Collins outlines how technology can be used to accelerate companies from good to great.[68] Table 1.3 shows how a few companies accomplished this move. Ultimately, it is not how much a company spends on information systems but how it makes and manages investments in technology. Companies can spend less and get more value.

Table 1.3

How Some Companies Used Technologies to Move from Good to Great

(Source: Data from Jim Collins, *Good to Great*, Harper Collins Books, 2001, p. 300.)

Company/Business Unit	Business	Competitive Use of Information Systems
Circuit City	Consumer electronics	Developed sophisticated sales and inventory-control systems to deliver a consistent experience to customers
Gillette	Shaving products	Developed advanced computerized manufacturing systems to produce high-quality products at low cost
Walgreens	Drug and convenience stores	Developed satellite communications systems to link local stores to centralized computer systems
Wells Fargo	Financial services	Developed 24-hour banking, ATMs, investments, and increased customer service using information systems

Factors That Lead Firms to Seek Competitive Advantage

A number of factors can lead to attaining a competitive advantage. Michael Porter, a prominent management theorist, suggested a now widely accepted competitive forces model, also called the **five-forces model**. The five forces include (1) the rivalry among existing competitors, (2) the threat of new entrants, (3) the threat of substitute products and services, (4) the bargaining power of buyers, and (5) the bargaining power of suppliers. The more these forces combine in any instance, the more likely firms will seek competitive advantage and the more dramatic the results of such an advantage will be.

five-forces model
A widely accepted model that identifies five key factors that can lead to attainment of competitive advantage, including (1) the rivalry among existing competitors, (2) the threat of new entrants, (3) the threat of substitute products and services, (4) the bargaining power of buyers, and (5) the bargaining power of suppliers.

Rivalry Among Existing Competitors
Typically, highly competitive industries are characterized by high fixed costs of entering or leaving the industry, low degrees of product differentiation, and many competitors. Although all firms are rivals with their competitors, industries with stronger rivalries tend to have more firms seeking competitive advantage. To gain an advantage over competitors, companies constantly analyze how they use their resources and assets. This *resource-based view* is an

approach to acquiring and controlling assets or resources that can help the company achieve a competitive advantage. For example, a transportation company might decide to invest in radio-frequency technology to tag and trace products as they move from one location to another.

Threat of New Entrants

A threat appears when entry and exit costs to an industry are low and the technology needed to start and maintain a business is commonly available. For example, a small restaurant is threatened by new competitors in the restaurant industry. Owners of small restaurants do not require millions of dollars to start the business, food costs do not decline substantially for large volumes, and food processing and preparation equipment is easily available. When the threat of new market entrants is high, the desire to seek and maintain competitive advantage to dissuade new market entrants is also usually high.

In the restaurant industry, competition is fierce because entry costs are low. Therefore, a small restaurant that enters the market can be a threat to existing restaurants.

(Source: © Sergio Pitamitz/Getty Images.)

Threat of Substitute Products and Services

Companies that offer one type of goods or services are threatened by other companies that offer similar goods or services. The more consumers can obtain similar products and services that satisfy their needs, the more likely firms are to try to establish competitive advantage. For example, consider the photographic industry. When digital cameras became popular, traditional film companies had to respond to stay competitive and profitable. Traditional film companies, such as Kodak and others, started to offer additional products and enhanced services, including digital cameras, the ability to produce digital images from traditional film cameras, and Web sites that could be used to store and view pictures.

Bargaining Power of Customers and Suppliers

Large customers tend to influence a firm, and this influence can increase significantly if the customers can threaten to switch to rival companies. When customers have a lot of bargaining power, companies increase their competitive advantage to retain their customers. Similarly, when the bargaining power of suppliers is strong, companies need to improve their competitive advantage to maintain their bargaining position. Suppliers can also help an organization gain a competitive advantage. Some suppliers enter into strategic alliances with firms and eventually act as a part of the company. Suppliers and companies can use telecommunications to link their computers and personnel to react quickly and provide parts or supplies as necessary to satisfy customers. Government agencies are also using strategic alliances. The investigative units of the U.S. Customs and Immigration and Naturalization Service entered into a strategic alliance to streamline investigations.

Strategic Planning for Competitive Advantage

To be competitive, a company must be fast, nimble, flexible, innovative, productive, economical, and customer oriented. It must also align its IS strategy with general business strategies and objectives.[69] Given the five market forces just mentioned, Porter and others have proposed a number of strategies to attain competitive advantage, including cost leadership, differentiation, niche strategy, altering the industry structure, creating new products and services, and improving existing product lines and services.[70] In some cases, one of these strategies becomes dominant. For example, with a cost leadership strategy, cost can be the key consideration, at the expense of other factors if need be.

- **Cost leadership.** Deliver the lowest possible price for products and services. Wal-Mart and other discount retailers have used this strategy for years. Cost leadership is often achieved by reducing the costs of raw materials through aggressive negotiations with suppliers, becoming more efficient with production and manufacturing processes, and reducing warehousing and shipping costs. Some companies use outsourcing to cut costs when making products or completing services.

Wal-Mart and other discount retailers have used a cost leadership strategy to deliver the lowest possible price for products and services.

(Source: © Jeff Zelevansky/Getty Images.)

- **Differentiation.** Deliver different products and services. This strategy can involve producing a variety of products, giving customers more choices, or delivering higher-quality products and services. Many car companies make different models that use the same basic parts and components, giving customers more options. Other car companies attempt to increase perceived quality and safety to differentiate their products and appeal to consumers who are willing to pay higher prices for these features. Companies that try to differentiate their products often strive to uncover and eliminate fake products produced and delivered by others.[71] Some believe counterfeit products cost companies about $600 billion annually. To distinguish their products from fakes, companies are inserting microscopic particles or other markers to allow them, government regulators, and law enforcement agencies to distinguish genuine products from bogus ones.
- **Niche strategy.** Deliver to only a small, niche market. Porsche, for example, doesn't produce inexpensive station wagons or large sedans. It makes high-performance sports cars and SUVs. Rolex only makes high-quality, expensive watches. It doesn't make inexpensive, plastic watches that can be purchased for $20 or less.

INFORMATION SYSTEMS @ WORK

Grand & Toy Seeks Competitive Advantage by Tracking Key Performance Indicators

The landscape of today's global business environment is one of mergers, acquisitions, and relationships. Companies combine and form partnerships to share benefits and extend their reach into new regions and markets. OfficeMax, a dominant global force in retail and commercial office supplies, with close to 1,000 super-stores worldwide, extended its reach into Canada by acquiring Grand & Toy, Canada's largest commercial office products company. Rather than rebranding and reorganizing Grand & Toy (named after James Grand and Samuel Toy), OfficeMax allowed Grand & Toy to operate independently, a wise move since Grand & Toy has proved itself to be a smart company, implementing cutting-edge information systems with impressive results.

Recently, Grand & Toy has been experimenting with new information management tools in hopes of better understanding the ebb and flow of market demand and weaknesses within its organization. Grand & Toy uses two tools from Clarity Systems—a Key Performance Indicator (KPI) reporting application that tracks important data in the system over time, and a tool called the Defec-tor Detector, which indicates when a customer reduces order quantity. Using these tools, Grand & Toy can closely monitor its customers and business to determine when and where it has successes and problems.

Grand & Toy's new information system from Clarity is an example of a trend in business management called corporate performance management (CPM). CPM is an extension of ERP that allows a business to use the information it collects to analyze and improve business practices and processes. Analyzing business processes often depends on analyzing key indicators—specific information within the system that indicates broader trends. For example, many senior managers look at growth, profitability, productivity, and satisfaction as methods of determin-ing the health of a business. Key indicator information can be com-bined to compile scores for each of these areas. Tracking these

areas over time help managers and employees evaluate the health of the company. Information systems are used to analyze these areas at a variety of levels.

Using corporate performance management tools decision-makers at Grand & Toy can evaluate the company's performance at many levels with little effort. Using the Clarity system, a manager can study sales for a specific geographic area, on a given day, or for the entire corporation for the previous year. Tracking key indicators and adjusting corporate performance accordingly is one way businesses are working to win a competitive advantage.

Discussion Questions

1. What types of data do you think go in to scoring the growth, profitability, productivity, and satisfaction of a company like Grand & Toy?
2. How can Grand & Toy maintain a competitive advantage over other office supply companies that use the same information system from Clarity?

Critical Thinking Questions

1. What key indicators would you watch if you managed an office supply business? Which are internal, and which are external?
2. Three hundred employees of Grand & Toy use the Clarity sys-tem on a daily basis. Why do you think it is important to provide access to the system to so many people in the company?

SOURCES: Ruffolo, Rafael, "Grand & Toy Gets Business Performance Clarity," *itWorldCanada*, July 6, 2007, *www.itworldcanada.com/Pages/Docbase/ ViewArticle.aspx?ID=idgml-36107f64- e342-4b25-8968-5ef53dd8433e&Portal=2e6e7040-2373-432d- b393-91e487ee7d70&ParaStart=0&ParaEnd=10&direction=next&Next=Next.* Staff, "Retail Case Study: Grand & Toy," Clarity Systems Case Studies, accessed December 26, 2007, *www.claritysystems.com/Resources/ CaseStudies.aspx.* Grand & Toy Web site, accessed December 26, 2007, *www.grandandtoy.com.*

Porsche is an example of a company with a niche strategy, producing only high-performance sports cars and SUVs.

(Source: © Sajjad Hussain/AFP/ Getty Images.)

strategic alliance (strategic partnership)
An agreement between two or more companies that involves the joint production and distribution of goods and services.

- **Altering the industry structure.** Change the industry to become more favorable to the company or organization. The introduction of low-fare airline carriers, such as Southwest Airlines, has forever changed the airline industry, making it difficult for traditional airlines to make high profit margins. To fight back, airlines such as Delta launched their own low-fare flights. Creating strategic alliances can also alter the industry structure. A **strategic alliance**, also called a **strategic partnership**, is an agreement between two or more companies that involves the joint production and distribution of goods and services.

- **Creating new products and services.** Introduce new products and services periodically or frequently. This strategy always helps a firm gain a competitive advantage, especially for the computer industry and other high-tech businesses. If an organization does not introduce new products and services every few months, the company can quickly stagnate, lose market share, and decline. Companies that stay on top are constantly developing new products and services. A large U.S. credit-reporting agency, for example, can use its information system to help it explore new products and services in different markets.

- **Improving existing product lines and service.** Make real or perceived improvements to existing product lines and services. Manufacturers of household products are always advertising new and improved products. In some cases, the improvements are more perceived than real refinements; usually, only minor changes are made to the existing product, such as to reduce the amount of sugar in breakfast cereal. Some direct-order companies are improving their service by using Radio Frequency Identification (RFID) tags to identify and track the location of their products as they are shipped from one location to another. Customers and managers can instantly locate products as they are shipped from suppliers to the company, to warehouses, and finally to customers.

- **Other strategies.** Some companies seek strong *growth* in sales, hoping that it can increase profits in the long run due to increased sales. Being the *first to market* is another competitive strategy. Apple Computer was one of the first companies to offer complete and ready-to-use personal computers. Some companies offer *customized* products and services to achieve a competitive advantage. Dell, for example, builds custom PCs for consumers. *Hire the best people* is another example of a competitive strategy. The assumption is that the best people will determine the best products and services to deliver to the market and the best approach to deliver these products and services. Companies can also combine one or more of these strategies. In addition to customization, Dell attempts to offer low-cost computers (cost leadership) and top-notch service (differentiation).

PERFORMANCE-BASED INFORMATION SYSTEMS

Businesses have passed through at least three major stages in their use of information systems. In the first stage, organizations focused on using information systems to reduce costs and improve productivity. The National ePrescribing Patient Safety Initiative, for example, offers powerful software to doctors used to reduce medication errors and costs.[72] United Airlines uses flight-planning software to help it schedule flights and compute fuel needs based on distance traveled and wind currents.[73] The company estimates that it saves almost $1,500 on fuel for many long-distance flights. In this stage, companies generally ignored the revenue potential, not looking for opportunities to use information systems to increase sales. The second stage was defined by Porter and others. It was oriented toward gaining a competitive advantage. In many cases, companies spent large amounts on information systems and downplayed the costs. Today, companies are shifting from strategic management to performance-based management of their information systems. In this third stage, companies carefully consider both strategic advantage and costs. They use productivity, return on investment (ROI), net present value, and other measures of performance to evaluate the contributions their information systems make to their businesses. Figure 1.18 illustrates these stages. This balanced approach attempts to reduce costs and increase revenues.

Figure 1.18

Three Stages in the Business Use of Information Systems

Productivity

Developing information systems that measure and control productivity is a key element for most organizations. **Productivity** is a measure of the output achieved divided by the input required. A higher level of output for a given level of input means greater productivity; a lower level of output for a given level of input means lower productivity. The numbers

productivity
A measure of the output achieved divided by the input required.

assigned to productivity levels are not always based on labor hours—productivity can be based on factors such as the amount of raw materials used, resulting quality, or time to produce the goods or service. The value of the productivity number is not as significant as how it compares with other time periods, settings, and organizations. A number of politicians and healthcare professionals hope that keeping electronic medical records (EMRs) on computerized databases will increase the productivity of doctors and healthcare professionals as well as reducing healthcare costs.[74]

$$Productivity = (Output \: / \: Input) \times 100\%$$

After a basic level of productivity is measured, an information system can monitor and compare it over time to see whether productivity is increasing. Then, a company can take corrective action if productivity drops below certain levels. An automotive company, for example, might use robots in assembling new cars to increase its labor productivity and reduce costs. In addition to measuring productivity, an information system can be used within a process to significantly increase productivity. Thus, improved productivity can result in faster customer response, lower costs, and increased customer satisfaction. A study of Canadian productivity increases, for example, showed that more than half of the country's productivity gains were caused by improvements in equipment and machinery.[75] 20 percent was caused by worker improvements.

Return on Investment and the Value of Information Systems

return on investment (ROI)
One measure of IS value that investigates the additional profits or benefits that are generated as a percentage of the investment in IS technology.

One measure of IS value is **return on investment (ROI)**. This measure investigates the additional profits or benefits that are generated as a percentage of the investment in IS technology. A small business that generates an additional profit of $20,000 for the year as a result of an investment of $100,000 for additional computer equipment and software would have a return on investment of 20 percent ($20,000/$100,000). Because of the importance of ROI, many computer companies provide ROI calculators to potential customers. ROI calculators are typically provided on a vendor's Web site and can be used to estimate returns. According to Megan Burns, an analyst for Forrester Research, "What ROI models allow you to do is run through the what if scenarios ..."[76]

Paris-based PPR, France's biggest clothing retailer, recently acquired Puma AG, Europe's second-biggest sporting-goods maker. The merger can help the company create a global brand that straddles sports and fashion.

(Source: Courtesy of AP Photo/ Christof Stache.)

Earnings Growth

Another measure of IS value is the increase in profit, or earnings growth, it brings. For instance, a mail-order company might install an order-processing system that generates a seven percent earnings growth compared with the previous year.

Market Share and Speed to Market

Market share is the percentage of sales that a product or service has in relation to the total market. If installing a new online catalog increases sales, it might help a company increase its market share by 20 percent. IS can also help organizations bring new products and services to customers in less time. This is often called speed to market. A music producer, for example, can bring a new song or record to the market by placing it on an online music site faster than it can produce CDs and ship them to retail music stores.

Customer Awareness and Satisfaction

Although customer satisfaction can be difficult to quantify, about half of today's best global companies measure the performance of their information systems based on feedback from internal and external users. Some companies and nonprofit organizations use surveys and questionnaires to determine whether the IS investment has increased customer awareness and satisfaction. University researchers at the University of Auckland, for example, developed surveys and other measurements of an electronic learning system, called CECIL, to determine the satisfaction and experience of students using the electronic learning approach.[77]

Total Cost of Ownership

Another way to measure the value of information systems was developed by the Gartner Group and is called the **total cost of ownership (TCO)**. TCO is the sum of all costs over the life of the information system and includes the cost to acquire the technology, technical support, administrative costs, end-user operations, etc. Market research groups often use TCO to compare products and services. For example, a survey of large global enterprises ranked messaging and collaboration software products using the TCO model.[78] TransUnion Interactive, a credit reporting company, uses TCO to rate and select hardware.[79] According to TransUnion's Chief Technology Officer, "Looking at the total cost of ownership and short implementation cycle, Azul's hardware was the best alternative for us, providing minimal downside risk." TCO is also used by many other companies to rate and select hardware, software, databases, and other computer-related components.

Return on investment, earnings growth, market share, customer satisfaction, and TCO are only a few measures that companies use to plan for and maximize the value of their IS investments. Regardless of the difficulties, organizations must attempt to evaluate the contributions that information systems make to assess their progress and plan for the future. Information systems and personnel are too important to leave to chance.

total cost of ownership (TCO)
The measurement of the total cost of owning computer equipment, including desktop computers, networks, and large computers.

Risk

In addition to the return-on-investment measures of a new or modified system, managers must also consider the risks of designing, developing, and implementing these systems. Information systems can sometimes be costly failures. Some companies, for example, have attempted to implement ERP systems and failed, costing them millions of dollars. In other cases, e-commerce applications have been implemented with little success. The costs of development and implementation can be greater than the returns from the new system. The risks of designing, developing, and implementing new or modified systems are covered in more detail in Chapter 8, which discuss systems development.

CAREERS IN INFORMATION SYSTEMS

Realizing the benefits of any information system requires competent and motivated IS personnel, and many companies offer excellent job opportunities. *Knowledge workers (KWs)* are people who create, use, and disseminate knowledge. They are usually professionals in science, engineering, business, and other areas who specialize in information systems. Numerous schools have degree programs with such titles as information systems, computer information systems, and management information systems. These programs are typically offered by information schools, business schools, and within computer science departments. Graduating students with degrees in information systems have attracted high starting salaries. In addition, students are increasingly completing business degrees with a global or international orientation. Here are some of the IS skills that some experts believe are important for IS workers to have.[80]

1. Machine learning
2. Mobilizing applications
3. Wireless networking
4. Human-computer interface
5. Project management
6. General networking skills
7. Network convergence technicians
8. Open-source programming
9. Business intelligence systems
10. Embedded security
11. Digital home technology integration
12. Languages, including C#, C++, and Java

The job market in the early 2000s was tight. Many jobs were lost in U.S. companies as firms merged, outsourced certain jobs overseas, or went bankrupt. Today, demand for IS personnel is on the rise, along with salaries.[81] The U.S. Department of Labor's Bureau of Labor Statistics predicts that many technology jobs will increase through 2012 or beyond. Today, the median salary for IS personnel is almost $80,000, while the average IS manager makes slightly more than $100,000.[82] Table 1.4 summarizes some of the best places to work as an IS professional.[83]

Table 1.4

Best Places to Work as an IS Professional

Source: Brandel, Mary, "Best Places to Work in IT 2007," *Computerworld,* June 18, 2007, p. 34.

Company	Description
Quick Loans	The Internet loan company offers many training opportunities and has excellent promotion practices, benefits, and high retention.
University of Miami	The university is highly rated for diversity. It has good training and career development.
Sharp Health Care	The company has a training budget for IS professionals of about $3,500 annually per person. About half of its managers are women.
The Capital Groups Companies	This investment management firm has a low turnover rate for IS professionals. The company has a reputation for business managers working closely with IS professionals.
The Mitre Corporation	This is a nonprofit organization started at the Massachusetts Institute of Technology (MIT). The organization has good work/life balance programs with flextime, part-time work, and telecommuting.
BAE Systems, Inc.	This large defense company offers many opportunities for IS professionals to work on the latest technologies and computer systems. The company has a training budget for IS professionals of about $30 million.
General Mills, Inc.	As a maker of such brands as Green Giant and Betty Crocker, the company has a successful "Women in IS" program that helps women advance in their IS careers. The company also has an IS Manager's forum that helps IS managers with global issues, project management, and recruitment issues.
University of Pennsylvania	The university offers an excellent child-support program for its IS workers, including its Baby Prep 101 seminar. The university also offers other programs and seminars, including retirement planning.
Anheuser-Busch Companies	This popular brewer offers good vacation programs for its IS professionals. Eligible employees also get free beer or nonalcoholic beverages every month.
Fairfax County Public Schools	This large public school system helps young IS professionals advance in their careers by offering the IT Leadership Development Cohort Program. The school also gives out a number of awards, like the Going The Extra Mile award.

Opportunities in information systems are also available to people from foreign countries, including Russia and India. The U.S. H-1B and L-1 visa programs seek to allow skilled employees from foreign lands into the United States.[84] These programs, however, are limited and in high demand. The L-1 visa program is often used for intracompany transfers for multinational companies. The H-1B program can be used for new employees. In the first few days that applications were available for the H-1B program in 2007, over 130,000 applications were filed for 65,000 positions.[85] The number of H-1B visas offered annually can be political and controversial.[86] Some fear that the H-1B program is being abused to replace high paid U.S. workers with less expensive foreign workers. Some believe that companies pretend to seek U.S. workers, while actually seeking less expensive foreign workers.[87] In 2007, two U.S. senators on the Senate Judiciary Subcommittee on Immigration sent letters of concern to a number of Indian firms that were using the H-1B program to staff their U.S. operations with IS personnel from foreign countries.[88] One IS professional, concerned about abuses in the H1-B program, got a law degree and is now suing companies that he feels are violating H-1B rules.[89] Others, however, believe the H-1B program and similar programs are invaluable to the U.S. economy and competitiveness.

Roles, Functions, and Careers in IS

There are many exciting and rewarding careers in IS as reported by the Bureau of Labor Statistics and research done by a variety of organizations.[90] Professionals with careers in information systems can work in an IS department or outside a traditional IS department as Web developers, computer programmers, systems analysts, computer operators, and many other positions. In addition to technical skills, they need skills in written and verbal communication, an understanding of organizations and the way they operate, and the ability to work with people and in groups. Today, many good information, business, and computer science schools require these business and communications skills of their graduates.

In general, IS professionals are charged with maintaining the broadest perspective on organizational goals. Most medium to large organizations manage information resources through an IS department. In smaller businesses, one or more people might manage information resources, with support from outsourced services. (Recall that outsourcing is also popular with larger organizations.) As shown in Figure 1.19, the IS organization has three primary responsibilities: operations, systems development, and support.

Operations

People in the operations component of a typical IS department work with information systems in corporate or business unit computer facilities. They tend to focus more on the *efficiency* of IS functions rather than their effectiveness.

System operators primarily run and maintain IS equipment, and are typically trained at technical schools or through on-the-job experience. They are responsible for starting, stopping, and correctly operating mainframe systems, networks, tape drives, disk devices, printers, and so on. Other operations include scheduling, hardware maintenance, and preparing input and output. Data-entry operators convert data into a form the computer system can use. They can use terminals or other devices to enter business transactions, such as sales orders and payroll data. Increasingly, data entry is being automated—captured at the source of the transaction rather than entered later. In addition, companies might have local area network and Web operators who run the local network and any Web sites the company has.

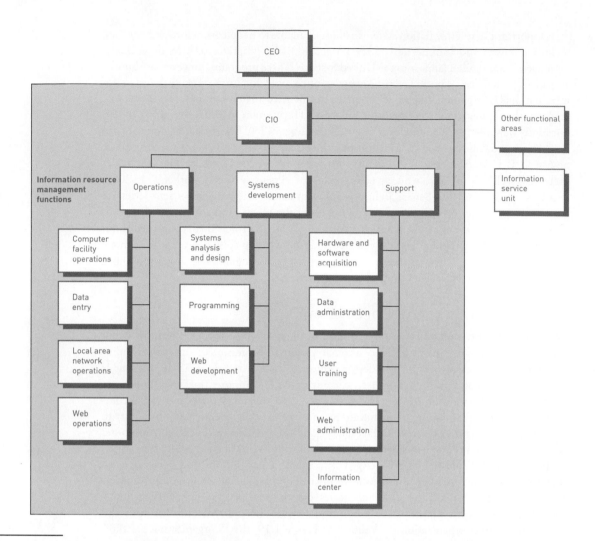

Figure 1.19

The Three Primary
Responsibilities of Information
Systems

Each of these elements—
operations, systems development,
and support—contains sub-
elements that are critical to the
efficient and effective performance
of the organization.

Systems Development

The systems development component of a typical IS department focuses on specific development projects and ongoing maintenance and review. Systems analysts and programmers, for example, address these concerns to achieve and maintain IS effectiveness. The role of a systems analyst is multifaceted. Systems analysts help users determine what outputs they need from the system and construct plans for developing the necessary programs that produce these outputs. Systems analysts then work with one or more programmers to make sure that the appropriate programs are purchased, modified from existing programs, or developed. A computer programmer uses the plans the systems analyst created to develop or adapt one or more computer programs that produce the desired outputs. In some cases, foreign companies are actively recruiting IS professionals in the U.S. Katrina Anderson, for example, was hired by Infosys, an Indian company, to train in India and return to the U.S. as a programmer.[91] According to Anderson, "The opportunity to train in India was eye-opening, as I came to realize how respected and prominent Infosys is within the country." To help businesses select the best analysts and programmers, companies such as TopCoder offer tests to evaluate the proficiency and competence of current IS employees or job candidates. TopCoder Collegiate Challenge allows programming students to compete with other programmers around the world.[92] Some companies, however, are skeptical of the usefulness of these types of tests.[93]

With the dramatic increase in the use of the Internet, intranets, and extranets, many companies have Web or Internet developers who create effective and attractive Web sites for customers, internal personnel, suppliers, stockholders, and others who have a business relationship with the company. The Internet is also being used to help with systems development projects.[94] Professor Luis von Ahn of Carnegie Mellon University, for example, has developed an Internet game that shows images to two or more players and asks them to type key words

System operators focus on the efficiency of IS functions, rather than their effectiveness. Their primary responsibility is to run and maintain IS equipment.

(Source: Courtesy iStockphoto.com.)

that describe the image. If the key words match, points are given in the game. Professor von Ahn uses the matches to develop key words to describe and categorize the images on the Internet. In other words, the game players are doing the work of developing names and categories for images on the Web, an important function that is difficult to do without human judgment and work.

Support

The support component of a typical IS department provides user assistance in hardware and software acquisition and use, data administration, user training and assistance, and Web administration. Increasingly, training is done using the Internet.[95] Microsoft, for example, offers free training in areas including time management, marketing, sales, and other areas (*office.microsoft.com/en-us/officelive/FX102119031033.aspx*). Other companies, such as Hewlett Packard (*www.hp.com/sbso*), also offer online training. In many cases, support is delivered through an information center.

IS personnel provide assistance in hardware and software acquisition, data administration, user training and assistance, and Web administration.

(Source: Courtesy of Christ Schmidt/iStockPhoto.com.)

Because IS hardware and software are costly, a specialized support group often manages computer hardware and software acquisitions. This group sets guidelines and standards for the rest of the organization to follow in making purchases. They must gain and maintain an understanding of available technology and develop good relationships with vendors.

A database administrator focuses on planning, policies, and procedures regarding the use of corporate data and information. For example, database administrators develop and disseminate information about the corporate databases for developers of IS applications. In addition, the database administrator monitors and controls database use.

User training is key to get the most from any information system, and the support area ensures that appropriate training is available. Training can be provided by internal staff or from external sources. For example, internal support staff can train managers and employees in the best way to enter sales orders, to receive computerized inventory reports, and to submit expense reports electronically. Companies also hire outside firms to help train users in other areas, including the use of word processing, spreadsheets, and database programs.

Web administration is another key area for support staff. With the increased use of the Internet and corporate Web sites, Web administrators are sometimes asked to regulate and monitor Internet use by employees and managers to make sure that it is authorized and appropriate. Web administrators also maintain the corporate Web site to keep it accurate and current, which can require substantial resources.

The support component typically operates the information center. An **information center** provides users with assistance, training, application development, documentation, equipment selection and setup, standards, technical assistance, and troubleshooting. Although many firms have attempted to phase out information centers, others have changed their focus from technical training to helping users find ways to maximize the benefits of the information resource.

Information Service Units

An **information service unit** is basically a miniature IS department attached and directly reporting to a functional area in a large organization. Notice the information service unit shown earlier in Figure 1.19. Even though this unit is usually staffed by IS professionals, the project assignments and the resources necessary to accomplish these projects are provided by the functional area to which it reports. Depending on the policies of the organization, the salaries of IS professionals staffing the information service unit might be budgeted to either the IS department or the functional area.

Typical IS Titles and Functions

The organizational chart shown in Figure 1.19 is a simplified model of an IS department in a typical medium or large organization. Many organizations have even larger departments, with increasingly specialized positions such as librarian or quality assurance manager. Smaller firms often combine the roles shown in Figure 1.19 into fewer formal positions.

The Chief Information Officer

The role of the chief information officer (CIO) is to employ an IS department's equipment and personnel to help the organization attain its goals.[96] The CIO is usually a vice president concerned with the overall needs of the organization, who sets corporate-wide policies, and plans, manages, and acquires information systems. In one survey, more than 60 percent of CIOs reported directly to the president of the company or the chief executive officer (CEO).[97] Some of the CIO's top concerns include integrating IS operations with corporate strategies, keeping up with the rapid pace of technology, and defining and assessing the value of systems development projects. According to a survey, almost 80 percent of CIOs are actively involved in or consulted on most major decisions.[98] Tom Shelman, CIO for Northrop Grumman Corporation, for example, changed his job description to be more strategic, meet with customers, and help win new business.[99] When asked about his new role, Mr. Shelman said, "It was a significant change."

Depending on the size of the IS department, several people might work in senior IS managerial levels. Some job titles associated with IS management are the CIO, vice president of information systems, manager of information systems, and chief technology officer (CTO). A central role of all these people is to communicate with other areas of the organization to determine changing needs. Often these employees are part of an advisory or steering committee that helps the CIO and other IS managers make decisions about the use of information systems. Together they can best decide what information systems will support corporate goals. The CTO, for example, typically works under a CIO and specializes in networks and related equipment and technology.

information center
A support function that provides users with assistance, training, application development, documentation, equipment selection and setup, standards, technical assistance, and troubleshooting.

information service unit
A miniature IS department.

A company's CIO is usually a vice president who sets corporate-wide policies, and plans, manages, and acquires information systems.

(Source: © Click Productions/Getty Images.)

LAN Administrators

Local area network (LAN) administrators set up and manage the network hardware, software, and security processes. They manage the addition of new users, software, and devices to the network. They also isolate and fix operations problems. LAN administrators are in high demand and often solve both technical and nontechnical problems.

Internet Careers

The bankruptcy of some Internet start-up companies in the early 2000s, called the *dot-gone era* by some, has resulted in layoffs for some firms. Executives of these bankrupt start-up Internet companies lost hundreds of millions of dollars in a few months. Yet, the use of the Internet to conduct business continues to grow and has stimulated a steady need for skilled personnel to develop and coordinate Internet usage. As shown in Figure 1.19, these careers are in the areas of Web operations, Web development, and Web administration. As with other areas in IS, many top-level administrative jobs are related to the Internet. These career opportunities are found in both traditional companies and those that specialize in the Internet.

Internet jobs within a traditional company include Internet strategists and administrators, Internet systems developers, Internet programmers, and Internet or Web site operators. Some companies suggest a new position, chief Internet officer, with responsibilities and a salary similar to the CIO's.

In addition to traditional companies, Internet companies offer exciting career opportunities. These companies include Amazon.com, Yahoo!, eBay, and many others. Systest, for example, specializes in finding and eliminating digital bugs that could halt the operation of a computer system.[100]

Often, the people filling IS roles have completed some form of certification. **Certification** is a process for testing skills and knowledge resulting in an endorsement by the certifying authority that an individual is capable of performing a particular job. Certification frequently involves specific, vendor-provided or vendor-endorsed coursework. Popular certification programs include Microsoft Certified Systems Engineer, Certified Information Systems Security Professional (CISSP), Oracle Certified Professional, Cisco Certified Security Professional (CCSP), and many others.[101]

certification
A process for testing skills and knowledge, which results in a statement by the certifying authority that states an individual is capable of performing a particular kind of job.

Other IS Careers

To respond to the increase in attacks on computers, new and exciting careers have developed in security and fraud detection and prevention. Today, many companies have IS security positions such as a chief information security officer or a chief privacy officer. Some universities offer degree programs in security or privacy. The National Insurance Crime Bureau, a nonprofit organization supported by roughly 1,000 property and casualty insurance companies, uses computers to join forces with special investigation units and law enforcement agencies, as well as to conduct online fraud-fighting training to investigate and prevent these types of crimes.[102] The University of Denver has offered a program in video-game development.[103] It is even possible to work from home in an IS field. Programmers, systems developers, and others are also working from home in developing new information systems.

In addition to working for an IS department in an organization, IS personnel can work for large consulting firms, such as Accenture (*www.accenture.com*), IBM (*www.ibm.com/services*), EDS (*www.eds.com*), and others.[104] Some consulting jobs can entail frequent travel because consultants are assigned to work on various projects wherever the client is. Such roles require excellent people and project management skills in addition to IS technical skills.

Other IS career opportunities include being employed by technology companies, such as Microsoft (*www.microsoft.com*), Google (*www.google.com*), Dell (*www.dell.com*), and many others. Such a role enables an individual to work on the cutting edge of technology, which can be extremely challenging and exciting. As some computer companies cut their services to customers, new companies are being formed to fill the need. With names such as Speak with a Geek and the Geek Squad, these companies are helping people and organizations with their computer-related problems that computer vendors are no longer solving.

Finding a Job in IS

There are many traditional approaches to finding a job in the information systems area, including on-campus visits from recruiters and referrals from professors, friends, and family members. Many colleges and universities have excellent programs to help students develop résumés and conduct job interviews. Developing an online resume can be critical to finding a good job. Many companies accept résumés online and use software to search for key words and skills used to screen job candidates.[105] Thus, having the right key words and skills can mean the difference between getting a job interview and not being considered.

Increasingly, students are using the Internet and other sources to find IS jobs. Many Web sites, such as Monster.com, post job opportunities for Internet careers and more traditional careers. Most large companies list job opportunities on their Web sites. These sites allow prospective job hunters to browse job opportunities, locations, salaries, benefits, and other factors. In addition, some sites allow job hunters to post their résumés. Many of the social networking sites, including MySpace and Facebook, can be used to help get job leads. Facebook and Jobster are investigating a joint venture to launch a job site available on Facebook's Internet site.[106] Corporate recruiters also use the Internet or Web logs (blogs) to gather information on existing job candidates or to locate new job candidates.[107] According to Ryan Loken, an executive at Wal-Mart Stores, Inc., "Blogs are a tool in the tool kit." Loken has used the Internet and blogs to fill over 100 corporate positions. To fill more than 5,000 intern positions and entry-level jobs, Ernst & Young LLP uses the Internet, blogs, and social networking sites such as Facebook and MySpace. The company has a page on the Facebook Internet site to attract job candidates.[108] "It's a very good thing for communicating with potential job seekers. You're reaching the student in their lair," says Mark Mehler of CareerXroads.

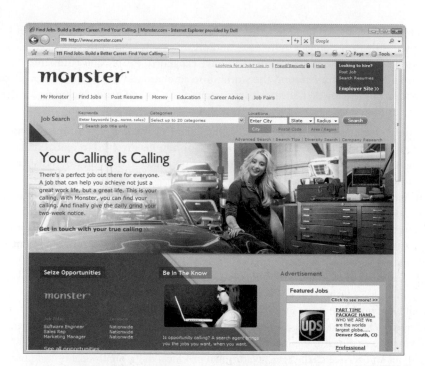

Internet job sites such as Monster.com allow job hunters to browse job opportunities and post their résumés.

(Source: *www.monster.com*.)

GLOBAL CHALLENGES IN INFORMATION SYSTEMS

Changes in society as a result of increased international trade and cultural exchange, often called globalization, have always had a big impact on organizations and their information systems. In his book, *The World Is Flat,* Thomas Friedman describes three eras of globalization.[109] See Table 1.5. According to Friedman, we have progressed from the globalization of countries to the globalization of multinational corporations and individuals. Today, people in remote areas can use the Internet to compete with and contribute to other people, the largest corporations, and entire countries. These workers are empowered by high-speed Internet access, making the world flatter. In the Globalization 3.0 era, designing a new airplane or computer can be separated into smaller subtasks and then completed by a person or small group that can do the best job. These workers can be located in India, China, Russia, Europe, and other areas of the world. The subtasks can then be combined or reassembled into the complete design. This approach can be used to prepare tax returns, diagnose a patient's medical condition, fix a broken computer, and many other tasks.

Era	Dates	Characterized by
Globalization 1.0	Late 1400–1800	Countries with the power to explore and influence the world
Globalization 2.0	1800–2000	Multinational corporations that have plants, warehouses, and offices around the world
Globalization 3.0	2000–today	Individuals from around the world who can compete and influence other people, corporations, and countries by using the Internet and powerful technology tools

Table 1.5

Eras of Globalization

Today's information systems have led to greater globalization. High-speed Internet access and networks that can connect individuals and organizations around the world create more international opportunities. Global markets have expanded. People and companies can get products and services from around the world, instead of around the corner or across town. These opportunities, however, introduce numerous obstacles and issues, including challenges involving culture, language, and many others.

- **Cultural challenges.** Countries and regional areas have their own cultures and customs that can significantly affect individuals and organizations involved in global trade.
- **Language challenges.** Language differences can make it difficult to translate exact meanings from one language to another.
- **Time and distance challenges.** Time and distance issues can be difficult to overcome for individuals and organizations involved with global trade in remote locations. Large time differences make it difficult to talk to people on the other side of the world. With long distance, it can take days to get a product, a critical part, or a piece of equipment from one location to another location.
- **Infrastructure challenges.** High-quality electricity and water might not be available in certain parts of the world. Telephone services, Internet connections, and skilled employees might be expensive or not readily available.
- **Currency challenges.** The value of different currencies can vary significantly over time, making international trade more difficult and complex.
- **Product and service challenges.** Traditional products that are physical or tangible, such as an automobile or bicycle, can be difficult to deliver to the global market. However, *electronic products (e-products)* and *electronic services (e-services)* can be delivered to customers electronically, over the phone, networks, through the Internet, or other electronic means. Software, music, books, manuals, and help and advice can all be delivered globally and over the Internet.
- **Technology transfer issues.** Most governments don't allow certain military-related equipment and systems to be sold to some countries. Even so, some believe that foreign companies are stealing the intellectual property, trade secrets, and copyrighted materials, and are counterfeiting products and services.
- **State, regional, and national laws.** Every state, region, and country has a set of laws that must be obeyed by citizens and organizations operating in the country. These laws can deal with a variety of issues, including trade secrets, patents, copyrights, protection of personal or financial data, privacy, and much more. Laws restricting how data enters or exits a country are often called *transborder data-flow* laws. Keeping track of these laws and incorporating them into the procedures and computer systems of multinational and transnational organizations can be very difficult and time consuming, requiring expert legal advice.
- **Trade agreements.** Countries often enter into trade agreements with each other. The North American Free Trade Agreement (NAFTA) and the Central American Free Trade Agreement (CAFTA) are examples. The European Union (EU) is another example of a group of countries with an international trade agreement.[110] The EU is a collection of mostly European countries that have joined together for peace and prosperity. In addition to NAFTA, CAFTA, and the EU, there are many other trade agreements. The Australia-United States Free Trade Agreement (AUSFTA) was signed into law in 2005. The Korean-United States Free Trade Agreement (KORUS-FTA) was signed into law in 2007.[111] There are free trade agreements between Bolivia and Mexico, Canada and Costa Rica, Canada and Israel, Chile and Korea, Mexico and Japan, the United States and Jordan, and many others.[112]

SUMMARY

Principle

The value of information is directly linked to how it helps decision makers achieve the organization's goals.

Data consists of raw facts; information is data transformed into a meaningful form. The process of defining relationships among data requires knowledge. Knowledge is an awareness and understanding of a set of information and the way that information can support a specific task. To be valuable, information must have several characteristics: It should be accurate, complete, economical to produce, flexible, reliable, relevant, simple to understand, timely, verifiable, accessible, and secure. The value of information is directly linked to how it helps people achieve their organization's goals.

Information systems are sets of interrelated elements that collect (input), manipulate and store (process), and disseminate (output) data and information. Input is the activity of capturing and gathering new data, processing involves converting or transforming data into useful outputs, and output involves producing useful information. Feedback is the output that is used to make adjustments or changes to input or processing activities.

Principle

Knowing the potential impact of information systems and having the ability to put this knowledge to work can result in a successful personal career and organizations that reach their goals.

Information systems play an important role in today's businesses and society. The key to understanding the existing variety of systems begins with learning their fundamentals. The types of systems used within organizations can be classified into four basic groups: (1) e-commerce and m-commerce, (2) TPS and ERP, (3) MIS and DSS, and (4) specialized business information systems.

E-commerce involves any business transaction executed electronically between parties such as companies (business-to-business), companies and consumers (business-to-consumer), business and the public sector, and consumers and the public sector. The major volume of e-commerce and its fastest-growing segment is business-to-business transactions that make purchasing easier for big corporations. E-commerce offers opportunities for small businesses by enabling them to market and sell at a low cost worldwide, thus enabling them to enter the global market. Mobile commerce (m-commerce) are transactions conducted anywhere, anytime. M-commerce relies on the use of wireless communications to allow managers and corporations to place orders and conduct business using handheld computers, portable phones, laptop computers connected to a network, and other mobile devices.

The most fundamental system is the transaction processing system (TPS). A transaction is any business-related exchange. The TPS handles the large volume of business transactions that occur daily within an organization. TPSs include order processing, purchasing, accounting, and related systems.

An enterprise resource planning (ERP) system is a set of integrated programs that is capable of managing a company's vital business operations for an entire multisite, global organization. Although the scope of an ERP system may vary from company to company, most ERP systems provide integrated software to support the manufacturing and finance business functions of an organization.

A management information system (MIS) uses the information from a TPS to generate information useful for management decision making. The focus of an MIS is primarily on operational efficiency. A decision support system (DSS) is an organized collection of people, procedures, databases, and devices used to support problem-specific decision making. The DSS differs from an MIS in the support given to users, the decision emphasis, the development and approach, and system components, speed, and output. The specialized business information systems include knowledge management systems, artificial intelligence systems, expert systems, and virtual reality systems. Knowledge management systems are organized collections of people, procedures, software, databases and devices used to create, store, share, and use the organization's knowledge and experience.

Principle

System users, business managers, and information systems professionals must work together to build a successful information system.

Systems development is the activity of creating or modifying existing business systems. The goal of the systems investigation is to gain a clear understanding of the problem to be solved or opportunity to be addressed. If the decision is to continue with the solution, the next step, systems analysis, defines the problems and opportunities of the existing system. Systems design determines how the new system will work to meet the business needs defined during systems analysis. Systems implementation involves creating or acquiring the various system components (hardware, software, databases, etc.) defined in the design step, assembling them, and putting the new system into operation. The purpose of systems maintenance and review is to check and modify the system so that it continues to meet changing business needs.

Principle

The use of information systems to add value to the organization can also give an organization a competitive advantage.

An organization is a formal collection of people and various other resources established to accomplish a set of goals. The primary goal of a for-profit organization is to maximize shareholder value. Nonprofit organizations include social groups, religious groups, universities, and other organizations that do not have profit as the primary goal. Organizations are systems with inputs, transformation mechanisms, and outputs.

Value-added processes increase the relative worth of the combined inputs on their way to becoming final outputs of the organization. The value chain is a series (chain) of activities that includes (1) inbound logistics, (2) warehouse and storage, (3) production, (4) finished product storage, (5) outbound logistics, (6) marketing and sales, and (7) customer service.

Supply chain management (SCM) helps determine what supplies are required, what quantities are needed to meet customer demand, how the supplies are to be processed (manufactured) into finished goods and services, and how the shipment of supplies and products to customers is to be scheduled, monitored, and controlled. Customer relationship management (CRM) programs help a company manage all aspects of customer encounters, including marketing and advertising, sales, customer service after the sale, and programs to help keep and retain loyal customers. CRM can help a company collect customer data, contact customers, educate customers on new products, and actively sell products to existing and new customers.

Organizations use information systems to support organizational goals. Because information systems typically are designed to improve productivity, methods for measuring the system's impact on productivity should be devised. In the late 1980s and early 1990s, overall productivity did not seem to increase with increases in investments in information systems. Often called the *productivity paradox*, this situation troubled many economists who were expecting to see dramatic productivity gains. In the early 2000s, however, productivity again seemed on the rise.

Organizational culture and change are important internal issues that affect most organizations. Organizational culture consists of the major understandings and assumptions for a business, a corporation, or an organization. Organizational change deals with how for-profit and nonprofit organizations plan for, implement, and handle change. Change can be caused by internal or external factors. Many European countries, for example, adopted the euro, a single European currency, which changed how financial companies do business and how they use their information systems.

User satisfaction with a computer system and the information it generates often depends on the quality of the system and the resulting information. A quality information system is usually flexible, efficient, accessible, and timely. The extent to which technology is used throughout an organization is a function of technology diffusion, infusion, and acceptance. Technology diffusion is a measure of how widely technology is in place throughout an organization. Technology infusion is the extent to which technology permeates an area or department. The technology acceptance model (TAM) investigates factors, such as perceived usefulness of the technology, ease of use of the technology, the quality of the information system, and the degree to which the organization supports the use of the information system, to predict IS usage and performance.

Competitive advantage is usually embodied in either a product or service that has the most added value to consumers and that is unavailable from the competition or in an internal system that delivers benefits to a firm not enjoyed by its competition. The five-forces model covers factors that lead firms to seek competitive advantage: rivalry among existing competitors, the threat of new market entrants, the threat of substitute products and services, the bargaining power of buyers, and the bargaining power of suppliers. Three strategies to address these factors and to attain competitive advantage include altering the industry structure, creating new products and services, and improving existing product lines and services.

The ability of an information system to provide or maintain competitive advantage should also be determined. Several strategies for achieving competitive advantage include enhancing existing products or services or developing new ones, as well as changing the existing industry or creating a new one.

Developing information systems that measure and control productivity is a key element for most organizations. A useful measure of the value of an IS project is return on investment (ROI). This measure investigates the additional profits or benefits that are generated as a percentage of the investment in IS technology. Total cost of ownership (TCO) can also be a useful measure.

Principle

Cooperation between business managers and IS personnel is the key to unlocking the potential of any new or modified system.

Information systems personnel typically work in an IS department that employs a chief information officer, systems analysts, computer programmers, computer operators, and a number of other people. The overall role of the chief information officer (CIO) is to employ an IS department's equipment and personnel in a manner that will help the organization attain its goals. Systems analysts help users determine what outputs they need from the system and construct the plans for developing the necessary programs that produce these outputs. Systems analysts then work with one or more programmers to make sure that the appropriate programs are purchased, modified from existing programs, or developed. The major responsibility of a computer programmer is to use the plans developed by the systems analyst to develop or adapt one or more computer programs that produce the desired outputs. Computer operators are responsible for starting, stopping, and correctly operating mainframe systems, networks, tape drives, disk devices, printers, and so on. LAN administrators set up and manage the network hardware, software, and security processes. Trained personnel are also increasingly needed to set up and manage a company's Internet site, including Internet strategists, Internet systems developers, Internet programmers, and Web site operators. Information systems personnel may also work in other functional departments or areas in a support capacity. In addition to technical skills, IS personnel also need skills in written and verbal communication, an understanding of organizations and the way they operate, and the ability to work with people (users). In general, IS personnel are charged with maintaining the broadest enterprise-wide perspective.

In addition to working for an IS department in an organization, IS personnel can work for one of the large consulting firms, such as Accenture, EDS, and others. Another IS career opportunity is to be employed by a hardware or software vendor developing or selling products.

Today's information systems have led to greater globalization. High-speed Internet access and networks that can connect individuals and organizations around the world create more international opportunities. Global markets have expanded. People and companies can get products and services from around the world, instead of around the corner or across town. These opportunities, however, introduce numerous obstacles and issues, including challenges involving culture, language, and many others.

CHAPTER 1: SELF-ASSESSMENT TEST

The value of information is directly linked to how it helps decision makers achieve the organization's goals.

1. A (An) _____ is a set of interrelated components that collect, manipulate, and disseminate data and information and provide a feedback mechanism to meet an objective.

2. Knowledge workers are usually professionals in science, engineering, business, and other areas. True or False?

Knowing the potential impact of information systems and having the ability to put this knowledge to work can result in a successful personal career and organizations that reach their goals.

3. A (An) _____ consists of hardware, software, databases, telecommunications, people, and procedures.

4. Computer programs that govern the operation of a computer system are called
    a. feedback
    b. feedforward
    c. software
    d. transaction processing system

5. What is an organized collection of people, procedures, software, databases and devices used to create, store, share, and use the organization's experience and knowledge?
    a. TPS (transaction processing system)
    b. MIS (management information system)
    c. DSS (decision support system)
    d. KMS (knowledge management system)

System users, business managers, and information systems professionals must work together to build a successful information system.

6. What involves creating or acquiring the various system components (hardware, software, databases, etc.) defined in the design step, assembling them, and putting the new system into operation?
    a. systems implementation
    b. systems review
    c. systems development
    d. systems design

7. _____ involves anytime, anywhere commerce that uses wireless communications.

8. _____ involves contracting with outside professional services to meet specific business needs.

The use of information systems to add value to the organization can also give an organization a competitive advantage.

9. _____ change can help an organization improve raw materials supply, the production process, and the products and services offered by the organization.

10. Technology infusion is a measure of how widely technology is spread throughout an organization. True or False?

Cooperation between business managers and IS personnel is the key to unlocking the potential of any new or modified system.

11. Who is involved in helping users determine what outputs they need and constructing the plans needed to produce these outputs?
    a. the CIO
    b. the applications programmer
    c. the systems programmer
    d. the systems analyst

12. An information center provides users with assistance, training, and application development. True or False?

13. The _____ is typically in charge of the information systems department or area in a company.

REVIEW QUESTIONS

1. What are the components of any information system?
2. How would you distinguish data and information? Information and knowledge?
3. Identify at least six characteristics of valuable information.
4. What is a computer-based information system? What are its components?
5. What are the most common types of computer-based information systems used in business organizations today? Give an example of each.
6. What is the difference between e-commerce and m-commerce?
7. Describe three applications of virtual reality.
8. What is a knowledge management system? Give an example.
9. What is the technology acceptance model (TAM)?
10. What is user satisfaction?
11. What are some general strategies employed by organizations to achieve competitive advantage?
12. Define the term *productivity*. Why is it difficult to measure the impact that investments in information systems have on productivity?
13. What is supply chain management?
14. What is the total cost of ownership?
15. What is the role of the systems analyst? What is the role of the programmer?
16. What is the operations component of a typical IS department?
17. What is the role of the chief information officer?

DISCUSSION QUESTIONS

1. Describe the "ideal" automated auto license plate renewal system for the drivers in your state. Describe the input, processing, output, and feedback associated with this system.
2. Describe how information systems are used at school or work.
3. You have decided to open an Internet site to buy and sell used music CDs to other students. Describe the value chain for your new business.
4. How is it that useful information can vary widely from the quality attributes of valuable information?
5. What is the difference between DSS and knowledge management?
6. Discuss the potential use of virtual reality to enhance the learning experience for new automobile drivers. How might such a system operate? What are the benefits and potential drawbacks of such a system?
7. Discuss how information systems are linked to the business objectives of an organization.
8. You have been hired to work in the IS area of a manufacturing company that is starting to use the Internet to order parts from its suppliers and offer sales and support to its customers. What types of Internet positions would you expect to see at the company?
9. How would you measure user satisfaction with a registration program at a college or university? What are the important features that would make students and faculty satisfied with the system?
10. You have been asked to participate in the preparation of your company's strategic plan. Specifically, your task is to analyze the competitive marketplace using Porter's five-forces model. Prepare your analysis, using your knowledge of a business you have worked for or have an interest in working for.
11. Based on the analysis you performed in the preceding discussion question, what possible strategies could your organization adopt to address these challenges? What role could information systems play in these strategies? Use Porter's strategies as a guide.
12. You have been hired as a sales representative for a sporting goods store. You would like the IS department to develop new software to give you reports on which customers are spending the most at your store. Describe your role in getting the new software developed. Describe the roles of the systems analysts and the computer programmers.
13. Imagine that you are the CIO for a large, multinational company. Outline a few of your key responsibilities.

14. You have decided to open an Internet site to buy and sell used music CDs to other students. Describe the supply chain for your new business.
15. What sort of IS position would be most appealing to you—working as a member of an IS organization, being a con-sultant, or working for an IS hardware or software vendor? Why?
16. What are your career goals and how can a computer-based information system be used to achieve them?

PROBLEM-SOLVING EXERCISES

1. Prepare a data disk and a backup disk for the problem-solving exercises and other computer-based assignments you will complete in this class. Create one directory for each chapter in the textbook (you should have 9 directories). As you work through the problem-solving exercises and complete other work using the computer, save your assignments for each chapter in the appropriate directory. On the label of each disk be sure to include your name, course, and section. On one disk write "Working Copy"; on the other write "Backup."

2. Do some research to obtain estimates of the rate of growth of the e-commerce and m-commerce. Use the plotting capabilities of your spreadsheet or graphics software to produce a bar chart of that growth over a number of years. Share your findings with the class.

3. Using a word processing program, write a detailed job description of a CIO for a medium sized manufacturing company. Use a graphics program to make a presentation on the requirements for the new CIO.

TEAM ACTIVITIES

1. Before you can do a team activity, you need a team! The class members may self-select their teams, or the instructor may assign members to groups. Once your group has been formed, meet and introduce yourselves to each other. You will need to find out the first name, hometown, major, and e-mail address and phone number of each member. Find out one interesting fact about each member of your team, as well. Come up with a name for your team. Put the information on each team member into a database and print enough copies for each team member and your instructor.

2. Have your team interview a company that recently introduced new technology. Write a brief report that describes the extent of technology infusion and diffusion.

3. With your team, interview one or more managers at a local manufacturing company. Describe the company's supply chain. How does the company manage handle customer relationship management?

WEB EXERCISES

1. Throughout this book, you will see how the Internet provides a vast amount of information to individuals and organizations. We will stress the World Wide Web, or simply the Web, which is an important part of the Internet. Most large universities and organizations have an address on the Internet, called a Web site or home page. The address of the Web site for the publisher of this text is *http://www.course.com*. You can gain access to the Internet through a browser, such as Internet Explorer or Netscape. Using an Internet browser, go to the Web site for this publisher. What did you find? Try to obtain information on this book. You may be asked to develop a report or send an e-mail message to your instructor about what you found.

2. Go to an Internet search engine, such as *www.yahoo.com*, or *www.google.com*, and search for information about virtual reality. Write a brief report that summarizes what you found and the companies that provide knowledge management products.

3. Use the Internet to search for information about user satisfaction. You can use a search engine, such as Yahoo!, or a database at your college or university. Write a brief report describing what you found.

CAREER EXERCISES

1. In the Career Exercises found at the end of every chapter, you will explore how material in the chapter can help you excel in your college major or chosen career. Write a brief report on the career that appeals to you the most. Do the same for two other careers that interest you.
2. Research careers in accounting, marketing, information systems, and two other career areas that interest you.

Describe the job opportunities, job duties, and the possible starting salaries for each career area in a report.
3. Pick the five best companies for your career. Describe how each company uses information systems to help achieve a competitive advantage.

CASE STUDIES

Case One

Yansha Leans on IS to Stay Competitive

More than ever before, Chinese retailers are facing local competition from foreign companies. China's highly regulated economy has long insulated businesses from competition. Now that China is loosening its regulations in an effort to benefit from international trade, its own businesses must work harder than ever to become more efficient and effective and keep customers.

Yansha is one of China's biggest retailers. It sells upscale designer clothes from around the world along with other fine merchandise. One of its largest retail stores occupies 215,000 square feet in Beijing's famous Youyi Shopping City.

Yansha has long experienced market leadership in China, but in recent years has felt increasing competition from international companies. Yansha's management team was aware that its methods of communication with suppliers—the placing and receiving of orders—was less than efficient. It was also aware of other inefficiencies in communication throughout the organization. For Yansha to maintain its leadership role in the market, it would need to cut the waste, and become lean and mean in its application of information system technology.

Yansha turned to IBM China Research Lab to evaluate its information systems and recommend the latest technologies to bring it up to date. IBM implemented a massive system upgrade across the entire enterprise, an enterprise resource planning (ERP) system. The ERP allows Yansha executives and managers to view real-time performance data, such as sales across all locations, in certain regions, or in one particular store. Using this system, managers could, for example, determine the success of a particular marketing approach. The new ERP interfaces with a new supply chain management (SCM) system that provides close communication between Yansha and its suppliers. These two systems working together, the ERP system and the SCM system, allows for Yansha and its suppliers to work as one tightly knit organization.

The new systems required a big investment of time and money for Yansha, but the benefits have vastly overshadowed the costs. The new systems reduced the time it takes for suppliers to ship merchandise to Yansha (order lead time) from 2.5 days to 4.5 hours. The order acknowledgment rate has increased from 80 percent to 99 percent. Order errors have been reduced from nine to one percent. The money saved by Yansha receiving the right merchandise at the right time has saved the company enough money to pay for its expensive new information systems within nine months of rolling them out. Achieving a return on investment (ROI) in such a short time is something any chief information officer (CIO) would be proud of.

Discussion Questions

1. China is experiencing a rapidly evolving economy. Why do you think most international businesses are looking to China as both an opportunity and a threat?
2. Yansha has raced to catch up with the latest information systems and technologies, but remains a local Chinese business. What might Yansha's next move be in order to increase its revenue? How might that move be most successfully taken?

Critical Thinking Questions

1. What benefits do local businesses such as Yansha have over foreign businesses. How can those advantages be used to maintain leadership in a market?
2. What benefits do international businesses have over smaller local businesses? How can they be used to infiltrate a new market and take over leadership?

SOURCES: Staff, "Yansha department store embraces supplier collaboration to streamline processes," IBM Success Stories, November 29, 2007, www-01.ibm.com/software/success/cssdb.nsf/CS/JSTS-79BMT3?OpenDocument&Site=corp&cty=en_us. Yansha's Web site, accessed December 21, 2007, www.yansha.com.cn.

Case Two

Customer Service Drives Information Systems at Volvo Cars Belgium

You can often gauge the role of information systems within a business by evaluating how the chief information officer (CIO), or head of information systems, interacts professionally with the president, chief executive officer (CEO), and other high-ranking decision makers. At Volvo Cars Belgium, the company's Brussels-based subsidiary, the head of information systems, Michelangelo Adamo, who goes by the title "IT supervisor," reports directly to the customer service manager. This is unique since most individuals in Michelangelo's position would report to the CEO, president, or even chief operating officer (COO) or chief financial officer (CFO).

The relationship of Volvo's IT supervisor to the customer service manager indicates Volvo's strong emphasis on customer service as a primary goal of the business. It also suggests a belief that information systems should always be designed in a manner that provides additional service to the customer. While Michelangelo is the IT supervisor for the 65-dealer network, his ultimate goal is to sell more cars, and to do so, he needs to make and keep customers happy. Consider the following example of customer service-driven information systems.

Recently Volvo Cars Belgium completed a massive information system implementation that connects the 65 Belgium Volvo dealers over a common network, and to Volvo headquarters in Goteborg, Sweden. The overarching goal of the project was to improve customer service. Through the new system, Volvo Cars Belgium is able to track a single car through its entire lifecycle; from the moment the order is placed, to delivery, to after-sale maintenance and customer service. This allows Volvo to see to it that customers get the right service at the right time, regardless of what dealership they visit. The system also speeds up repairs by allowing a dealership to order parts for a vehicle from Volvo suppliers at the same time the service appointment is made.

The system also allows Volvo Cars Belgium to easily collect sales and service data and apply standardized performance metrics across the entire dealer network. This data is passed on to Volvo headquarters where it is used to target dealerships that need improvement and reward dealerships that excel.

Volvo Cars Belgium's new integrated information system illustrates the emphasis that the company places on customer service. The time and investment in the new system has paid off, with customer-approval ratings jumping for the first time in years.

Discussion Questions

1. Compare how the goals of information systems may differ if driven by the CEO, CIO, customer service, or equally by all business areas?
2. How did joining the dealerships with a common data network help Volvo Cars Belgium improve customer satisfaction?

Critical Thinking Questions

1. What other information systems could assist an auto dealer in improving customer service?
2. How can an auto dealer judge customer satisfaction, and how can information systems be used to compile an overall customer approval rating for a dealership?

SOURCES: Briody, Dan, "How SOA Boosts Customer Service," CIO Insight, March 7, 2007, *www.cioinsight.com/article2/0,1540,2104339,00.asp*. Volvo Cars Belgium Web site, accessed December 27, 2007, *www.volvocars.be*.

Questions for Web Case

See the Web site for this book to read about the Whitmann Price Consulting case for this chapter. The following questions cover this Web case.

Whitman Price Consulting: A New Systems Initiative

Discussion Questions

1. What advantages would the proposed Advanced Mobile Communications and Information System provide for Whitman Price Consulting? What problems might it assist in eliminating?
2. Why do you think Josh and Sandra have been asked to interview the managers of the six business units within WPC as a first step? As IT professionals, Josh, Sandra, and their boss Matt know much more about technology and information systems than the heads of the business units. Shouldn't they be able to design the system without suggestions from amateurs? Including more people in the planning stage is sure to complicate the process.

Critical Thinking Questions

1. If you were Josh or Sandra, what questions would you ask the heads of the six business units?
2. If you were Josh or Sandra, what additional research might you request of your IT staff at this point?

NOTES

Sources for the opening vignette: Staff, "Fossil integrates retail and wholesale operations with IBM and SAP," IBM Success Stories, *www-01.ibm.com/software/success/cssdb.nsf/CS/STRD- 795MNK? OpenDocument&Site=corp&cty=en_us.* Fossil Web site, accessed December 26, 2007, *www.fossil.com.* SAP Web site, accessed December 26, 2007, *www.sap.com/about/press/press.epx?pressid=4514.*

1   Mills, Steven, "News," *Computerworld,* June 4, 2007, p. 20.

2   Dhar, V. and Sundararajan, A., "Information Technologies in Business: A Blueprint for Education and Research," *Information Systems Research,* June 2007, p. 125.

3   Smith, Frederick, "A Budding Network," *Forbes,* May 7, 2007, p. 64.

4   Chaker, Anne Marie, "Yale on $0 Per Day," *The Wall Street Journal,* February 15, 2007, p. D1.

5   Hoffman, Thomas, "Saving Lives Via Video at Sutter Health's eICU," *Computerworld,* June 25, 2007, p. 28.

6   Oh, Wonseok and Pinsonneault, Alian, "On the Assessment of the Strategic Value of Information Technologies," *MIS Quarterly,* June 2007, p. 239.

7   Peppard, J. and Daniel, E., "Managing The Realization of Business Benefits from IT Investments," *MIS Quarterly Executive,* June 2007, p. 1.

8   Ayres, I. and Nalebuff, B., "Experiment," *Forbes,* September 3, 2007, p. 130.

9   McDonald, Ian, "Making Paperless Trails at Lloyd's," *The Wall Street Journal,* August 13, 2007, p. B1.

10 Wingfield, Nick, "Hide the Button," *The Wall Street Journal,* July 25, 2007, p. D1.

11 Lawton, Christopher, "Dumb Terminals Can Be a Smart Move," *The Wall Street Journal,* January 30, 2007, p. B3.

12 Clark, Don, "Intel Scores Speed Breakthrough," *The Wall Street Journal,* July 25, 2007, p. B4.

13 Wildstrom, Stephen, "How Flash Will Change PCs," *Business Week,* June 11, 2007.

14 Gonsalves, Antone, "A Supercomputer on a Chip," *Information Week,* June 25, 2007, p. 26.

15 Einhorn, Bruce, "Intel Inside the Third World," *Business Week,* July 9, 2007, p. 38.

16 Fab Lab, *http://fab.cba.mit.edu/.* accessed August 25, 2007.

17 Spanbauer, Scott, "The Right Operating System for You," *PC World,* April 2007, p. 102.

18 Dunn, Scott, "Sync Your PC's Tunes with Windows Mobile Devices," *PC World,* March 2007, p. 124.

19 Arar, Yardena, "Say So Long to Shrink-Wrapped Software," *PC World,* February 2007, p. 37.

20 "Adobe Creative Suite 3," *www.adobe.com/products/creativesuite/,* accessed August 20, 2007.

21 Nickum, Chris, "The BI Prescription," *Optimize Magazine,* April 2007, p. 45.

22 Havenstein, Heather, "Aetna Clients to Get Access to Online Health Data," *Computerworld,* January 15, 2007, p. 7.

23 Greenemeier, Larry, "Data on Loan," *Information Week,* April 23, 2007, p. 33.

24 Schupak, Amanda, "90 Years of Networks," *Forbes,* May 7, 2007, p. 106.

25 Malykhina, E. and Gardner, D., "A $48 Billion Upgrade," *Information Week,* April 2, 2007, p. 22.

26 Gomes, Lee, "Paradoxes Abound in Telecommuting," *The Wall Street Journal,* January 23, 2007, p. B3.

27 Malykhina, Elena, "China Picks Google," *Information Week,* January 8, 2007, p. 17.

28 McBride, Sarah, "Online-Radio Fight Reaches New Pitch," *The Wall Street Journal,* May 31, 2007, p. A13.

29 Ali, Sarmad, "When I Was Your Age, We Didn't Have Sites for Writing Our Bios," *The Wall Street Journal,* May 31, 2007, p. B1.

30 Hardy, Quentin, "Better Than YouTube," *Forbes,* May 21, 2007, p. 72.

31 McMillan, Robert, "Is Web 2.0 Safe?" *PC World,* July 2007, p. 18.

32 Borrows, P. and Crockett, R., "Turning Cell Phones on Their Ear," *Business Week,* January 22, 2007, p. 40.

33 McCartney, Scott, "WiFi in the Sky," *The Wall Street Journal,* April 3, 2007, p. B9.

34 Green, Heather, "Don't Quit Your Day Job, Podcasters," *Business Week,* April 9, 2007, p. 72.

35 Buckman, Rebecca, "Backers Thrive on B2B Firms," *The Wall Street Journal,* July 16, 2007, p. C1.

36 Holahan, Catherine, "Going, Going, Everywhere," *Business Week,* June 18, 2007, p. 62.

37 Buckman, Rebecca, "Just Charge It to Your Cell Phone," *The Wall Street Journal,* June 21, 2007, p. B3.

38 Ma, M. and Agarwal, R., "Through a Glass Darkly," *Information Systems Research,* March 2007, p. 42.

39 Fahey, Jonathan, "Car Talk," *Forbes,* January 29, 2007, p. 52.

40 Greenemeier, Larry, "A Cancer-Zapping Heroine," *Information Week,* June 4, 2007, p. 22.

41 Staff, "Boeing Simulates and Manufactures 787 Dreamliner," *www.3ds.com/news-events/press-room/release/1357/1,* accessed August 16, 2007.

42 Peng, Tina, "Why Trunk Shows Are Going Virtual," *The Wall Street Journal,* August 11, 2007, p. P1.

43 Anthes, Gary, "Split-Second Trading," *Computerworld,* May 21, 2007, p. 40.

44 Vijayan, J., "Feds Delay Database for Money Transfer Records," *Computerworld,* January 22, 2007, p. 1.

45 Imse, Ann and Gathright, Alan, "Governor Seeks to Bring Order to Computer Chaos," *Rocky Mountain News,* July 23, 2007, p. 4.

46 Bartels, Lynn, "State Might Scrap DMV Computers," *Rocky Mountain News,* August 14, 2007, p. 5.

47 Nelson, Ryan, "IT Project Management: Infamous Failures, Classic Mistakes, and Best Practices," *MIS Quarterly Executive,* June 2007, p. 67.

48 Rajeev, Sharma and Yetton, Philip, "The Contingent Effects of Training, Technical Complexity, and Task Interdependence on Successful Information Systems Implementation," *MIS Quarterly,* June 2007, p. 219.

49 Flandez, Raymond, "In Search of Help," *The Wall Street Journal,* March 19, 2007, p. R8.

50 McKenzie, P. and Jayanthi, S., "Ball Aerospace Explores Operational and Financial Trade-Offs in Batch Sizing in Implementing JIT," *Interfaces,* March 2007, p. 108.

51 Craighead, C., et al., "The Severity of Supply Chain Disruptions," *Decision Sciences,* February 2007, p. 131.

52 Abboud, Leila, "Global Suppliers Play Catch-Up in Information Age," *The Wall Street Journal,* January 4, 2007, p. B3.

53 Staff, "Average Customer Relationship Management Program Takes 16.2 Months to Develop," *Science Letter,* May 1, 2007, p. 600.

54 Livari, J. and Huisman, M., "The Relationship Between Organizational Culture and the Deployment of Systems Development Methodologies," *MIS Quarterly,* March 2007, p. 35.

55 Mehta, M. and Hirschheim, R., "Strategic Alignment in Mergers and Acquisitions," *Journal of the Association of Information Systems,* March 2007, p. 143.

56 Christensen, Clayton, *The Innovator's Dilemma*, Harvard Business School Press, 1997, p. 225 and *The Inventor's Solution*, Harvard Business School Press, 2003.

57 Wixom, Barbara and Todd, Peter, "A Theoretical Integration of User Satisfaction and Technology Acceptance," *Information Systems Research,* March 2005, p. 85.

58 Bailey, J. and Pearson, W., "Development of a Tool for Measuring and Analyzing Computer User Satisfaction," *Management Science,* 29(5), 1983, p. 530.

59 Chaparro, Barbara, et al., "Using the End-User Computing Satisfaction Instrument to Measure Satisfaction with a Web Site," *Decision Sciences,* May 2005, p. 341.

60 Schwarz, A. and Chin, W., "Toward an Understanding of the Nature and Definition of IT Acceptance," *Journal of the Association for Information Systems,* April 2007, p. 230.

61 Davis, F., "Perceived Usefulness, Perceived Ease of Use, and User Acceptance of Information Technology," *MIS Quarterly,* 13(3) 1989, p. 319. Kwon, et al., "A Test of the Technology Acceptance Model," *Proceedings of the Hawaii International Conference on System Sciences*, January 4–7, 2000.

62 Barki, H., et al., "Information System Use-Related Activity," *Information Systems Research,* June 2007, p. 173.

63 Loch, Christoph and Huberman, Bernardo, "A Punctuated-Equilibrium Model of Technology Diffusion," *Management Science*, February 1999, p. 160.

64 Tornatzky, L and Fleischer, M., "The Process of Technological Innovation," *Lexington Books,* Lexington, MA, 1990; Zhu, K.and Kraemer, K., "Post-Adoption Variations in Usage and Value of E-Business by Organizations," *Information Systems Research,* March 2005, p. 61.

65 Armstrong, Curtis and Sambamurthy, V., "Information Technology Assimilation in Firms," *Information Systems Research*, April 1999, p. 304.

66 Agarwal, Ritu and Prasad, Jayesh, "Are Individual Differences Germane to the Acceptance of New Information Technology?" *Decision Sciences*, Spring 1999, p. 361.

67 Squeri, Steve, "What's Next for IT," *The Wall Street Journal,* July 30, 2007, p. R6.

68 Collins, Jim, *Good to Great*, Harper Collins Books, 2001, p. 300.

69 Porter, M. E., *Competitive Advantage: Creating and Sustaining Superior Performance,* New York: Free Press, 1985; *Competitive Strategy: Techniques for Analyzing Industries and Competitors*, The Free Press, 1980; and *Competitive Advantage of Nations*, The Free Press, 1990.

70 Porter, M. E. and Millar, V., "How Information Systems Give You Competitive Advantage," *Journal of Business Strategy*, Winter 1985. *See also* Porter, M. E., *Competitive Advantage* (New York: Free Press, 1985).

71 Staff, "Technologies to Spot Fakes Are Short Lived," *The Wall Street Journal,* May 26, 2007, p. A6.

72 Havenstein, Heather, "U.S. Doctors Offered Free Prescribing Software," *Computerworld,* January 22, 2007, p. 6.

73 Carey, Susan, "Calculating Costs in the Clouds," *The Wall Street Journal,* March 6, 2007, p. B1.

74 McGee, M., "Urgent Care," *Information Week,* May 26, 2007, p. 40.

75 Beauchesne, Eric, "Machines, Not People, Boosted Productivity," *The Gazette (Montreal),* June 26, 2007, p. B5.

76 Pratt, Mary, "Web Site ROI," *Computerworld,* May 28, 2007, p. 30.

77 Davis, Robert and Wong, Don, "Conceptualizing and Measuring the Optimal Experience of the eLearning Environment," *Decision Sciences Journal of Innovation Education,* January 2007, p. 97.

78 Staff, "Detailed Comparison of a Range of Factors Affecting Total Cost of Ownership (TCO) for Messaging and Collaboration," *M2 Presswire*, September 5, 2003.

79 Metzer, Scott, "The Allure of New IT," *Information Week,* August 20, 2007, p. CIO5.

80 Brandel, Mary, "12 IT Skills That Employers Can't Say No To," *Computerworld,* July 11, 2007.

81 Levina, N. and Xin, M., "Comparing IT Workers' Compensation Across Country Contexts," *Information Systems Research,* June 2007, p. 193.

82 Murphy, Chris, "Six-Figure Club," *Business Week,* May 7, 2007, p. 86.

83 Brandel, Mary, "Best Places to Work in IT 2007," *Computerworld,* June 18, 2007, p. 34.

84 McDougall, P. and Murphy, C., "Another Round of H-1B," *Information Week,* March 26, 2007, p. 17.

85 McGee, Marianne, "The H-1B Limit," *Information Week,* April 9, 2007, p. 29.

86 Thibodeau, P., "Another H-1B Fight Looms in Congress," *Computerworld,* January 8, 2007, p. 9.

87 Herbst, Moira, "Americans Need Not Apply," *Business Week,* July 9, 2007, p. 10.

88 McGee, M. and Murphy, C., "Feds Cast Wary Eye on Indian Outsourcers' Use of H-1B," *Information Week,* May 21, 2007, p. 32.

89 Miano, John, "Founder of Programmers Guild," *Information Week,* February 5, 2005, p. 15.

90 The Bureau of Labor Statistics, *www.bls.gov*, accessed January 11, 2008

91 Brandel, Mary, "Three IT Pros Talk About the Flip Side of Outsourcing," *Computerworld,* May 28, 2007, p. 33.

92 Top Coder Collegiate Challenge, *www.topcoder.com*, accessed September 2, 2007.

93 TopCoder, *www.topcoder.com*, accessed August 29, 2007.

94 Gomes, Lee, "Computer Scientists Pull a Tom Sawyer to Finish Grunt Work," *The Wall Street Journal,* June 27, 2007, p. B1.

95 Flandez, Raymund, "Firms Go Online to Train Employees," *The Wall Street Journal,* August 14, 2007, p. B4.

96 Martin, Richard, "The CIO Dilemma," *Information Week,* March 26, 2007, p. 38.

97 Gillooly, Brian, "CIO Role Revs Up," *Optimize Magazine,"* June, 2007, p. 22.

98 Gillooly, Brian, "CIO: Time To Step Up," *Information Week,* June 4, 2007, p. 45.

99 Tam, P., "CIO Jobs Morph from Tech Support into Strategy," *The Wall Street Journal,* February 20, 2007, p. B1.

100 *www.systest.com*, accessed November 9, 2007.

101 Kelly, C.J., "Getting Certified and Just a Bit Certifiable," *Computerworld,* January 29, 2007, p. 29.

102 *www.nicb.org*, accessed November 9, 2007.

103 *www.du.edu/newsroom/releases/media/2007-08-01-gamecamp.html*, accessed November 9, 2007.

104 *www.ibm.com/services*, *www.eds.com*, and *www.accenture.com*, accessed November 9, 2007

105 MacMillan, Douglas, "The Art of the Online Résumé," *Business Week,* May 7, 2007, p. 86.

106 Athavaley, A., "Facebook, Jobster to Partner on Site for Job-Seeker Profiles," *The Wall Street Journal,* February 8, 2007, p. D2.

107 Needleman, Sarah, "How Blogging Can Help You Get a New Job," *The Wall Street Journal,* April 10, 2007, p. B1.

108 White, Erin, "Employers Are Putting New Face on Web Recruiting," *The Wall Street Journal,* January 8, 2007, p. B3.

109 Friedman, Thomas, "The World Is Flat," *Farrar, Straus and Giroux,* 2005, p. 488.

110 *www.europa.eu.int*, accessed August 15, 2007

111 *www.ustr.gov/Document_Library/Press_Releases/2007/June/United_States_the_Republic_of_Korea_Sign_Lmark_Free_Trade_Agreement.html*, accessed November 8, 2007.

112 *www.sice.oas.org/tradee.asp*, accessed August 15, 2007

PART
· 2 ·

Technology

Hardware and Software

PRINCIPLES	LEARNING OBJECTIVES
▪ Computer hardware must be carefully selected to meet the evolving needs of the organization and its supporting information systems.	▪ Identify and discuss the role of the essential hardware components of a computer system. ▪ Identify the characteristics of and discuss the usage of various classes of single-user and multiuser computer systems.
▪ Systems and application software are critical in helping individuals and organizations achieve their goals.	▪ Identify and briefly describe the functions of the two basic kinds of software. ▪ Outline the role of the operating system and identify the features of several popular operating systems.
▪ Organizations should not develop proprietary application software unless doing so will meet a compelling business need that can provide a competitive advantage.	▪ Discuss how application software can support personal, workgroup, and enterprise business objectives. ▪ Identify three basic approaches to developing application software and discuss the pros and cons of each.
▪ Organizations should choose a programming language whose functional characteristics are appropriate for the task at hand, considering the skills and experience of the programming staff.	▪ Outline the overall evolution and importance of programming languages and clearly differentiate among the generations of programming languages.
▪ The software industry continues to undergo constant change; users need to be aware of recent trends and issues to be effective in their business and personal life.	▪ Identify several key software issues and trends that have an impact on organizations and individuals.

Information Systems in the Global Economy
UB Spirits, India
UB Spirits Serves up Success

The UB Group is a successful and growing international conglomerate based in Bangalore, India. The bright future of the conglomerate is illustrated by the rising steel and glass structures in the heart of Bangalore and known as UB City—a 7-acre high-rise business, shopping, and living community. The UB Group is invested in several industries including aviation, fertilizers, engineering, information technology, pharmaceuticals, and alcoholic beverages, or spirits. While this may sound like an unusual and even dangerous combination, the diverse business portfolio is making the UB Group billions of dollars.

The alcoholic beverage division of UB, called UB Spirits, is itself a successful international corporation made up of several successful distilleries: McDowell, Herbertsons, and Triumph Distillers and Vintners. Combined, these companies produce 140 brands, 15 of which are top-shelf classics. The brands are produced in 75 locations across India. UB Spirits dominates the Indian marketplace and ranks as the world's second largest distilled spirits corporation, with sales that exceed 60 million cases a year.

To comply with laws that control the distribution of alcoholic beverages in India and around the world, UB Spirits often must negotiate a selling price for their products with a national government. Because of this constraint, UB Spirits has to improve its profits by minimizing overhead and streamlining operations. One obvious waste of resources in the business was its outdated and bloated computer-based information systems. Each UB Spirits facility across India used its own information systems on its own servers—111 Microsoft Windows-based servers in all, which required constant attention. At the close of each month, the data from the disparate systems would be merged in a costly and lengthy process to produce corporate reports. It was clear to the decision makers at UB Spirits that if the company was going to grow, it would have to invest in a new system with state-of-the-art hardware.

As with many system overhauls, UB Spirits began by choosing an ERP software package before it decided on hardware. This is smart because different ERP solutions have varying hardware requirements. Working with IBM, the company selected SAP R/3, which includes information systems for every area of the business.

After selecting an ERP system, the company built the infrastructure on which the system would run. The main goal of the overhaul was to reduce the work required to maintain and administer the system. The company selected an IBM System i550 because it is powerful and simple to manage. The i550 server has multiple processors, up to 64 GB of memory, and up to 77 TB of disk storage. The i550 easily supports 400 UB Spirits users and 1.5 TB of corporate data. It can run all of the corporate systems and be accessed by 64 UB Spirits offices over a network.

By switching to a centralized server system, UB Spirits was able to replace 111 servers spread across India with one server located in its home office. The new system assisted UB Spirits in lowering manufacturing costs. It standardized business processes across the enterprise and accelerated monthly financial reporting. UB Spirits also reassigned 20 of its IT personnel from server administration to more productive projects.

UB Spirits found that the path to higher profits lay in centralizing its operations and investing in large, powerful servers. Other companies have found that they can accomplish more by distributing systems over many workstations in what is called grid computing. The hardware in which a business invests is intimately linked to the type of information system it implements. Creating an underlying infrastructure from hardware should support the evolving needs of the business and its information systems.

As you read this chapter, consider the following:

- How does the type of hardware and software a company acquires affect the way the company operates?
- Businesses are in constant fluctuation—growing, diversifying, acquiring, and always working to reduce costs. How do these conditions and requirements affect the purchase of the firm's hardware and software?

Why Learn About Hardware and Software?

Organizations invest in computer hardware and software to improve worker productivity, increase revenue, reduce costs, and provide better customer service. Those that don't may be stuck with outdated hardware and software that is unreliable and cannot take advantage of the latest advances. As a result, obsolete hardware and software can place an organization at a competitive disadvantage. Managers, no matter what their career field and educational background, are expected to know enough about their business needs to be able to ask tough questions of those recommending the hardware and software to meet those needs. Cooperation and sharing of information between business managers and IT managers is needed to make wise IT investments that yield real business results. Managers in marketing, sales, and human resources often help IS specialists assess opportunities to apply hardware and software and evaluate the various options and features. Managers in finance and accounting especially must also keep an eye on the bottom line, guarding against overspending, yet be willing to invest in computer hardware and software when and where business conditions warrant it.

Today's use of technology is practical—it's intended to yield real business benefits, as demonstrated by UB Spirits. Employing information technology and providing additional processing capabilities can increase employee productivity, expand business opportunities, and allow for more flexibility. This chapter discusses the hardware and software components of a computer-based information system (CBIS), beginning with a definition of hardware.

hardware
Any machinery (most of which uses digital circuits) that assists in the input, processing, storage, and output activities of an information system.

Hardware consists of any machinery (most of which uses digital circuits) that assists in the input, processing, storage, and output activities of an information system. When making hardware decisions, the overriding consideration of a business should be how hardware can support the objectives of the information system and the goals of the organization.

COMPUTER SYSTEMS: INTEGRATING THE POWER OF TECHNOLOGY

To assemble an effective and efficient system, you should select and organize components while understanding the trade-offs between overall system performance and cost, control, and complexity. For instance, in building a car, manufacturers try to match the intended use of the vehicle to its components. Racing cars, for example, require special types of engines, transmissions, and tires. Selecting a transmission for a racing car requires balancing how much engine power can be delivered to the wheels (efficiency and effectiveness) with how expensive the transmission is (cost), how reliable it is (control), and how many gears it has (complexity). Similarly, organizations assemble computer systems so that they are effective, efficient, and well suited to the tasks that need to be performed.

Because the business needs and their importance vary at different companies, the IS solutions they choose can be quite different.

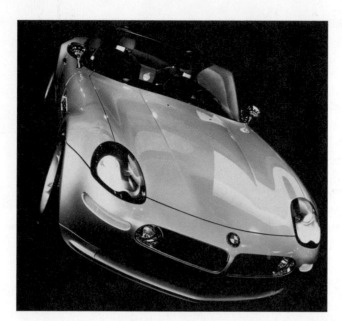

People involved in selecting their organization's computer hardware must clearly understand current and future business requirements so they can make informed acquisition decisions. Consider the following examples of applying business knowledge to reach critical hardware decisions.

- Bosch Security Products provides and maintains physical security systems to control access and detect intrusions for the secure locations of large organizations such as the Dutch Army. For a single customer, Bosch might perform preventive maintenance on thousands of components to ensure that its system is working properly. Service technicians used to rely on a paper-intensive process that required them to fill out forms on the status of each device. The process was error prone and could not provide customers with the comprehensive reports they needed to verify that all necessary repairs and replacements had been completed. Maintenance technicians and their managers defined the requirements for a new and improved solution to meet the business needs. They then consulted with IS experts and chose a PDA-based device for data entry instead of a laptop device because it was lighter, easier to handle, and more resistant to rough handling.[1]

- The Iowa Health System is a network of physicians, hospitals, civic leaders, and local volunteers who serve more than 100 communities. To support patient care, it operates a multifacility Picture Archiving and Communications System in which they store and manage image data such as magnetic resonance images (MRIs). The storage capacity requirements of this system are rapidly growing to 1 million exams per year. Staff and administrators identified additional data storage needs for secure and redundant storage of patient data and to meet the United States Health Insurance Portability and Accountability Act of 1996 (HIPAA) requirements for data integrity. Taking these requirements, the IS staff selected and implemented an appropriate data storage solution that met all needs in a cost-effective manner.[2]

As these examples demonstrate, choosing the right computer hardware requires understanding its relationship to the information system and the needs of the organization. Furthermore, hardware objectives are subordinate to, but supportive of, the information system and the current and future needs of the organization.

HARDWARE COMPONENTS

Computer system hardware components include devices that perform the functions of input, processing, data storage, and output (see Figure 2.1).

central processing unit (CPU)
Part of the computer that consists of three associated elements: the arithmetic/logic unit, the control unit, and the register areas.

arithmetic/logic unit (ALU)
Part of the CPU that performs mathematical calculations and makes logical comparisons.

control unit
Part of the CPU that sequentially accesses program instructions, decodes them, and coordinates the flow of data in and out of the ALU, the registers, primary storage, and even secondary storage and various output devices.

The ability to process (organize and manipulate) data is a critical aspect of a computer system, in which processing is accomplished by an interplay between one or more of the central processing units and primary storage. Each **central processing unit (CPU)** consists of two primary elements: the arithmetic/logic unit and the control unit. The **arithmetic/logic unit (ALU)** performs mathematical calculations and makes logical comparisons. The **control unit** sequentially accesses program instructions, decodes them, and coordinates the flow of data in and out of the ALU, primary storage, and even secondary storage and various output devices. Primary memory, which holds program instructions and data, is closely associated with the CPU.

Now that you have learned about the basic hardware components and the way they function, you are ready to examine processing power, speed, and capacity. These three attributes determine the capabilities of a hardware device.

PROCESSING AND MEMORY DEVICES: POWER, SPEED, AND CAPACITY

The components responsible for processing—the CPU and memory—are housed together in the same box or cabinet, called the *system unit*. All other computer system devices, such as the monitor and keyboard, are linked either directly or indirectly into the system unit housing. As discussed previously, achieving IS objectives and organizational goals should be the primary consideration in selecting processing and memory devices. In this section, we investigate the characteristics of these important devices.

Processing Characteristics and Functions

Because efficient processing and timely output are important, organizations use a variety of measures to gauge processing speed. These measures include the time it takes to complete a machine cycle, clock speed, and others.

Clock Speed

Each CPU produces a series of electronic pulses at a predetermined rate, called the **clock speed**, which affects machine cycle time. The control unit executes an instruction in accordance with the electronic cycle, or pulses of the CPU "clock." Each instruction takes at least the same amount of time as the interval between pulses. The shorter the interval between pulses, the faster each instruction can be executed. The clock speed for personal computers is in the multiple gigahertz (GHz), or billions of cycles per second, range.

Physical Characteristics of the CPU

CPU speed is also limited by physical constraints. Most CPUs are collections of digital circuits imprinted on silicon wafers, or chips, each no bigger than the tip of a pencil eraser. To turn a digital circuit within the CPU on or off, electrical current must flow through a medium (usually silicon) from point A to point B. The speed at which it travels between points can be increased by either reducing the distance between the points or reducing the resistance of the medium to the electrical current.

Memory Characteristics and Functions

Located physically close to the CPU (to decrease access time), memory provides the CPU with a working storage area for program instructions and data. The chief feature of memory is that it rapidly provides the data and instructions to the CPU.

Storage Capacity

Like the CPU, memory devices contain thousands of circuits imprinted on a silicon chip. Each circuit is either conducting electrical current (on) or not (off). Data is stored in memory as a combination of on or off circuit states. Usually 8 bits are used to represent a character, such as the letter *A*. Eight bits together form a **byte (B)**. In most cases, storage capacity is measured in bytes, with 1 byte equivalent to one character of data. The contents of the Library of Congress, with over 126 million items and 530 miles of bookshelves, would require about 20 petabytes of digital storage. Table 2.1 lists units for measuring computer storage.

clock speed
A series of electronic pulses produced at a predetermined rate that affects machine cycle time.

byte (B)
Eight bits that together represent a single character of data.

Name	Abbreviation	Number of Bytes
Byte	B	1
Kilobyte	KB	2^{10} or approximately 1,024 bytes
Megabyte	MB	2^{20} or 1,024 kilobytes (about 1 million)
Gigabyte	GB	2^{30} or 1,024 megabytes (about 1 billion)
Terabyte	TB	2^{40} or 1,024 gigabytes (about 1 trillion)
Petabyte	PB	2^{50} or 1,024 terabytes (about 1 quadrillion)
Exabyte	EB	2^{60} or 1,024 petabytes (about 1 billion billion, or 1 quintillion)

Table 2.1

Units for Measuring Computer Storage

Types of Memory

Several forms of memory are available. Instructions or data can be temporarily stored in **random access memory (RAM)**. RAM is temporary and volatile—RAM chips lose their contents if the current is turned off or disrupted (as in a power surge, brownout, or electrical noise generated by lightning or nearby machines). RAM chips are mounted directly on the computer's main circuit board or in chips mounted on peripheral cards that plug into the computer's main circuit board. These RAM chips consist of millions of switches that are sensitive to changes in electric current.

Read-only memory (ROM), another type of memory, is usually nonvolatile. In ROM, the combination of circuit states is fixed, and therefore its contents are not lost if the power is removed. ROM provides permanent storage for data and instructions that do not change, such as programs and data from the computer manufacturer, including the instructions that tell the computer how to start up when power is turned on.

Multiprocessing

There are a number of forms of **multiprocessing**, which involves the simultaneous execution of two or more instructions.

Multicore Microprocessor

A **multicore microprocessor** combines two or more independent processors into a single computer so that they can share the workload and boost processing capacity. "A dual-core processor is like a four-lane highway—it can handle up to twice as many cars as its two-lane predecessor without making each car drive twice as fast." In addition, a dual-core processor enables people to perform multiple tasks simultaneously such as playing a game and burning a CD. AMD and Intel are battling for leadership in the multicore processor marketplace.

Both Intel and AMD have improved on dual processors by introducing new quad-core chips. However, a major need for basic research in computer science is to develop software that can actually take advantage of four processors. Multicore systems are most effective when they run programs that can split their workload among multiple CPUs. Such applications include working with large databases and multimedia. Intel has introduced Viiv (rhymes with five), a marketing initiative that combines Intel products including the Core 2 Quad processor with additional hardware and software to build an extremely powerful multimedia computer capable of running the highly processing-intensive applications associated with high-definition entertainment.

Parallel Computing

Another form of multiprocessing, called **parallel processing**, speeds processing by linking several processors to operate at the same time, or in parallel. The most frequent business uses for parallel processing are modeling, simulation, and analysis of large amounts of data. In response to higher fuel prices and the desire to reduce carbon dioxide emissions, auto manufacturers are introducing smaller models. However, consumers are concerned about the safety of these smaller cars. To address these concerns, automobile engineers use finite element modeling and massively parallel processing computer systems to simulate crashes. Such simulations are much less expensive than using actual cars and crash dummies. Also, the speed and accuracy of the computer simulations allows for many more crash tests at an earlier stage in the design than using physical crashes with crash dummies. As a result, engineers gain confidence earlier in the design process that their car will pass federal safety standards, enabling them to bring the car to market sooner.[3]

Grid Computing

Grid computing is the use of a collection of computers, often owned by many people or organizations, to work in a coordinated manner to solve a common problem. Grid computing is one low-cost approach to parallel processing. The grid can include dozens, hundreds, or even thousands of computers that run collectively to solve extremely large parallel processing problems. Key to the success of grid computing is a central server that acts as the grid leader and traffic monitor. This controlling server divides the computing task into subtasks and

---

random access memory (RAM)
A form of memory in which instructions or data can be temporarily stored.

read-only memory (ROM)
A nonvolatile form of memory.

multiprocessing
The simultaneous execution of two or more instructions at the same time.

multicore microprocessor
Microprocessor that combines two or more independent processors into a single computer so they can share the workload and deliver a big boost in processing capacity.

parallel processing
The simultaneous execution of the same task on multiple processors in order to obtain results faster.

grid computing
The use of a collection of computers, often owned by multiple individuals or organizations, to work in a coordinated manner to solve a common problem.

assigns the work to computers on the grid that have (at least temporarily) surplus processing power. The central server also monitors the processing, and if a member of the grid fails to complete a subtask, it will restart or reassign the task. When all the subtasks are completed, the controlling server combines the results and advances to the next task until the whole job is completed.

Folding@home is a grid computing project with more than one million people around the world downloading and running software to form one of the largest grids in the world. To carry out their various functions, proteins self-assemble into a particular shape in a process called "folding." The goal of the Folding@home project is to research protein folding and misfolding and gain an understanding of how this protein behavior is related to diseases such as Alzheimer's, Parkinson's, and many forms of cancer. It takes a single computer about one day to simulate a nanosecond (1/1,000,000th of a second) in the life of a protein. The folding process takes about 10,000 nanoseconds. Thus, 10,000 days (30 years) are required to simulate a single folding! The Folding@home group has developed ways to speed up the simulation of protein folding by dividing the work among over 100,000 processors.[4] In September 2007, with more than half a million PlayStation 3 consoles participating on the grid, the combined computing power exceeded 1×10^{15} floating-point operations per second—more than twice the speed of the world's fastest stand-alone supercomputer.[5]

By installing the Folding@home screen saver program on their personal computers, more than one million people around the world contribute their idle CPU time to research into diseases such as Alzheimer's, Parkinson's, and many forms of cancer.

(Source: *http://folding.stanford.edu*)

Cloud Computing

Cloud computing involves using a giant cluster of computers that serves as a host to run applications that require high-performance computing. Cloud computing supports a wider variety of applications than grid computing and pools computing resources so they can be managed primarily by software rather than people. IBM and Google have provided hardware, software, and services to many universities so that students and faculty can explore cloud computing and massively parallel computing. For example, University of Washington students use the Google cluster to scan millions of changes made to the online encyclopedia Wikipedia to identify spam.[6] Amazon.com offers the Elastic Compute Cloud as a service that provides users with a highly expandable computing capacity for Web site development and operations.[7] Some industry observers think that Google is planning a bold move into the delivery of applications for businesses and consumers. Indeed, CEO Eric Schmidt has stated that "Most people who run small businesses would like to throw out their infrastructure and use ours for $50 per year." Companies are asking whether Google, IBM, Amazon.com, or others can provide a cheaper, more reliable, more secure, more flexible, and more powerful alternative through cloud computing.[8]

cloud computing
Using a giant cluster of computers to serve as a host to run applications that require high-performance computing.

SECONDARY STORAGE AND INPUT AND OUTPUT DEVICES

As you have seen, memory is an important factor in determining overall computer system power. However, memory provides only a small amount of storage area for the data and instructions the CPU requires for processing. Computer systems also need to store larger amounts of data, instructions, and information more permanently than main memory allows. Secondary storage, also called *permanent storage*, serves this purpose.

Compared with memory, secondary storage offers the advantages of nonvolatility, greater capacity, and greater economy. Most forms of secondary storage are considerably less expensive than memory (see Table 2.2). Because of the electromechanical processes involved in using secondary storage, however, it is considerably slower than memory. The selection of secondary storage media and devices requires understanding their primary characteristics—access method, capacity, and portability.

Table 2.2

Cost Comparison for Various Forms of Data Storage

(Source: Office Depot Web site, *www.officedepot.com*, January 18, 2008.)

Description	Cost Per GB	Storage Capacity (GB)	Cost per GB
72 GB DAT 72 data cartridge	$14.95	72	$0.21
10 - 4.7 GB DVD+R discs	$9.95	47	$0.21
20 GB 4 MM backup data tape	$16.99	20	$0.85
120 GB portable hard drive	$139.99	120	$1.16
25 GB Rewritable Blu-ray disc	$29.99	25	$1.20
9.1 GB Write Once Read Many optical disc	$69.95	9.1	$7.69
1 GB flash drive	$7.99	1	$7.99
512 MB DDR2 SDRAM memory upgrade	$24.99	0.512	$48.81

All forms of secondary storage cost considerably less per megabyte of capacity than SDRAM, although they have slower access times. A data tape cartridge costs about $.16 per gigabyte, while SDRAM can cost over $40 per gigabyte.

Access Methods

sequential access
Retrieval method in which data must be accessed in the order in which it is stored.

Data and information access can be either sequential or direct. **Sequential access** means that data must be accessed in the order in which it is stored. For example, inventory data stored sequentially may be stored by part number, such as 100, 101, 102, and so on. If you want to retrieve information on part number 125, you need to read and discard all the data relating to parts 001 through 124.

direct access
Retrieval method in which data can be retrieved without the need to read and discard other data.

Direct access means that data can be retrieved directly, without having to pass by other data in sequence. With direct access, it is possible to go directly to and access the needed data—such as part number 125—without reading through parts 001 through 124. For this reason, direct access is usually faster than sequential access. The devices used to sequentially access secondary storage data are simply called **sequential access storage devices (SASDs)**; those used for direct access are called **direct access storage devices (DASDs)**.

sequential access storage device (SASD)
Device used to sequentially access secondary storage data.

Secondary Storage Devices

direct access storage device (DASD)
Device used for direct access of secondary storage data.

The most common forms of secondary storage include magnetic tapes, magnetic disks, and optical discs. Some of these media (magnetic tape) allow only sequential access, while others (magnetic and optical discs) provide direct and sequential access.

Magnetic Tapes

One common secondary storage medium is **magnetic tape**. Similar to the kind of tape found in audio- and videocassettes, magnetic tape is a Mylar film coated with iron oxide. Portions of the tape are magnetized to represent bits. Magnetic tape is a sequential access storage medium. Although access is slower, magnetic tape is usually less expensive than disk storage. Magnetic tape is often used to back up disk drives and to store data off-site for recovery in case of disaster. Technology is improving to provide tape storage devices with greater capacities and faster transfer speeds. Large, bulky tape drives have been replaced with much smaller tape cartridge devices measuring a few millimeters in diameter that take up much less floor space and allow hundreds of tape cartridges to be stored in a small area. The U.S. federal government is the largest user of magnetic tape in the world, buying over 1 million reels of tape each year for use by such organizations as the Internal Revenue Service, National Oceanic and Atmospheric Administration, the Federal Reserve Bank, and the various branches of the military.[9]

Magnetic Disks

Magnetic disks are also coated with iron oxide; they can be thin metallic platters (hard disks, see Figure 2.2) or Mylar film (diskettes). As with magnetic tape, magnetic disks represent bits using small magnetized areas. Magnetic disks are direct-access storage devices that enable fast data retrieval and are used by companies that need to respond quickly to customer requests. For example, if a manager needs information on the credit history of a customer or the seat availability on a particular flight, the information can be obtained in seconds if the data is stored on a direct access storage device.

RAID

Companies' data storage needs are expanding rapidly. Today's storage configurations routinely entail many hundreds of gigabytes. However, putting the company's data online involves a serious business risk—the loss of critical business data can put a corporation out of operation. The concern is that the most critical mechanical components inside a disk storage device—the disk drives, the fans, and other input/output devices—can break.

Organizations now require their data storage devices to be fault tolerant—they can continue with little or no loss of performance in the event of a failure of one or more key components. **Redundant array of independent/inexpensive disks (RAID)** is a method of storing data so that if a hard drive fails, the lost data on that drive can be rebuilt. With this approach, data is stored redundantly on different physical disk drives using a technique called *stripping* to evenly distribute the data. Medkinetics is a small (12 employees) business that automates collecting and submitting for approval information about a doctor's qualifications. Jim Cox, founder and president of Medkinetics, says: "The high availability of data is also really important to us." The firm employs terabytes of inexpensive but secure RAID storage.[10]

SAN

A **storage area network (SAN)** uses computer servers, distributed storage devices, and networks to tie everything together, as shown in Figure 2.3. To increase the speed of storing and retrieving data, high-speed communications channels are often used. Although SAN technology is relatively new, a number of companies are using SAN to successfully and efficiently

magnetic tape
Secondary storage medium; Mylar film coated with iron oxide with portions of the tape magnetized to represent bits.

magnetic disk
Common secondary storage medium, with bits represented by magnetized areas.

Figure 2.2

Hard Disk

Hard disks give direct access to stored data. The read/write head can move directly to the location of a desired piece of data, dramatically reducing access times, as compared with magnetic tape.

(Source: Courtesy of Seagate Technology.)

redundant array of independent/inexpensive disks (RAID)
Method of storing data that generates extra bits of data from existing data, allowing the system to create a "reconstruction map" so that if a hard drive fails, the system can rebuild lost data.

storage area network (SAN)
Technology that provides high-speed connections between data-storage devices and computers over a network.

store critical data. The Bombay Company designs and markets home furnishings and decorative accessories via 422 retail outlets, catalogs, and the Internet. The firm implemented 1.5 TB of SAN data storage to effectively hold its inventory data.[11]

compact disc read-only memory (CD-ROM)

A common form of optical disc on which data, once it has been recorded, cannot be modified.

digital video disc (DVD)

A storage medium used to store digital video or computer data.

Figure 2.4

Digital Video Disc and Player

DVDs look like CDs but have a greater storage capacity and can transfer data at a faster rate.

(Source: Courtesy of Toshiba America Information Systems.)

Optical Discs

A common optical disc is the **compact disc read-only memory (CD-ROM)** with a storage capacity of 740 MB of data. After data is recorded on a CD-ROM, it cannot be modified—the disc is "read-only." A CD burner, the informal name for a CD recorder, is a device that can record data to a compact disc. *CD-recordable (CD-R)* and *CD-rewritable (CD-RW)* are the two most common types of drives that can write CDs, either once (in the case of CD-R) or repeatedly (in the case of CD-RW). CD-rewritable (CD-RW) technology allows PC users to back up data on CDs.

Digital Video Disc

A **digital video disc (DVD)** is a five-inch diameter CD-ROM look-alike with the ability to store about 135 minutes of digital video or several gigabytes of data (see Figure 2.4). Software programs, video games, and movies are common uses for this storage medium.

DVDs have replaced recordable and rewritable CD discs (CD-R and CD-RW) as the preferred format for sharing movies and photos. Whereas a CD can hold about 740 MB of data, a single-sided DVD can hold 4.7 GB, with double-sided DVDs having a capacity of 9.4 GB. Unfortunately, DVD manufacturers haven't agreed on a recording standard, so several types of recorders and discs are currently in use. Recordings can be made on record-once discs (DVD-R and DVD+R) or on rewritable discs (DVD-RW, DVD+RW, and DVD-RAM). Not all types of rewritable DVDs are compatible with other types.

The Blu-ray high-definition video-disc format based on blue-laser technology stores at least three times as much data as a DVD now holds. The primary use for this new format is in home entertainment equipment to store high definition video, though this format can also store computer data.

Flash Memory

Flash memory is a silicon computer chip that, unlike RAM, is nonvolatile and keeps its memory when the power is shut off. It gets its name from the fact that the microchip is organized so that a section of memory cells (called a *block*) is erased or reprogrammed in a single action, or "flash." Solid-state-memory disks (SSDs) that use flash memory are supplementing or replacing traditional hard drives that employ power-hungry spinning platters with mobile read/write heads near data surfaces. The result is longer laptop battery life and more protection for data. Digital music players and cameras use flash memory to hold music and photos. Compared with other types of secondary storage, flash memory can be accessed more quickly and consumes less power and storage space. Flash memory chips cost much more per megabyte than a traditional hard disk.

The overall trend in secondary storage is toward more direct-access methods, higher capacity, and increased portability. The business needs and needs of individual users should be considered when selecting a specific type of storage. In general, the ability to store large amounts of data and information and access it quickly can increase organizational effectiveness and efficiency.

Input Devices

Your first experience with computers is usually through input and output devices. These devices are the gateways to the computer system—you use them to provide data and instructions to the computer and receive results from it. Input and output devices are part of a computer's user interface, which includes other hardware devices and software that allow you to interact with a computer system.

As with other computer system components, an organization should keep their business goals in mind when selecting input and output devices. For example, many restaurant chains use handheld input devices or computerized terminals that let waiters enter orders efficiently and accurately. These systems have also cut costs by helping to track inventory and market to customers.

Literally hundreds of devices can be used for data input, ranging from special-purpose devices used to capture specific types of data to more general-purpose input devices. We will now discuss several.

Personal Computer Input Devices

A keyboard and a computer mouse are the most common devices used for entry of data such as characters, text, and basic commands. Some companies are developing newer keyboards that are more comfortable, adjustable, and faster to use. These keyboards, such as the split keyboard by Microsoft and others, are designed to avoid wrist and hand injuries caused by hours of keyboarding. Using the same keyboard, you can enter sketches on the touchpad and text using the keys.

flash memory
A silicon computer chip that, unlike RAM, is nonvolatile and keeps its memory when the power is shut off.

A keyboard and mouse are two of the most common devices for computer input. Wireless mice and keyboards are now readily available.

(Source: Courtesy of Hewlett-Packard Company.)

You use a computer mouse to point to and click symbols, icons, menus, and commands on the screen. The computer takes a number of actions in response, such as placing data into the computer system.

Speech-Recognition Technology

speech-recognition technology
Enables a computer equipped with a source of speech input such as a microphone to interpret human speech as an alternative means of providing data or instructions to the computer.

Speech-recognition technology enables a computer equipped with a source of speech input such as a microphone to interpret human speech as an alternative means of providing data or instructions to the computer. The most basic systems require you to train the system to recognize your speech patterns or are limited to a small vocabulary of words. More advanced systems can recognize continuous speech without requiring you to break up your speech into discrete words. Very advanced systems used by the government and military can interpret a voice it has never heard and understand a rich vocabulary.

Companies that must constantly interact with customers are eager to reduce their customer support costs while improving the quality of their service. One company, Dial Directions, offers a free cell phone direction service. Users dial 347-328-4667 and tell the voice-activated service their originating location and desired destination and receive instant text messages with MapQuest driving directions on their cell phone.[12]

Digital Cameras

digital camera
Input device used with a PC to record and store images and video in digital form.

Digital cameras record and store images or video in digital form. When you take pictures, the images are electronically stored in the camera. You can download the images to a computer either directly or by using a flash memory card. After you store the images on the computer's hard disk, you can edit them, send them to another location, paste them into another application, or print them. For example, you can download a photo of your project team captured by a digital camera and then post it on a Web site or paste it into a project status report. Digital cameras have eclipsed film cameras used by professional photographers for photo quality and features such as zoom, flash, exposure controls, special effects, and even video-capture capabilities. With the right software, you can add sound and handwriting to the photo.

The primary advantage of digital cameras is saving time and money by eliminating the need to process film. In fact, digital cameras that can easily transfer images to DVDs have made the consumer film business of Kodak and Fujitsu nearly obsolete. Until film-camera users switch to digital cameras, Kodak is allowing photographers to have it both ways. When you want to develop print film, Kodak offers the option of placing pictures on a DVD in addition to the traditional prints. After the photos are stored on the DVD, they can be edited, placed on a Web site, or sent electronically to business associates or friends around the world.

Organizations use digital cameras for research as well as for business purposes. Microsoft chairman Bill Gates and philanthropist Charles Simonyi donated $30 million to build the Large Synoptic Survey Telescope on a mountain in Chile. When operational in 2014, the 8.4 meter telescope will include a 3,200 megapixel digital camera that captures up to 30 TB of image data per night. The images from deep space will be loaded onto the Web and made available to the public.[13]

Touch-Sensitive Screens

Advances in screen technology allow display screens to function as input as well as output devices. By touching certain parts of a sensitive screen, you can execute a program or cause the computer to take an action. Touch-sensitive screens are frequently used at gas stations for customers to select grades of gas and request a receipt, at fast-food restaurants for order clerks to enter customer choices, at information centers in hotels to allow guests to request facts about local eating and drinking establishments, and at amusement parks to provide directions to patrons. They also are used in kiosks at airports and department stores.

Optical Data Readers

You can use a special scanning device called an *optical data reader* to scan documents. The two categories of optical data readers are for optical mark recognition (OMR) and optical character recognition (OCR). You use OMR readers for test scoring and other purposes when test takers use pencils to fill in boxes on OMR paper, which is also called a "mark sense form." OMR systems are used in standardized tests, including the SAT and GMAT tests, and are being considered as a means to capture voters' choices on Election Day. In comparison, most OCR readers use reflected light to recognize and scan various characters. With special software, OCR readers can convert handwritten or typed documents into digital data. After being entered, this data can be shared, modified, and distributed over computer networks to hundreds or thousands of people.

Con-way Inc. is a $4.7 billion company that offers freight transportation and logistics services. The firm converted to an OCR-based system so that the timesheets can be processed and scanned at any of 38 locations in the United States and Canada, and then forwarded electronically to the payroll office. Initially, the OCR scans were 85 percent accurate, but over time and with a few improvements, the scans are now 99.9 percent accurate. Con-way eliminated the cost of shipping the forms to Portland along with cost of three full-time positions in the payroll department. In addition, the OCR system increased the speed of the entire process and made it more reliable.[14]

Magnetic Ink Character Recognition (MICR) Devices

In the 1950s, the banking industry became swamped with paper checks, loan applications, bank statements, and so on. To remedy this overload and process documents more quickly, the industry developed *magnetic ink character recognition (MICR)*, a system for reading this data quickly. With MICR, data to help clear and route checks is placed on the bottom of a check or other form using a special magnetic ink. Data printed with this ink using a special character set can be read by both people and computers.

Pen Input Devices

By touching the screen with a pen input device, you can activate a command or cause the computer to perform a task, enter handwritten notes, and draw objects and figures. Pen input requires special software and hardware. Handwriting recognition software can convert handwriting on the screen into text. The Tablet PC from Microsoft and its hardware partners can transform handwriting into typed text and store the "digital ink" just the way a person writes it. Users can use a pen to write and send e-mail, add comments to Word documents, mark up PowerPoint presentations, and even hand-draw charts in a document. That data can then be moved, highlighted, searched, and converted into text. If perfected, this interface is likely to become widely used. Pen input is especially attractive if you are uncomfortable using a keyboard. The success of pen input depends on how accurately handwriting can be read and translated into digital form and at what cost.

Radio Frequency Identification

The purpose of a **Radio Frequency Identification (RFID)** system is to transmit data by a mobile device, called a tag, which is read by an RFID reader and processed according to the needs of an information system program (Figure 2.5). One popular application of RFID is to place a microchip on retail items and install in-store readers that track the inventory on the shelves to determine when shelves should be restocked. Recall that the RFID tag chip includes a special form of erasable programmable read-only memory (EPROM) that holds data about the item to which the tag is attached. A radio-frequency signal can update this memory as the status of the item changes. The data transmitted by the tag might provide identification, location information, or details about the product tagged, such as date manufactured, retail price, color, or date of purchase.

Boekhandels Groep Nederland (BGN) is a major book retailer with 40 stores in the Netherlands that sell to roughly 30,000 customers per day. BGN implemented item-level RFID tagging to track the movement of books along with new software to create a tightly integrated warehouse-to-consumer supply chain. With this solution, BGN simplified the inventory process, reduced errors in inventory, and improved the entire supply chain process.[15]

Radio Frequency Identification (RFID)

A technology that employs a microchip with an antenna that broadcasts its unique identifier and location to receivers.

RFID tag

Output Devices

Computer systems provide output to decision makers at all levels of an organization so they can solve a business problem or capitalize on a competitive opportunity. In addition, output from one computer system can provide input into another computer system. The desired form of this output might be visual, audio, or even digital. Whatever the output's content or form, output devices are designed to provide the right information to the right person in the right format at the right time.

Display Monitors

The display monitor is a TV-screen-like device on which output from the computer is displayed. Because traditional monitors use a cathode ray tube to display images, they are sometimes called *CRTs*. The monitor works in much the same way as a traditional TV screen—one or more electron beams are generated from cathode ray tubes. As the beams strike a phosphorescent compound (phosphor) coated on the inside of the screen, a dot on the screen called a *pixel* lights up. The electron beam sweeps back and forth across the screen so that as the phosphor starts to fade, it is struck and lights up again.

With today's wide selection of monitors, price and overall quality can vary tremendously. The quality of a screen is often measured by the number of horizontal and vertical pixels used to create it. More pixels per square inch means a higher resolution, or clarity and sharpness of the image. The distance between one pixel on the screen and the next nearest pixel is known as dot pitch. The common range of dot pitch is from .25 mm to .31 mm. The smaller the number, the better the picture with a dot pitch of .28 mm or smaller considered good.

The characteristics of screen color depend on the quality of the monitor, the amount of RAM in the computer system, and the monitor's graphics adapter card. The Color/Graphics Adapter (CGA) was one of the first technologies to display color images on the screen. Today, Super Video Graphics Array (SVGA) displays are standard, providing vivid colors and high resolution. Digital Video Interface (DVI) is designed to maximize the visual quality of digital display devices such as flat-panel LCD computer displays.

Liquid Crystal Displays (LCDs)

LCD displays are flat displays that use liquid crystals—organic, oil-like material placed between two polarizers—to form characters and graphic images on a backlit screen. These displays are easier on your eyes than CRTs because they are flicker-free, brighter, and don't emit the type of radiation that makes some CRT users worry. In addition, LCD monitors take up less space and use less than half of the electricity required to operate a comparably sized CRT monitor. *Thin-film transistor (TFT) LCDs* are a type of liquid crystal display that assigns a transistor to control each pixel, resulting in higher resolution and quicker response to changes on the screen. TFT LCD monitors have displaced the older CRT technology and are commonly available in sizes from 12 to 30 inches.

Read the Ethical and Societal Issues special feature to learn about the various approaches being taken to capture votes in an accurate and verifiable manner.

Collecting Accurate and Verifiable Data Where It Counts

Imagine having to design or choose an input device that will satisfy every person's needs: the young, elderly, intelligent, illiterate, sighted, or blind. Then imagine that this device has to provide a 100 percent guarantee that it is easy to use for all and collects accurate data—exactly what the user wanted to enter. Sound challenging? That's the struggle that countries around the world are facing as they continue to create the perfect voting machine.

As the technology revolution races ahead, those responsible for voting systems are trying to harness technology to streamline the voting process. Submitting paper ballots now seems prehistoric in this day of movie downloads and cell phone text messaging. It was only natural that the touch screen would make its way into the voting process—with disastrous results. Touch screen machines, also called Direct Recording Electronic (DRE) units, allow the voter to press the name of the person for whom they want to vote. Each vote is either stored in the machine's storage device to be collected later and batch processed, or sent directly to a central database over a private network.

The use of touch screen machines has led to numerous questionable elections and accusations of scandal. The most prominent are the 2000 and 2004 United States presidential elections, where the close results were questioned due to voting irregularities caused by electronic voting machines.

To overcome the problems with touch screen voting machines, many experts feel that a paper backup of a citizen's vote should be generated along with the electronic vote. By providing a "paper trail" of votes, questionable elections can be easily checked. At the time of this writing, 12 U.S. states still have no paper record requirements.

Many voting administrators have given up on touch screen voting systems altogether. In the 2008 primary presidential elections, the state of New Hampshire relied on optical mark recognition (OMR) technology for their voting. Voters fill in the circle next to a candidate's name on a card. The voter's card is scanned to record the votes, and is filed away as a backup in case a recount is needed. Some precincts in New Hampshire provide voters with simple paper ballots that are counted by hand at the polling place. Visually impaired and disabled can use a touch-tone phone to place their votes. In this way, New Hampshire uses several methods to collect votes.

Other states are experimenting with other systems. Oregon holds its votes by mail. Citizens do not have to travel to a precinct center to cast their votes; instead, they simply mark their ballots, stamp them, and put them in the mailbox. The state claims record voting turnouts and little strife.

Some states seem committed to touch screen systems. Despite a report describing several methods of compromising the vote records of its voting machines, the Crawford County, Ohio, county commissioner tells the citizens that there is nothing to worry about. Since only officials from the county are provided with access codes to the inner workings of the machines, the system should be secure.

The voting machine debate extends beyond the United States to every other voting country. In Germany, a group of computer experts collected signatures to request that a court grant an injunction stopping the use of electronic voting machines. They wanted the system switched to a paper ballot system. They argue that the system had security flaws that allowed a hacker to manipulate voting outcomes. The group contended that the government didn't have the technical understanding to ensure an accurate vote count using the electronic system.

As the search for the perfect voting machine continues, one thing is clear: Collecting data into a system that is verifiably accurate, using easy and fast methods, can be a challenge. Security experts put forth three requirements for touch screen machines: They should produce a voter-verifiable paper trail, use software that is open to examination by the public, and provide verifiable ballots to safeguard against machine failure.

Discussion Questions

1. Do you think that one method of collecting data into a voting system can satisfy all the different types of voters? Or are multiple methods required?
2. What would be your concerns about elections by mail, such as the system used in Oregon?

Critical Thinking Questions

1. Of the systems described in this feature, which would you most like to use? In other words, describe your ideal voting method.
2. What are the security risks of your ideal voting method?

SOURCES: Weiss, Tod, "As primary season ramps up, an e-voting snapshot," *Computerworld*, January 8, 2008, *www.computerworld.com/action/article.do? command=viewArticleBasic&articleId=9056 098*. Smith, Jane, "Officials confident voting machines pose no problems," The Meadville Tribune, January 12, 2008, *www.meadvilletribune.com/local/local_story_009222956.html*. Kirk, Jeremy, "German activists move to block e-voting," *NetworkWorld*, January 8, 2008, *www.networkworld.com/news/2008/010808-german-activists-move-to-block.html?fsrc=rss-security*. Kim, Myung, "Most clerks pushing for mail ballots," Rocky Mountain News, January 12, 2008, *www.rockymountainnews.com/news/2008/jan/12/most-clerks-pushing-for-mail-ballots*.

Organic Light-Emitting Diodes

Organic light-emitting diode (OLED) technology is based on research by Eastman Kodak Company and is appearing on the market in small electronic devices. OLEDs use the same base technology as LCDs, with one key difference: Whereas LCD screens contain a fluorescent backlight and the LCD acts as a shutter to selectively block that light, OLEDs directly emit light. OLEDs can provide sharper and brighter colors than LCDs and CRTs, and because they don't require a backlight, the displays can be half as thick as LCDs and used in flexible displays. Another big advantage is that OLEDs don't break when dropped. OLED technology can also create three-dimensional (3-D) video displays by taking a traditional LCD monitor and then adding layers of transparent OLED films to create the perception of depth without the need for 3-D glasses or laser optics.[16]

Printers and Plotters

Hard copy is paper output from a device called a printer. Printers with different speeds, features, and capabilities are available. Some can be set up to accommodate different paper forms such as blank check forms, invoice forms, and so forth. Newer printers allow businesses to create customized printed output for each customer from standard paper and data input using full color.

The speed of the printer is typically measured by the number of pages printed per minute (ppm). Like a display screen, the quality, or resolution, of a printer's output depends on the number of dots printed per inch. A 600-dpi (dots-per-inch) printer prints more clearly than a 300-dpi printer. A recurring cost of using a printer is the inkjet or laser cartridge that is used as pages are printed. Figure 2.6 shows an inkjet printer.

Figure 2.6

Hewlett-Packard CM8060 Inkjet Printer

Inkjet printers, available in a wide variety of speeds and price ranges, have many features, including color capabilities. They are a popular solution for printing hard copies of information.

(Source: Courtesy of Hewlett-Packard.)

Laser printers are generally faster than inkjet printers and can handle more volume than inkjet printers. Laser printers print 15 to 50 pages per minute (ppm) for black and white and 4 to 20 ppm for color. Inkjet printers print 10 to 30 ppm for black and white and 2 to 10 ppm for color.

Plotters are a type of hard-copy output device used for general design work. Businesses typically use these devices to generate paper or acetate blueprints, schematics, and drawings of buildings or new products onto paper or transparencies. Standard plot widths are 24 inches and 36 inches, and the length can be whatever meets the need—from a few inches to several feet.

Digital Audio Player

A **digital audio player** is a device that can store, organize, and play digital music files. **MP3** (MPEG-1 Audio Layer-3) is a popular format for compressing a sound sequence into a very small file while preserving the original level of sound quality when it is played. By compressing the sound file, it requires less time to download the file and less storage space on a hard drive.

You can use many different music devices about the size of a cigarette pack to download music from the Internet and other sources. These devices have no moving parts and can store hours of music. Apple expanded into the digital music market with an MP3 player (the iPod) and the iTunes Music Store, which allows you to find music online, preview it, and download it in a way that is safe, legal, and affordable. The Apple iPod has a 2.5-inch screen and can play video, including selected TV shows you can download from the iTunes Music Store. Other MP3 manufacturers include Dell, Sony, Samsung, Iomega, and Motorola, whose Rokr product is the first iTunes-compatible phone.

The Apple iPod Touch 3.5-inch widescreen lets the user watch movies and TV shows and view photos and album art. The display automatically adjusts the view when it's rotated from portrait to landscape. An ambient light sensor adjusts brightness to match the current lighting conditions. It also supports wireless networking so that the user can access the Internet, view YouTube videos, and purchase music from the iTunes Wi-Fi Music Store.

digital audio player
A device that can store, organize, and play digital music files.

MP3
A standard format for compressing a sound sequence into a small file.

The Apple's iPod Touch
(Source: Courtesy of Apple.)

COMPUTER SYSTEM TYPES

Computer systems can range from desktop (or smaller) portable computers to massive supercomputers that require housing in large rooms. Let's examine the types of computer systems in more detail. Table 2.3 shows general ranges of capabilities for various types of computer systems.

Handheld Computers

Handheld computers are single-user computers that provide ease of portability because of their small size—some are as small as a credit card. These systems often include a variety of software and communications capabilities. Most are compatible with and can communicate with desktop computers over wireless networks. Some even add a built-in GPS receiver with

handheld computer
A single-user computer that provides ease of portability because of its small size.

Factor	Single-User Systems					Multiuser System			
	Handheld	Ultra Laptop	Portable	Thin Client	Desktop	Work-station	Server	Mainframe	Supercomputer
Cost Range	$90 to $900	$700 to $2250	$500 to $3,000	$300 to $900	$400 to $2,500	$3,000 to $40,000	$500 to $50,000	>$100,000	>$250,000
Weight	<24 oz.	<3 lbs.	<7 lbs.	<15 lbs.	<25 lbs.	<25 lbs.	>25 lbs.	>200 lbs.	>200 lbs.
Typical Size	Palm size	Size of a notebook	Size of a notebook	Fits on desktop	Fits on desktop	Fits on desktop	Three-drawer filing cabinet	Refrigerator	Refrigerator and larger
Typical Use	Organize personal data	Improve productivity of highly mobile worker	Improve worker productivity	Enter data and access the Internet	Improve worker productivity	Perform engineering, CAD, and software development	Perform network and Internet applications	Perform computing tasks for large organizations and provide massive data storage	Run scientific applications; perform intensive number crunching
Example	HP iPAQ Pocket PC	Fujitsu Lifebook Q2010	Dell Inspiron T5450	Wyse V90LE Thin Client	Mac Pro	Sun Ultra 40 M2 workstation	Hewlett-Packard HP ProLiant BL	Unisys Clear Path	IBM RS/6000 SP

Table 2.3

Types of Computer Systems

software that can integrate the location data into the application. For example, if you click an entry in an electronic address book, the device displays a map and directions from your current location. Such a computer can also be mounted in your car and serve as a navigation system. One of the shortcomings of handheld computers is that they require lots of power relative to their size.

The Coca-Cola field sales force uses a Pocket PC (a handheld computer that runs the Microsoft Windows Mobile operating system) to automate the collection of information about sales calls, customers, and prospects. They chose a Pocket PC over a laptop because of the cost savings and because it is easier to point and click using radio buttons and drop-down menus on the Pocket PC than to fumble with a keyboard and mouse on a much heavier laptop.[17]

Ultra Laptop Computer

An ultra laptop is a laptop computer weighing less than 3 pounds (1.4 kg) and is usually targeted for use by business travelers. Such laptops typically have a screen that measures 12 inches (30 cm) or less diagonally with a less than full-size keyboard. Many ultra laptops come with extended battery life and energy-efficient CPUs. Popular ultra laptop computers are made by Fujitsu, Lenovo, Sony, Samsung, and Apple. (See Figure 2.7.) Some of the devices use tablet PC technology with a built-in keyboard and can accept handwritten notes on its computer screen. These devices cost between $700 and $2,250 as of January 2008.[18]

Figure 2.7

The MacBook Air

The MacBook Air is an ultraportable laptop that measures 0.76 inches deep at the back and tapers down to 0.16 inches at the front. It weighs 3 pounds and includes a 13-inch LED screen and full-size keyboard.

(Source: Courtesy of Apple.)

Portable Computers

A variety of **portable computers,** those that can be carried easily, are now available. A *laptop computer* is a small, lightweight PC about the size of a three-ring notebook. The even smaller and lighter *notebook* and *subnotebook* computers offer similar computing power. Some notebook and subnotebook computers fit into docking stations of desktop computers to provide additional storage and processing capabilities.

Tablet computers are especially popular and useful in the healthcare, retail, insurance, and manufacturing industries because of their versatility. DT Research provides portable tablet personal computers that come with optional input devices including an integrated bar-code scanner, a card reader, and a camera. The bar-code scanner can capture data from retail, patient, or shipping labels. The card reader can capture data from any card with a magnetic stripe, such as a credit card or driver's license.[19] CSX Transportation, one of the nation's largest railroads, uses DT Research's WebDT 360 to enable train conductors to monitor systems while onboard and communicate with stations for real-time updates. The WebDT 360 has improved operations efficiency and worker productivity.[20]

Low Cost Laptops The mission of the nonprofit One Laptop per Child (OLPC) association is to provide children around the world with new opportunities to explore, experiment, and express themselves with the help of a low-cost laptop priced at about $100. OLPC was founded by Nicholas Negroponte of the Massachusetts Institute of Technology and includes a wide variety of members from academia, the arts, business, and the information technology industry. Negroponte states: "It's an education project, not a laptop project."[21] OLPC launched a "give one, get one" campaign in North America, asking consumers to pay $400 for an XO for themselves and a "free" XO to be given to a child in a developing country. The bright green computer is designed to be extremely rugged and durable with child-friendly features including an easy-to-use interface. See Figure 2.8.

portable computer
Computer small enough to be carried easily.

Figure 2.8

The OLPC XO Laptop Computer
(Source: Courtesy of fuseproject.)

Thin Client

A **thin client** is a low-cost, centrally managed computer with no extra drives, such as a CD or DVD drive, or expansion slots. These computers have limited capabilities and perform only essential applications, so they remain "thin" in terms of the client applications they include. These stripped-down versions of desktop computers do not have the storage capacity or computing power of typical desktop computers, nor do they need it for the role they play. With no hard disk, they never pick up viruses or suffer a hard disk crash. Unlike personal computers, thin clients download software from a network when needed, making support, distribution, and updating of software applications much easier and less expensive. Thin-client manufacturers include Hewlett-Packard, Wyse, BOSaNOVA, and DTR Research.

thin client
A low-cost, centrally managed computer with essential but limited capabilities and no extra drives, such as a CD or DVD drive, or expansion slots.

Desktop Computers

Desktop computers are relatively small, inexpensive single-user computer systems that are highly versatile. Named for their size—the parts are small enough to fit on or beside an office desk—*desktop computers* can provide sufficient memory and storage for most business computing tasks. Desktop computers have become standard business tools; more than 30 million are in use in large corporations.

desktop computer
A relatively small, inexpensive single-user computer that is highly versatile.

The Nokia 770 Internet Tablet is designed for wireless Internet browsing and electronic mail, and includes software such as Internet radio, an RSS news reader, and audio and video players.

(Source: Courtesy of Nokia.)

In addition to traditional PCs that use Intel processors and Microsoft software, there are other options. One of the most popular is the iMac by Apple Computer.

Workstations

Workstations are more powerful than personal computers but still small enough to fit on a desktop. They are used to support engineering and technical users who perform heavy mathematical computing, computer-aided design (CAD), and other applications requiring a high-end processor. Such users need very powerful CPUs, large amounts of main memory, and extremely high-resolution graphic displays.

Servers

A computer **server** is a computer used by many users to perform a specific task, such as running network or Internet applications. Servers typically have large memory and storage capacities, along with fast and efficient communications abilities. A Web server is used to handle Internet traffic and communications. An Internet caching server stores Web sites that are frequently used by a company. An enterprise server stores and provides access to programs that meet the needs of an entire organization. A file server stores and coordinates program and data files. A transaction server is used to process business transactions. Server systems consist of multiuser computers including supercomputers, mainframes, and servers.

Servers offer great **scalability**, the ability to increase the processing capability of a computer system so that it can handle more users, more data, or more transactions in a given period. Scalability is increased by adding more, or more powerful, processors.

Mainframe Computers

A **mainframe computer** is a large, powerful computer shared by dozens or even hundreds of concurrent users connected to the machine over a network. The mainframe computer must reside in a data center with special heating, ventilating, and air-conditioning (HVAC) equipment to control temperature, humidity, and dust levels. In addition, most mainframes are kept in a secure data center with limited access to the room. The construction and maintenance of a controlled-access room with HVAC can add hundreds of thousands of dollars to the cost of owning and operating a mainframe computer.

The new role of the mainframe is as a large information-processing and data-storage utility for a corporation—running jobs too large for other computers, storing files and databases too large to be stored elsewhere, and storing backups of files and databases created elsewhere. The mainframe can handle the millions of daily transactions associated with airline, automobile, and hotel/motel reservation systems. It can process the tens of thousands of daily queries necessary to provide data to decision support systems. Its massive storage and input/output capabilities enable it to play the role of a video computer, providing full-motion video to multiple, concurrent users.

IBM mainframe computer customers include the top 25 banks and the top 25 retailers in the world who use the machines for processing large amounts of transactions. For example, the Bank of China houses 350 million accounts with three billion transaction histories and processes 30 million transactions in less than an hour using IBM's System z mainframe computer.[22]

workstation
A more powerful personal computer that is used for technical computing, such as engineering, but still fits on a desktop.

server
A computer designed for a specific task, such as network or Internet applications.

scalability
The ability to increase the capability of a computer system to process more transactions in a given period by adding more, or more powerful, processors.

mainframe computer
Large, powerful computer often shared by hundreds of concurrent users connected to the machine via terminals.

Supercomputers

Supercomputers are the most powerful computer systems with the fastest processing speeds and highest performance. They are *special-purpose machines* designed for applications that require extensive and rapid computational capabilities. Originally, supercomputers were used primarily by government agencies to perform the high-speed number crunching needed in weather forecasting and military applications. With recent reductions in the cost of these machines, they are now used more broadly for commercial purposes by oil and gas companies, financial institutions and animated film makers.

supercomputers
The most powerful computer systems, with the fastest processing speeds.

IBM's Blue Gene/L System at the Lawrence Livermore National Laboratory is the fastest supercomputer in the world and can perform 596 trillion floating-point operations per second.

(Source: Courtesy of IBM Corporation.)

We now turn to the other critical component of effective computer systems—software. Like hardware, software has made great technological leaps in a relatively short time span.

OVERVIEW OF SOFTWARE

As you learned in Chapter 1, software consists of computer programs that control the workings of computer hardware. **Computer programs** are sequences of instructions for the computer. **Documentation** describes the program functions to help the user operate the computer system. The program displays some documentation on screen, while other forms appear in external resources, such as printed manuals. People using commercially available software are usually asked to read and agree to End-User License Agreements (EULAs). After reading the EULA, you normally have to click an "I agree" button before you can use the software, which can be one of two basic types: systems software and application software.

computer programs
Sequences of instructions for the computer.

documentation
Text that describes the program functions to help the user operate the computer system.

Application software has the greatest potential to affect processes that add value to a business because it is designed for specific organizational activities and functions.

(Source: © Jim West / Alamy.)

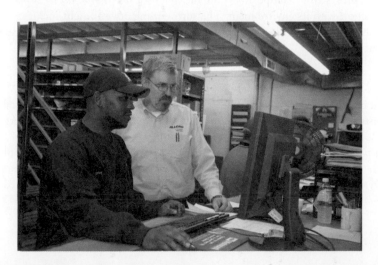

systems software
The set of programs designed to coordinate the activities and functions of the hardware and various programs throughout the computer system.

application software
The programs that help users solve particular computing problems.

Systems software is the set of programs designed to coordinate the activities and functions of the hardware and various programs throughout the computer system. Each type of systems software is designed for a specific CPU design and class of hardware. **Application software** consists of programs that help users solve particular computing problems. In most cases, application software resides on the computer's hard disk before it is brought into the computer's main memory and run. Application software can also be stored on CDs, DVDs, and even flash or keychain storage devices that plug into a USB port. Before a person, group, or enterprise decides on the best approach for acquiring application software, they should analyze their goals and needs carefully.

Supporting Individual, Group, and Organizational Goals

Every organization relies on the contributions of individuals, groups, and the entire enterprise to achieve business objectives. To help them achieve these objectives, the organization provides them with specific application software and information systems. One useful way of classifying the many potential uses of information systems is to identify the scope of the problems and opportunities addressed by a particular organization, called the sphere of influence. These spheres of influence are personal, workgroup, and enterprise, as shown in Table 2.4.

Table 2.4

Classifying Software by Type and Sphere of Influence

Software	Personal	Workgroup	Enterprise
Systems software	Personal computer and workstation operating systems	Network operating systems	Midrange computer and mainframe operating systems
Application software	Word processing, spreadsheet, database, graphics	Electronic mail, group scheduling, shared work, collaboration	General ledger, order entry, payroll, human resources

Information systems that operate within the *personal sphere of influence* serve the needs of an individual user. These information systems enable users to improve their personal effectiveness, increasing the amount of work that can be done and its quality. Such software is often referred to as *personal productivity software.*

A *workgroup* is two or more people who work together to achieve a common goal. A workgroup may be a large, formal, permanent organizational entity such as a section or department or a temporary group formed to complete a specific project. An information system that operates in the *workgroup sphere of influence* supports a workgroup in the attainment of a common goal. Users of such applications must be able to communicate, interact, and collaborate to be successful.

Information systems that operate within the *enterprise sphere of influence* support the firm in its interaction with its environment. The surrounding environment includes customers, suppliers, shareholders, competitors, special-interest groups, the financial community, and government agencies.

Installing and Removing New Software

Before you can use any type of software, it must be installed on a computer. Installing new software usually involves only a few setup steps. Software for personal computers typically comes on CDs or is downloaded from the Web.

When possible, it is best to remove software using an add/remove software utility that comes with the operating system or that is part of some utility software such as Norton System Works and McAfee QuickClean. This will help ensure that all elements of unwanted software are removed.

SYSTEMS SOFTWARE

Controlling the operations of computer hardware is one of the most critical functions of systems software. Systems software also supports the application programs' problem-solving capabilities. Different types of systems software include operating systems and utility programs.

Operating Systems

An operating system (OS) is a set of computer programs that control the computer hardware and act as an interface with application programs (see Figure 2.9). Operating systems can control one computer or multiple computers, or they can allow multiple users to interact with one computer. The various combinations of OSs, computers, and users include the following:

- **Single computer with a single user.** This system is commonly used in a personal computer or a handheld computer that allows one user at a time.
- **Single computer with multiple users.** This system is typical of larger, mainframe computers that can accommodate hundreds or thousands of people, all using the computer at the same time.
- **Multiple computers.** This system is typical of a network of computers, such as a home network with several computers attached or a large computer network with hundreds of computers attached around the world.
- **Special-purpose computers.** This system is typical of a number of computers with specialized functions, such as those that control sophisticated military aircraft, the space shuttle, and some home appliances.

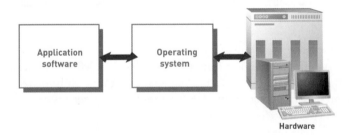

Figure 2.9

The Role of Operating Systems

The role of the operating system and other systems software is as an interface or buffer between users and application software and hardware.

The OS, which plays a central role in the functioning of the complete computer system, is usually stored on disk. After a computer system is started, or "booted up," portions of the OS are transferred to memory as they are needed. You can also boot a computer from a CD, DVD, or even a thumb drive that plugs into a USB port. A storage device that contains some or all of the OS is often called a "rescue disk" because you can use it to start the computer if you have problems with the primary hard disk.

The collection of programs that make up the operating system performs a variety of activities, including the following:

- Performing common computer hardware functions
- Providing a user interface and input/output management
- Providing a degree of hardware independence
- Managing system memory
- Managing processing tasks
- Providing networking capability
- Controlling access to system resources
- Managing files

Common Hardware Functions

All applications must perform certain hardware-related tasks, such as the following:

- Get input from the keyboard or another input device
- Retrieve data from disks
- Store data on disks
- Display information on a monitor or printer

Each of these tasks requires a detailed set of instructions. The OS converts a basic request into the set of detailed instructions that the hardware requires. In effect, the OS acts as an intermediary between the application and the hardware. The typical OS performs hundreds of such tasks, translating each into one or more instructions for the hardware. The OS notifies the user if input or output devices need attention, if an error has occurred, and if anything abnormal happens in the system.

User Interface and Input/Output Management

One of the most important functions of any OS is providing a **user interface**. A user interface allows people to access and command the computer system. The first user interfaces for mainframe and personal computer systems were command based.

A **command-based user interface** requires text commands to be given to the computer to perform basic activities. For example, the command ERASE 00TAXRTN would cause the computer to erase or delete a file called 00TAXRTN. RENAME and COPY are other examples of commands used to rename files and copy files from one location to another.

A **graphical user interface (GUI)** uses pictures called *icons* and menus displayed on screen to send commands to the computer system. Many people find that GUIs are easier to use because users intuitively grasp the functions. Today, the most widely used graphical user interface is Windows by Microsoft. As the name suggests, Windows is based on the use of a window, or a portion of the display screen dedicated to a specific application. The screen can display several windows at once. The use of GUIs has contributed greatly to the increased use of computers because users no longer need to know command-line syntax to accomplish a task.

Hardware Independence

To run, applications request services from the OS through a defined **application program interface (API)**, as shown in Figure 2.10. Programmers can use APIs to create application software without understanding the inner workings of the operating system.

user interface

Element of the operating system that allows you to access and command the computer system.

command-based user interface

A user interface that requires you to give text commands to the computer to perform basic activities.

graphical user interface (GUI)

An interface that uses icons and menus displayed on screen to send commands to the computer system.

application program interface (API)

Interface that allows applications to make use of the operating system.

Figure 2.10

Application Program Interface Links Application Software to the Operating System

Memory Management

The OS also controls how memory is accessed and maximizes available memory and storage. Newer OSs typically manage memory better than older OSs. The memory-management feature of many OSs allows the computer to execute program instructions effectively and to speed processing. One way to increase the performance of an old computer is to upgrade to a newer OS and increase the amount of memory.

Most OSs support virtual memory, which allocates space on the hard disk to supplement the immediate, functional memory capacity of RAM. Virtual memory works by swapping programs or parts of programs between memory and one or more disk devices—a concept called paging. This reduces CPU idle time and increases the number of jobs that can run in a given time span.

Processing Tasks

The task-management features of today's OSs manage all processing activities. Task management allocates computer resources to make the best use of each system's assets. Task-management software can permit one user to run several programs or tasks at the same time (multitasking) and allow several users to use the same computer at the same time (time-sharing).

An OS with multitasking capabilities allows a user to run more than one application at the same time. Without having to exit a program, you can work in one application, easily pop into another, and then jump back to the first program, picking up where you left off. Better still, while you're working in the *foreground* in one program, one or more other applications can be churning away, unseen, in the *background*, sorting a database, printing a document, or performing other lengthy operations that otherwise would monopolize your computer and leave you staring at the screen unable to perform other work. Multitasking can save users a considerable amount of time and effort.

Time-sharing allows more than one person to use a computer system at the same time. For example, 15 customer service representatives might be entering sales data into a computer system for a mail-order company at the same time. In another case, thousands of people might be simultaneously using an online computer service to get stock quotes and valuable business news.

The ability of the computer to handle an increasing number of concurrent users smoothly is called *scalability*. This feature is critical for systems expected to handle a large number of users such as a mainframe computer or a Web server. Because personal computer OSs are usually oriented toward single users, they do not need to manage multiple-user tasks often.

Networking Capability

Most operating systems include networking capabilities so that computers can join together in a network to send and receive data and share computing resources. PCs running Mac, Windows, or Linux operating systems allow users to easily set up home or business networks for sharing Internet connections, printers, storage, and data. Operating systems for larger server computers are designed specifically for computer networking environments.

Access to System Resources and Security

Because computers often handle sensitive data that can be accessed over networks, the OS needs to provide a high level of security against unauthorized access to the users' data and programs. Typically, the OS establishes a logon procedure that requires users to enter an identification code, such as a user name, and a matching password. If the identification code is invalid or if the password does not match the identification code, the user cannot gain access to the computer. Some OSs require that user passwords change frequently—such as every 20 to 40 days. If the user is successful in logging on to the system, the OS restricts access to only portions of the system for which the user has been authorized. The OS records who is using the system and for how long and reports any attempted breaches of security.

File Management

The OS manages files to ensure that files in secondary storage are available when needed and that they are protected from access by unauthorized users. Many computers support multiple users who store files on centrally located disks or tape drives. The OS keeps track of where each file is stored and who can access it. The OS must determine what to do if more than one user requests access to the same file at the same time. Even on stand-alone personal computers with only one user, file management is needed to track where files are located, what size they are, when they were created, and who created them.

Current Operating Systems

Early OSs were very basic. Today, however, more advanced OSs have been developed, incorporating sophisticated features and impressive graphics effects. Table 2.5 classifies a number of current OSs by sphere of influence.

Table 2.5

Popular Operating Systems Cross All Three Spheres of Influence

Personal	Workgroup	Enterprise
Microsoft Windows Vista, Windows XP, Windows Mobile, Windows Automotive, and Windows Embedded	Microsoft Windows Server 2003 and Server 2008	Microsoft Windows Server 2003 and Server 2008
Mac OS X	Mac OS X Server	
UNIX	UNIX	UNIX
Solaris	Solaris	Solaris
Linux	Linux	Linux
Red Hat Linux	Red Hat Linux	Red Hat Linux
Palm OS	Netware	
	IBM i5/OS and z/OS	IBM i5/OS and z/OS
	HP-UX 11i	HP-UX 11i

Microsoft PC Operating Systems

Since a small company called Microsoft developed PC-DOS and MS-DOS to support the IBM personal computer introduced in the 1980s, personal computer OSs have steadily evolved. *PC-DOS* and *MS-DOS* had command-driven interfaces that were difficult to learn and use. Each new version of OS has improved the ease of use, processing capability, reliability, and ability to support new computer hardware devices.

Windows XP (XP reportedly stands for the wonderful e*xp*erience that you will have with your personal computer) was released in fall 2001. Previous consumer versions of Windows were notably unstable and crashed frequently, requiring frustrating and time-consuming reboots. With XP, Microsoft sought to bring reliability to the consumer.

Microsoft released Windows Vista in 2007 with the goal of providing a more secure and stable operating system. See Figure 2.11 The new operating system includes a number of new features. The most advanced versions of Vista include a 3-D graphics interface called Aero. However, the system requirements for Windows Vista with Aero require many users to purchase new, more powerful PCs. Another issue was that some software and hardware designed for Windows XP would not run on Vista. Today, Microsoft has over 90 percent of the PC OS market, Apple has roughly seven percent, Linux publishers, and other companies account for the rest of the PC OS market.

Apple Computer Operating Systems

Although IBM system platforms traditionally use one of the Windows OSs and Intel microprocessors (often called *Wintel* for this reason), Apple computers have used non-Intel microprocessors designed by Apple, IBM, and Motorola and a proprietary Apple OS—the Mac OS. Newer Apple computers, however, use Intel chips. Although Wintel computers hold the largest share of the business PC market, Apple computers are also quite popular, especially in the fields of publishing, education, graphic arts, music, movies, and media.

The Apple OSs have also evolved over a number of years and often provide features not available from Microsoft. Starting in July 2001, the Mac OS X was installed on all new

Figure 2.11

Microsoft Windows Vista

Macs. It includes an entirely new user interface, which provides a new visual appearance for users—including luminous and semitransparent elements, such as buttons, scroll bars, windows, and fluid animation to enhance the user's experience.

Since its first release, OS X has upgraded several times from Jaguar (OS X.2) to Panther (OS X.3) to Tiger (OS X.4) to Leopard, the most recent version of OS X, released in 2007 to compete with Windows Vista.

OS X Leopard includes an attractive 3-D graphical user interface that Apple claims is more intuitive than Windows. Leopard includes Time Machine, a powerful backup tool that allows users to view their system as it looked in the past and resurrect deleted files. Leopard also includes multiple desktops, a video chat program that allows users to pose in front of imaginary landscapes, a powerful system search utility, and other updated software. Because Mac OS X runs on Intel processors, Mac users can set up their PC to run both Windows Vista and Mac OS X and select which platform they want to work with when they boot their PC. Macs are also considered very secure, with no widespread virus or spyware infections to date.

Linux

Linux is an OS developed by Linus Torvalds in 1991 as a student in Finland. The OS is distributed under the GNU General Public License, and its source code is freely available to everyone. It is, therefore, called an open-source operating system. This doesn't mean, however, that Linux and its assorted distributions are necessarily free—companies and developers can charge money for a distribution as long as the source code remains available. Linux is actually only the kernel of an OS, the part that controls hardware, manages files, separates processes, and so forth. Several combinations of Linux are available, with various sets of capabilities and applications to form a complete OS. Each of these combinations is called a *distribution* of Linux. Many distributions are available as free downloads.

Linux is available on the Internet and from other sources, including Red Hat Linux and Caldera OpenLinux. Many people and organizations use Linux. In addition, several large computer vendors, including IBM, Hewlett-Packard, and Intel, support the Linux operating system. CIOs within many organizations are considering switching to Linux and open-source software because of security concerns with Microsoft software. Linux is also making inroads to the consumer PC market with their GUI distributions. Both Dell and Lenovo sell notebook computers running Ubuntu and SuSE Linux.[23]

Radio station KRUU, "the Voice of Fairfield" in Iowa, is a nonprofit, community-based radio station. It broadcasts locally every day, 24 hours a day, and online to 30 countries (*www.kruufm.com*). The station supports about 100 hosts and 75 programs, broadcasting programs ranging from bedtime stories to death-metal music. When shopping for an operating system to use in the studio, KRUU selected Linux Ubuntu.[24] "Our requirements were quite complex and our decision to go with Ubuntu was based on three factors and Ubuntu won hands down," stated Sundar Raman, a presenter at the station. The three factors were: (1) Ubuntu looks good and is simple for both Windows and Mac users to use, (2) Ubuntu is reliable and easy to manage both locally and remotely, and (3) Ubuntu software supports professional audio editing and mixing software and hardware. One benefit of using Linux Ubuntu is the user community support. KRUU found all the answers it needed regarding running a professional studio on Linux from the "Ubuntu-Studio" community. Communicating with other Linux professionals helped the station use Linux computers for all of their computing tasks, including recording and mixing consoles.

Workgroup Operating Systems

To keep pace with user demands, the technology of the future must support a world in which network usage, data-storage requirements, and data-processing speeds increase at a dramatic rate. This rapid increase in communications and data-processing capabilities pushes the boundaries of computer science and physics. Powerful and sophisticated OSs are needed to run the servers that meet these business needs for workgroups. Small businesses, for example, often use workgroup OSs to run networks and perform critical business tasks.

Windows Server

Windows Server 2008 is a powerful Web server management system that enables organizations to deploy data and services throughout the organization using state-of-the-art Web technologies. It includes virtualization tools that allow multiple and various operating systems to run on a single server. It also provides advanced security features, improved reliability, simple management tools, ease of installation, troubleshooting tools, and remote administration.

UNIX

UNIX is a powerful OS originally developed by AT&T for minicomputers. UNIX can be used on many computer system types and platforms, from personal computers to mainframe systems. UNIX also makes it much easier to move programs and data among computers or to connect mainframes and personal computers to share resources. There are many variants of UNIX—including HP/UX from Hewlett-Packard, AIX from IBM, UNIX SystemV from UNIX Systems Lab, Solaris from Sun Microsystems, and SCO from Santa Cruz Operations. Sun Microsystems hopes that its open-source Solaris will attract developers to make the software even better.

NetWare

NetWare is a network OS sold by Novell that can support users on Windows, Macintosh, and UNIX platforms. NetWare provides directory software to track computers, programs, and people on a network, helping large companies to manage complex networks. NetWare users can log on from any computer on the network and use their own familiar desktop with all their applications, data, and preferences.

Red Hat Linux

Red Hat Software offers a Linux network OS that taps into the talents of tens of thousands of volunteer programmers who generate a steady stream of improvements for the Linux OS. The *Red Hat Linux* network OS is very efficient at serving Web pages and can manage a cluster of up to eight servers. Linux environments typically have fewer virus and security problems than other OSs. Distributions such as SuSE and Red Hat have proven Linux to be a very stable and efficient OS.

Mac OS X Server

The *Mac OS X Server* provides UNIX-style process management. Protected memory puts each service in its own well-guarded chunk of dynamically allocated memory, preventing a single process from going awry and bringing down the system or other services. Under pre-emptive multitasking, a computer OS uses some criteria to decide how long to allocate to any one task before giving another task a turn to use the OS. Preempting is the act of taking control of the OS from one task and giving it to another. A common criterion for preempting is simply elapsed time or certain applications can be given higher priority than other applications, giving the higher-priority programs longer processing times. Preemptive multitasking ensures that each process gets the right amount of CPU time and the system resources it needs for optimal efficiency and responsiveness.

Enterprise Operating Systems

New mainframe computers provide the computing and storage capacity to meet massive data processing requirements and provide a large number of users with high performance and excellent system availability, strong security, and scalability. In addition, a wide range of application software has been developed to run in the mainframe environment, making it possible to purchase software to address almost any business problem. As a result, mainframe computers remain the computing platform of choice for mission-critical business applications for many companies. *z/OS* from IBM, *HP-UX* from Hewlett-Packard, and *Linux* are examples of mainframe operating systems.

z/OS

The *z/OS* is IBM's first 64-bit enterprise OS. It supports IBM's z900 and z800 lines of mainframes that can come with up to sixteen 64-bit processors. (The z stands for zero downtime.) The OS provides several new capabilities to make it easier and less expensive for users to run large mainframe computers. The OS has improved workload management and advanced e-commerce security. The IBM zSeries mainframe, like previous generations of IBM mainframes, lets users subdivide a single computer into multiple smaller servers, each of which can run a different application. In recognition of the widespread popularity of a competing OS, z/OS allows partitions to run a version of the Linux OS. This means that a company can upgrade to a mainframe that runs the Linux OS.

HP-UX and Linux

HP-UX is a robust UNIX-based OS from Hewlett-Packard designed to handle a variety of business tasks, including online transaction processing and Web applications. It supports Internet database, and a variety of business applications on server and mainframe enterprise system. It can work with Java programs and Linux applications. HP-UX supports Hewlett-Packard's computers and those designed to run Intel's Itanium processors. *Red Hat Enterprise Linux* for IBM mainframe computers is another example of an enterprise operating system.

Operating Systems for Small Computers, Embedded Computers, and Special-Purpose Devices

New OSs and other software are changing the way we interact with personal digital assistants (PDAs), smart phones, cell phones, digital cameras, TVs, and other appliances. These OSs are also called *embedded operating systems* because they are typically embedded within a device, such as an automobile, TV recorder, or other device. Embedded software is a multibillion dollar industry. Some of these OSs allow handheld devices to be synchronized with PCs using cradles, cables, and wireless connections. Cell phones also use embedded OSs (see Figure 2.12). In addition, some OSs have been developed for special-purpose devices, such as TV set-top boxes, computers on the space shuttle, computers in military weapons, and computers in some home appliances.

An IT group within the United States Department of Agriculture recently deployed BlackBerries to their IT staff.[25] The high-speed network connection between BlackBerry and

the organization's private network allowed system support staff to troubleshoot problems on
Linux, UNIX, and Microsoft servers located in the home office from any location.

Palm OS

ACCESS Systems makes the Palm operating system that is used in over 30 million handheld
computers and smartphones manufactured by Palm, Inc. and other companies. Palm also
develops and supports applications, including business, multimedia, games, productivity,
reference and education, hobbies and entertainment, travel, sports, utilities, and wireless
applications. Today, the smartphone market is overtaking the PDA market, as mobile users
prefer to combine phone and information services in one device. OSs for this market are also
provided by Research in Motion, Microsoft, Symbian, Apple (for the iPhone), and others.

Windows Embedded

Windows Embedded is a family of Microsoft OSs included with or embedded into small
computer devices. Windows Embedded includes several versions that provide computing
power for TV set-top boxes, automated industrial machines, media players, medical devices,
digital cameras, PDAs, GPS receivers, ATMs, gaming devices, and business devices such as
cash registers. Microsoft Auto provides a computing platform for automotive software such
as Ford Sync. The Ford Sync system uses an in-dashboard display and wireless networking
technologies to link automotive systems with cell phones and portable media players. See
Figure 2.13.

Windows Mobile

Windows Mobile is an operating system designed for smartphones and PDAs. Different versions of Windows Mobile support either a touch screen interface or a menu-driven interface.
In addition to supporting typical cellular services, Windows Mobile provides handwriting
recognition, instant messaging technology, support for more secure Internet connections,
and the ability to beam information to other devices. Dozens of phones provided by all of
the major carriers run Windows Mobile.

Figure 2.13

Microsoft Auto and Ford Sync

The Ford Sync system, developed on the Microsoft Auto operating system, allows drivers to wirelessly connect cell phones and media devices to automotive systems.

(Source: Courtesy of Microsoft Corporation and Ford Motor Company.)

UTILITY PROGRAMS

Utility programs help to perform maintenance or correct problems with a computer system. For example, some utility programs merge and sort sets of data, keep track of computer jobs being run, compress files of data before they are stored or transmitted over a network (thus saving space and time), and perform other important tasks. Some utility programs can help computer systems run better and longer without problems.

Utility programs can also help to secure and safeguard data. For example, the recording and motion picture industry uses digital rights management (DRM) technologies to prevent copyright-protected movies and music from being unlawfully copied. Music and media files are encoded so that software running on players recognizes and plays only legally obtained copies. DRM has been criticized for infringing on the freedom and rights of customers. Record companies are experimenting with DRM-free music to see if it increases sales.

Although many PC utility programs come installed on computers, you can also purchase utility programs separately. Table 2.6 provides examples of some common types of utilities.

utility programs
Programs that help to perform maintenance or correct problems with a computer system.

Table 2.6

Examples of Utility Programs

Personal	Workgroup	Enterprise
Software to compress data so that it takes less hard disk space	Software to provide detailed reports of workgroup computer activity and status of user accounts	Software to archive contents of a database by copying data from disk to tape
Screen saver	System management software that enables a support person to monitor the growing number of desktop computers attached to the server	Virtualization software that enables one computer to simulate other computers or multiple computers.
Antivirus and antispyware software	Software that reports unsuccessful user logon attempts	Software that reports the status of a particular computer job

APPLICATION SOFTWARE

Application software applies the power of a computer to give individuals, workgroups, and the entire enterprise the ability to solve problems and perform specific tasks. Application programs interacts with systems software, and the systems software directs the computer hardware to perform the necessary tasks.

Applications help you perform common tasks, such as create and format text documents, perform calculations, or manage information, though some applications are more specialized. A pharmaceutical company, for example, has developed application software to detect the early signs of Parkinson's disease. Application software is used throughout the medical profession to save and prolong lives.

The functions performed by application software are diverse, and range from personal productivity to business analysis. For example, application software can help sales managers track sales of a new item in a test market. Software from IntelliVid monitors video feeds from store security cameras and notifies security when a shopper is behaving suspiciously.[26] Most of the computerized business jobs and activities discussed in this book involve application software. We begin by investigating the types and functions of application software.

Types and Functions of Application Software

The key to unlocking the potential of any computer system is application software. A company can either develop a one-of-a-kind program for a specific application (called **proprietary software**) or purchase and use an existing software program (sometimes called **off-the-shelf software**). It is also possible to modify some off-the-shelf programs, giving a blend of off-the-shelf and customized approaches. These different sources of software are shown in Figure 2.14. The relative advantages and disadvantages of proprietary software and off-the-shelf software are summarized in Table 2.7.

proprietary software
One-of-a-kind program developed for a specific application.

off-the-shelf software
An existing software program that can be purchased.

Table 2.7

A Comparison of Proprietary and Off-the-Shelf Software

Proprietary Software		Off-the-Shelf Software	
Advantages	**Disadvantages**	**Advantages**	**Disadvantages**
You can get exactly what you need in terms of features, reports, and so on.	It can take a long time and significant resources to develop required features.	The initial cost is lower because the software firm can spread the development costs over many customers.	An organization might have to pay for features that are not required and never used.
Being involved in the development offers control over the results.	In-house system development staff may become hard pressed to provide the required level of ongoing support and maintenance because of pressure to move on to other new projects.	The software is likely to meet the basic business needs—you can analyze existing features and the performance of the package before purchasing.	The software might lack important features, thus requiring future modification or customization. This can be very expensive because users must adopt future releases of the software as well.
You can modify features that you might need to counteract an initiative by competitors or to meet new supplier or customer demands. A merger with or acquisition of another firm also requires software changes to meet new business needs.	The features and performance of software that has yet to be developed presents more potential risk.	The package is likely to be of high quality because many customer firms have tested the software and helped identify its bugs.	The software might not match current work processes and data standards.

Figure 2.14

Types of Application Software

Some off-the-shelf software can be customized to suit user needs.

Software Helps Target Radiation Treatment for Cancer

Doctors have been using radiation therapy as a treatment for cancer since the 1940s. The treatment has saved countless lives, yet has been somewhat imprecise until recently. The original method of treating a tumor with radiation used a linear accelerator that delivered radiation in rectangular beams. Doctors used lead blocks to prevent the beams from harming healthy tissue. The process was cumbersome and only partially effective. Surrounding tissue was often destroyed along with the tumor.

In the 1980s, a machine called an MLC, for multileaf collimator, was invented. The MLC had motorized leaves to disrupt the beam of radiation and focus it more closely on where it was needed. Still, the treatment was imprecise, lacking real-time control of the radiation intensity and direction.

Until the mid-1990s, most of the development of radiation treatment technologies focused on hardware. Varian Medical Systems decided that devising a more effective system would require a heavy investment in software development. Computing processors and hardware were advanced enough to precisely control beams of radiation, but the software to empower the hardware had yet to be developed. Varian transformed itself from a hardware company to a software company to get the job done.

Varian hired experts in programming embedded controls, user interfaces, treatment planning, and databases. It proceeded incrementally over many years to develop a trustworthy and powerful system called the SmartBeam IMRT (for Intensity Modulated Radiation Therapy), which is now in use at thousands of medical facilities around the world.

The SmartBeam IMRT combines an x-ray and radiation technology into one device that rotates around the patient delivering radiation at precise intensities from any angle. The machine is the first that allows physicians to examine and treat a tumor at the same time. The on-board imager produces "high-resolution images of tumors and tracks changes in a tumor's shape, size, and position... that when coupled with SmartBeam IMRT, allows clinicians to be even more precise when targeting tumors," according to *Computerworld*. The magazine awarded Varian the top prize for information systems in manufacturing in its 2007 Computerworld Honors Program.

Discussion Questions

1. What role does software play in the SmartBeam IMRT medical system?
2. Why couldn't Varian produce the SmartBeam IMRT before it did?

Critical Thinking Questions

1. What additional safeguards must be programmed into the software that runs the SmartBeam IMRT that aren't necessary in typical PC software?
2. How do you think the development of the SmartBeam IMRT launched Varian to the top of the market in cancer treatment systems?

SOURCES: Pratt, Mary K., "Software Helps Target Radiation Treatment for Cancer," *Computerworld*, December 3, 2007, *www.computerworld.com/action/article.do?command=viewArticleBasic&articleId=3048 65& pageNumber=1*. Varian Medical Systems Web site, *www.varian.com*, accessed February 3, 2007.

Many companies use off-the-shelf software to support business processes. Key questions for selecting off-the-shelf software include: (1) Will the software run on the OS and hardware you have selected? (2) Does the software meet the essential business requirements that have been defined? (3) Is the software manufacturer financially solvent and reliable? and (4) Does the total cost of purchasing, installing, and maintaining the software compare favorably to the expected business benefits?

Some off-the-shelf programs can be modified, in effect blending the off-the-shelf and customized approaches. For example, a software developer might write a collection of programs to be used in an auto body shop and include features to generate estimates, order parts, and process insurance. Body shops of all types have these needs. Designed properly—and with provisions for minor tailoring for each user—the same software package can be sold to many users. However, because each body shop has slightly different requirements, some modifications to the software might be needed. As a result, software vendors often provide a wide range of services, including installing their standard software, modifying the software as the customer requires, training users, and providing other consulting services.

Another approach to obtaining a customized software package is to use an application service provider. An **application service provider** (ASP) is a company that can provide the software, support, and computer hardware on which to run the software from the user's facilities over a network. An ASP can also simplify a complex corporate software package so that it is easier for the users to set up and manage. ASPs provide contract customization of off-the-shelf software, and they speed deployment of new applications while helping IS managers avoid implementation headaches, reducing the need for many skilled IS staff members and decreasing project start-up expenses. Such an approach allows companies to devote more time and resources to more important tasks. For example, Avanax, a Silicon Valley company that develops intelligent photonic solutions for optical networks, uses a Product Lifecycle Management system provided by SAP. The system runs on SAP servers which has helped Avanax reduce costs and provide much higher levels of service.[27]

Using an ASP involves some risks—sensitive information could be compromised in a number of ways, including unauthorized access by employees or computer hackers; the ASP might not be able to keep its computers and network up and running as consistently as necessary; or a disaster could disable the ASP's data center, temporarily putting an organization out of business. These are legitimate concerns that an ASP must address.

The high overhead of an ASP designing, running, managing, and supporting many customized applications for many businesses has led to a new form of software distribution known as software as a service. **Software as a service** (SaaS) allows businesses to subscribe to Web-delivered business application software by paying a monthly service charge or a per-use fee. Like ASP, SaaS providers maintain software on their own servers and provide access to it over the Internet. SaaS usually uses a Web browser-based user interface. SaaS can reduce expenses by sharing its running applications among many businesses. For example, Sears, JCPenney, and Wal-Mart might use customer relationship management software provided by a common SaaS provider. Providing one high-quality SaaS application to thousands of businesses is much more cost-effective than custom designing software for each business.

Personal Application Software

Hundreds of computer applications can help individuals at school, home, and work. The features of personal application software are summarized in Table 2.8. In addition to these general-purpose programs, there are literally thousands of other personal computer applications to perform specialized tasks: to help you do your taxes, get in shape, lose weight, get medical advice, write wills and other legal documents, make repairs to your computer, fix your car, write music, and edit your pictures and videos (see Figure 2.15). This type of software, often called *user software* or *personal productivity software*, includes the general-purpose tools and programs that support individual needs.

Type of Software	Explanation	Example	Vendor
Word processing	Create, edit, and print text documents	Word WordPerfect Google Docs Pages Writer	Microsoft Corel Google Apple Sun
Spreadsheet	Provide a wide range of built-in functions for statistical, financial, logical, database, graphics, and date and time calculations	Excel Lotus 1-2-3 Spreadsheet Numbers Calc	Microsoft Lotus/IBM Google Apple Sun
Database	Store, manipulate, and retrieve data	Access Approach dBASE Base	Microsoft Lotus/IBM Borland Sun
Graphics	Develop graphs, illustrations, and drawings	Illustrator FreeHand	Adobe Macromedia
Project management	Plan, schedule, allocate, and control people and resources (money, time, and technology) needed to complete a project according to schedule	Project for Windows On Target Project Schedule Time Line	Microsoft Symantec Scitor Symantec
Financial management	Provide income and expense tracking and reporting to monitor and plan budgets (some programs have investment portfolio management features)	Quicken Money	Intuit Microsoft
Desktop publishing (DTP)	Use with personal computers and high-resolution printers to create high-quality printed output, including text and graphics; various styles of pages can be laid out; art and text files from other programs can also be integrated into "published" pages	QuarkXPress Publisher PageMaker Ventura Publisher Pages	Quark Microsoft Adobe Corel Apple
Creativity	Generate innovative and creative ideas and problem solutions. The software does not propose solutions, but provides a framework conducive to creative thought. The software takes users through a routine, first naming a problem, then organizing ideas and "wishes," and offering new information to suggest different ideas or solutions	Organizer Notes	Macromedia Lotus

Table 2.8

Examples of Personal Productivity Software

Word Processing

If you write reports, letters, or term papers, word processing applications can be indispensable. The majority of personal computers in use today have word processing applications installed. Such applications can be used to create, edit, and print documents. Most come with a vast array of features, including those for checking spelling, creating tables, inserting formulas, creating graphics, and much more. This book (and most like it) was entered into a word processing application using a personal computer.

Figure 2.15

Personal Productivity Software

PIM software assists individuals, groups, and organizations with organizing appointments, schedules, contacts, and to-do lists.

A team of people can use a word processing program to collaborate on a project. The authors and editors who developed this book, for example, used the Track Changes and Reviewing features of Microsoft Word to track and make changes to chapter files. You can add comments or make revisions to a document that a coworker can review and either accept or reject.

Spreadsheet Analysis

People use spreadsheets to prepare budgets, forecast profits, analyze insurance programs, summarize income tax data, and analyze investments. Whenever numbers and calculations are involved, spreadsheets should be considered. Features of spreadsheets include graphics, limited database capabilities, statistical analysis, built-in business functions, and much more (see Figure 2.16). The business functions include calculation of depreciation, present value, internal rate of return, and the monthly payment on a loan, to name a few. Optimization is another powerful feature of many spreadsheet programs. *Optimization* allows the spreadsheet to maximize or minimize a quantity subject to certain constraints. For example, a small furniture manufacturer that produces chairs and tables might want to maximize its profits. The constraints could be a limited supply of lumber, a limited number of workers that can assemble the chairs and tables, or a limited amount of various hardware fasteners that might be required. Using an optimization feature, such as Solver in Microsoft Excel, the spreadsheet can determine what number of chairs and tables to produce with labor and material constraints to maximize profits.

Figure 2.16

Spreadsheet Program

Spreadsheet programs should be considered when calculations are required.

Database Applications

Database applications are ideal for storing, manipulating, and retrieving data. These applications are particularly useful when you need to manipulate a large amount of data and produce reports and documents. Database manipulations include merging, editing, and sorting data. The uses of a database application are varied. You can keep track of a CD collection, the items in your apartment, tax records, and expenses. A student club can use a database to store names, addresses, phone numbers, and dues paid. In business, a database application can help process sales orders, control inventory, order new supplies, send letters to customers, and pay employees. Database management systems can be used to track orders, products, and customers, analyze weather data to make forecasts for the next several days, and summarize medical research results. A database can also be a front end to another application. For example, you can use a database application to enter and store income tax information, then export the stored results to other applications, such as a spreadsheet or tax-preparation application (see Figure 2.17).

Figure 2.17

Database Program

After being entered into a database application, information can be manipulated and used to produce reports and documents.

Graphics Programs

With today's graphics programs, it is easy to develop attractive graphs, illustrations, and drawings. Graphics programs can be used to develop advertising brochures, announcements, and full-color presentations. If you are asked to make a presentation at school or work, you can use a graphics program to develop and display slides while you are making your talk. A graphics program can be used to help you make a presentation, a drawing, or an illustration. See Figure 2.18. Most presentation graphics programs come with many pieces of *clip art*, such as drawings and photos of people meeting, medical equipment, telecommunications equipment, entertainment, and much more.

Personal Information Managers

Personal information managers (PIMs) help individuals, groups, and organizations store useful information, such as a list of tasks to complete or a list of names and addresses. They usually provide an appointment calendar and a place to take notes. In addition, information in a PIM can be linked. For example, you can link an appointment with a sales manager that appears in the calendar with information on the sales manager in the address book. When you click the appointment in the calendar, information on the sales manager from the address

Presentation Graphics Program

Graphics programs can help you make a presentation at school or work.

software suite

A collection of single application programs packaged in a bundle.

book is automatically opened and displayed on the computer screen. Microsoft Outlook is an example of a PIM software package.

Software Suites and Integrated Software Packages

A **software suite** is a collection of single application programs packaged in a bundle. Software suites can include word processors, spreadsheets, database management systems, graphics programs, communications tools, organizers, and more. Some suites support the development of Web pages, note taking, and speech recognition, where applications in the suite can accept voice commands and record dictation. Software suites offer many advantages. The software programs have been designed to work similarly, so after you learn the basics for one application, the other applications are easy to learn and use. Buying software in a bundled suite is cost-effective; the programs usually sell for a fraction of what they would cost individually.

Microsoft Office, Corel's WordPerfect Office, Lotus SmartSuite, and Sun Microsystems's StarOffice are examples of popular general-purpose software suites for personal computer users. Microsoft Office has the largest market share. The Free Software Foundation offers software similar to Sun Microsystems's StarOffice that includes word processing, spreadsheet, database, presentation graphics, and e-mail applications for the Linux OS. OpenOffice is another Office suite for Linux.[36] Wine can run any Windows application, including those in Microsoft Office, on Linux, although some features might not work as well as with a Microsoft OS. Each of these software suites includes a spreadsheet program, word processor, database program, and graphics package with the ability to move documents, data, and diagrams among them (see Table 2.9). Thus, a user can create a spreadsheet and then cut and paste that spreadsheet into a document created using the word processing application.

More than a hundred million people worldwide use the Microsoft Office software suite, with Office 2007 representing the latest version of the productivity software. Microsoft Office goes beyond its role as a mainstream package of ready-to-run applications with the extensive custom development facilities of Visual Basic for Applications (VBA)—a built-in facility that is part of every Office application. Using VBA, users can enhance off-the-shelf applications to tailor the programs for special tasks.

Some companies are offering Web-based productivity software suites that require no installation, only a Web browser. Zoho, Google, and Thinkfree offer free online word processing, spreadsheet, presentation, and other software that require no installation on the

Personal Productivity Function	Microsoft Office	Lotus Symphony	Corel WordPerfect Office	Sun Star Office	Apple iWork	Google
Word Processing	Word	Documents	WordPerfect	Writer	Pages	Docs
Spreadsheet	Excel	Spreadsheets	Quattro Pro	Calc	Numbers	Spreadsheet
Presentation Graphics	PowerPoint	Presentations	Presentations	Impress	Keynote	Presentation
Database	Access		Paradox	Base		

Table 2.9

Major Components of Leading Software Suites

PC. Documents created with the software can be stored on the Web server. Currently these online applications are not as powerful and robust as installed software such as Microsoft Office. However, it is likely that as the technology becomes more powerful, and network connection speeds increase, users will need to install less software on their own PCs and turn instead to using software online.

In addition to suites, some companies produce *integrated application packages* that contain several programs. For example, *Microsoft Works* is one program that contains basic word processing, spreadsheet, database, address book, calendar, and other applications. Although not as powerful as stand-alone software included in software suites, integrated software packages offer a range of capabilities for less money. Some integrated packages cost about $100.

Workgroup Application Software

Workgroup application software is designed to support teamwork, whether people are in the same location or dispersed around the world. This support can be accomplished with software known as *groupware* that helps groups of people work together effectively. Microsoft Exchange Server, for example, has groupware and e-mail features. Also called *collaborative software*, the approach allows a team of managers to work on the same production problem, letting them share their ideas and work via connected computer systems. The "Three Cs" rule for successful implementation of groupware is summarized in Table 2.10.

workgroup application software
Software that supports teamwork, whether in one location or around the world.

Quality	Description
Convenient	If it's too hard to use, it's not used; it should be as easy to use as the telephone.
Content	It must provide a constant stream of rich, relevant, and personalized content.
Coverage	If it isn't easy to access, it might never be used.

Table 2.10

Ernst & Young's "Three Cs" Rule for Groupware

Examples of workgroup software include group scheduling software, electronic mail, and other software that enables people to share ideas. Lotus Notes from IBM, for example, lets companies use one software package and one user interface to integrate many business processes. Lotus Notes can allow a global team to work together from a common set of documents, have electronic discussions using threads of discussion, and schedule team meetings. As the program matured, Lotus added services to it and renamed it Domino (Lotus Notes is now the name of the e-mail package), and now an entire third-party market has emerged to build collaborative software based on Domino.

The Web-based software described in the previous section is ideal for group use. Because documents are stored on an Internet server, anyone with an Internet connection can access them easily. Google provides options in its online applications that allow users to share documents, spreadsheets, presentations, calendars, and notes with other specified users or everyone on the Web (see Figure 2.19). This makes it convenient for several people to contribute to a document without concern for software compatibility or storage.

Figure 2.19

Google's Online Applications

Google applications are designed to share documents, presentations, spreadsheets, calendars, and notes with specific users or everyone on the Web.

Enterprise Application Software

Software that benefits an entire organization can also be developed or purchased. Some software vendors, such as SAP, specialize in developing software for enterprises. A fast-food chain, for example, might develop a materials ordering and distribution program to make sure that each of its franchises gets the necessary raw materials and supplies during the week. This materials ordering and distribution program can be developed internally using staff and resources in the IS department or purchased from an external software company. Boeing and DaimlerChrysler use enterprise software to design new airplanes and automotive products. The software simulates the effectiveness and safety of designs, allowing the companies to save time and money compared to developing physical prototypes of airplanes and vehicles. iSite from geoVue (*www.geovue.com*) is site-selection software that lets companies analyze factors to help determine the location of new stores. Table 2.11 lists examples of enterprise application software. Many organizations are moving to integrated enterprise software that supports supply chain management (movement of raw materials from suppliers through shipment of finished goods to customers), as shown in Figure 2.20.

Table 2.11

Examples of Enterprise Application Software

Type of Software	Description
Accounts receivable	Sales ordering
Accounts payable	Order entry
Airline industry operations	Payroll
Automatic teller systems	Human resource management
Cash-flow analysis	Check processing
Credit and charge card administration	Tax planning and preparation
Manufacturing control	Receiving
Distribution control	Restaurant management
General ledger	Retail operations
Stock and bond management	Invoicing
Savings and time deposits	Shipping
Inventory control	Fixed asset accounting

Figure 2.20

Use of Integrated Supply Chain Management Software

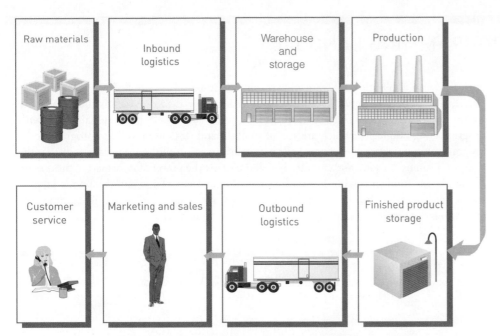

Integrated enterprise software to support supply chain management

Organizations can no longer respond to market changes using nonintegrated information systems based on overnight processing of yesterday's business transactions, conflicting data models, and obsolete technology. As a result, many corporations are turning to **enterprise resource planning (ERP) software**, a set of integrated programs that manage a company's vital business operations for an entire multisite, global organization. Thus, an ERP system must be able to support multiple legal entities, multiple languages, and multiple currencies. Although the scope of an ERP system may vary from vendor to vendor, most ERP systems provide integrated software to support manufacturing and finance. Additional functions supported might include human resources, sales, and distribution. The primary benefits of implementing ERP include eliminating inefficient systems, easing adoption of improved work processes, improving access to data for operational decision making, standardizing technology vendors and equipment, and enabling the implementation of supply chain management. Even small businesses can benefit from enterprise application software. Intuit's QuickBooks and Microsoft's Office Small Business Accounting are accounting and record keeping programs for small businesses and organizations.

enterprise resource planning (ERP) software
A set of integrated programs that manage a company's vital business operations for an entire multisite, global organization.

Application Software for Information, Decision Support, and Specialized Purposes

Specialized application software for information, decision support, and other purposes is available in every industry. Genetic researchers, for example, are using software to visualize and analyze the human genome. Music executives use decision support software to help pick the next hit song. Sophisticated decision support software is also being used to increase the cure rate for cancer by analyzing about 100 scans of a cancerous tumor to create a 3-D view of the tumor. Software can then consider thousands of angles and doses of radiation to determine the best program of radiation therapy. The software analysis takes only minutes, but the results can save years or decades of life for the patient. As you will see in future chapters, information, decision support, and specialized systems are used in businesses of all sizes and types to increase profits or reduce costs. But how are all these systems actually developed or built? The answer is through the use of programming languages, discussed next.

PROGRAMMING LANGUAGES

programming languages
Sets of keywords, symbols, and a system of rules for constructing statements by which humans can communicate instructions to be executed by a computer.

Both OSs and application software are written in coding schemes called *programming languages*. The primary function of a programming language is to provide instructions to the computer system so that it can perform a processing activity. IS professionals work with **programming languages**, which are sets of symbols and rules used to write program code. Programming involves translating what a user wants to accomplish into instructions that the computer can understand and execute. The desire to use the power of information processing efficiently in problem solving has pushed the development of literally thousands of programming languages, but only a few dozen are commonly used today. A brief summary of the various programming language generations is provided in Table 2.12.

Table 2.12

The Evolution of Programming Languages

Generation	Language	Approximate Development Date	Sample Statement or Action
First	Machine language	1940s	00010101
Second	Assembly language	1950s	MVC
Third	High-level language	1960s	READ SALES
Fourth	Query and database languages	1970s	PRINT EMPLOYEE NUMBER IF GROSS PAY>1000
Beyond Fourth	Natural and intelligent languages	1980s	IF gross pay is greater than 40, THEN pay the employee overtime pay

Although many programming languages are used to write new business applications, more lines of code are written in COBOL in existing business applications than any other programming language. Today, programmers often use visual and object-oriented languages. In the future, they will likely be using artificial intelligence languages to a greater extent. In general, these languages are easier for nonprogrammers to use compared with older generation languages.

SOFTWARE ISSUES AND TRENDS

Because software is such an important part of today's computer systems, issues such as software bugs, licensing, and global software support have received increased attention.

Software Bugs

A software bug is a defect in a computer program that keeps it from performing as it is designed to perform. Some software bugs are obvious and cause the program to terminate unexpectedly. Other bugs are more subtle and allow errors to creep into your work. Computer and software vendors say that as long as people design and program hardware and software, bugs are inevitable. In fact, according to the Pentagon and the Software Engineering Institute at Carnegie Mellon University, there are typically 5 to 15 bugs in every 1,000 lines of code. The following list summarizes tips for reducing the impact of software bugs.

- Register all software so that you receive bug alerts, fixes, and patches.
- Check the manual or read-me files for work-arounds.
- Access the support area of the manufacturer's Web site for patches.
- Install the latest software updates.

- Before reporting a bug, make sure that you can re-create the circumstances under which it occurs.
- When you can re-create the bug, call the manufacturer's tech support line.
- Avoid buying the latest release of software for several months or a year until the software bugs have been discovered and removed.

Copyrights and Licenses

Most software products are protected by law using copyright or licensing provisions. Those provisions can vary, however. In some cases, you are given unlimited use of software on one or two computers. This is typical with many applications developed for personal computers. In other cases, you pay for your usage—if you use the software more, you pay more. This approach is becoming popular with software placed on networks or larger computers. Most of these protections prevent you from copying software and giving it to others without restrictions. Some software now requires that you *register* or *activate* it before it can be fully used. Registration and activation sometimes put software on your hard disk that monitors activities and changes to your computer system.

Software Upgrades

Software companies revise their programs and sell new versions periodically. In some cases, the revised software offers new and valuable enhancements. In other cases, the software uses complex program code that offers little in terms of additional capabilities. In addition, revised software can contain bugs or errors. When software companies stop supporting older software versions or releases, some customers feel forced to upgrade to the newer software. Deciding whether to purchase the newest software can be a problem for corporations and people with a large investment in software. Should the newest version be purchased when it is released? Some users do not always get the most current software upgrades or versions, unless it includes significant improvements or capabilities. Instead, they might upgrade to newer software only when it offers vital new features. Software upgrades usually cost much less than the original purchase price.

Global Software Support

Large, global companies have little trouble persuading vendors to sell them software licenses for even the most far-flung outposts of their company. But can those same vendors provide adequate support for their software customers in all locations? Supporting local operations is one of the biggest challenges IS teams face when putting together standardized, company-wide systems. In slower technology growth markets, such as Eastern Europe and Latin America, there may be no official vendor presence at all. Instead, large vendors such as Sybase, IBM, and Hewlett-Packard typically contract out support for their software to local providers.

One approach that has been gaining acceptance in North America is to outsource global support to one or more third-party distributors. The software-user company may still negotiate its license with the software vendor directly, but it then hands over the global support contract to a third-party supplier. The supplier acts as a middleman between software vendor and user, often providing distribution, support, and invoicing. American Home Products Corporation handles global support for both Novell NetWare and Microsoft Office applications this way—throughout the 145 countries in which it operates. American Home Products negotiated the agreements directly with the vendors for both purchasing and maintenance, but fulfillment of the agreement is handled exclusively by Philadelphia-based Softsmart, an international supplier of software and services.

In today's computer systems, software is an increasingly critical component. Whatever approach individuals and organizations take to acquire software, it is important for everyone to be aware of the current trends in the industry. Informed users are wiser consumers, and they can make better decisions.

SUMMARY

Principle

Computer hardware must be carefully selected to meet the evolving needs of the organization and its supporting information systems.

Hardware devices work together to perform input, processing, data storage, and output. Processing is performed by an interplay between the central processing unit (CPU) and memory. Primary storage, or memory, provides working storage for program instructions and data to be processed and provides them to the CPU. Together, a CPU and memory process data and execute instructions.

Processing that uses several processing units is called multiprocessing. A multicore processor combines two or more independent processors into a single computer so that they can share the workload and boost processing capacity. Parallel processing involves linking several processors to work together to solve complex problems. Grid computing is the use of a collection of computers, often owned by multiple individuals or organizations, to work in a coordinated manner to solve a common problem.

Computer systems can store large amounts of data and instructions in secondary storage, which is less volatile and has greater capacity than memory. Storage media can be either sequential access or direct access. Common forms of secondary storage include magnetic tape, magnetic disk, optical disc storage, and flash memory. Redundant array of independent/inexpensive disks (RAID) is a method of storing data that allows the system to more easily recover data in the event of a hardware failure. Storage area network (SAN) uses computer servers, distributed storage devices, and networks to provide fast and efficient storage.

Input and output devices allow users to provide data and instructions to the computer for processing and allow subsequent storage and output. These devices are part of a user interface through which humans interact with computer systems. Input and output devices vary widely, but they share common characteristics of speed and functionality.

A keyboard and computer mouse are the most common devices used for entry of data. Speech recognition technology enables a computer to interpret human speech as an alternative means of providing data and instructions. Digital cameras record and store images or video in digital form. Handwriting recognition software can convert handwriting on the screen into text. Radio-frequency identification (RFID) technology employs a microchip, called a tag, to transmit data which is read by an RFID reader. The data transmitted could include facts such as item identification number, location information, or other details about the item tagged.

Output devices provide information in different forms, from hard copy to sound to digital format. Display monitors are standard output devices; monitor quality is determined by size, number of colors that can be displayed, and resolution. Other output devices include printers and plotters.

The main computer system types are handheld computers, ultra portable computers, portable computers, thin clients, desktop computers, workstations, servers, mainframe computers, and supercomputers. Personal computers (PCs) are small, inexpensive computer systems. Handheld (palmtop) computers are increasingly popular for portable computing and communications needs. Portable computers range from laptops, to notebooks, to subnotebooks, to tablet computers. A thin client is a low-cost, centrally managed computer with limited capabilities. Desktop computers are relatively small, inexpensive single-user computer systems that are highly versatile. Workstations are advanced PCs with greater memory, processing, and graphics abilities. A computer server is a computer used by many users to perform a specific task, such as running network or Internet applications. A mainframe computer is a large, powerful computer shared by dozens or even hundreds of concurrent users connected to the computer over a network. Supercomputers are extremely fast computers used to solve the most intensive computing problems.

Grid computing is the use of a collection of computers, often owned by many people or organizations, to work in a coordinated manner to solve a common problem. Grid computing is one low-cost approach to parallel processing.

Principle

Systems and application software are critical in helping individuals and organizations achieve their goals.

Software consists of programs that control the workings of the computer hardware. The two main categories of software are systems software and application software. Systems software is a collection of programs that interacts between hardware and application software. Application software can be proprietary or off the shelf, and enables people to solve problems and perform specific tasks.

An operating system (OS) is a set of computer programs that controls the computer hardware to support users' computing needs. An OS converts an instruction from an application into a set of instructions needed by the hardware. This intermediary role allows hardware independence. An OS also manages memory, which involves controlling storage access and use by converting logical requests into physical locations and by placing data in the best storage space, perhaps virtual memory.

An OS manages tasks to allocate computer resources through multitasking and time-sharing. With multitasking, users can run more than one application at a time. Time-sharing allows more than one person to use a computer system at the same time.

The ability of a computer to handle an increasing number of concurrent users smoothly is called scalability, a feature critical for systems expected to handle a large number of users.

An OS also provides a user interface, which allows users to access and command the computer. A command-based user interface requires text commands to send instructions; a graphical user interface (GUI), such as Windows, uses icons and menus.

Software applications use the OS by requesting services through a defined application program interface (API). Programmers can use APIs to create application software without having to understand the inner workings of the OS. APIs also provide a degree of hardware independence so that the underlying hardware can change without necessarily requiring a rewrite of the software applications.

Over the years, several popular OSs have been developed. These include several proprietary OSs used primarily on mainframes. MS-DOS is an early OS for IBM-compatibles. Older Windows OSs are GUIs used with DOS. Newer versions, such as Windows Vista and XP, are fully functional OSs that do not need DOS. Apple computers use proprietary OSs such as the Mac OS and Mac OS X. UNIX is a powerful OS that can be used on many computer system types and platforms, from personal computers to mainframe systems. UNIX makes it easy to move programs and data among computers or to connect mainframes and personal computers to share resources. Linux is the kernel of an OS whose source code is freely available to everyone. Several variations of Linux are available, with sets of capabilities and applications to form a complete OS, for example, Red Hat Linux. z/OS and HP-UX are OSs for mainframe computers. Some OSs have been developed to support consumer appliances such as Palm OS, Windows CE.Net, Windows XP Embedded, Pocket PC, and variations of Linux.

Principle

Organizations should not develop proprietary application software unless doing so will meet a compelling business need that can provide a competitive advantage.

Application software applies the power of the computer to solve problems and perform specific tasks. One useful way of classifying the many potential uses of information systems is to identify the scope of problems and opportunities addressed by a particular organization or its sphere of influence. For most companies, the spheres of influence are personal, workgroup, and enterprise.

User software, or personal productivity software, includes general-purpose programs that enable users to improve their personal effectiveness, increasing the quality and amount of work that can be done. Software that helps groups work together is often called workgroup application software, and includes group scheduling software, electronic mail, and other software that enables people to share ideas. Enterprise

software that benefits the entire organization can also be developed or purchased. Many organizations are turning to enterprise resource planning software, a set of integrated programs that manage a company's vital business operations for an entire multisite, global organization.

Three approaches to developing application software are to build proprietary application software, buy existing programs off the shelf, or use a combination of customized and off-the-shelf application software. Building proprietary software (in-house or on contract) has the following advantages: the organization will get software that more closely matches its needs; by being involved with the development, the organization has further control over the results; and the organization has more flexibility in making changes. The disadvantages include the following: it is likely to take longer and cost more to develop, the in-house staff will be hard pressed to provide ongoing support and maintenance, and there is a greater risk that the software features will not work as expected or that other performance problems will occur.

Purchasing off-the-shelf software has many advantages. The initial cost is lower, there is a lower risk that the software will fail to work as expected, and the software is likely to be of higher quality than proprietary software. Some disadvantages are that the organization might pay for features it does not need, the software might lack important features requiring expensive customization, and the system might require process reengineering.

Some organizations have taken a third approach—customizing software packages. This approach usually involves a mixture of the preceding advantages and disadvantages and must be carefully managed.

An application service provider (ASP) is a company that can provide the software, support, and computer hardware on which to run the software from the user's facilities over a network. ASPs provide contract customization of off-the-shelf software, and they speed deployment of new applications while helping IS managers avoid implementation headaches. Use of ASPs reduces the need for many skilled IS staff members and also lowers a project's start-up expenses.

Software as a service, or SaaS allows business to subscribe to Web-delivered business application software by paying a monthly service charge or a per use fee.

Although hundreds of computer applications can help people at school, home, and work, the primary applications are word processing, spreadsheet analysis, database, graphics, and online services. A software suite, such as SmartSuite, WordPerfect, StarOffice, or Office, offers a collection of powerful programs.

Principle

Organizations should choose a programming language whose functional characteristics are appropriate for the task at hand, considering the skills and experience of the programming staff.

All software programs are written in coding schemes called programming languages, which provide instructions to a computer to perform some processing activity. The several classes of programming languages include machine, assembly, high-level, query and database, and natural and intelligent languages.

Programming languages have changed since their initial development in the early 1950s. In the first generation, computers were programmed in machine language, and the second generation of languages used assembly languages. The third generation consists of many high-level programming languages that use English-like statements and commands. Fourth-generation languages include database and query languages such as SQL.

Users frequently use fourth generation and higher level programming languages to develop their own simple programs.

Principle

The software industry continues to undergo constant change; users need to be aware of recent trends and issues to be effective in their business and personal life.

Software bugs, software licensing and copyrighting, software upgrades, and global software support are all important software issues and trends.

A software bug is a defect in a computer program that keeps it from performing in the manner intended. Software bugs are common, even in key pieces of business software.

Software upgrades are an important source of increased revenue for software manufacturers and can provide useful new functionality and improved quality for software users.

Global software support is an important consideration for large, global companies putting together standardized, company-wide systems. A common solution is outsourcing global support to one or more third-party software distributors.

CHAPTER 2: SELF-ASSESSMENT TEST

Computer hardware must be carefully selected to meet the evolving needs of the organization and its supporting information systems.

1. Organizations typically make a one-time investment in the computer hardware necessary to meet their needs with little need for future changes and upgrades. True or False?

2. Executing an instruction by the CPU involves two phases: the instruction phase and the _____ phase.

3. Which of the following components performs mathematical calculations and makes logical comparisons?
   a. control unit
   b. register
   c. ALU
   d. main memory

Systems and application software are critical in helping individuals and organizations achieve their goals.

4. Application software like Microsoft Excel manipulates the computer hardware directly. True or False?

5. Today's operating systems support _____, the ability to run multiple processes seemingly simultaneously.

Organizations should not develop proprietary application software unless doing so will meet a compelling business need that can provide a competitive advantage.

6. Software that enables users to improve their personal effectiveness, increasing the amount of work they can do and its quality is called _____.
   a. personal productivity software
   b. operating system software
   c. utility software
   d. graphics software

7. Optimization can be found in which type of application software?
   a. spreadsheets
   b. word processing programs
   c. database programs
   d. graphics software

8. A program to detect and eliminate viruses is an example of what type of software?
   a. personal productivity software
   b. operating system software
   c. utility software
   d. application software

Organizations should choose a programming language whose functional characteristics are appropriate to the task at hand, taking into consideration the skills and experience of the programming staff.

9. A class of application software that helps groups work together and collaborate is called _____ .

10. More lines of code for current applications are written in the _____ programming language than in any other language.

The software industry continues to undergo constant change; users need to be aware of recent trends and issues to be effective in their business and personal life.

11. _____ are an important source of increased revenue for software manufacturers and can provide useful new functionality and improved quality for software users.

REVIEW QUESTIONS

1. When determining the appropriate hardware components of a new information system, what role must the end user of the system play?
2. Identify two basic characteristics of RAM and ROM.
3. What is RFID technology? How does it work?
4. Identify the three components of the CPU and explain the role of each.
5. What is RAID storage technology? What advantages does it offer?
6. Identify and briefly describe the various classes of computers.
7. Give four examples of personal productivity software.
8. What is software as a service? What advantages does it provide for meeting an organization's software needs?
9. What are the two basic types of software? Briefly describe the role of each.
10. What is multiprocessing? What is multitasking?
11. What is an application service provider? What issues arise in considering the use of one?
12. Define the term utility software and give two examples.
13. What does the acronym API stand for? What is the role of an API?
14. Describe the term *enterprise resource planning* (*ERP*) system. What functions does such a system perform?

DISCUSSION QUESTIONS

1. Briefly discuss the advantages and disadvantages of frequent software upgrades from the perspective of the user of that software. How about from the perspective of the software manufacturer?
2. What would be the advantages for a university computer lab to install thin clients rather than standard desktop personal computers? Can you identify any disadvantages?
3. Which would you rather have—a hand held computer or a tablet computer? Why?
4. If cost were not an issue, describe the characteristics of your ideal computer. What would you use it for? Would you choose a handheld, portable, desktop, or workstation computer? Why? Which operating system would you want it to run?
5. Identify the three spheres of influence and briefly discuss the software needs of each.
6. Identify the two fundamental sources for obtaining application software. Discuss the advantages and disadvantages of each source.
7. Define the software as a service. Discuss some of the pros and cons of using software as a service. What precautions might you take to minimize the risk of using one?
8. In what ways is an operating system for a mainframe computer different from the operating system for a laptop computer? In what ways are they similar?
9. How can application software improve the effectiveness of a large enterprise? What are some of the benefits associated with implementation of an enterprise resource planning system? What are some issues that could keep the use of an enterprise resource planning system from being successful?

PROBLEM-SOLVING EXERCISES

1. Develop a spreadsheet that compares the features, initial purchase price, and ongoing operating costs for three laser printers. Now do the same for three inkjet printers. Write a brief memo on which printer you would choose and why. Cut and paste the spreadsheet into a document.
2. Use word processing software to document what your needs are as a computer user and your justification for selecting either a desktop or laptop computer. Find a Web site that allows you to order and customize a computer and select those options that meet your needs in a cost-effective manner. Assume that you have a budget of $1500. Enter the computer specifications into an Excel spreadsheet that you cut and paste into the document defining your needs. E-mail the document to your instructor.
3. Use a database program to enter five software products you are likely to use at work. List the name, vendor or manufacturer, cost, and features in the columns of a database table. Use a word processor to write a report on the software. Copy the database table into the word processing program.

TEAM ACTIVITIES

1. With two or three of your classmates, visit a computer retail store and identify the most popular ultra laptop computers. Interview members of the sales staff to find out why they think this particular laptop is popular.
2. Form a group of three or four classmates. Identify and contact an employee of a local firm. Interview the individual and describe the application software the company uses and the importance of the software to the organization. Write a brief report summarizing your findings.
3. Divide your team into two groups. The first group should prepare a report using a word processing program. Make sure to include a large number of spelling, grammatical, and similar errors in the document. The second group should use the word processing program's features to locate and eliminate the errors. The entire team should write a report on the advantages and limitations of the spelling and grammar checking features of the word processing program you used. What additional features would you like to see in future word processing programs?

WEB EXERCISES

1. Use the Web to research four productivity software suites from various vendors (see *http://en.wikipedia.org/wiki/Office_Suite*). Create a table in a word processing document to show what applications are provided by the competing suites. Write a few paragraphs on which suite you think best matches your needs and why.
2. Do research on the Web and develop a two-page report summarizing the latest consumer appliance OSs. Which one seems to be gaining the most widespread usage? Why do you think this is the case?
3. Do research on the Web about application software that is used in an industry and is of interest to you. Write a brief report describing how the application software can be used to increase profits or reduce costs.

CAREER EXERCISES

1. Imagine that you are going to buy a single handheld device to improve your communication and organizational abilities. What tasks do you need it to perform? What features would you look for in this device? Visit a computer store or a consumer electronics store and see whether you can purchase such a device for under $400.

2. Think of your ideal job. Describe five application software packages that could help you advance in your career. If the software package doesn't exist, describe the kinds of software packages that could help you in your career.

CASE STUDIES

Case One
Advance America Implements Grid Computing

Chances are you have seen places that offer payday loans in your town. Payday loans are short-term loans designed for people that run out of money before payday, but can repay the loan when their paycheck arrives. Advance America is the leading payday loan company in the United States. It includes 3,000 centers in 37 states, and employs nearly 7,000 people, according to its Web site. Advance America is big, and growing bigger every day. Its growth in recent years is straining the capabilities of its client-server information system infrastructure and holding the company back from further growth.

Advance America used a system in which each center was equipped with an independent hardware and software environment. Installation and maintenance costs were high, and compiling data for all centers was time consuming and difficult. Each night the thousands of centers would upload their data to the main server for consolidation. With the growing number of centers, there wasn't enough time in the night to process all of the incoming data. Advance America's system had run up against a wall. It was time for a change.

Advance America decided to invest in a new system based on a grid computing architecture. They installed thin client machines to run in each center, connecting via the Web to a fault-tolerant server cluster running Oracle database software. The server cluster consists of a four-node cluster of IBM P5 series servers, which include four processors per node for a total of 16 processors. The servers in the cluster work as a grid by sharing the work load of the entire organization equally among them. A pair of Cisco load balancers make sure that processing is distributed evenly among the servers for maximum performance. The new system includes a 2 TB storage area network (SAN) that uses an IBM disk array controlled by the Oracle Automatic Storage Management (ASM) software.

System IT managers at headquarters use a central grid-management console to oversee the entire nationwide network. Problems are easily identified and fixed through the centrally managed system. So far, the system has provided 100 percent uptime at the cash-advance centers.

Advance America took a chance with its $3.8 million investment in this new technology, but it has paid off. Center managers can now tap into "a continuously updated central database and generate reports in near real time." The new system has decreased the time it takes to open a new Advance America center. Managers are getting information much more quickly, making it easier for them to analyze business performance and customer trends. The new system is also easy to expand as the business grows. It is estimated that the new system will provide total net benefits of almost $3 million over five years for an ROI of 131 percent.

Discussion Questions

1. How does grid computing provide Advance America managers with faster access to data?
2. How did grid computing assist Advance America in breaking through the wall that held it back from growth?

Critical Thinking Questions

1. Why is the new grid computing system at Advance America much easier to install, manage, and maintain than its old system?
2. How might Advance America expand its system as the company outgrows it?

SOURCES: Staff, "Advance America Grows with Oracle Enterprise Grid," *Computerworld Honors Program*, 2007, *www.cwhonors.org/viewCaseStudy.asp?NominationID=104*. Advance America Web Site, *www.advanceamerica.net/values.php*, accessed January 12, 2008. Staff, "Integrated Data Infrastructure Pays Off for Advance America, Cash Advance Centers, Inc," Oracle Customer Snapshot, *www.oracle.com/customers/snapshots/advance_america.pdf*, accessed January 12, 2008.

Case Two
Systems Management Software Helps Fight Crime

The York regional police protect 1,800 square kilometers north of Toronto, Canada. Until recently, the force has been challenged trying to keep the rugged Panasonic laptops in its 200 cruisers, boats, and helicopters up-to-date, secure, and synchronized. Although the notebooks were wirelessly networked, the data, software, and systems were not necessarily well synchronized.

To apply system updates and patches every few months, officers were required to check in at the main station where they had to wait for a few hours while the notebook was updated. With over 200 laptops on the force, this cost the department hundreds of working hours every few months.

Not only were human resources wasted, but the internal IT department was pushed past its limit. Staff members sometimes unknowingly worked to solve the same problems. They spent too much time coordinating applications, running backups, and trying to keep up with new law enforcement applications.

Recently the York police installed system management software from Microsoft called System Center Configuration Manager 2007. The software allows system administrators at the main station to access notebooks remotely over the network for system upgrades, patch management, software distribution, and hardware and software inventory. No longer do officers need to spend hours waiting on their PC updates. PCs are updated as needed over the wireless network. New software and system changes are pushed out to all notebooks simultaneously so all officers have the same information and services at all times.

The new systems software allows the department to come close to its paperless ideal. An e-ticketing system allows officers to swipe a driver's license, run a background check, and issue a ticket in minutes. Officers receive daily briefings online and submit reports directly from their notebooks, which allows them to stay on the road rather than at a desk. The new system has freed up the IT staff to concentrate on delivering new and useful services rather than just maintaining the old services.

The York regional police are looking forward to the next edition of System Center Configuration Manager, which promises to support streaming media. They would like to use it to stream video from the helicopter to cruisers on the road.

Discussion Questions

1. What unique challenges did the York regional police IT staff have to overcome?
2. How did Microsoft System Center Configuration Manager resolve the issues for the York regional police?

Critical Thinking Questions

1. What other types of industries would benefit from products like Microsoft System Center Configuration Manager? Why?
2. What general lesson regarding information system administration can you take from this case?

SOURCES: Smith, Briony, "Cops roll out remote patch updates, e-ticketing," *ITWorld Canada*, November 29, 2007, *www.itworldcanada.com/a/Enterprise-Business-Applications/50132083-0699-401d-b7ce-d6f1c16193b5.html*. Microsoft System Center Configuration Manager Web site, *www.microsoft.com/smserver*, accessed February 3, 2008. York Regional Police Web site, *www.yrp.on.ca*, accessed February 3, 2008.

Questions for Web Case

See the Web site for this book to read about the Whitmann Price Consulting case for this chapter. Following are questions concerning this Web case.

Whitmann Price Consulting: Choosing Hardware

Discussion Questions

1. What considerations led Josh and Sandra to lean towards a Blackberry as the handheld device on which to run the AMCIS?
2. How did Josh and Sandra organize their hardware considerations for the new system?

Critical Thinking Questions

1. What role does system compatibility play in Josh and Sandra's decision?
2. What device(s) might have been chosen if the system requirements called for a 12-inch display and the ability to take handwritten notes and communicate through voice, text, and video?

Whitmann Price Consulting: Software Considerations

Discussion Questions

1. What three types of software made the BlackBerry ideal for meeting the needs of Whitmann Price?
2. How did the choice of hardware affect options for software solutions? If Sandra and Josh picked a newly developed handheld device unknown in industry, how would that change the solutions?

Critical Thinking Questions

1. For software other than the BlackBerry software, should Sandra and Josh look to BlackBerry Alliance Program members or should they plan to have their own software engineers develop the software? What are the benefits and drawbacks of either option?
2. What process would you use to evaluate software for the Advanced Mobile Communications and Information System?

NOTES

1 "Bosch Security Projects – The 2007 Computerworld Honors Program," *Computerworld, www.cwhonors.org.viewCaseStudy.asp?NominationID=65*, accessed January 27, 2008.

2 "Iowa Health System – The 2007 ComputerWorld Honors Program," *Computerworld, www.cwhonors.org/viewCaseStudy.asp?NominatonID=347*, accessed January 27, 2008.

3 Brown, Stuart F., "Crashing Cars When They're Still a Gleam in the Designer's Eye," *The New York Times*, June 17, 2007.

4 "The Science Behind Folding@home," Folding@home Web site, *folding.stanford.edu/English/Science*, accessed January 4, 2008.

5 Blue Gene," IBM Research Web site, *domino.reseaarch.ibm.co/comm./research_projectsw.nsf/pages/bluegene.index.html*, accessed January 4, 2008.

6 "IBM and Google Float in Parallel Cloud," Grant Gross Blog, *Computerworld*, October 9, 2007.

7 "Amazon Elastic Compute Cloud (Amazon EC2) – Beta," *www.amazon.com/gp/browse.html?node=201590011*, accessed January 14, 2008.

8 Carr, Nicholas, Rough Type blog, "Google's Cloud," October 26, 2007, *www.roughtype.com/archives/2007/10/googles_cloud.php*, accessed January 15, 2008.

9 "Government Sale of Used Magnetic Tape Storage Not a Big Security Risk, GAO Reports," *Network World*, September 21, 2007.

10 Quain, John, R., "2007 Small Business Awards: The Winners," *PC Magazine*, September 26, 2007.

11 "Remodeling the Data Center: A Centralized Data Storage Management Model," *whitepapers.zdNet.com*, accessed January 30, 2008.

12 Jones, K.C., "Free Cell Phone Direction Service Gets Voice Recognition Boost," *InformationWeek*, November 13, 2007.

13 McDougall, Paul, "Gates, Simonyi Give $30 Million to Build Giant Telescope," *InformationWeek*, January 4, 2008.

14 Martin, Richard, "InformationWeek 500: Con-way Earns No. 1 Ranking in Information Week 500," *InformationWeek*, September 18, 2007.

15 "Boekhandels Groep Nederland (BGN)" The 2007 Computerworld Honors Program, *www.cwhonors.org/viewCaseStudy.asp?NominationID=119*, accessed January 16, 2008.

16 Criswell, Chad, "OLED Monitors and 3-D Display Tech," *Suite 101.com*, December 4, 2007.

17 "Pocket PC Development Provides Sales Force Tool for Coca-Cola," Syware Web site, *www.syware.com*, accessed January 31, 2008.

18 Dejean, Davi, "Review: Three Ultralight Laptops Get the Job Done," *InformationWeek*, March 17, 2007.

19 Malykhina, Elena, "DT Research Adds Barcode Scanners, Card Readers, Cameras to Tablet PCs," *InformationWeek*, November 16, 2007.

20 Industry Solutions – CSX, DT Research Web site, *www.dtresearch.com/industry/case.htm*, accessed January 31, 2008.

21 OLPC Web site, *laptop.org/vision/index.shtml*, accessed January 31, 2008.

22 Malykhina, Elena, "IBM: The Mainframe Is Alive and Going Strong," *InformationWeek*, June 21, 2007.

23 Gruener, Wolfgang, "Linux Returns to Thinkpads," *tgdaily*, January 15, 2008, *www.tgdaily.com/content/view/35615/145*.

24 Ubuntu success stories, "Ubuntu proved the optimal choice for a non-profit radio station," *www.ubuntu.com/products/casestudies/KRUU*, accessed June 14, 2007.

25 BlackBerry Customer Success, "IT Group Deploys Remote Systems Management on BlackBerry for Business-Critical Systems," *http://na.BlackBerry.com/eng/solutions/types/government/#tab_tab_customersuccess*, accessed June 14, 2007.

26 Adams, Steve, "Boston-area firm offers video intelligence to merchants," *The Patriot Ledger*, November 17, 2007, *http://ledger.southofboston.com/articles/2007/11/17/business/biz02.txt*.

27 SAP Case Study, "Avanex Inc.," Accessed February 15, 2008, *www.sap.com/industries/hightech/midsize/customers/stories/avanex.ex*p.

CHAPTER · 3 ·

Database Systems and Business Intelligence

PRINCIPLES	LEARNING OBJECTIVES
▪ Data management and modeling are key aspects of organizing data and information.	▪ Define general data management concepts and terms, highlighting the advantages of the database approach to data management. ▪ Describe the relational database model and outline its basic features.
▪ A well-designed and well-managed database is an extremely valuable tool in supporting decision making.	▪ Identify the common functions performed by all database management systems, and identify popular database management systems.
▪ The number and types of database applications will continue to evolve and yield real business benefits.	▪ Identify and briefly discuss current database applications.

Information Systems in the Global Economy Wal-Mart, United States

Warehousing and Mining Data on a Grand Scale

One company that really needs to know how to manage data is Wal-Mart. With a total of over 800 million transactions per day in over 7,000 stores around the world, Wal-Mart produces more data in a day than many businesses produce in a lifetime. No matter what size the business, databases and the systems that manage them provide the foundation on which business decisions are made.

Wal-Mart is successful due to its ability to learn from the data it collects. In a nutshell, Wal-Mart owes its success to its databases and business intelligence tools—software tools that manipulate data to provide useful information to Wal-Mart decision-makers.

At Wal-Mart headquarters in Arkansas, massive amounts of data are collected every day from its stores around the world, and stored in a data warehouse over a petabyte in size—which is a quadrillion bytes or a million gigabytes. A data warehouse is a large database that collects data from many sources, which can then be analyzed to guide business decisions. Wal-Mart uses HP's Neoview technology for its data warehouse. The system integrates data warehousing hardware, software, and services to manage large amounts of data. It is ideal for a company looking for a powerful database tool that is easy to administer.

Neoview offers "next generation business intelligence" features that embed useful information mined from the data directly into the systems that executives, managers, and employees use every day. "A lot of people in the company are asking for quicker and easier access to data. We want to make sure it's readable and usable by internal customers," says Jim Scantlin, the director of enterprise information management at Wal-Mart.

Specifics about how Wal-Mart uses business intelligence are corporate secrets that the company works hard to keep from its competitors. Clearly one of the top goals of the system is to determine which products are selling well at various locations so that Wal-Mart can manage inventory and promotions. When asked about the role of business intelligence in Wal-Mart's business strategies, Wal-Mart CTO Nancy Stewart says, "Business intelligence is huge. It is huge." Without sophisticated data analyses, making decisions regarding business strategy would be like running through the woods wearing a blindfold. A data warehouse not only allows a company to navigate through current market conditions, but in many cases provides information that allows the business to predict and plan for the future.

Wal-Mart uses its databases, data warehouse, and business intelligence tools to collect, analyze, and disseminate massive amounts of data across its networks every day. Top-level executives, regional managers, store managers, and associates are provided with custom-designed reports, charts, and graphs presented in easy-to-read dashboard software that lets users understand the state of the business at any time so they can do their jobs more effectively. As a pilot watches and analyzes the gauges and meters on the control panel of a jumbo jet to provide a smooth flight, Wal-Mart executives and managers watch and analyze the dashboard of Wal-Mart's data warehouse to keep the business running smoothly.

As you read this chapter, consider the following:

- What role do databases and business intelligence systems play in the overall effectiveness of information systems?
- What techniques do businesses use to maximize the value of the information provided from databases and business intelligence systems?

Why Learn About Database Systems and Business Intelligence?

A huge amount of data is entered into computer systems every day. Where does all this data go and how is it used? How can it help you on the job? In this chapter, you will will learn about database systems and business intelligence tools that can help you make the most effective use of information. If you become a marketing manager, you can access a vast store of data on existing and potential customers from surveys, their Web habits, and their past purchases. This information can help you sell products and services. If you become a corporate lawyer, you will have access to past cases and legal opinions from sophisticated legal databases. This information can help you win cases and protect your organization legally. If you become a human resource (HR) manager, you will be able to use databases and business intelligence tools to analyze the impact of raises, employee insurance benefits, and retirement contributions on long-term costs to your company. Regardless of your field of study in school, using database systems and business intelligence tools will likely be a critical part of your job. In this chapter, you will see how you can use data mining to extract valuable information to help you succeed. This chapter starts by introducing basic concepts of database management systems.

A database is an organized collection of data. Like other components of an information system, a database should help an organization achieve its goals. A database can contribute to organizational success by providing managers and decision makers with timely, accurate, and relevant information based on data. For example, at Creative Artists Agency (CAA), a successful Hollywood talent agency, a database helps agents organize information about clients.[1] With clients such as Tom Cruise, Julia Roberts, and Brad Pitt, a talent agency must prevent mistakes and misunderstandings. CAA's database can store various types of information about each client. For example, the database informs agents about movies in which Tom Cruise is acting, movies he is producing, products he is endorsing, and any other pertinent information about the actor's career. Using the database, an agent could find all clients that are associated with a particular product or film, or all the products and films associated with one client. Databases also help companies generate information to reduce costs, increase profits, track past business activities, and open new market opportunities. In some cases, organizations collaborate in creating and using international databases. Six organizations, including the Organization of Petroleum Exporting Countries (OPEC), International Energy Agency (IEA), and the United Nations, use a database to monitor the global oil supply.

database management system (DBMS)

A group of programs that manipulate the database and provide an interface between the database and the user of the database and other application programs.

database administrator (DBA)

A skilled IS professional who directs all activities related to an organization's database.

A **database management system** (**DBMS**) consists of a group of programs that manipulate the database and provide an interface between the database and its users and other application programs. Usually purchased from a database company, a DBMS provides a single point of management and control over data resources, which can be critical to maintaining the integrity and security of the data. A database, a DBMS, and the application programs that use the data make up a database environment. A **database administrator** (**DBA**) is a skilled and trained IS professional who directs all activities related to an organization's database, including providing security from intruders. A security breach at an Ivy League college provided an intruder with access to a database that stored student's private information.[2] Such data breaches have become commonplace for businesses and organizations because many databases are now accessible from the Internet. Data quality and accuracy also continue to be important issues for DBAs. A database error in the United Kingdom left 400,000 people without paychecks in March, 2007.[3]

Databases and database management systems are becoming even more important to businesses as they deal with increasing amounts of digital information. A report from IDC, called "The Diverse and Exploding Digital Universe," estimates the size of the digital universe to be 281 exabytes, or 281 billion gigabytes. By 2011, there will be 1,800 exabytes of electronic data in existence, or 1.8 zettabytes.[4] If a tennis ball were one byte of information, a zettabyte-sized ball would be around the size of one Earth. IDC recommends that businesses and organizations move now to create policies, tools, and standards to accommodate the approaching tidal wave of digital data and information.[5]

DATA MANAGEMENT

Without data and the ability to process it, an organization could not successfully complete most business activities. It could not pay employees, send out bills, order new inventory, or produce information to assist managers in decision making. Recall that data consists of raw facts, such as employee numbers and sales figures. For data to be transformed into useful information, it must first be organized in a meaningful way.

The Hierarchy of Data

Data is generally organized in a hierarchy that begins with the smallest piece of data used by computers (a bit) and progresses through the hierarchy to a database. A bit (a binary digit) represents a circuit that is either on or off. Bits can be organized into units called *bytes*. A byte is typically eight bits. Each byte represents a **character**, which is the basic building block of information. A character can be an uppercase letter (A, B, C... Z), lowercase letter (a, b, c... z), numeric digit (0, 1, 2... 9), or special symbol (., !, [+], [-], /, ...).

Characters can be combined to form a field. A **field** is typically a name, number, or combination of characters that describes an aspect of a business object (such as an employee, a location, or a truck) or activity (such as a sale). In addition to being entered into a database, fields can be computed from other fields. *Computed fields* include the total, average, maximum, and minimum value. A collection of related data fields is a **record**. By combining descriptions of the characteristics of an object or activity, a record can provide a complete description of the object or activity. For instance, an employee record is a collection of fields about one employee. One field includes the employee's name, another field contains the address, and still others the phone number, pay rate, earnings made to date, and so forth. A collection of related records is a **file**—for example, an employee file is a collection of all company employee records. Likewise, an inventory file is a collection of all inventory records for a particular company or organization. Some database software refers to files as tables.

At the highest level of this hierarchy is a *database*, a collection of integrated and related files. Together, bits, characters, fields, records, files, and databases form the **hierarchy of data** (see Figure 3.1). Characters are combined to make a field, fields are combined to make a record, records are combined to make a file, and files are combined to make a database. A database houses not only all these levels of data but also the relationships among them.

Data Entities, Attributes, and Keys

Entities, attributes, and keys are important database concepts. An **entity** is a generalized class of people, places, or things (objects) for which data is collected, stored, and maintained. Examples of entities include employees, inventory, and customers. Most organizations organize and store data as entities.

An **attribute** is a characteristic of an entity. For example, employee number, last name, first name, hire date, and department number are attributes for an employee (see Figure 3.2). The inventory number, description, number of units on hand, and location of the inventory item in the warehouse are attributes for items in inventory. Customer number, name, address, phone number, credit rating, and contact person are attributes for customers. Attributes are usually selected to reflect the relevant characteristics of entities such as employees or customers. The specific value of an attribute, called a **data item**, can be found in the fields of the record describing an entity.

character
A basic building block of information, consisting of uppercase letters, lowercase letters, numeric digits, or special symbols.

field
Typically a name, number, or combination of characters that describes an aspect of a business object or activity.

record
A collection of related data fields.

file
A collection of related records.

hierarchy of data
Bits, characters, fields, records, files, and databases.

entity
A generalized class of people, places, or things for which data is collected, stored, and maintained.

attribute
A characteristic of an entity.

data item
The specific value of an attribute.

Employee #	Last name	First name	Hire date	Dept. number
005-10-6321	Johns	Francine	10-07-1997	257
549-77-1001	Buckley	Bill	02-17-1979	632
098-40-1370	Fiske	Steven	01-05-1985	598

KEY FIELD

ATTRIBUTES (fields)

ENTITIES (records)

Most organizations use attributes and data items. Many governments use attributes and data items to help in criminal investigations. The United States Federal Bureau of Investigation is building the "world's largest computer database of peoples' physical characteristics."[6] At a cost of $1 billion, the database management system named Next Generation Identification will catalog digital images of faces, fingerprints, and palm prints of U.S. citizens and visitors. Each person in the database is an entity, each biometric category is an attribute, and each image is a data item. The information will be used as a forensics tool and to increase homeland security.

As discussed, a collection of fields about a specific object is a record. A **key** is a field or set of fields in a record that identifies the record. A **primary key** is a field or set of fields that uniquely identifies the record. No other record can have the same primary key. The primary key is used to distinguish records so that they can be accessed, organized, and manipulated. For an employee record, such as the one shown in Figure 3.2, the employee number is an example of a primary key.

Locating a particular record that meets a specific set of criteria might be easier and faster using a combination of secondary keys. For example, a customer might call a mail-order company to place an order for clothes. If the customer does not know the correct primary key (such as a customer number), a secondary key (such as last name) can be used. In this case, the order clerk enters the last name, such as Adams. If several customers have a last name of Adams, the clerk can check other fields, such as address, first name, and so on, to find the correct customer record. After locating the correct customer record, the order can be completed and the clothing items shipped to the customer.

key
A field or set of fields in a record that is used to identify the record.

primary key
A field or set of fields that uniquely identifies the record.

The Database Approach

At one time, applications used only particular files. For example, a payroll application would use a payroll file. In other words, each application used files dedicated to that application. This approach to data management, whereby separate data files are created and stored for each application program, is called the **traditional approach to data management**.

Today, most organizations use the **database approach to data management**, where multiple application programs share a pool of related data. A database offers the ability to share data and information resources. Federal databases, for example, often include the results of DNA tests as an attribute for convicted criminals. The information can be shared with law enforcement officials around the country.

To use the database approach to data management, additional software—a database management system (DBMS)—is required. As previously discussed, a DBMS consists of a group of programs that can be used as an interface between a database and the user of the database and application programs. Typically, this software acts as a buffer between the application programs and the database itself. Figure 3.3 illustrates the database approach.

traditional approach to data management
An approach whereby separate data files are created and stored for each application program.

database approach to data management
An approach whereby a pool of related data is shared by multiple application programs.

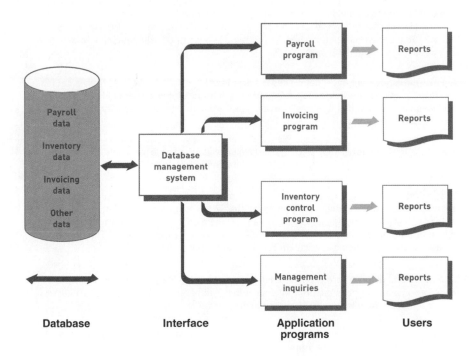

Figure 3.3

The Database Approach to Data Management

Table 3.1 lists some of the primary advantages of the database approach, and Table 3.2 lists some disadvantages.

Many modern databases serve entire enterprises, encompassing much of the data of the organization. Often, distinct yet related databases are linked to provide enterprisewide databases. For example, many Wal-Mart stores include in-store medical clinics for customers. Wal-Mart uses a centralized electronic health records database that stores the information of all patients across all stores.[7] The database is interconnected with the main Wal-Mart database to provide information about customer's interactions with the clinics and stores. The Ethical and Societal Issues box provides more information about databases used for electronic health record systems.

Advantages	Explanation
Improved strategic use of corporate data	Accurate, complete, up-to-date data can be made available to decision makers where, when, and in the form they need it. The database approach can also give greater visibility to the organization's data resource.
Reduced data redundancy	Data is organized by the DBMS and stored in only one location. This results in more efficient use of system storage space.
Improved data integrity	With the traditional approach, some changes to data were not reflected in all copies of the data kept in separate files. The database approach prevents this problem because no separate files contain copies of the same piece of data.
Easier modification and updating	The DBMS coordinates data modifications and updates. Programmers and users do not have to know where the data is physically stored. Data is stored and modified once. Modification and updating is also easier because the data is commonly stored in only one location.
Data and program independence	The DBMS organizes the data independently of the application program, so the application program is not affected by the location or type of data. Introduction of new data types not relevant to a particular application does not require rewriting that application to maintain compatibility with the data file.
Better access to data and information	Most DBMSs have software that makes it easy to access and retrieve data from a database. In most cases, users give simple commands to get important information. Relationships between records can be more easily investigated and exploited, and applications can be more easily combined.
Standardization of data access	A standardized, uniform approach to database access means that all application programs use the same overall procedures to retrieve data and information.
A framework for program development	Standardized database access procedures can mean more standardization of program development. Because programs go through the DBMS to gain access to data in the database, standardized database access can provide a consistent framework for program development. In addition, each application program need address only the DBMS, not the actual data files, reducing application development time.
Better overall protection of the data	Accessing and using centrally located data is easier to monitor and control. Security codes and passwords can ensure that only authorized people have access to particular data and information in the database, thus ensuring privacy.
Shared data and information resources	The cost of hardware, software, and personnel can be spread over many applications and users. This is a primary feature of a DBMS.

Table 3.1

Advantages of the Database Approach

Table 3.2

Disadvantages of the Database Approach

Disadvantages	Explanation
More complexity	DBMSs can be difficult to set up and operate. Many decisions must be made correctly for the DBMS to work effectively. In addition, users have to learn new procedures to take full advantage of a DBMS.
More difficult to recover from a failure	With the traditional approach to file management, a failure of a file affects only a single program. With a DBMS, a failure can shut down the entire database.
More expensive	DBMSs can be more expensive to purchase and operate. The expense includes the cost of the database and specialized personnel, such as a database administrator, who is needed to design and operate the database. Additional hardware might also be required.

Web-Based Electronic Health Record Systems

The United States Federal government is pushing for most Americans to have their medical records stored in electronic form by 2014. Electronic health record (EHR) systems store patient records in a central database that can be accessed by many physicians at more than one location. Such a system eliminates problems caused by duplicate records at different physician offices, avoids having to fill out a new patient history with each new physician visited by the patient, and reduces errors made by incorrectly deciphering handwritten notes and prescriptions. Electronic records can make for a better and healthier world. However, the cost of moving to electronic systems is prohibitive, especially for small medical practices. At this point, only ten percent of small medical offices and five percent of solo practitioners have moved to EHR systems.

Although the government is introducing financial incentives to encourage physicians to use EHR systems, some big companies that aren't typically associated with health care are becoming involved—particularly Microsoft and Google. Approximately 52 percent of adults look to the Web when seeking health advice. Google and Microsoft believe that they can better assist health consumers by providing them with a robust tool for managing their health records. Microsoft's tool is named HealthVault, while Google's is named Google Health. The companies see their EHR systems as a solution to the government's problem for finding a low-cost records system designed for both physicians and patients.

John D. Halamka, a doctor and CIO of the Harvard Medical School, thinks systems in which the patient manages the information, such as those proposed by Microsoft and Google, are the inevitable future. "Patients will ultimately be the stewards of their own information," Halamka stated. "In the future, health care will be a much more collaborative process between patients and doctors."

Google agrees that patients should be in charge. A statement at Google Health's welcome page reads, "At Google, we feel patients should be in charge of their health information, and they should be able to grant their health care providers, family members, or whomever they choose, access to this information. Google Health was developed to meet this need."

But just how private and secure will our medical records be when stored in Web-accessible databases, protected only by one password? Privacy and security concerns are raised both by corporate access to private records by Microsoft and Google and outsider access by hackers. It is likely that both companies will use automated systems to target advertising at individuals based on medical records, just as Google's Gmail places ads next to e-mail messages based on the message contents. Unauthorized users might also be able to access records stored on a network that billions of users around the world use.

Another problem that complicates Google and Microsoft's involvement is that third-party medical record services are not covered by the Health Insurance Portability and Accountability Act (HIPAA). HIPAA provides strict standards for keeping medical records private. If a patient chooses to use Microsoft or Google to store medical records, those records would no longer be protected by the standards imposed by HIPAA in its current form.

As in similar cases, patients should weigh the costs in terms of privacy and security against the benefits of convenience and data reliability. Meanwhile, the software vendors need to work to build higher levels of security, privacy assurances, and customer trust.

Discussion Questions

1. Why does the U.S. Federal government want to move health records to electronic systems?
2. What benefits and risks are offered by Web-based health records management systems like Google Health?

Critical Thinking Questions

1. How might Google and Microsoft reassure users about the privacy and security issues posed in this sidebar?
2. Would you consider registering for Google Health? Why or why not?

Sources: Lohr, Steve, "Google and Microsoft Look to Change Health Care," *New York Times*, August 14, 2007, *www.nytimes.com/2007/08/14/technology/14healthnet.html*, AP Staff, "Google ventures into health records biz," *CNN.com*, February 21, 2008, *www.cnn.com/2008/TECH/02/21/google.records.ap*.

DATA MODELING AND DATABASE CHARACTERISTICS

Because today's businesses have so many elements, they must keep data organized so that it can be used effectively. A database should be designed to store all data relevant to the business and provide quick access and easy modification. Moreover, it must reflect the business processes of the organization. When building a database, an organization must carefully consider these questions:

- **Content.** What data should be collected and at what cost?
- **Access.** What data should be provided to which users and when?
- **Logical structure.** How should data be arranged so that it makes sense to a given user?
- **Physical organization.** Where should data be physically located?

Data Modeling

Key considerations in organizing data in a database include determining what data to collect in the database, who will have access to it, and how they might want to use the data. After determining these details, an organization can create a database. Building a database requires two different types of designs: a logical design and a physical design. The *logical design* of a database is an abstract model of how the data should be structured and arranged to meet an organization's information needs. The logical design involves identifying relationships among the data items and grouping them in an orderly fashion. Because databases provide both input and output for information systems throughout a business, users from all functional areas should assist in creating the logical design to ensure that their needs are identified and addressed. *Physical design* starts from the logical database design and fine-tunes it for performance and cost considerations (such as improved response time, reduced storage space, and lower operating cost). For example, the database administrator at Intermountain Healthcare in Salt Lake City, Utah, combined the databases of 21 hospitals and 100 clinics into one integrated system, saving the organization the cost of dozens of servers, and providing new and improved services.[8] The person who fine-tunes the physical design must have an in-depth knowledge of the DBMS. For example, the logical database design might need to be altered so that certain data entities are combined, summary totals are carried in the data records rather than calculated from elemental data, and some data attributes are repeated in more than one data entity. These are examples of **planned data redundancy**, which improves the system performance so that user reports or queries can be created more quickly.

One of the tools database designers use to show the logical relationships among data is a data model. A **data model** is a diagram of entities and their relationships. Data modeling usually involves understanding a specific business problem and analyzing the data and information needed to deliver a solution. When done at the level of the entire organization, this is called enterprise data modeling. **Enterprise data modeling** is an approach that starts by investigating the general data and information needs of the organization at the strategic level, and then examines more specific data and information needs for the various functional areas and departments within the organization. Various models have been developed to help managers and database designers analyze data and information needs. An entity-relationship diagram is an example of such a data model.

Entity-relationship (ER) diagrams use basic graphical symbols to show the organization of and relationships between data. In most cases, boxes in ER diagrams indicate data items or entities contained in data tables, and diamonds show relationships between data items and entities. In other words, ER diagrams show data items in tables (entities) and the ways they are related.

ER diagrams help ensure that the relationships among the data entities in a database are correctly structured so that any application programs developed are consistent with business operations and user needs. In addition, ER diagrams can serve as reference documents after a database is in use. If changes are made to the database, ER diagrams help design them. Figure 3.4 shows an ER diagram for an order database. In this database design, one salesperson serves many customers. This is an example of a one-to-many relationship, as indicated by

planned data redundancy
A way of organizing data in which the logical database design is altered so that certain data entities are combined, summary totals are carried in the data records rather than calculated from elemental data, and some data attributes are repeated in more than one data entity to improve database performance.

data model
A diagram of data entities and their relationships.

enterprise data modeling
Data modeling done at the level of the entire enterprise.

entity-relationship (ER) diagrams
Data models that use basic graphical symbols to show the organization of and relationships between data.

the one-to-many symbol (the "crow's-foot") shown in Figure 3.4. The ER diagram also shows that each customer can place one-to-many orders, each order includes one-to-many line items, and many line items can specify the same product (a many-to-one relationship). This database can also have one-to-one relationships. For example, one order generates one invoice.

Figure 3.4

An Entity-Relationship (ER) Diagram for a Customer Order Database

Development of ER diagrams helps ensure that the logical structure of application programs is consistent with the data relationships in the database.

The Relational Database Model

Although there are a number of different database models, including flat files, hierarchical, and network models, the **relational model** has become the most popular, and use of this model will continue to increase. The relational model describes data using a standard tabular format. In a database structured according to the relational model, all data elements are placed in two-dimensional tables, called *relations*, which are the logical equivalent of files. The tables in relational databases organize data in rows and columns, simplifying data access and manipulation. It is normally easier for managers to understand the relational model (see Figure 3.5) than other database models.

Databases based on the relational model include IBM DB2, Oracle, Sybase, Microsoft SQL Server, Microsoft Access, and MySQL. Oracle is currently the market leader in general-purpose databases, with over 40 percent of the $16.5 billion database market. IBM comes in second with about 21 percent, and Microsoft third with about 19 percent.[9]

In the relational model, each row (or record) of a table represents a data entity, with the columns (or fields) of the table representing attributes. Each attribute can accept only certain values. The allowable values for these attributes are called the **domain**. The domain for a particular attribute indicates what values can be placed in each column of the relational table. For instance, the domain for an attribute such as gender would be limited to male or female. A domain for pay rate would not include negative numbers. In this way, defining a domain can increase data accuracy.

Manipulating Data

After entering data into a relational database, users can make inquiries and analyze the data. Basic data manipulations include selecting, projecting, and joining. **Selecting** involves eliminating rows according to certain criteria. Suppose a project table contains the project number, description, and department number for all projects a company is performing. The president of the company might want to find the department number for Project 226, a sales manual project. Using selection, the president can eliminate all rows but the one for Project 226 and see that the department number for the department completing the sales manual project is 598.

relational model
A database model that describes data in which all data elements are placed in two-dimensional tables, called *relations*, which are the logical equivalent of files.

domain
The allowable values for data attributes.

selecting
Manipulating data to eliminate rows according to certain criteria.

Figure 3.5

A Relational Database Model

In the relational model, all data elements are placed in two-dimensional tables, or relations. As long as they share at least one common element, these relations can be linked to output useful information. Note that some organizations might use employee number instead of Social Security number (SSN) in Data Tables 2 and 3.

Data Table 1: Project Table

Project	Description	Dept. number
155	Payroll	257
498	Widgets	632
226	Sales manual	598

Data Table 2: Department Table

Dept.	Dept. name	Manager SSN
257	Accounting	005-10-6321
632	Manufacturing	549-77-1001
598	Marketing	098-40-1370

Data Table 3: Manager Table

SSN	Last name	First name	Hire date	Dept. number
005-10-6321	Johns	Francine	10-07-1997	257
549-77-1001	Buckley	Bill	02-17-1979	632
098-40-1370	Fiske	Steven	01-05-1985	598

projecting
Manipulating data to eliminate columns in a table.

joining
Manipulating data to combine two or more tables.

linking
Data manipulation that combines two or more tables using common data attributes to form a new table with only the unique data attributes.

Projecting involves eliminating columns in a table. For example, a department table might contain the department number, department name, and Social Security number (SSN) of the manager in charge of the project. A sales manager might want to create a new table with only the department number and the Social Security number of the manager in charge of the sales manual project. The sales manager can use projection to eliminate the department name column and create a new table containing only department number and SSN.

Joining involves combining two or more tables. For example, you can combine the project table and the department table to create a new table with the project number, project description, department number, department name, and Social Security number for the manager in charge of the project.

As long as the tables share at least one common data attribute, the tables in a relational database can be **linked** to provide useful information and reports. Being able to link tables to each other through common data attributes is one of the keys to the flexibility and power of relational databases. Suppose the president of a company wants to find out the name of the manager of the sales manual project and the length of time the manager has been with the company. Assume that the company has the manager, department, and project tables shown in Figure 3.5. A simplified ER diagram showing the relationship between these tables is shown in Figure 3.6. Note the crow's-foot by the project table. This indicates that a department can have many projects. The president would make the inquiry to the database, perhaps via a personal computer. The DBMS would start with the project description and search the project table to find out the project's department number. It would then use the department number to search the department table for the manager's Social Security number. The department number is also in the department table and is the common element that links the project table to the department table. The DBMS uses the manager's Social Security number to search the manager table for the manager's hire date. The manager's Social Security number is the common element between the department table and the manager table. The final result is that the manager's name and hire date are presented to the president as a response to the inquiry (see Figure 3.7).

Figure 3.6

A Simplified ER Diagram Showing the Relationship Between the Manager, Department, and Project Tables

```
          Manager
             |
         Supervises
             |
        Department
             |
         Performs
             |
          Project
```

Figure 3.7

Linking Data Tables to Answer an Inquiry

In finding the name and hire date of the manager working on the sales manual project, the president needs three tables: project, department, and manager. The project description (Sales manual) leads to the department number (598) in the project table, which leads to the manager's SSN (098-40-1370) in the department table, which leads to the manager's name (Fiske) and hire date (01-05-1985) in the manager table. Again, note that some organizations might use employee number instead of Social Security number (SSN).

Data Table 1: Project Table

Project number	Description	Dept. number
155	Payroll	257
498	Widgets	632
226	Sales manual	598

Data Table 2: Department Table

Dept. number	Dept. name	Manager SSN
257	Accounting	005-10-6321
632	Manufacturing	549-77-1001
598	Marketing	098-40-1370

Data Table 3: Manager Table

SSN	Last name	First name	Hire date	Dept. number
005-10-6321	Johns	Francine	10-07-1997	257
549-77-1001	Buckley	Bill	02-17-1979	632
098-40-1370	Fiske	Steven	01-05-1985	598

One of the primary advantages of a relational database is that it allows tables to be linked, as shown in Figure 3.7. This linkage is especially useful when information is needed from multiple tables. For example, the manager's Social Security number is maintained in the manager table. If the Social Security number is needed, it can be obtained by linking to the manager table.

The relational database model is by far the most widely used. It is easier to control, more flexible, and more intuitive than other approaches because it organizes data in tables. As shown in Figure 3.8, a relational database management system, such as Access, provides tips and tools for building and using database tables. In this figure, the database displays information about data types and indicates that additional help is available. The ability to link relational tables also allows users to relate data in new ways without having to redefine complex relationships. Because of the advantages of the relational model, many companies use it for large corporate databases, such as those for marketing and accounting. The relational model can also be used with personal computers and mainframe systems. A travel reservation company, for example, can develop a fare-pricing system by using relational database technology that can handle millions of daily queries from online travel companies, such as Expedia, Travelocity, and Orbitz.

Figure 3.8

Building and Modifying a Relational Database

Relational databases provide many tools, tips, and shortcuts to simplify the process of creating and modifying a database.

(Source: Courtesy of Microsoft Corporation.)

DATABASE MANAGEMENT SYSTEMS

Creating and implementing the right database system ensures that the database will support both business activities and goals. But how do we actually create, implement, use, and update a database? The answer is found in the database management system. As discussed earlier, a DBMS is a group of programs used as an interface between a database and application programs or a database and the user. The capabilities and types of database systems, however, vary considerably. For example, visitors to the Baseball Hall of Fame in Cooperstown New York use a DBMS to search baseball highlight films from famous games and plays.[10] DBMSs are used to manage all kinds of data for all kinds of purposes.

Overview of Database Types

Database management systems can range from small, inexpensive software packages to sophisticated systems costing hundreds of thousands of dollars. The following sections discuss a few popular alternatives. See Figure 3.9 for one example.

Flat File

A flat file is a simple database program whose records have no relationship to one another. Flat file databases are often used to store and manipulate a single table or file, and do not use any of the database models discussed previously, such as the relational model. Many spreadsheet and word processing programs have flat file capabilities. These software packages can sort tables and make simple calculations and comparisons. Microsoft OneNote is designed to let people put ideas, thoughts, and notes into a computer file. In OneNote, each note can be placed anywhere on a page or in a box on a page, called a *container*. Pages are organized into sections and subsections that appear as colored tabs. After you enter a note, you can retrieve, copy, and paste it into other applications, such as word processing and spreadsheet programs. Microsoft uses OneNote as the primary technology for its management training

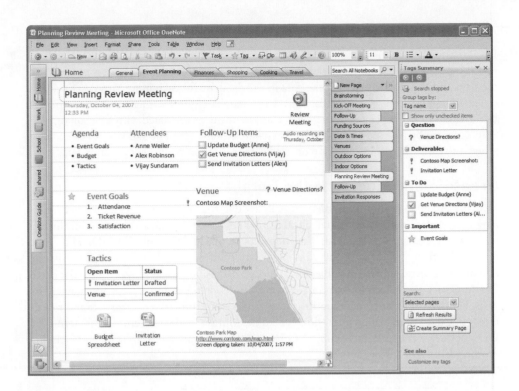

Figure 3.9

Microsoft OneNote

Microsoft OneNote lets you gather any type of information and then retrieve, copy, and paste the information into other applications, such as word processing and spreadsheet programs.

classes. OneNote allows managers-in-training to collect photos, handwritten notes, online content, and audio recordings in one flat file.[11] OneNote enables Microsoft to offer training to a larger number of managers, while saving $360,000 per year in printed training materials.

Similar to OneNote, Evernote is a free database that can store notes and other pieces of information. Considering the amount of information today's high-capacity hard disks can store, the popularity of databases that can handle unstructured data will continue to grow.

Single User

A database installed on a personal computer is typically meant for a single user. Microsoft Office Access and FileMaker Pro are designed to support single-user implementations. Microsoft InfoPath is another example of a database program that supports a single user. This software is part of the Microsoft Office suite, and it helps people collect and organize information from a variety of sources. InfoPath has built-in forms that can be used to enter expense information, time-sheet data, and a variety of other information.

Multiple Users

Small, midsize, and large businesses need multiuser DBMSs to share information throughout the organization over a network. These more powerful, expensive systems allow dozens or hundreds of people to access the same database system at the same time. Popular vendors for multiuser database systems include Oracle, Microsoft, Sybase, and IBM. Many single-user databases, such as Microsoft Access, can be implemented for multiuser support over a network, though they often are limited in the amount of users they can support.

All DBMSs share some common functions, such as providing a user view, physically storing and retrieving data in a database, allowing for database modification, manipulating data, and generating reports. These DBMSs can handle the most complex data-processing tasks, and because they are accessed over a network, one database can serve many locations around the world. For example, Surya Roshni Ltd is a major manufacturer of lighting products based in New Delhi, India, with a global reach. One Oracle database stored on servers in New Delhi provides corporate information to associates around the world.[12]

Providing a User View

Because the DBMS is responsible for access to a database, one of the first steps in installing and using a large database involves telling the DBMS the logical and physical structure of the data and relationships among the data in the database for each user. This description is called a **schema** (as in schematic diagram). Large database systems, such as Oracle, typically use schemas to define the tables and other database features associated with a person or user. A schema can be part of the database or a separate schema file. The DBMS can reference a schema to find where to access the requested data in relation to another piece of data.

schema
A description of the entire database.

Creating and Modifying the Database

Schemas are entered into the DBMS (usually by database personnel) via a data definition language. A **data definition language** (DDL) is a collection of instructions and commands used to define and describe data and relationships in a specific database. A DDL allows the database's creator to describe the data and relationships that are to be contained in the schema. In general, a DDL describes logical access paths and logical records in the database. Figure 3.10 shows a simplified example of a DDL used to develop a general schema. The *Xs* in Figure 3.10 reveal where specific information concerning the database should be entered. File description, area description, record description, and set description are terms the DDL defines and uses in this example. Other terms and commands can be used, depending on the particular DBMS employed.

data definition language (DDL)
A collection of instructions and commands used to define and describe data and relationships in a specific database.

Figure 3.10

Using a Data Definition Language to Define a Schema

```
SCHEMA DESCRIPTION
SCHEMA NAME IS XXXX
AUTHOR        XXXX
DATE          XXXX
FILE DESCRIPTION
      FILE NAME IS XXXX
        ASSIGN XXXX
      FILE NAME IS XXXX
        ASSIGN XXXX
AREA DESCRIPTION
      AREA NAME IS XXXX
RECORD DESCRIPTION
      RECORD NAME IS XXXX
      RECORD ID IS XXXX
      LOCATION MODE IS XXXX
      WITHIN XXXX AREA FROM XXXX THRU XXXX
SET DESCRIPTION
      SET NAME IS XXXX
      ORDER IS XXXX
      MODE IS XXXX
      MEMBER IS XXXX
      .
      .
      .
```

data dictionary
A detailed description of all the data used in the database.

Another important step in creating a database is to establish a **data dictionary**, a detailed description of all data used in the database. The data dictionary contains the following data:

- Name of the data item
- Aliases or other names that may be used to describe the item
- Range of values that can be used
- Type of data (such as alphanumeric or numeric)
- Amount of storage needed for the item
- Notation of the person responsible for updating it and the various users who can access it
- List of reports that use the data item

A data dictionary can also include a description of data flows, the way records are organized, and the data-processing requirements. Figure 3.11 shows a typical data dictionary entry.

```
                    NORTHWESTERN MANUFACTURING

    PREPARED BY:          D. BORDWELL
    DATE:                 04 AUGUST 2007
    APPROVED BY:          J. EDWARDS
    DATE:                 13 OCTOBER 2007
    VERSION:              3.1
    PAGE:                 1 OF 1


    DATA ELEMENT NAME:    PARTNO
    DESCRIPTION:          INVENTORY PART NUMBER
    OTHER NAMES:          PTNO
    VALUE RANGE:          100 TO 5000
    DATA TYPE:            NUMERIC
    POSITIONS:            4 POSITIONS OR COLUMNS
```

Figure 3.11

A Typical Data Dictionary Entry

For example, the information in a data dictionary for the part number of an inventory item can include the following:

- Name of the person who made the data dictionary entry (D. Bordwell)
- Date the entry was made (August 4, 2007)
- Name of the person who approved the entry (J. Edwards)
- Approval date (October 13, 2007)
- Version number (3.1)
- Number of pages used for the entry (1)
- Part name (PARTNO)
- Other part names that might be used (PTNO)
- Range of values (part numbers can range from 100 to 5,000)
- Type of data (numeric)
- Storage required (four positions are required for the part number)

A data dictionary is valuable in maintaining an efficient database that stores reliable information with no redundancy, and makes it easy to modify the database when necessary. Data dictionaries also help computer and system programmers who require a detailed description of data elements stored in a database to create the code to access the data.

Storing and Retrieving Data

One function of a DBMS is to be an interface between an application program and the database. When an application program needs data, it requests the data through the DBMS. Suppose that to calculate the total price of a new car, an auto dealer pricing program needs price data on the engine option—six cylinders instead of the standard four cylinders. The application program requests this data from the DBMS. In doing so, the application program follows a logical access path. Next, the DBMS, working with various system programs, accesses a storage device, such as disk drives, where the data is stored. When the DBMS goes to this storage device to retrieve the data, it follows a path to the physical location (physical access path) where the price of this option is stored. In the pricing example, the DBMS might go to a disk drive to retrieve the price data for six-cylinder engines. This relationship is shown in Figure 3.12.

This same process is used if a user wants to get information from the database. First, the user requests the data from the DBMS. For example, a user might give a command, such as LIST ALL OPTIONS FOR WHICH PRICE IS GREATER THAN 200 DOLLARS. This

Figure 3.12

Logical and Physical Access
Paths

is the logical access path (LAP). Then, the DBMS might go to the options price section of a disk to get the information for the user. This is the physical access path (PAP).

Two or more people or programs attempting to access the same record in the same database at the same time can cause a problem. For example, an inventory control program might attempt to reduce the inventory level for a product by ten units because ten units were just shipped to a customer. At the same time, a purchasing program might attempt to increase the inventory level for the same product by 200 units because more inventory was just received. Without proper database control, one of the inventory updates might be incorrect, resulting in an inaccurate inventory level for the product. **Concurrency control** can be used to avoid this potential problem. One approach is to lock out all other application programs from access to a record if the record is being updated or used by another program.

concurrency control
A method of dealing with a situation in which two or more people need to access the same record in a database at the same time.

Manipulating Data and Generating Reports

After a DBMS has been installed, employees, managers, and consumers can use it to review reports and obtain important information. For example, the Food Allergen and Consumer Protection Act, effective in 2006, requires that food manufacturing companies generate reports on the ingredients, formulas, and food preparation techniques for the public. Using a DBMS, a company can easily manage this requirement.

Some databases use *Query-by-Example (QBE)*, which is a visual approach to developing database queries or requests. Like Windows and other GUI operating systems, you can perform queries and other database tasks by opening windows and clicking the data or features you want (see Figure 3.13).

In other cases, database commands can be used in a programming language. For example, C++ commands can be used in simple programs that will access or manipulate certain pieces of data in the database. Here's another example of a DBMS query: SELECT * FROM EM-PLOYEE WHERE JOB_CLASSIFICATION = "C2". The * tells the program to include all columns from the EMPLOYEE table. In general, the commands that are used to manipulate the database are part of the **data manipulation language** (DML). This specific language, provided with the DBMS, allows managers and other database users to access, modify, and make queries about data contained in the database to generate reports. Again, the application programs go through schemas and the DBMS before actually getting to the physically stored data on a device such as a disk.

**data manipulation language
(DML)**
The commands that are used to manipulate the data in a database.

In the 1970s, D. D. Chamberlain and others at the IBM Research Laboratory in San Jose, California, developed a standardized data manipulation language called *Structured Query Language (SQL)*, pronounced like the word *sequel* or spelled out as *SQL*. The EMPLOYEE query shown earlier is written in SQL. In 1986, the American National

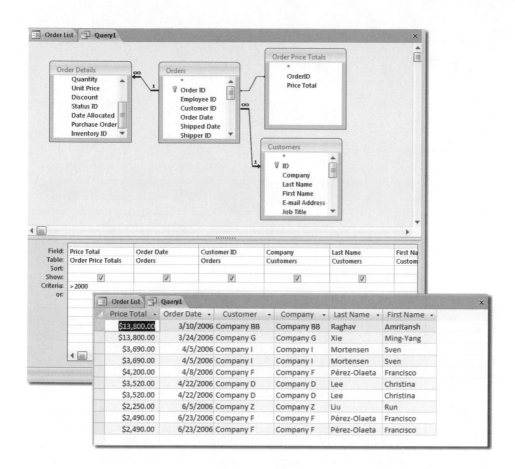

Figure 3.13

Query-by-Example

Some databases use Query-by-Example (QBE) to generate reports and information.

Standards Institute (ANSI) adopted SQL as the standard query language for relational databases. Since ANSI's acceptance of SQL, interest in making SQL an integral part of relational databases on both mainframe and personal computers has increased. SQL has many built-in functions, such as average (AVG), the largest value (MAX), the smallest value (MIN), and others. Table 3.3 contains examples of SQL commands.

SQL Command	Description
SELECT ClientName, Debt FROM Client WHERE Debt > 1000	This query displays all clients (ClientName) and the amount they owe the company (Debt) from a database table called Client for clients who owe the company more than $1,000 (WHERE Debt > 1000).
SELECT ClientName, ClientNum, OrderNum FROM Client, Order WHERE Client.ClientNum=Order.ClientNum	This command is an example of a join command that combines data from two tables: the client table and the order table (FROM Client, Order). The command creates a new table with the client name, client number, and order number (SELECT ClientName, ClientNum, OrderNum). Both tables include the client number, which allows them to be joined. This is indicated in the WHERE clause, which states that the client number in the client table is the same as (equal to) the client number in the order table (WHERE Client.ClientNum = Order.ClientNum).
GRANT INSERT ON Client to Guthrie	This command is an example of a security command. It allows Bob Guthrie to insert new values or rows into the Client table.

Table 3.3

Examples of SQL Commands

SQL lets programmers learn one powerful query language and use it on systems ranging from PCs to the largest mainframe computers (see Figure 3.14). Programmers and database users also find SQL valuable because SQL statements can be embedded into many programming languages, such as the widely used C++ and COBOL languages. Because SQL uses standardized and simplified procedures for retrieving, storing, and manipulating data in a database system, the popular database query language can be easy to understand and use.

Figure 3.14

Structured Query Language

Structured Query Language (SQL) has become an integral part of most relational databases, as shown by this screen from Microsoft Office Access 2007.

After a database has been set up and loaded with data, it can produce desired reports, documents, and other outputs (see Figure 3.15). These outputs usually appear in screen displays or hard-copy printouts. The output-control features of a database program allow you to select the records and fields to appear in reports. You can also make calculations specifically for the report by manipulating database fields. Formatting controls and organization options (such as report headings) help you to customize reports and create flexible, convenient, and powerful information-handling tools.

Figure 3.15

Database Output

A database application offers sophisticated formatting and organization options to produce the right information in the right format.

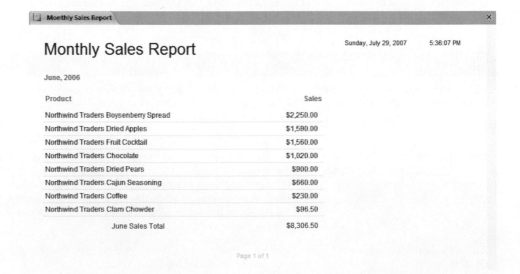

A DBMS can produce a wide variety of documents, reports, and other output that can help organizations achieve their goals. The most common reports select and organize data to present summary information about some aspect of company operations. For example, accounting reports often summarize financial data such as current and past-due accounts. Many companies base their routine operating decisions on regular status reports that show the progress of specific orders toward completion and delivery.

Databases can also provide support to help executives and other people make better decisions. A database by Intellifit, for example, can be used to help shoppers make better decisions and get clothes that fit when shopping online. The database contains true sizes of apparel from various clothing companies that do business on the Web. The process starts when a customer's body is scanned into a database at one of the company's locations, typically in a shopping mall. About 200,000 measurements are taken to construct a 3-D image of the person's body shape. The database then compares the actual body dimensions with sizes given by Web-based clothing stores to get an excellent fit.[13]

Database Administration

Database systems require a skilled database administrator. A DBA is expected to have a clear understanding of the fundamental business of the organization, be proficient in the use of selected database management systems, and stay abreast of emerging technologies and new design approaches. The role of the DBA is to plan, design, create, operate, secure, monitor, and maintain databases. Typically, a DBA has a degree in computer science or management information systems and some on-the-job training with a particular database product or more extensive experience with a range of database products. See Figure 3.16.

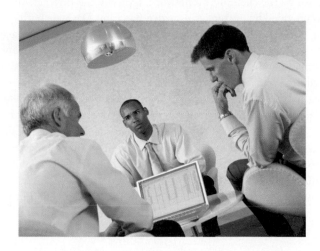

Database Administrator

The role of the database administrator (DBA) is to plan, design, create, operate, secure, monitor, and maintain databases.

(Source: BananaStock / Alamy.)

The DBA works with users to decide the content of the database—to determine exactly what entities are of interest and what attributes are to be recorded about those entities. Thus, personnel outside of IS must have some idea of what the DBA does and why this function is important. The DBA can play a crucial role in the development of effective information systems to benefit the organization, employees, and managers.

The DBA also works with programmers as they build applications to ensure that their programs comply with database management system standards and conventions. After the database is built and operating, the DBA monitors operations logs for security violations. Database performance is also monitored to ensure that the system's response time meets users' needs and that it operates efficiently. If there is a problem, the DBA attempts to correct it before it becomes serious.

Some organizations have also created a position called the *data administrator*, a nontechnical, but important role that ensures that data is managed as an important organizational resource. The **data administrator** is responsible for defining and implementing consistent principles for a variety of data issues, including setting data standards and data definitions that apply across all the databases in an organization. For example, the data administrator would ensure that a term such as "customer" is defined and treated consistently in all corporate databases. This person also works with business managers to identify who should have read or update access to certain databases and to selected attributes within those databases. This information is then communicated to the database administrator for implementation. The data administrator can be a high-level position reporting to top-level managers.

data administrator
A nontechnical position responsible for defining and implementing consistent principles for a variety of data issues.

Popular Database Management Systems

Some popular DBMSs for single users include Microsoft Access and FileMaker Pro. The complete database management software market encompasses software used by professional programmers and that runs on midrange, mainframe, and supercomputers. The entire market generates billions of dollars per year in revenue, including IBM, Oracle, and Microsoft. Although Microsoft rules in the desktop PC software market, its share of database software on larger computers is small.

Like other software products, a number of open-source database systems are available, including PostgreSQL and MySQL. Open-source software was described in Chapter 4. In addition, many traditional database programs are now available on open-source operating systems. The popular DB2 relational database from IBM, for example, is available on the Linux operating system. The Sybase IQ database and other databases are also available on the Linux operating system.

A new form of database system is emerging that some refer to as *Database as a Service* (*DaaS*) and others as Database 2.0. DaaS is similar to software as a service (SaaS). Recall that a SaaS system is one in which the software is stored on a service provider's servers and accessed by the client company over a network. In DaaS, the database is stored on a service provider's servers and accessed by the client over a network, typically the Internet. In DaaS, database administration is provided by the service provider. SaaS and DaaS are both part of the larger cloud computing trend. Recall from Chapter 3 that cloud computing uses a giant cluster of computers that serves as a host to run applications that require high performance computing. In cloud computing, all information systems and data are maintained and managed by service providers and delivered over the Internet. Businesses and individuals are freed from having to install, service, maintain, upgrade, and safeguard their systems.

More than a dozen companies are moving in the DaaS direction. They include Google, Microsoft, Intuit, Serran Tech, MyOwnDB, and Trackvia.[14] XM Radio, Google, JetBlue Airways, Bank of America, Southwest Airlines, and others use QuickBase from service provider Intuit to manage their databases out of house.[15] JetBlue, for example, uses a DaaS from Intuit to organize and manage IT projects.[16] Because the database and DBMS are available from any Internet connection, those involved in managing and implementing systems development projects can record their progress and check on other's progress from any location.

Special-Purpose Database Systems

In addition to the popular database management systems just discussed, some specialized database packages are used for specific purposes or in specific industries. For example, the Israeli Holocaust Database (*www.yadvashem.org*) is a special-purpose database available through the Internet and contains information on about three million people in 14 languages. A unique special-purpose DBMS for biologists called Morphbank (*www.morphbank.net*) allows researchers from around the world to continually update and expand a library of over 96,000 biological images to share with the scientific community and the public. The iTunes Store music and video catalog is a special-purpose database system. When you search for your favorite artist, you are querying the database.

Selecting a Database Management System

The database administrator often selects the best database management system for an organization. The process begins by analyzing database needs and characteristics. The information needs of the organization affect the type of data that is collected and the type of database management system that is used. Important characteristics of databases include the following:

- **Database size.** The number of records or files in the database
- **Database cost.** The purchase or lease costs of the database
- **Concurrent users.** The number of people who need to use the database at the same time (the number of concurrent users)

- **Performance.** How fast the database is able to update records
- **Integration.** The ability to be integrated with other applications and databases
- **Vendor.** The reputation and financial stability of the database vendor

The Web-based Morphbank database allows scientists from around the world to upload and share biological and microscopic photographs and descriptions which support research in many areas.

(Source: *www.morphbank.net*)

For many organizations, database size doubles about every year or two. With the increasing use of digital media—images, video, and audio—data storage demands are growing exponentially. In fact, the volume of data being created has surpassed the world's available storage capacity.[17] The growing need for data storage has not escaped the notice of large technology companies such as Google and Microsoft, who are buying hundreds of acres of land and building huge data centers to support the world's data storage needs.[18] Meanwhile, many businesses and government agencies are working to consolidate data dispersed across the organization into smaller, more efficient centralized systems.

Using Databases with Other Software

Database management systems are often used with other software or the Internet. A DBMS can act as a front-end application or a back-end application. A *front-end application* is one that directly interacts with people or users. Marketing researchers often use a database as a front end to a statistical analysis program. The researchers enter the results of market questionnaires or surveys into a database. The data is then transferred to a statistical analysis program to determine the potential for a new product or the effectiveness of an advertising campaign. A *back-end application* interacts with other programs or applications; it only indirectly interacts with people or users. When people request information from a Web site, the Web site can interact with a database (the back end) that supplies the desired information. For example, you can connect to a university Web site to find out whether the university's library has a book you want to read. The Web site then interacts with a database that contains a catalog of library books and articles to determine whether the book you want is available.

DATABASE APPLICATIONS

Today's database applications manipulate the content of a database to produce useful information. Common manipulations are searching, filtering, synthesizing, and assimilating the data contained in a database, using a number of database applications. These applications allow users to link the company databases to the Internet, set up data warehouses and marts, use databases for strategic business intelligence, place data at different locations, use online processing and open connectivity standards for increased productivity, develop databases with the object-oriented approach, and search for and use unstructured data, such as graphics, audio, and video.

Linking the Company Database to the Internet

Linking databases to the Internet is one reason the Internet is so popular. A large percentage of corporate databases are accessed over the Internet through a standard Web browser. Being able to access bank account data, student transcripts, credit card bills, product catalogs, and a host of other data online is convenient for individual users, and increases effectiveness and efficiency for businesses and organizations. Amazon.com, Apple's iTunes store, eBay, and others have made billions of dollars by combining databases, the Internet, and smart business models.

As discussed in the Ethical and Societal Issues sidebar, Google is rolling out a DBMS that will provide patients and physicians with one storage location for all medical records, accessed through a Web browser.[19] Access to private medical information over the public Web has some privacy advocates concerned. However, the convenience that the system offers by dramatically reducing the amount of paper forms to fill out and store, along with the reduction of clerical errors through streamlined data management procedures, has most in the field supporting the move to a centralized system. Google protects patient records with encryption and authentication technologies.

Developing a seamless integration of traditional databases with the Internet is often called a *semantic Web*. A semantic Web allows people to access and manipulate a number of traditional databases at the same time through the Internet. The World Wide Web Consortium has established standards for a semantic Web in hopes of some day evolving the Web into one big database that is easy to manage and traverse. Yahoo has recently announced its commitment to complying with the standards for a semantic Web.[20]

Although the semantic Web standards have not been embraced by all businesses, many software vendors—including IBM, Oracle, Microsoft, Macromedia, and Inline Internet Systems—are incorporating the Internet into their products. Such databases allow companies to create an Internet-accessible catalog, which is a database of items, descriptions, and prices. As evidenced by the Web, most companies are utilizing these tools to take their business online.

In addition to the Internet, organizations are gaining access to databases through networks to find good prices and reliable service. Connecting databases to corporate Web sites and networks can lead to potential problems, however. A recent study found that nearly half a million database servers were vulnerable to attack over the Internet due to the lack of proper security measures.[21]

Data Warehouses, Data Marts, and Data Mining

The raw data necessary to make sound business decisions is stored in a variety of locations and formats. This data is initially captured, stored, and managed by transaction processing systems that are designed to support the day-to-day operations of the organization. For decades, organizations have collected operational, sales, and financial data with their online transaction processing (OLTP) systems. The data can be used to support decision making using data warehouses, data marts, and data mining.

Data Warehouses

A **data warehouse** is a database that holds business information from many sources in the enterprise, covering all aspects of the company's processes, products, and customers. The data warehouse provides business users with a multidimensional view of the data they need to analyze business conditions. Data warehouses allow managers to *drill down* to get more detail or *roll up* to take detailed data and generate aggregate or summary reports. A data warehouse is designed specifically to support management decision making, not to meet the needs of transaction processing systems. A data warehouse stores historical data that has been extracted from operational systems and external data sources (see Figure 3.17). This operational and external data is "cleaned up" to remove inconsistencies and integrated to create a new information database that is more suitable for business analysis.

data warehouse
A database that collects business information from many sources in the enterprise, covering all aspects of the company's processes, products, and customers.

Figure 3.17

Elements of a Data Warehouse

Data warehouses typically start out as very large databases, containing millions and even hundreds of millions of data records. As this data is collected from the various production systems, a historical database is built that business analysts can use. To keep it fresh and accurate, the data warehouse receives regular updates. Old data that is no longer needed is purged from the data warehouse. Updating the data warehouse must be fast, efficient, and automated, or the ultimate value of the data warehouse is sacrificed. It is common for a data warehouse to contain from three to ten years of current and historical data. Data-cleaning tools can merge data from many sources into one database, automate data collection and verification, delete unwanted data, and maintain data in a database management system. Data warehouses can also get data from unique sources. Oracle's Warehouse Management software, for example, can accept information from Radio Frequency Identification (RFID) technology, which is being used to tag products as they are shipped or moved from one location to another. Instead of recalling hundreds of thousands of cars because of a possible defective part, automotive companies could determine exactly which cars had the defective parts and only recall the 10,000 cars with the bad parts using RFID. The savings would be huge.

1-800-flowers.com uses a data warehouse to reference customer historical data. DBMS software accessed over the corporate intranet gives marketing professionals the information they need to determine customer interests based on past interactions.[22]

Table 3.4

Comparison of OLTP and Data Warehousing

The primary advantage of data warehousing is the ability to relate data in innovative ways. However, a data warehouse can be extremely difficult to establish, with the typical cost exceeding $2 million. Table 3.4 compares online transaction processing (OLTP) and data warehousing.

Characteristic	OLTP Database	Data Warehousing
Purpose	Support transaction processing	Support decision making
Source of data	Business transactions	Multiple files, databases—data internal and external to the firm
Data access allowed users	Read and write	Read only
Primary data access mode	Simple database update and query	Simple and complex database queries with increasing use of data mining to recognize patterns in the data
Primary database model employed	Relational	Relational
Level of detail	Detailed transactions	Often summarized data
Availability of historical data	Very limited—typically a few weeks or months	Multiple years
Update process	Online, ongoing process as transactions are captured	Periodic process, once per week or once per month
Ease of process	Routine and easy	Complex, must combine data from many sources; data must go through a data cleanup process
Data integrity issues	Each transaction must be closely edited	Major effort to "clean" and integrate data from multiple sources

Data Marts

A **data mart** is a subset of a data warehouse. Data marts bring the data warehouse concept—online analysis of sales, inventory, and other vital business data that has been gathered from transaction processing systems—to small and medium-sized businesses and to departments within larger companies. Rather than store all enterprise data in one monolithic database, data marts contain a subset of the data for a single aspect of a company's business—for example, finance, inventory, or personnel. In fact, a specific area in the data mart might contain more detailed data than the data warehouse would provide.

Data marts are most useful for smaller groups who want to access detailed data. A warehouse contains summary data that can be used by an entire company. Because data marts typically contain tens of gigabytes of data, as opposed to the hundreds of gigabytes in data warehouses, they can be deployed on less powerful hardware with smaller secondary storage devices, delivering significant savings to an organization. Although any database software can be used to set up a data mart, some vendors deliver specialized software designed and priced specifically for data marts. Already, companies such as Sybase, Software AG, Microsoft, and others have announced products and services that make it easier and cheaper to deploy these scaled-down data warehouses. The selling point: Data marts put targeted business information into the hands of more decision makers. For example, the Defense Acquisition University (DAU), which is responsible for continuing education and career management for employees of the U.S. Department of Defense, uses data marts to provide administrators, instructors, and staff with domain-specific information.[23] A data warehouse is used to combine information from more than 50 disconnected sources, and the DBMS then organizes the information into area-specific data marts, which produce reports accessible through an online dashboard application. The system is estimated to save DAU personnel three to five years of labor.

data mart
A subset of a data warehouse.

Data Mining

Data mining is an information-analysis tool that involves the automated discovery of patterns and relationships in a data warehouse. Like gold mining, data mining sifts through mountains of data to find a few nuggets of valuable information. The University of Maryland has developed a data-mining technique to "forecast terrorist behavior based on past actions."[24] The system uses a real-time data extraction tool called T-REX to scour an average of 128,000 articles a day and forecast future activities of over 110 terrorist groups.

data mining
An information-analysis tool that involves the automated discovery of patterns and relationships in a data warehouse.

MySpace.com mines the data of all of its members to determine which ads should be displayed for each member to attract the maximum attention and hits.[25]

Data mining's objective is to extract patterns, trends, and rules from data warehouses to evaluate (i.e., predict or score) proposed business strategies, which will improve competitiveness, increase profits, and transform business processes. It is used extensively in marketing to improve customer retention; cross-selling opportunities; campaign management; market, channel, and pricing analysis; and customer segmentation analysis (especially one-to-one marketing). In short, data-mining tools help users find answers to questions they haven't thought to ask.

E-commerce presents another major opportunity for effective use of data mining. Attracting customers to Web sites is tough; keeping them can be next to impossible. For example, when retail Web sites launch deep-discount sales, they cannot easily determine how many first-time customers are likely to come back and buy again. Nor do they have a way of understanding which customers acquired during the sale are price sensitive and more likely to jump on future sales. As a result, companies are gathering data on user traffic through their Web sites and storing the data in databases. This data is then analyzed using data-mining techniques to personalize the Web site and develop sales promotions targeted at specific customers.

predictive analysis
A form of data mining that combines historical data with assumptions about future conditions to predict outcomes of events, such as future product sales or the probability that a customer will default on a loan.

Predictive analysis is a form of data mining that combines historical data with assumptions about future conditions to predict outcomes of events, such as future product sales or the probability that a customer will default on a loan. Retailers use predictive analysis to upgrade occasional customers into frequent purchasers by predicting what products they will buy if offered an appropriate incentive. Genalytics, Magnify, NCR Teradata, SAS Institute, Sightward, SPSS, and Quadstone have developed predictive analysis tools. Predictive analysis software can be used to analyze a company's customer list and a year's worth of sales data to find new market segments that could be profitable.

The City of Richmond Police Department uses predictive analysis to predict "when and where crimes were likely to occur, so officers can be on hand to prevent their occurrence."[26]

(Source: Courtesy of Mitch Kezar.)

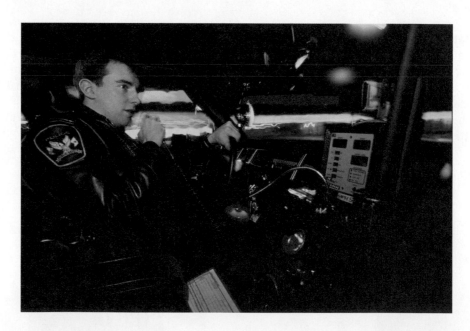

Traditional DBMS vendors are well aware of the great potential of data mining. Thus, companies such as Oracle, Sybase, Tandem, and Red Brick Systems are all incorporating data-mining functionality into their products. Table 3.5 summarizes a few of the most frequent applications for data mining.

Business Intelligence

business intelligence
The process of gathering enough of the right information in a timely manner and usable form and analyzing it to have a positive impact on business strategy, tactics, or operations.

The use of databases for business-intelligence purposes is closely linked to the concept of data mining. **Business intelligence (BI)** involves gathering enough of the right information in a timely manner and usable form and analyzing it so that it can have a positive effect on business strategy, tactics, or operations. IMS Health, for example, provides a BI system designed to assist businesses in the pharmaceutical industry with custom marketing to

Application	Description
Branding and positioning of products and services	Enable the strategist to visualize the different positions of competitors in a given market using performance (or other) data on dozens of key features of the product and then to condense all that data into a perceptual map of only two or three dimensions.
Customer churn	Predict current customers who are likely to switch to a competitor.
Direct marketing	Identify prospects most likely to respond to a direct marketing campaign (such as a direct mailing).
Fraud detection	Highlight transactions most likely to be deceptive or illegal.
Market basket analysis	Identify products and services that are most commonly purchased at the same time (e.g., nail polish and lipstick).
Market segmentation	Group customers based on who they are or on what they prefer.
Trend analysis	Analyze how key variables (e.g., sales, spending, promotions) vary over time.

physicians, pharmacists, nurses, consumers, government agencies, and nonprofit healthcare organizations.[27] Business intelligence turns data into useful information that is then distributed throughout an enterprise. It provides insight into the causes of problems, and when implemented can improve business operations and sometimes even save lives. For example, BI software at the Sahlgrenska University Hospital in Gothenburg, Sweden, has helped neurosurgeons save lives by identifying complications in patient conditions after cranial surgery.[28] The Information Systems at Work box shows how business intelligence is used in the utilities industry.

Competitive intelligence is one aspect of business intelligence and is limited to information about competitors and the ways that knowledge affects strategy, tactics, and operations. Competitive intelligence is a critical part of a company's ability to see and respond quickly and appropriately to the changing marketplace. Competitive intelligence is not espionage—the use of illegal means to gather information. In fact, almost all the information a competitive-intelligence professional needs can be collected by examining published information sources, conducting interviews, and using other legal, ethical methods. Using a variety of analytical tools, a skilled competitive-intelligence professional can by deduction fill the gaps in information already gathered.

The term **counterintelligence** describes the steps an organization takes to protect information sought by "hostile" intelligence gatherers. One of the most effective counterintelligence measures is to define "trade secret" information relevant to the company and control its dissemination.

Table 3.5

Common Data-Mining Applications

competitive intelligence
One aspect of business intelligence limited to information about competitors and the ways that knowledge affects strategy, tactics, and operations.

counterintelligence
The steps an organization takes to protect information sought by "hostile" intelligence gatherers.

Yangtze Power Harnesses the Power

Perhaps you've heard of the Yangtze River in China, and the enormous Three Gorges Dam being erected to harness the river's force for hydroelectric power. Due to be completed in 2011, the Three Gorges Dam will generate 22,500 megawatts of electricity, more than any other hydroelectric facility in the world. The company that will operate the dam is Yangtze Power, China's largest publicly listed utility company.

For years, Yangtze Power has managed the Gezhouba Power Station and six commissioned generating units. It has maintained business data in five databases, supporting its five divisions: Power Generation Management, Finance, Human Resources, Contract Management, and Safety and Control Management. Keeping data in siloed systems—separate, unconnected systems—limited information transfer through the enterprise. If a manager from Human Resources wanted to evaluate data from Contract Management, he would have to e-mail someone in that department to have a report generated and transferred. As Yangtze Power looked ahead to growth and the addition of the world's largest hydroelectric power generator, the company knew that its information would need to flow more freely through the enterprise in order for it to make the best business decisions in a timely fashion.

After evaluating products from Business Objects, Cognos, Informatica, MicroStrategy, and Oracle, Yangtze Power decided to go with Oracle to design one centralized database for all of its information because it was the only company that could provide one integrated system.

In March 2007, Yangtze Power's technology team worked with Oracle to develop a needs analysis and begin data preparation. Requirements were defined to cover six major areas of the business, including 65 performance indices and 370 reports. Through extensive preparation and testing, the system was up and running by November 2007.

Oracle's business intelligence tools allow senior managers to analyze performance on a daily basis, highlight areas for improvement, and monitor the results of business strategies. Each morning, reports on the previous day's critical activities are waiting on managers' desks. The new database stores three years of data, so that managers can draw on historical data when analyzing business performance. Communication between departments has improved, since everyone accesses the same data from a central system, and reports can easily be generated tailored to meet any business need.

Oracle's BI tools are used to create customized reports and charts including pie charts, broken curve diagrams, histograms, and radar maps. Being able to visualize data and trends in data enables a deeper analysis of the organization's business performance.

Yangtze Power has gained control over its flow of information through the enterprise. Now it is working to gain similar control over the raging waters of the Yangtze River.

Discussion Questions

1. What was wrong with Yangtze Power's previous database system, and how was it affecting the business?
2. What solution did Yangtze opt for, and how did it improve business?

Critical Thinking Questions

1. How does a centralized database improve communications within an organization?
2. In what situations might one centralized database not be practical for an enterprise?

Sources: Oracle Success Stories, "Yangtze Power Improves Business Intelligence with Integrated Database and Analysis Tools," 2008, *www.oracle.com/customers/snapshots/yangtze-power-case-study.pdf*, Yangtze River Web site, *www.yangtzeriver.org*, accessed April 2, 2008, Oracle Database and BI Tools, *www.oracle.com/database*, accessed April 2, 2008.

Distributed Databases

Distributed processing involves placing processing units at different locations and linking them via telecommunications equipment. A **distributed database**—a database in which the data can be spread across several smaller databases connected through telecommunications devices—works on much the same principle. A user in the Milwaukee branch of a clothing manufacturer, for example, might make a request for data that is physically located at corporate headquarters in Milan, Italy. The user does not have to know where the data is physically stored (see Figure 3.18).

distributed database

A database in which the data can be spread across several smaller databases connected via telecommunications devices.

Figure 3.18

The Use of a Distributed Database

For a clothing manufacturer, computers might be located at corporate headquarters, in the research and development center, in the warehouse, and in a company-owned retail store. Telecommunications systems link the computers so that users at all locations can access the same distributed database no matter where the data is actually stored.

Warehouse

Research and development

STORE

Retail outlet

User request

Distributed databases give corporations and other organizations more flexibility in how databases are organized and used. Local offices can create, manage, and use their own databases, and people at other offices can access and share the data in the local databases. Giving local sites more direct access to frequently used data can improve organizational effectiveness and efficiency significantly. The New York City Police Department, for example, has thousands of officers searching for information located on servers in offices around the city.

Despite its advantages, distributed processing creates additional challenges in integrating different databases (information integration), maintaining data security, accuracy, timeliness, and conformance to standards. Distributed databases allow more users direct access at different sites; thus, controlling who accesses and changes data is sometimes difficult. Also, because distributed databases rely on telecommunications lines to transport data, access to data can be slower.

replicated database
A database that holds a duplicate set of frequently used data.

To reduce telecommunications costs, some organizations build a replicated database. A **replicated database** holds a duplicate set of frequently used data. The company sends a copy of important data to each distributed processing location when needed or at predetermined times. Each site sends the changed data back to update the main database on an update cycle that meets the needs of the organization. This process, often called *data synchronization*, is used to make sure that replicated databases are accurate, up to date, and consistent with each other. A railroad, for example, can use a replicated database to increase punctuality, safety, and reliability. The primary database can hold data on fares, routings, and other essential information. The data can be continually replicated and downloaded on a read-only basis from the master database to hundreds of remote servers across the country. The remote locations can send back the latest figures on ticket sales and reservations to the main database.

Online Analytical Processing (OLAP)

**online analytical processing
(OLAP)**
Software that allows users to explore data from a number of perspectives.

For nearly two decades, multidimensional databases and their analytical information display systems have provided flashy sales presentations and trade show demonstrations. All you have to do is ask where a certain product is selling well, for example, and a colorful table showing sales performance by region, product type, and time frame appears on the screen. Called **online analytical processing (OLAP)**, these programs are now being used to store and deliver data warehouse information efficiently. The leading OLAP software vendors include Microsoft, Cognos, SAP, Business Objects, MicroStrategy, Applix, Infor, and Oracle. Lufthansa Cargo depends on OLAP to deliver up-to-the-minute company statistics that help the company compete in the growing global air-freight market.[29] The market is growing by six percent annually, and competitors are emerging all around the world to get a piece of the action. Lufthansa Cargo uses OLAP to analyze its data to provide the fastest service to its customers and the lowest rates.

The value of data ultimately lies in the decisions it enables. Powerful information-analysis tools in areas such as OLAP and data mining, when incorporated into a data warehousing architecture, bring market conditions into sharper focus and help organizations deliver greater competitive value. OLAP provides top-down, query-driven data analysis; data mining provides bottom-up, discovery-driven analysis. OLAP requires repetitive testing of user-originated theories; data mining requires no assumptions and instead identifies facts and conclusions based on patterns discovered. OLAP, or multidimensional analysis, requires a great deal of human ingenuity and interaction with the database to find information in the database. A user of a data-mining tool does not need to figure out what questions to ask; instead, the approach is, "Here's the data, tell me what interesting patterns emerge." For example, a data-mining tool in a credit card company's customer database can construct a profile of fraudulent activity from historical information. Then, this profile can be applied to all incoming transaction data to identify and stop fraudulent behavior, which might otherwise go undetected. Table 3.6 compares the OLAP and data-mining approaches to data analysis.

Table 3.6

Comparison of OLAP and Data Mining

Characteristic	OLAP	Data Mining
Purpose	Supports data analysis and decision making	Supports data analysis and decision making
Type of analysis supported	Top-down, query-driven data analysis	Bottom-up, discovery-driven data analysis
Skills required of user	Must be very knowledgeable of the data and its business context	Must trust in data mining tools to uncover valid and worthwhile hypotheses

Object-Relational Database Management Systems

An **object-oriented database** uses the same overall approach of objected-oriented programming that was discussed in Chapter 4. With this approach, both the data and the processing instructions are stored in the database. For example, an object-oriented database could store monthly expenses and the instructions needed to compute a monthly budget from those expenses. A traditional DBMS might only store the monthly expenses. The King County Metro Transit system in the state of Washington uses an object-oriented database in a system supplied by German vendor Init to manage the routing and accounting of its bus line.[30] Object-oriented databases are useful when a database contains complex data that needs to be processed quickly and efficiently.

In an object-oriented database, a *method* is a procedure or action. A sales tax method, for example, could be the procedure to compute the appropriate sales tax for an order or sale—for example, multiplying the total amount of an order by five percent, if that is the local sales tax. A *message* is a request to execute or run a method. For example, a sales clerk could issue a message to the object-oriented database to compute sales tax for a new order. Many object-oriented databases have their own query language, called *object query language (OQL)*, which is similar to SQL, discussed previously.

An object-oriented database uses an **object-oriented database management system** **(OODBMS)** to provide a user interface and connections to other programs. Computer vendors who sell or lease OODBMSs include Versant and Objectivity. Many organizations are selecting object-oriented databases for their processing power. Versant's OODBMS, for example, is being used by companies in the telecommunications, defense, online gaming, and healthcare industries, and by government agencies. The *Object Data Standard* is a design standard created by the *Object Database Management Group (www.odmg.org)* for developing object-oriented database systems.

An **object-relational database management system (ORDBMS)** provides a complete set of relational database capabilities plus the ability for third parties to add new data types and operations to the database. These new data types can be audio, images, unstructured text, spatial, or time series data that require new indexing, optimization, and retrieval features. Each of the vendors offering ORDBMS facilities provides a set of application programming interfaces to allow users to attach external data definitions and methods associated with those definitions to the database system. They are essentially offering a standard socket into which users can plug special instructions. DataBlades, Cartridges, and Extenders are the names applied by Oracle and IBM to describe the plug-ins to their respective products. Other plug-ins serve as interfaces to Web servers.

Visual, Audio, and Other Database Systems

In addition to raw data, organizations are increasingly finding a need to store large amounts of visual and audio signals in an organized fashion. Credit card companies, for example, enter pictures of charge slips into an image database using a scanner. The images can be stored in the database and later sorted by customer name, printed, and sent to customers along with their monthly statements. Image databases are also used by physicians to store x-rays and transmit them to clinics away from the main hospital. Financial services, insurance companies, and government branches are using image databases to store vital records and replace paper documents. Drug companies often need to analyze many visual images from laboratories. Chesapeake Energy maintains a database filled with scanned images of terrain and drilling locations.[31] Visual databases can be stored in some object-relational databases or special-purpose database systems. Many relational databases can also store graphic content.

Combining and analyzing data from different databases is an increasingly important challenge. Global businesses, for example, sometimes need to analyze sales and accounting data stored around the world in different database systems. Companies such as IBM are developing *virtual database systems* to allow different databases to work together as a unified database system. Banc of America Securities Prime Brokerage, for example, turned to database virtualization to address management and performance problems. Since its implementation,

object-oriented database
A database that stores both data and its processing instructions.

object-oriented database management system (OODBMS)
A group of programs that manipulate an object-oriented database and provide a user interface and connections to other application programs.

object-relational database management system (ORDBMS)
A DBMS capable of manipulating audio, video, and graphical data.

the virtual database system has reduced storage administration by 95 percent and decreased the need for more storage capacity by 50 percent.[32]

In addition to visual, audio, and virtual databases, other special-purpose database systems meet particular business needs. *Spatial data technology* involves using a database to store and access data according to the locations it describes and to permit spatial queries and analysis. MapInfo software from Pitney Bowes allows businesses such as Home Depot, Sonic Restaurants, CVS Corporation, and Chico's to choose the optimal location for new stores and restaurants based on geospatial demographics.[33] The software provides information about local competition, populations, and traffic patterns to predict how a business will fare in a particular location. Builders and insurance companies use spatial data to make decisions related to natural hazards. Spatial data can even be used to improve financial risk management with information stored by investment type, currency type, interest rates, and time.

Spatial data technology is used by NASA to store data from satellites and Earth stations. Location-specific information can be accessed and compared.

(Source: Courtesy of NASA.)

SUMMARY

Principle

Data management and modeling are key aspects of organizing data and information.

Data is one of the most valuable resources that a firm possesses. It is organized into a hierarchy that builds from the smallest element to the largest. The smallest element is the bit, a binary digit. A byte (a character such as a letter or numeric digit) is made up of eight bits. A group of characters, such as a name or number, is called a field (an object). A collection of related fields is a record; a collection of related records is called a file. The database, at the top of the hierarchy, is an integrated collection of records and files.

An entity is a generalized class of objects for which data is collected, stored, and maintained. An attribute is a characteristic of an entity. Specific values of attributes—called data items—can be found in the fields of the record describing an entity. A data key is a field within a record that is used to identify the record. A primary key uniquely identifies a record, while a secondary key is a field in a record that does not uniquely identify the record.

Traditional file-oriented applications are often characterized by program-data dependence, meaning that they have data organized in a manner that cannot be read by other programs. To address problems of traditional file-based data management, the database approach was developed. Benefits of this approach include reduced data redundancy, improved data consistency and integrity, easier modification and updating, data and program independence, standardization of data access, and more-efficient program development.

One of the tools that database designers use to show the relationships among data is a data model. A data model is a map or diagram of entities and their relationships. Enterprise data modeling involves analyzing the data and information needs of an entire organization. Entity-relationship (ER) diagrams can be employed to show the relationships between entities in the organization.

The relational model places data in two-dimensional tables. Tables can be linked by common data elements, which are used to access data when the database is queried. Each row represents a record. Columns of the tables are called attributes, and allowable values for these attributes are called the domain. Basic data manipulations include selecting, projecting, and joining. The relational model is easier to control, more flexible, and more intuitive than the other models because it organizes data in tables.

Principle

A well-designed and well-managed database is an extremely valuable tool in supporting decision making.

A DBMS is a group of programs used as an interface between a database and its users and other application programs. When an application program requests data from the database, it follows a logical access path. The actual retrieval of the data follows a physical access path. Records can be considered in the same way: A logical record is what the record contains; a physical record is where the record is stored on storage devices. Schemas are used to describe the entire database, its record types, and their relationships to the DBMS.

A DBMS provides four basic functions: providing user views, creating and modifying the database, storing and retrieving data, and manipulating data and generating reports. Schemas are entered into the computer via a data definition language, which describes the data and relationships in a specific database. Another tool used in database management is the data dictionary, which contains detailed descriptions of all data in the database.

After a DBMS has been installed, the database can be accessed, modified, and queried via a data manipulation language. A more specialized data manipulation language is the query language, the most common being Structured Query Language (SQL). SQL is used in several popular database packages today and can be installed on PCs and mainframes.

Popular single-user DBMSs include Corel Paradox and Microsoft Access. IBM, Oracle, and Microsoft are the leading DBMS vendors. Database as a Service (DaaS), or Database 2.0, is a new form of database service in which clients lease use of a database on a service provider's site.

Selecting a DBMS begins by analyzing the information needs of the organization. Important characteristics of databases include the size of the database, the number of concurrent users, its performance, the ability of the DBMS to be integrated with other systems, the features of the DBMS, the vendor considerations, and the cost of the database management system.

Principle

The number and types of database applications will continue to evolve and yield real business benefits.

Traditional online transaction processing (OLTP) systems put data into databases very quickly, reliably, and efficiently, but they do not support the types of data analysis that today's businesses and organizations require. To address this need, organizations are building data warehouses, which are relational database management systems specifically designed to support management decision making. Data marts are subdivisions of data warehouses, which are commonly devoted to specific purposes or functional business areas.

Data mining, which is the automated discovery of patterns and relationships in a data warehouse, is emerging as a practical approach to generating hypotheses about the patterns and anomalies in the data that can be used to predict future behavior.

Predictive analysis is a form of data mining that combines historical data with assumptions about future conditions to forecast outcomes of events such as future product sales or the probability that a customer will default on a loan.

Business intelligence is the process of getting enough of the right information in a timely manner and usable form and analyzing it so that it can have a positive effect on business strategy, tactics, or operations. Competitive intelligence is one aspect of business intelligence limited to information about competitors and the ways that information affects strategy, tactics, and operations. Competitive intelligence is not espionage—the use of illegal means to gather information. Counterintelligence describes the steps an organization takes to protect information sought by "hostile" intelligence gatherers.

With the increased use of telecommunications and networks, distributed databases, which allow multiple users and different sites access to data that may be stored in different physical locations, are gaining in popularity. To reduce telecommunications costs, some organizations build replicated databases, which hold a duplicate set of frequently used data.

Multidimensional databases and online analytical processing (OLAP) programs are being used to store data and allow users to explore the data from a number of different perspectives.

An object-oriented database uses the same overall approach of objected-oriented programming, first discussed in Chapter 4. With this approach, both the data and the processing instructions are stored in the database. An object-relational database management system (ORDBMS) provides a complete set of relational database capabilities, plus the ability for third parties to add new data types and operations to the database. These new data types can be audio, video, and graphical data that require new indexing, optimization, and retrieval features.

In addition to raw data, organizations are increasingly finding a need to store large amounts of visual and audio signals in an organized fashion. A number of special-purpose database systems are also being used.

CHAPTER 3: SELF-ASSESSMENT TEST

Data management and modeling are key aspects of organizing data and information.

1. A group of programs that manipulate the database and provide an interface between the database and the user of the database and other application programs is called a(n) _____.
   a. GUI
   b. operating system
   c. DBMS
   d. productivity software

2. A(n) _____ is a skilled and trained IS professional who directs all activities related to an organization's database.

3. Data redundancy is a desirable quality in a database. True or False?

4. A(n) _____ is a field or set of fields that uniquely identifies a database record.
   a. attribute
   b. data item
   c. key
   d. primary key

5. A(n) _____ uses basic graphical symbols to show the organization of and relationships between data.

6. What database model places data in two-dimensional tables?
   a. relational
   b. network
   c. normalized
   d. hierarchical

A well-designed and well-managed database is an extremely valuable tool in supporting decision making.

7. _____ involves combining two or more database tables.

8. After data has been placed into a relational database, users can make inquiries and analyze data. Basic data manipulations include selecting, projecting, and optimizing. True or False?

9. Because the DBMS is responsible for providing access to a database, one of the first steps in installing and using a database involves telling the DBMS the logical and physical structure of the data and relationships among the data in the database. This description of an entire database is called a(n) _____.

10. The commands used to access and report information from the database are part of the _____.
    a. data definition language
    b. data manipulation language
    c. data normalization language
    d. schema

11. Access is a popular DBMS for _____.
    a. personal computers
    b. graphics workstations
    c. mainframe computers
    d. supercomputers

12. A new trend in database management, known as Database as a Service, places the responsibility of storing and managing a database on a service provider. True or False?

The number and types of database applications will continue to evolve and yield real business benefits.

13. A(n) _____ holds business information from many sources in the enterprise, covering all aspects of the company's processes, products, and customers.

14. An information-analysis tool that involves the automated discovery of patterns and relationships in a data warehouse is called _____.
    a. a data mart
    b. data mining
    c. predictive analysis
    d. business intelligence

15. _____ allows users to predict the future based on database information from the past and present.

CHAPTER 3: SELF-ASSESSMENT TEST ANSWERS

(1) c (2) database administrator (3) False (4) d (5) entity-relationship diagram (6) a (7) Joining (8) False (9) schema (10) b (11) a (12) True (13) data warehouse (14) b (15) Predictive analysis

REVIEW QUESTIONS

1. What is an attribute? How is it related to an entity?
2. Define the term *database*. How is it different from a database management system?
3. What is the hierarchy of data in a database?
4. What is a flat file?
5. What is the purpose of a primary key? How can it be useful in controlling data redundancy?
6. What is the purpose of data cleanup?
7. What are the advantages of the database approach?
8. What is data modeling? What is its purpose? Briefly describe three commonly used data models.
9. What is a database schema, and what is its purpose?
10. How can a data dictionary be useful to database administrators and DBMS software engineers?
11. Identify important characteristics in selecting a database management system.
12. What is the difference between a data definition language (DDL) and a data manipulation language (DML)?
13. What is the difference between projecting and joining?
14. What is a distributed database system?
15. What is a data warehouse, and how is it different from a traditional database used to support OLTP?
16. What is meant by the "front end" and the "back end" of a DBMS?
17. What is data mining? What is OLAP? How are they different?
18. What is an ORDBMS? What kind of data can it handle?
19. What is business intelligence? How is it used?
20. In what circumstances might a database administrator consider using an object-oriented database?

DISCUSSION QUESTIONS

1. You have been selected to represent the student body on a project to develop a new student database for your school. What actions might you take to fulfill this responsibility to ensure that the project meets the needs of students and is successful?

2. Your company wants to increase revenues from its existing customers. How can data mining be used to accomplish this objective?
3. You are going to design a database for your cooking club to track its recipes. Identify the database characteristics

most important to you in choosing a DBMS. Which of the database management systems described in this chapter would you choose? Why? Is it important for you to know what sort of computer the database will run on? Why or why not?

4. Make a list of the databases in which data about you exists. How is the data in each database captured? Who updates each database and how often? Is it possible for you to request a printout of the contents of your data record from each database? What data privacy concerns do you have?

5. If you were the database administrator for the iTunes store, how might you use predictive analysis to determine which artists and movies will sell most next year?

6. You are the vice president of information technology for a large, multinational consumer packaged goods company (such as Procter & Gamble or Unilever). You must make a presentation to persuade the board of directors to invest $5 million to establish a competitive-intelligence organization—including people, data-gathering services, and software tools. What key points do you need to make in favor of this investment? What arguments can you anticipate that others might make?

7. Briefly describe how visual and audio databases can be used by companies today.

8. Identity theft, where people steal your personal information, continues to be a threat. Assume that you are the database administrator for a corporation with a large database. What steps would you implement to help prevent people from stealing personal information from the corporate database?

9. What roles do databases play in your most favorite online activities and Web sites?

PROBLEM-SOLVING EXERCISES

1. Develop a simple data model for the music you have on your MP3 player or in your CD collection, where each row is a song. For each row, what attributes should you capture? What will be the unique key for the records in your database? Describe how you might use the database.

2. A video movie rental store is using a relational database to store information on movie rentals to answer customer questions. Each entry in the database contains the following items: Movie ID No. (primary key), Movie Title, Year Made, Movie Type, MPAA Rating, Number of Copies on Hand, and Quantity Owned. Movie types are comedy, family, drama, horror, science fiction, and western. MPAA ratings are G, PG, PG-13, R, NC-17 and NR (not rated). Use a single-user database management system to build a data-entry screen to enter this data. Build a small database with at least ten entries.

3. To improve service to their customers, the salespeople at the video rental store have proposed a list of changes being considered for the database in the previous exercise. From this list, choose two database modifications and modify the data-entry screen to capture and store this new information. Proposed changes:

   a. Add the date that the movie was first available to help locate the newest releases.
   b. Add the director's name.
   c. Add the names of three primary actors in the movie.
   d. Add a rating of one, two, three, or four stars.
   e. Add the number of Academy Award nominations.

4. Your school maintains information about students in several interconnected database files. The student_contact file contains student contact information. The student_grades file contains student grade records, and the student_financial file contains financial records including tuition and student loans. Draw a diagram of the fields these three files might contain, which field is a primary key in each file, and which fields serve to relate one file to another. Use Figure 3.7 as a guide.

TEAM ACTIVITIES

1. In a group of three or four classmates, communicate with the person at your school that supervises information systems. Find out how many databases are used by your school and for what purpose. Also find out what policies and procedures are in place to protect the data stored from identity thieves and other threats.

2. As a team of three or four classmates, interview business managers from three different businesses that use databases to help them in their work. What data entities and data attributes are contained in each database? How do they access the database to perform analysis? Have they received training in any query or reporting tools? What do they like about their database and what could be improved? Do any of them use data-mining or OLAP techniques? Weighing the information obtained, select one of these databases as

being most strategic for the firm and briefly present your selection and the rationale for the selection to the class.

3. Imagine that you and your classmates are a research team developing an improved process for evaluating auto loan applicants. The goal of the research is to predict which applicants will become delinquent or forfeit their loan. Those who score well on the application will be accepted; those who score exceptionally well will be considered for lower-rate loans. Prepare a brief report for your instructor addressing these questions:

   a. What data do you need for each loan applicant?
   b. What data might you need that is not typically requested on a loan application form?

   c. Where might you get this data?
   d. Take a first cut at designing a database for this application. Using the chapter material on designing a database, show the logical structure of the relational tables for this proposed database. In your design, include the data attributes you believe are necessary for this database, and show the primary keys in your tables. Keep the size of the fields and tables as small as possible to minimize required disk drive storage space. Fill in the database tables with the sample data for demonstration purposes (ten records). After your design is complete, implement it using a relational DBMS.

WEB EXERCISES

1. Use a Web search engine to find information on specific products for one of the following topics: business intelligence, object-oriented databases, or database as a service. Write a brief report describing what you found, including a description of the database products and the companies that developed them.

2. List your five favorite Web sites. Consider the services that they provide. For each site, suggest how one or more databases might be used on the back end to supply information to visitors.

CAREER EXERCISES

1. What type of data is stored by businesses in a professional field that interests you? How many databases might be used to store that data? How would the data be organized within each database?

2. How could you use business intelligence (BI) to do a better job at work? Give some specific examples of how BI can give you a competitive advantage.

CASE STUDIES

Case One
The Getty Vocabularies

J. Paul Getty was an American industrialist who made his fortune in the oil business. He made his first million at age 25 in 1916, and later became the world's first billionaire. Getty viewed art as a "civilizing influence in society, and strongly believed in making art available to the public for its education and enjoyment." To that end, he created an art museum in Los Angeles, California, and established the J. Paul Getty Trust, commonly referred to as the Getty.

The Getty includes four branches: the Getty Museum, a research institute, a conservation institute, and a foundation.

In the 1980s, the Getty discovered a need within the art research community. Researchers lacked a common vocabulary with which to discuss art and artists' work. Establishing a scientific vocabulary with which to describe artwork, style, and technique would allow the study and appreciation of artwork to flourish. To meet this need, the Getty created and published the *Art and Architecture Thesaurus* (AAT) in 1990. The three-volume tome, which includes a thesaurus of geographic names and the Union List of Artist Names, has become a priceless resource for art historical research. It provides tools, standards, and best practices for documenting works of art, just as the Library of Congress provides a standard cataloging tool for libraries.

However, the massive AAT is difficult to search and is expensive to edit and update. Recognizing that a digital version of the resource would provide many benefits, the Getty recently began porting the AAT and associated volumes into a database that can be electronically searched and edited over the Web. To do so, the Getty had to first select a database technology in which to house the information, and a DBMS for use in searching and editing the contents.

One challenge of building an online AAT was that the various components of the resource were stored using different proprietary technologies. The first task was to collect them into one common technology, which required a custom-designed system. Technicians within the Getty opted to use Oracle databases and a product called PowerBuilder from Sybase, Inc., for the user interface. Custom coding was done in Perl and SQR programming languages to merge the components into a cohesive system. The result is a system called the Vocabulary Coordination System (VCS). The VCS is used to collect analyze, edit, merge, and distribute the terminology managed by the Getty vocabularies. A special Web-based interface was developed that made searching the volumes easy enough for anyone to manage. You can try it yourself at *www.getty.edu/research/conducting_research/vocabularies*.

The resulting system was so impressive that it won the Getty the Computerworld Honors Award in Media, Arts & Entertainment for innovative use of technology. The system makes it easy for scholars to update information in the vocabularies, and for everyone from school children to professional art historians to research and learn about art and art history. The Getty online vocabularies are an ideal realization of J. Paul Getty's original philosophy of promoting human civility through cultural awareness, creativity, and aesthetic enjoyment.

Discussion Questions

1. What purpose do the Getty vocabularies serve, and how are they supported through database technology?
2. How does using the Web as a front end to this database further support J. Paul Getty's vision?

Critical Thinking Questions

1. What concerns do you think the designers of the database had when making this valuable resource available online to the general public?
2. Why did the database designers need to use custom-designed code to collect the original data?

Sources: Pratt, Mary K., "The Getty makes art accessible with online database." *Computerworld*, March 10, 2008, *www.computerworld.com/ action/article.do?command=viewArticleBasic& taxonomyName=Databases& articleId=310236&taxonomyId=173&intsrc=kc_li_story*; Staff, "The Computerworld Honors Program: Web-Based Global Art Resources: The Getty Vocabularies," *Computerworld*, 2007, *www.cwhonors.org/viewCaseStudy.asp?NominationID=112*; The Getty Web site, *www.getty.edu*, accessed April 1, 2008.

Case Two

ETAI Manages Auto Parts Overload with Open Source Database

If you need a hard-to-find automobile part for a European import, you could probably find it in a catalog published by the ETAI Group in France. The ETAI catalog includes over 30 million parts for over 50,000 European car models manufactured during the past 15 years. The catalog is updated 100 times each year to stay current with the latest models.

While maintaining an average auto parts catalog might not seem a daunting task, this one is an exception. ETAI collects auto parts information from nine databases provided by parts manufacturers. Each database uses a unique design with different formats for parts numbers and varying amounts and types of fields for each part record. Over many years, ETAI had developed a system for collating the data using a variety of programming languages and platforms. The entire process required 15 steps and two to three weeks. It was so complicated that if ETAI's database administrator were to leave, his replacement would have a difficult time learning how the complicated system worked.

Philippe Bobo, the director of software and information systems at ETAI, knew it was time to improve the system. He and his team tested products from a variety of vendors over a five-week period, and eventually decided to work with Talend Open Data Solutions, based in Los Altos, California. Talend specializes in open-source database management systems that integrate data from various types of systems into a single target system—exactly what ETAI needed.

Talend designed a system for ETAI using a single standard programming language that queries the nine auto parts databases and streams the results into one data warehouse. It then cleans the data and standardizes it for output to a catalog format. The 15-step, three-week process is now reduced to one step and two days.

Philippe likes the open-source nature of Talend's solution because it makes it possible for his own software engineers to work with and adjust the software over time to accommodate new needs in the system. Updating the DBMS has reduced labor costs and production time, and made it possible for ETAI to expand into other types of catalogues and service manuals.

Discussion Questions

1. What challenges did ETAI face that made creating their catalog a three-week-long ordeal?
2. How did the solution provided by Talend reduce the job time by 90 percent?

Critical Thinking Questions

1. What benefits were provided by the open-source solution?
2. Why couldn't ETAI standardize the data formats in the nine databases?

Sources: Weiss, Todd R., "ETAI avoids data traffic jam with open source," *Computerworld*, December 17, 2007, *www.computerworld.com/action/ article.do?command=viewArticleBasic&articleId=90531 61&intsrc=news_list*; Weiss, Todd R., "ETAI Rides Open Source to Ease Data Traffic Jam," *Computerworld*, December 31, 2007, *www.computerworld.com/action/article.do? command=viewArticleBasic&articleId=30982 1*; Talend Open Data Solutions Web site, *www.talend.com*, accessed March 31, 2008; ETAI Web site, *www.etai.fr/g_instit/atout.htm*, accessed March 31, 2008.

Questions for Web Case

See the Web site for this book to read about the Whitmann Price Consulting case for this chapter. Following are questions concerning this Web case.

Whitmann Price Consulting: Database Systems and Business Intelligence

Discussion Questions

1. How will Whitmann Price consultants and the company itself benefit from their ability to call up corporate information in an instant anywhere and at any time?
2. Why will the database itself not require a change to support the new advanced mobile communications and information system?

Critical Thinking Questions

1. The Web has acted as a convenient standard for accessing all types of information from various types of computing platforms. How will this benefit the systems developers of Whitmann Price in developing forms and reports for the new mobile system?
2. What are the suggested limitations of using a BlackBerry device for accessing and interacting with corporate data?

NOTES

Sources for opening vignette: Havenstein, Heather, "Wal-Mart CTO details HP data warehouse move," *ITWorld Canada*, August 3, 2007, *www.itworldcanada.com/a/Enterprise-Business-Applications/ efb96e0a-18de-47e6-ac61-ddab5cc55b5b.html*; Wal-Mart Corporate Fact Sheet, accessed March 30, 2008; Hayes Weier, Mary, "Wal-Mart Speaks Out On HP Neoview Decision," *Information Week*, August 3, 2007, *www.informationweek.com/management/showArticle.jhtml? articleID=201203010, www.walmartstores.com/media/factsheets/fs_ 2230.pdf*; HP Neoview Enterprise Data Warehouse Web site, *http:// h20331.www2.hp.com/enterprise/cache/414444-0-0-225-121.html*, accessed March 30, 3008.

1 Wailgum, Thomas, "Hollywood agency updates systems to woo talent," *Computerworld*, January 19, 2007, *www.computerworld.com/ action/article.do?command=viewArticleBasic&taxonomyName= Business_Intelligence&articleId=9008545&taxonomyId=9&intsrc= kc_li_story.*
2 Vijayan, Jaikumar, "Harvard grad students hit in computer intrusion," *Computerworld*, March 13, 2008, *www.computerworld.com/action/ article.do?command=viewArticleBasic&articleId=9068221&source= rss_news10.*
3 Staff, "Bacs database fault leaves 400,000 without pay," *Computerworld UK*, March 30, 2007, *www.itworldcanada.com/a/Information- Architecture/94cb5e9e-79c5-4f29-a909-c7025be0d0b4.html.*
4 Mearian, Lucas, "Study: Digital universe and its impact bigger than we thought," *Computerworld*, March 11, 2008, *www.computerworld.com/action/article.do?command= viewArticleBasic&articleId=9067639&source=rss_news10.*
5 Koman, Richard, "Exploding Digital Data Growth Is a Challenge for IT," *Top Tech News*, March 11, 2008, *www.toptechnews.com/ story.xhtml?story_id=58752.*
6 Nakashima, Ellen, "FBI Prepares Vast Database of Biometrics," *Washington Post*, December 22, 2007, *www.washingtonpost.com/ wp-dyn/content/article/2007/12/21/AR2007122102544_pf.html.*
7 Kolbasuk McGee, Marianne, "Wal-Mart Requires In-Store Clinics To Use E-Health Records System," *Information Week*, February 16, 2008, *www.informationweek.com/story/showArticle.jhtml? articleID=206504257&cid=RSSfeed_IWK_All.*

8 Pratt, Mary, "Steven Barlow: Master of Data Warehousing," *Computerworld*, July 9, 2007, *www.computerworld.com/action/article.do? command=viewArticleBasic&taxonomyName=data_warehousing&a rticleId=297032&taxonomyId=55&intsrc=kc_feat.*
9 Mullins, Craig, "The Database Report - July 2007," *TDNA*, July 10, 2007, *www.tdan.com/view-featured-columns/5603.*
10 Fonseca, Brian, "Baseball Hall of Fame on deck for archive format change?" *Computerworld*, July 24, 2007, *www.computerworld.com/ action/article.do?command=viewArticleBasic&taxonomyName= servers_and_data_center&articleId=9027849&taxonomyId= 154&intsrc=kc_top.*
11 Microsoft Staff, "Microsoft Transforms Management Training into an Interactive, On-the-Job Experience," Microsoft Case Studies, August 30, 2007, *www.microsoft.com/casestudies/casestudy.aspx?cases- tudyid=4000000613.*
12 Oracle Staff, "Lighting Manufacturer Surya Roshni Streamlines Supply Chain with Bright Results," Oracle Customer Snapshot, 2008, *www.oracle.com/customers/snapshots/surya-roshni-snapshot.pdf.*
13 Staff, "INTELLIFIT Moves from Virtual Fitting (match-to-order) to True Mass Customization: Custom-made jeans with a high-tech twist," *Mass Customization & Open Innovation News*, February 15, 2008, *http://mass-customization.blogs.com/mass_customization_ open_i/2008/02/intellifit-move.html.*
14 IT Redux Web site, *http://itredux.com/office-20/database/? family=Database*, accessed March 23, 2008.
15 Lai, Eric, "Cloud database vendors: What, us worry about Microsoft?" *Computerworld*, March 12, 2008, *www.computerworld.com/action/ article.do?command=viewArticleBasic&articleId= 9067979&pageNumber=1.*
16 Intuit Staff, "Jet Blue – The Challenge," Intuit Case Studies, *http://quickbase.intuit.com/customers/jetblue_v4.pdf*, accessed April 10, 2008.
17 Arellano, Nestor, "Information created in 2007 will exceed storage capacity says IDC," *ITWorld Canada*, March 23, 2008, *www.itworldcanada.com/a/Information-Architecture/844a38f1- ccaf-445f-867f-498f041587c1.html.*
18 Gittlen, Sandra, "Data center land grab: How to get ready for the rush," *Computerworld*, March 12, 2007, *www.computerworld.com/*

action/article.do?command=viewArticleBasic&articleId=9012963.

19 AP Staff, "Google ventures into health records biz," CNN.com, February 21, 2008, *www.cnn.com/2008/TECH/02/21/ google.records.ap.*

20 Dixon, Guy, "Yahoo embraces semantic Web standards," *vnunet.com*, March 18, 2008, *www.vnunet.com/vnunet/news/ 2212249/yahoo-embraces-semantic-web.*

21 McMillan, Robert, "Researcher: Half a million database servers have no firewall," *Computerworld*, November 14, 2007, *www.computerworld.com/action/article.do?command= viewArticleBasic&articleId=9046821&source=rss_news10.*

22 SAS Staff, "1-800-FLOWERS.COM Gathers a Bouquet of CRM Capabilities," SAS Success Stories, *www.sas.com/success/ 1800flowers_IT.html*, accessed March 23, 2008.

23 Computerworld Staff, "The Computerworld Honors Program: Defense Acquisition University," *Computerworld*, 2007, *www.cwhonors.org/viewCaseStudy.asp?NominationID=313.*

24 Vijayan, Jaikumar, "Univ. of Md. launches data mining portal for counter-terrorism research," *Computerworld*, February 26, 2008, *www.computerworld.com/action/article.do? command=viewArticleBasic&articleId=9064938.*

25 Stone, Brad, "MySpace to Discuss Effort to Customize Ads," *New York Times*, September 18, 2007, *www.nytimes.com/2007/09/18/technology/18myspace.html?_r=1&oref=slogin.*

26 Computerworld Staff, "The Computerworld Honors Program: City of Richmond Police Department," *Computerworld*, 2007,

www.cwhonors.org/viewCaseStudy.asp?NominationID=303.

27 Nickum, Chris, "The BI Prescription," *OptimizeMag.com*, April 2007, Page 45.

28 Computerworld Staff, "The Computerworld Honors Program: QlikTech International," *Computerworld*, 2007, *www.cwhonors.org/ viewCaseStudy.asp?NominationID=155.*

29 Cognos Staff, "Lufthansa Cargo," Lufthansa Cargo Case Study, *www.cognos.com/pdfs/success_stories/ss_lufthansa-cargo.pdf*, accessed April 10, 2008.

30 Staff, "init wins contract worth 25 mill. US Dollar from Seattle," Init News Release, March 13, 2007, *www.init-ka.de/en_news/ PR_AH_2007/AH_070313_Seattle_en.php.*

31 Babcock, Charles, "Oracle Touts Virtualization, Apps – and 11g Of Course," *Information Week*, November 19, 2007, page 30.

32 Fonseca, Brian, "Virtualization Cutting Storage Costs for Some Large Firms," *Computerworld*, April 23, 2007, *www.computerworld.com/ action/article.do?command=viewArticleBasic&taxonomyId= 154&articleId=290230&intsrc=hm_topic.*

33 Staff, "MapInfo Customer Testimonials," MapInfo Web site, *www.mapinfo.com/location/integration? txtTopNav=836a2545d8a37f00dev- vcm100001a031dc7____&txtLeftNav=6e5819cd57f47f00dev- vcm100001a031dc7____&txtExtNav=391495ac2a771110Vgn- VCM10000021021dc7____*, accessed March 23, 2008.

CHAPTER
· 4 ·

Telecommunications, the Internet, Intranets, and Extranets

PRINCIPLES	LEARNING OBJECTIVES
▪ A telecommunications system has many fundamental components that must be carefully selected and work together effectively to enable people to meet personal and organization objectives.	▪ Identify and describe the fundamental components of a telecommunications system including media and hardware devices. ▪ Identify several network types and describe the uses and limitations of each. ▪ Name three basic processing alternatives for organizations that require two or more computer systems and discuss their fundamental features.
▪ The Internet and the Web provide a wide range of services, some of which are effective and practical for use today, others are still evolving, and still others will fade away from lack of use.	▪ Briefly describe how the Internet works, including alternatives for connecting to it and the role of Internet service providers. ▪ Describe how the World Wide Web works and the use of Web browsers, search engines, and other Web tools. ▪ Identify and briefly describe several applications associated with the Internet and the Web.
▪ Because use of the Internet and World Wide Web is becoming universal in the business environment, management, service and speed, privacy, and security issues must continually be addressed and resolved.	▪ Outline a process and identify tools used to create Web content. ▪ Define the terms *intranet* and *extranet* and discuss how organizations are using them. ▪ Identify several issues associated with the use of networks.

Information Systems in the Global Economy
Lamborghini, Italy
Web Portal and Online Collaboration Shrink Distances

The Internet has enabled thousands of businesses to extend their reach beyond borders to become global competitors. For a global business to succeed, all of its offices and personnel spread around the world must stay in sync. This is typically accomplished by being in close communications with headquarters. The Internet and Web allow employees separated by thousands of miles to operate as though they were seated across the table from each other. Automobile manufacturer Lamborghini, with headquarters in Sant'Agata Bolognese, Italy, recently discovered the benefits of employing the latest technologies to shrink the distances between its more than 100 dealerships scattered around the world.

The famous manufacturer of elite sports and racing cars of the highest quality had anything but high quality communications between dealerships and headquarters. Until recently, Lamborghini dealers relied on telephone, e-mail, and snail mail to order new cars and spare parts, and to learn about the latest marketing programs and business procedures. The multiple forms of inefficient communications were difficult to manage at headquarters and allowed important tasks to fall through the cracks. Executives at the company realized that their problems could be remedied by a more efficient use of Internet and Web technologies.

Lamborghini decided to create a Web portal, a custom designed Web page, that would provide dealerships with direct access to corporate databases and information systems over the Internet, resources that were previously only accessible by employees at headquarters. Secured by password authentication, and encryption, dealers could log in to the portal to check inventory, place orders for cars and parts, read daily announcements, and access procedural instructions. The new portal dramatically cuts down on paperwork and manual processing of orders at headquarters. Also, errors have been reduced by eliminating the two-stage system that used to require headquarters staff to interpret and type in order data into the system.

Lamborghini executives were so impressed by the savings and improvements offered by the new Web portal that they looked for ways to further enhance the system. Observing the benefits of online social networks and Web 2.0 technologies, Lamborghini looked for ways to implement these technologies into its own Web portal system. Social features were added that allowed dealers and personnel to post announcements and requests to be fielded by others on the network. For example, if a dealership needs a part for a vintage Lamborghini that is no longer manufactured, this system could get results, by finding someone within the network that had the part. A chat utility and file sharing features were also added to allow dealers around the world to collaborate on promotions and sales tactics.

Ultimately, Lamborghini created an intranet—a secure private network accessible over the Internet—so that dealers could have expanded access to the information systems and data stored on servers at headquarters. Using this system, the Lamborghini dealer in Orange County, California can access the same information, systems, databases, and services as the VP of sales in Sant'Agata Bolognese, Italy. The dealer could also develop relationships with the other hundred dealers around the world for more effective business practices.

Finally, Lamborghini developed a state of the art Web site that looks more like a top rate motion picture, or video game, than an automobile Web site. The site utilizes the latest technologies to thoroughly impress visitors and reflect the high quality of the corporation and its products.

As you read this chapter, consider the following:

- What role does telecommunications play in connecting businesses and growing the global economy?
- In what ways are the Internet and Web used by individuals to improve our quality of life and by businesses to improve the bottom line?

Why Learn About Telecommunications and Networks?

Today's decision makers need to access data wherever it resides. They must be able to establish fast, reliable connections to exchange messages, upload and download data and software, route business transactions to processors, connect to databases and network services, and send output to printers. Regardless of your chosen major or future career field, you will need the communications capabilities provided by telecommunications and networks including the Internet, especially if your work involves the supply chain. Among all business functions, supply chain management might use telecommunications and networks the most because it requires cooperation and communications among workers in inbound logistics, warehouse and storage, production, finished product storage, outbound logistics, and most importantly, with customers, suppliers, and shippers. Many supply chain organizations make use of the Web to purchase raw materials, parts, and supplies at competitive prices. All members of the supply chain must work together effectively to increase the value perceived by the customer, so partners must communicate well. Other employees in human resources, finance, research and development, marketing, and sales positions must also use communications technology to communicate with people inside and outside the organization. To be a successful member of any organization, you must be able to take advantage of the capabilities that these technologies offer you. This chapter begins by discussing the importance of effective communications.

In today's high-speed global business world, organizations need always-on, always-connected computing for traveling employees and for network connections to their key business partners and customers. As we saw in the opening vignette, a forward-thinking company such as Lamborghini is able to improve communications among its many dealers, leading to cost reductions and increases in sales. Here are additional examples of organizations using telecommunications and networks to move ahead.

- Wal-Mart, the world's largest retailer with $345 billion in sales, plans to include RFID tags on products in its 4,068 North American stores to improve inventory accuracy, thereby reducing lost sales and all but eliminating lost or missing merchandise. The net result is a savings of $287 million per year. Telecommunications between the RFID chips and scanners on forklift trucks and between the trucks and in-store computers is an essential component of achieving these savings.[1]
- Procter & Gamble (P&G) has implemented 13 of its planned 40 Video Collaboration Studios using the Cisco TelePresence system to empower the P&G community of more than 138,000 employees working in over 80 countries worldwide. The Studios foster a great degree of communication and collaboration without requiring members of a team to physically travel to meet. Use of this telecommunications technology is credited with helping P&G to bring its products to market faster and compete more effectively. Aflac, BT, McKesson, SAP, Verizon, and more than 100 other Cisco customers are also experimenting with the use of the technology.[2]
- Thousands of companies are employing Webcasts to inform and educate potential customers about their products and services.
- High-technology companies such as Boeing use a wide range of telecommunications technologies to support their business and collaborate with people from inside and outside the company. Boeing has created the LabNet to connect the various Boeing Labs and customers that test concepts and features under development. The LabNet enables all participants to visualize live, simulated, and computer-generated fighter jets as they demonstrate their performance under various test scenarios.[3]

AN OVERVIEW OF TELECOMMUNICATIONS

Telecommunications refers to the electronic transmission of signals for communications, by such means as telephone, radio, and television. Telecommunications is creating profound changes in business because it lessens the barriers of time and distance. Telecommunications not only is changing the way businesses operate, but the nature of commerce itself. As networks are connected with one another and transmit information more freely, a competitive marketplace demands excellent quality and service from all organizations.

Figure 4.1 shows a general model of telecommunications. The model starts with a sending unit (1), such as a person, a computer system, a terminal, or another device, that originates the message. The sending unit transmits a signal (2) to a modem (3) that can perform many tasks, which can include converting the signal into a different form or from one type to another. The modem then sends the signal through a medium (4). A **telecommunications medium** is any material substance that carries an electronic signal to support communications between a sending and receiving device. Another modem (5) connected to the receiving device (6) receives the signal. The process can be reversed, and the receiving unit (6) can send another message to the original sending unit (1). An important characteristic of telecommunications is the speed at which information is transmitted, which is measured in bits per second (bps). Common speeds are in the range of thousands of bits per second (Kbps) to millions of bits per second (Mbps) and even billions of bits per second (Gbps).

telecommunications medium
Anything that carries an electronic signal and serves as an interface between a sending device and a receiving device.

Figure 4.1

Elements of a Telecommunications System

Telecommunications devices relay signals between computer systems and transmission media.

Advances in telecommunications technology allow us to communicate rapidly with clients and coworkers almost anywhere in the world. Telecommunications also reduces the amount of time needed to transmit information that can drive and conclude business actions.

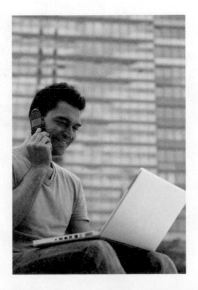

Telecommunications technology enables businesspeople to communicate with coworkers and clients from remote locations.

(Source: © BananaStock / Alamy.)

Channel Bandwidth

Telecommunications professionals consider the capacity of the communications path or channel when they recommend transmission media for a business. **Channel bandwidth** refers to the rate at which data is exchanged, usually measured in bits per second (bps)—the broader the bandwidth, the more information can be exchanged at one time. **Broadband communications** is a relative term but generally means a telecommunications system that can exchange data very quickly. For example, for wireless networks, broadband lets you send data at a rate greater than 1.5 Mbps. In general, today's organizations need more bandwidth for increased transmission speed to carry out their daily functions.

Communications Media

In designing a telecommunications system, the transmission media selected depends on the amount of information to be exchanged, the speed at which data must be exchanged, the level of concern for data privacy, whether or not the users are stationary or mobile, and many other business requirements. Transmission media can be divided into two broad categories: *guided transmission media*, in which communications signals are guided along a solid medium, and *wireless*, in which the communications signal is broadcast over airwaves as a form of electromagnetic radiation.

Guided Transmission Media Types

There are many different guided transmission media types. Table 4.1 summarizes the guided media types by physical media type. Common guided transmission media types are shown in Figure 4.2.

channel bandwidth
The rate at which data is exchanged over a communication channel, usually measured in bits per second (bps).

broadband communications
A telecommunications system in which a very high rate of data exchange is possible.

Table 4.1

Guided Transmission Media Types

Guided Media Types			
Media Type	**Description**	**Advantages**	**Disadvantages**
Twisted-pair wire	Twisted pairs of copper wire, shielded or unshielded	Used for telephone service; widely available	Transmission speed and distance limitations
Coaxial cable	Inner conductor wire surrounded by insulation	Cleaner and faster data transmission than twisted-pair wire	More expensive than twisted-pair wire
Fiber-optic cable	Many extremely thin strands of glass bound together in a sheathing; uses light beams to transmit signals	Diameter of cable is much smaller than coaxial; less distortion of signal; capable of high transmission rates	Expensive to purchase and install
Broadband over Power Lines	Data is transmitted over standard high-voltage power lines	Can provide Internet service to rural areas where cable and phone service may be non-existent	Can be expensive and may interfere with ham radios and police and fire communications

Figure 4.2

Types of Guided Transmission Media

Twisted-pair wire (left), coaxial cable (middle), fiber-optic cable (right)

(Source: © Greg Pease/Getty Images.)

Many utilities, cities, and organizations are experimenting with *broadband over power lines (BPL)* to provide network connections over standard high-voltage power lines. The Transportation Security Administration is testing the use of BPL at selected airports to connect airport passenger and other screening systems, cameras at ticket counters, and passport readers.[4] To access the Internet, BPL users connect their computer to a special hardware device that plugs into any electrical wall socket. A potential issue with BPL is that it can interfere with both ham radio broadcasts and police and fire radios. However, BPL can provide Internet service in rural areas where broadband access has lagged because electricity is more prevalent in homes than cable or even telephone lines.

Wireless Technologies

Wireless communications coupled with the Internet is revolutionizing how and where we gather and share information, collaborate in teams, listen to music or watch video, and stay in touch with our families and co-workers while on the road. With wireless capability, a coffee shop can become our living room or the bleachers at a ball park can become our office. The many advantages and freedom provided by wireless communications are causing many organizations to consider moving to an all wireless environment. Shanista, a transportation service serving Los Angeles, California, made the decision to move to wireless after tiring of the hassles of moving wired devices.[5]

Wireless telecommunications involves the broadcast of communications in one of three frequency ranges: microwave, radio, and infrared as shown in Table 4.2.

Technology	Description	Advantages	Disadvantages
Radio frequency range	Operates in the 3KHz–300 MHz range	Supports mobile users; costs are dropping	Signal highly susceptible to interception
Microwave—terrestrial and satellite frequency range	High-frequency radio signal (300 MHz–300 GHz) sent through atmosphere and space (often involves communications satellites)	Avoids cost and effort to lay cable or wires; capable of high-speed transmission	Must have unobstructed line of sight between sender and receiver; signal highly susceptible to interception
Infrared frequency range	Signals in the 300 GHz to 400 THz frequency range sent through air as light waves	Lets you move, remove, and install devices without expensive wiring	Must have unobstructed line of sight between sender and receiver; transmission effective only for short distances

Table 4.2

Frequency Ranges Used for Wireless Communications

Some of the more widely used wireless communications options are discussed next.

Near Field Communication (NFC) is a very short-range wireless connectivity technology designed for cell phones and credit cards. With NFC, consumers can swipe their credit cards or even cell phones within a few inches of point-of-sale terminals to pay for purchases. Consumers are using the technology in Germany and Austria, and pilot projects are being conducted in London, Singapore, the Netherlands, and Finland. In the United States, MasterCard and Visa are testing devices with embedded NFC and are looking for business partners to help evaluate the technology.[6]

Bluetooth is a wireless communications specification that describes how cell phones, computers, personal digital assistants, printers, and other electronic devices can be interconnected over distances of 10–30 feet at a rate of about 2 Mbps. Bluetooth enables users of multifunctional devices to synchronize with information in a desktop computer, send or receive faxes, print, and, in general, coordinate all mobile and fixed computer devices. The Bluetooth technology is named after the tenth century Danish King Harald Blatand, or Harold Bluetooth in English. He had been instrumental in uniting warring factions in parts of what is now Norway, Sweden, and Denmark—just as the technology named after him is designed to allow collaboration between differing devices such as computers, phones, and other electronic devices.

Near Field Communication (NFC)

A very short-range wireless connectivity technology designed for cell phones and credit cards.

Bluetooth

A wireless communications specification that describes how cell phones, computers, faxes, personal digital assistants, printers, and other electronic devices can be interconnected over distances of 10–30 feet at a rate of about 2 Mbps.

ultra wideband (UWB)
A wireless communications technology that transmits large amounts of digital data over short distances of up to 30 feet using a wide spectrum of frequency bands and very low power.

Ultra wideband (UWB) is a wireless communications technology that transmits large amounts of digital data over short distances of up to 30 feet using a wide spectrum of frequency bands and very low power. Ultra wideband has the potential to replace Bluetooth's 2 Mbps transmission speed with 400 Mbps rates for wirelessly connecting printers and other devices to desktop computers or enabling completely wireless home multimedia networks.[7] The manufacturers of electronic entertainment devices are particularly interested in the use of UWB. With UWB, a digital camcorder could play a just-recorded video on an HDTV without anyone having to fiddle with wires. A portable MP3 player could stream audio to high-quality surround-sound speakers anywhere in the room. A mobile computer user could wirelessly connect to a digital projector in a conference room to deliver a presentation.

Wi-Fi
A wireless telecommunication technology brand owned by the Wi-Fi Alliance that enables wireless devices to connect to a wireless access pint which has a wired connection to the Internet.

Wi-Fi is a wireless telecommunications technology brand owned by the Wi-Fi Alliance, which consists of about 300 technology companies including AT&T, Dell, Microsoft, Nokia, and Qualcomm. The alliance exists to improve the interoperability of wireless local area network products.

With a Wi-Fi wireless network, the user's computer, smartphone, or personal digital assistant has a wireless adapter that translates data into a radio signal and transmits it using an antenna. A wireless access point, which consists of a transmitter with an antenna, receives the signal and decodes it. The access point then sends the information to the Internet over a wired connection (see Figure 4.3). When receiving data, the wireless access point takes the information from the Internet, translates it into a radio signal, and sends it to the device's wireless adapter. These devices typically come with built-in wireless transmitters and software to enable them to alert the user to the existence of a Wi-Fi network. The area covered by one or more interconnected wireless access points is called a "hot spot." Current Wi-Fi access points have a maximum range of about 300 feet outdoors and 100 feet within a dry-walled building. Wi-Fi has proven so popular that hot spots are popping up in places such as airports, coffee shops, college campuses, libraries, and restaurants. Many cities have implemented Wi-Fi based networks for their citizens and government workers.

Figure 4.3

Wi-Fi Network

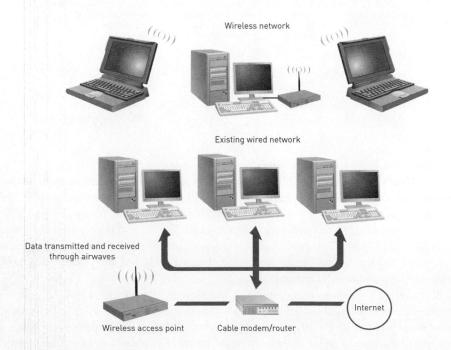

wireless mesh
A way to route communications between network nodes (computers or other devices) by allowing for continuous connections and reconfiguration around blocked paths by "hopping" from node to node until a connection can be established.

Wireless mesh uses multiple Wi-Fi access points to link a series of interconnected local area networks to form a wide area network capable of serving a large campus or entire city. Communications are routed among network nodes by allowing for continuous connections and reconfiguration around blocked paths by "hopping" from node to node until a connection can be established. Mesh networks are very robust: If one node fails, all the other nodes can still communicate with each other, directly or through one or more intermediate nodes.

The city of Tempe, Arizona implemented a mesh network to provide broadband wireless access for Tempe residents, visitors, students, and mobile workers on their laptop, PDA, or smartphone. The network has not been widely accepted—it has 1,000 outdoor access points but only 500 subscribers.[8]

3G wireless communication was developed by the International Telecommunications Union (ITU) who set out to establish a single standard for cellular networks in 1999. The goal was to standardize future digital wireless communications and allow global roaming with a single handset. Referred to as 3G, this standard provides for faster transmission speeds in the range of 2–4 Mbps. Originally, 3G was supposed to be a single, unified, worldwide standard, but the 3G standards effort split into several different standards. One standard is the Universal Mobile Telephone System (UMTS), which is the preferred solution for European countries that use Global System for Mobile (GSM) communications. GSM is the *de facto* wireless telephone standard in Europe with more than 120 million users worldwide in 120 countries. Another 3G-based standard is Code-Division Multiple Access (CDMA), which is used in Australia, Canada, China, India, Israel, Mexico, South Korea, the United States, and Venezuela. The wide variety of 3G cellular communications standards can support many business applications. The challenge is to enable these standards to intercommunicate and support fast, reliable, global wireless communications.

3G wireless communication is useful for business travelers, people on the go, and people who need to get or stay connected. Although Wi-Fi is an option, 3G is preferable to mobile users concerned about the availability, cost, and security associated with the use of public Wi-Fi networks.

4G wireless communication stands for fourth-generation broadband mobile wireless, which is expected to deliver more advanced versions of enhanced multimedia, smooth streaming video, universal access, portability across all types of devices, and eventually, worldwide roaming capability. 4G will also provide increased data transmission rates in the 20–40 Mbps range.

Pine Cellular, Pine Telephone, and Choctaw Electric are deploying Nortel 4G technology to provide homes and businesses in southeastern Oklahoma with reliable, wireless high-speed Internet. The network will provide low-cost, broad coverage and deliver wireless services to rural areas where construction of a wired network is less economical. The 4G services will be provided at no charge to the local police and fire departments and public schools.[9]

Worldwide Interoperability for Microwave Access (WiMAX) is the common name for a set of IEEE 802.16 wireless metropolitan-area network standards that support various types of communications access. In many respects, WiMAX operates like Wi-Fi, only over greater distances (up to 30 miles of separation) and at faster transmission speeds (up to 70 Mbps). Thus fewer WiMAX base stations are required to cover the same geographical area than when Wi-Fi technology is used. Mobile WiMAX provides both fixed and mobile access over the same network infrastructure. Fixed WiMAX is designed to deliver communications to homes and offices but cannot support mobile users. WiMAX is considered a 4G service.

In mid-2008, Sprint Nextel combined its wireless broadband unit with Clearwire to create a new communications company whose goal is to build the first national WiMAX network bringing coverage to 120 million people by the end of 2010. To supply the necessary phones, computer chips, and other equipment, Sprint is working with Intel, Motorola, Nokia, and Samsung to provide WiMAX-capable PC cards, gaming devices, laptops, cameras, and even phones.[10]

Telecommunications Hardware

Telecommunications hardware devices include modems, multiplexers, PBXs and front-end processors that enable electronic communications to occur or occur more efficiently. Switches, bridges, routers and gateways are devices for sending packets of data through one or more networks. Table 4.3 summarizes these common telecommunications devices.

Worldwide Interoperability for Microwave Access (WiMAX)
The common name for a set of IEEE 802.16 wireless metropolitan-area network standards that support different types of communications access.

Device	Function
Modem	Translates data from a digital form (as it is stored in the computer) into an analog signal that can be transmitted over ordinary telephone lines.
Fax modem	Facsimile devices, commonly called fax devices, allow businesses to transmit text, graphs, photographs, and other digital files via standard telephone lines. A fax modem is a very popular device that combines a fax with a modem, giving users a powerful communications tool.
Multiplexer	Allows several telecommunications signals to be transmitted over a single communications medium at the same time, thus saving expensive long-distance communications costs.
PBX	A communications system that manages both voice and data transfer within a building and to outside lines. In a PBX system, switching PBXs can be used to connect hundreds of internal phone lines to a few phone company lines.
Front-end processor	Special-purpose computer that manages communications to and from a computer system serving many people.
Switch	Uses the physical device address in each incoming message on the network to determine which output port it should forward the message to reach another device on the same network
Bridge	Connects one LAN to another LAN that uses the same telecommunications protocol.
Router	Forwards data packets across two or more distinct networks toward their destinations through a process known as routing. Often an Internet service provider (ISP) installs a router in a subscriber's home that connects the ISP's network to the network within the home
Gateway	A network device that serves as an entrance to another network.

Services

Telecommunications carriers organize communications channels, networks, hardware, software, people, and business procedures to provide valuable communications services.

Digital Subscriber Line (DSL)

digital subscriber line (DSL)
A telecommunications service that delivers high-speed Internet access to homes and small businesses over the existing phone lines of the local telephone network.

A **digital subscriber line** (DSL) is a telecommunications service that delivers high-speed Internet access to homes and small businesses over the existing phone lines of the local telephone network (see Figure 4.4). Most home and small business users are connected to an *asymmetric DSL (ADSL)* line designed to provide a connection speed from the Internet to the user (download speed) that is three to four times faster than the connection from the user back to the Internet (upload speed). ADSL does not require an additional phone line and yet provides "always-on" Internet access. A drawback of ADSL is that the farther the subscriber is from the local telephone office, the poorer the signal quality and the slower the transmission speed. ADSL provides a dedicated connection from each user to the phone company's local office, so the performance does not decrease as new users are added. Cable-modem users generally share a network loop that runs through a neighborhood so that adding users means lowering the actual transmission speeds.

Voice over Internet (VoIP) Services

Voice over Internet Protocol (VoIP)
A collection of technologies and communications protocols that enables your voice to be converted into packets of data that can be sent over a data network such as the Internet, a WAN or LAN.

Voice over Internet Protocol (VoIP) is a collection of technologies and communications protocols that enables voice conversations to be converted into packets of data that can be sent over a data network such as the Internet, a WAN, or LAN. You can use VoIP to make a call directly from a computer equipped with appropriate software and a microphone, a special VoIP phone, or an ordinary phone connected to an analog telephone adapter that converts the analog voice signal into data packets (see Figure 4.5). Table 4.4 lists the advantages and disadvantages of VoIP.

Telecommunications networks require state-of-the-art computer software technology to continuously monitor the flow of voice, data, and image transmission over billions of circuit miles worldwide.

(Source: © Roger Tully/Getty Images.)

Digital Subscriber Line (DSL)

At the local telephone company's central office, a DSL Access Multiplexer (DSLAM) takes connections from many customers and aggregates them onto a single, high-capacity connection to the Internet. Subscriber phone calls can be routed through a switch at the local telephone central office to the public telephone network.

VoIP Options

Table 4.4

Advantages and Disadvantages
of VoIP

Advantages	Disadvantages
A single data network can be used to carry both voice and data traffic, thus reducing total telecommunications expenses.	VoIP users may not be able to place calls if power is lost either locally or at the broadband carrier itself because broadband modems and other VoIP equipment depend on electricity from the power company.
VoIP enables online retailers to provide "click to talk" customer service—online customers needing additional help can click a hyperlink to create a VoIP phone connection to a live customer service representative.	Some VoIP calls might have lower quality than conventional phone calls if data packets are lost or delayed at any point in the network between VoIP users.
Some VoIP providers permit users to call anywhere in the world at a low cost per minute.	The support for the sending of faxes over VoIP is limited.
	Not all VoIP services connect directly to emergency services through 911.
	Not all VoIP providers offer directory assistance or white page listings.

Merrill Lynch, the global financial management and advisory company, implemented an advanced VoIP voice trading system. Over 4,000 IQ/MAX next-generation trading desktops were installed on its two largest trading floors in New York and London plus other trading floors around the globe.[11]

NETWORKS AND DISTRIBUTED PROCESSING

computer network
The communications media, devices, and software needed to connect two or more computer systems and/or devices.

A **computer network** consists of communications media, devices, and software needed to connect two or more computer systems or devices. The computers and devices on the networks are also called *network nodes*. After they are connected, the nodes can share data, information, work processes and allow employees to collaborate on projects. If a company uses networks effectively, it can grow into an agile, powerful, and creative organization, giving it a long-term competitive advantage. Organizations can use networks to share hardware, programs, and databases. Networks can transmit and receive information to improve organizational effectiveness and efficiency. They enable geographically separated workgroups to share documents and opinions, which fosters teamwork, innovative ideas, and new business strategies.

Network Types

Depending on the physical distance between nodes on a network and the communications and services it provides, networks can be classified as personal area, local area, metropolitan area and wide area network.

Personal Area Networks

personal area network (PAN)
A network that supports the interconnection of information technology within a range of 33 feet or so.

A **personal area network (PAN)** is a wireless network that connects information technology devices within a range of 33 feet or so. One device serves as the controller during wireless PAN initialization, and this controller device mediates communication within the PAN. The controller broadcasts a beacon that synchronizes all devices and allocates time slots for the devices. With a PAN, you can connect a laptop, digital camera, and portable printer without physical cables. You could download digital image data from the camera to the laptop and then print it on a high-quality printer—all wirelessly. Bluetooth is the industry standard for PAN communications.

Local Area Networks

A network that connects computer systems and devices within a small area such as an office, home, or several floors in a building is a **local area network (LAN)**. Typically, LANs are wired into office buildings and factories (see Figure 4.6).

local area network (LAN)
A network that connects computer systems and devices within a small area like an office, home, or several floors in a building.

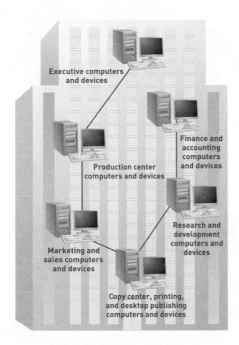

Figure 4.6

A Typical LAN

All LAN users within an office building can connect to each other's devices for rapid communication. For instance, a user in research and development could send a document from her computer to be printed at a printer located in the desktop publishing center.

Disneyland's House of the Future in Tomorrowland features information technology designed to enhance everyday living. A LAN that senses the presence of people throughout the home is the key to making this all work. The lights, temperature, and even the paintings on the wall adjust to preset personal preferences as people enter and leave the rooms. If someone clicks on the remote, the network dims the lights, shuts off any music, and draws the shades in preparation for the TV to turn on. The network also enables people to easily transfer music, photos, and videos among computers and TVs throughout the home.[12]

With more people working at home, connecting home computing devices and equipment into a unified network is on the rise. Small businesses are also connecting their systems and equipment. A home or small business can connect network, computers, printers, scanners, and other devices. A person working on one computer, for example, can use data and programs stored on another computer's hard disk. In addition, several computers on the network can share a single printer. To make home and small business networking a reality, many companies are offering standards, devices, and procedures.

Metropolitan Area Networks

A **metropolitan area network (MAN)** is a telecommunications network that connects users and their computers in a geographical area that spans a campus or city. Most MANs have a range of roughly 30 to 90 miles. For example, a MAN might redefine the many networks within a city into a single larger network or connect several LANs into a single campus LAN. The Miami-Dade Police Department consists of 3,000 officers and 1,500 civilians who serve and protect more than two million citizens over a 2,100 square mile area. The department implemented a MAN to enable its officers to gain easy access to the data they need while staying mobile on the streets rather than behind a desk.[13]

metropolitan area network (MAN)
A telecommunications network that connects users and their devices in a geographical area that spans a campus or city.

Wide Area Networks

A **wide area network (WAN)** is a telecommunications network that connects large geographic regions. A WAN might be privately owned or rented and includes public (shared users) networks. When you make a long-distance phone call or access the Internet, you are using a WAN. WANs usually consist of computer equipment owned by the user, together with data

wide area network (WAN)
A telecommunications network that ties together large geographic regions.

communications equipment and telecommunications links provided by various carriers and service providers (see Figure 4.7).

A Wide Area Network

WANs are the basic long-distance networks used around the world. The actual connections between sites, or nodes (shown by dashed lines), might be any combination of satellites, microwave, or cabling. When you make a long-distance telephone call or access the Internet, you are using a WAN.

North America

Jenny Craig, the weight management company, uses a WAN to capture and send daily the private information about its customers from some 500 locations in North America to its headquarters in Carlsbad, CA. Each location is in the process of setting up Wi-Fi access points and a DSL connection to the company's headquarters.[14]

WANs often provide communications across national borders, which involves national and international laws regulating the electronic flow of data across international boundaries, often called *transborder data flow*. Some countries have strict laws limiting the use of telecommunications and databases, making normal business transactions such as payroll costly, slow, or even impossible.

Read the Information Systems @ Work special feature to learn more about how one organization is using a WAN creatively to reduce travel-related times and costs and speed up the development of new products.

Telepresence Eliminates Travel and Saves Valuable Human Resources

Derek Chan, head of digital operations at DreamWorks Studios, is ecstatic. DreamWorks used to release an animated film every 18 to 36 months. "Now we're doing a show in May, another in November, and then May. We're reaching a scale no one else has been able to do. When we ended up building these systems, it changed the landscape for us," says Chan. The systems Chan is using are telepresence systems.

DreamWorks Animation SKG, creators of many popular films including "Shrek," "Madagascar," "Chicken Run," and "Bee Movie," is well known for its high-quality 3-D animation. Creating these movies takes the combined effort of many top animators located around the world. For DreamWorks, a great deal of collaboration takes place between its home studios in Glendale, California, and its subsidiary, Pacific Data Images, 400 miles north in Redwood City.

The long distance between sites was causing important DreamWorks executives, artists, and directors to waste time traveling rather than creating. DreamWorks tried a variety of network conferencing systems, but none provided a smooth stream of communication—in-person visits were much more productive. DreamWorks partnered with Hewlett-Packard (HP) to create a videoconferencing system that allows people around the world to communicate as though they were sitting around a conference table. The result is an HP product called HP Halo, a telepresence and videoconferencing system. The technology has proven successful and similar systems are being offered by other vendors under the general title of telepresence systems.

Telepresence participants sit at a long, one-sided conference table facing a wall covered with large displays. The room is equipped with unobtrusive video cameras and a high-resolution document camera. When connected to the telepresence studio at another location, the displays show the other meeting participants seated as if across the table. A large display above the participants shows documents that people want to share using the document camera or directly from a PC. Microphones and speakers allow participants to converse in a natural voice. People outside the telepresence environment can dial into the system to join in on audio.

Network connections are the most important component of the telepresence system. DreamWorks' HP Halo system provides a dedicated high-bandwidth network line between its Glendale studios and its Redwood City offices. Users describe the so-called tele-immersive environment as being stunningly lifelike. It's as though the participants are physically together.

For DreamWorks, this means a fundamental change in the way it does business. Teams can work together by sharing and discussing documents, images, and video, while cutting the time and cost of travel. Animators and producers use the system to collaborate from disperse geographic locations, developing storyboards, reviewing artwork, and adjusting character designs in real time. The document camera allows artists to sketch ideas to share with the group. The network is fast enough to transfer video clips from motion pictures while it is transferring live action video of participants.

Using HP Halo, DreamWorks became the first company to release two animated films in one year. Executives that previously traveled overseas once every three weeks now travel once every three months. Trips between Glendale and Redwood City have been reduced by as much as 80 percent.

Discussion Questions

1. What network considerations are involved when introducing a telepresence system? How might DreamWorks' requirements vary from a retail company such as Home Depot?
2. How did telepresence allow DreamWorks employees to be more productive and efficient?

Critical Thinking Questions

1. Although some workforces are becoming more mobile, others such as DreamWorks find it more effective to stay put. What types of business activities are best carried out through travel? What kinds of professionals benefit from avoiding travel? How do telecommunications assist both?
2. In your chosen career, do you anticipate a lot of travel or a little? What types of activities will you perform that require telecommunications?

Sources: King, Julia, "Premier 100 IT Leaders 2008," *Computerworld*, December 10, 2007, *www.computerworld.com/action/article.do? command=viewArticleBasic&articleId=305908&pageNumber=1*; HP Staff, "DreamWorks speeds film production with HP Halo Collaboration Studio," HP Case Study, *http://h20219.www2.hp.com/enterprise/downloads/Case% 20Study_DreamWorks%20hi-res_7_17_07.pdf*, accessed April 28, 2008; HP Halo Web site, *http://h20219.www2.hp.com/enterprise/cache/ 570006-0-0-0-121.html*, accessed April 28, 2008.

Distributed Processing

When an organization needs to use two or more computer systems, it can implement one of three basic processing alternatives: centralized, decentralized, or distributed. With **centralized processing**, all processing occurs in a single location or facility. This approach offers the highest degree of control because a single centrally managed computer performs all data processing. The Ticketmaster reservation service is an example of a centralized system. One central computer with a database stores information about all events and records the purchases of seats. Ticket clerks at various ticket selling locations can enter order data and print the results, or customers can place orders directly over the Internet.

With **decentralized processing**, processing devices are placed at various remote locations. Each processing device is isolated and does not communicate with any other processing device. Decentralized systems are suitable for companies that have independent operating units, such as 7-Eleven, where each of its 5,800 U.S. stores is managed to meet local retail conditions. Each store has a computer that runs over 50 business applications such as cash register operations, gasoline pump monitoring, and merchandising.

With **distributed processing**, computers are placed at remote locations but connected to each other via telecommunications devices. One benefit of distributed processing is that managers can allocate data to the locations that can process it most efficiently. For example, Kroger operates over 2,400 supermarkets, each with its own computer to support store operations such as customer checkout and inventory management. These computers are connected to a network so that sales data gathered by each store's computer can be sent to a huge data repository on a mainframe computer for efficient analysis by marketing analysts and product supply chain managers.

The September 11, 2001 terrorist attacks and the current relatively high level of natural disasters such as hurricane Katrina sparked many companies to distribute their workers, operations, and systems much more widely, a reversal of the recent trend toward centralization. The goal is to minimize the consequences of a catastrophic event at one location while ensuring uninterrupted systems availability.

Client/Server Systems

Users can share data through file server computing, which allows authorized users to download entire files from certain computers designated as file servers. After downloading data to a local computer, a user can analyze, manipulate, format, and display data from the file.

In **client/server** architecture, multiple computer platforms are dedicated to special functions such as database management, printing, communications, and program execution. These platforms are called *servers*. Each server is accessible by all computers on the network. Servers can be computers of all sizes; they store both application programs and data and are equipped with operating system software to manage the activities of the network. The server distributes programs and data to the other computers (clients) on the network as they request them. An application server holds the programs and data files for a particular application, such as an inventory database. The client or the server can do the processing.

Communications Protocols

A **communications protocol** is a set of rules that govern the exchange of information over a communications channel. The goal is to ensure fast, efficient, error-free communications over an imperfect communication channel. Protocols govern several levels of a telecommunications network. For example, some protocols determine data interchange at the hardware device level, and other protocols determine data interchange at the application program level.

---

centralized processing
Processing alternative in which processing occurs at a single location or facility.

decentralized processing
Processing alternative in which processing devices are placed at remote locations.

distributed processing
Processing alternative in which computers are placed at remote locations but are connected to each other via a network.

client/server
An architecture in which multiple computer platforms are dedicated to special functions such as database management, printing, communications, and program execution.

communications protocol
A set of rules that govern the exchange of information over a communications channel.

Communications Software

A **network operating system (NOS)** is systems software that controls the computer systems and devices on a network and allows them to communicate with each other. An NOS performs the same types of functions for the network as operating system software does for a computer, such as memory and task management and coordination of hardware. When network equipment (such as printers, plotters, and disk drives) is required, the NOS makes sure that these resources are used correctly. In most cases, companies that produce and sell networks provide the NOS. For example, NetWare is the NOS from Novell, a popular network environment for personal computer systems and equipment. Windows 2000, Windows 2003 and Windows 2008 are other common network operating systems. MySpace, the popular social networking Web site that offers an interactive, user-submitted network of friends, personal profiles, blogs, music and videos internationally, was one of the first very busy Web sites to adopt the use of Windows Server 2008.

Software tools and utilities are available for managing networks. With **network-management software**, a manager on a networked personal computer can monitor the use of individual computers and shared hardware (such as printers), scan for viruses, and ensure compliance with software licenses. Network-management software also simplifies the process of updating files and programs on computers on the network—a manager can make changes through a communications server instead of on individual computers. In addition, network-management software protects software from being copied, modified, or downloaded illegally and performs error control to locate telecommunications errors and potential network problems. Some of the many benefits of network-management software include fewer hours spent on routine tasks (such as installing new software), faster response to problems, and greater overall network control.

T-Mobile Austria GmbH is a subsidiary of T-Mobile International and serves about one-third of all mobile users in Austria. Its infrastructure is highly diverse and includes a mix of hardware from Alcatel, Cisco, Ericsson, Hewlett-Packard, and Siemens using the Microsoft Windows NT and 2000, Solaris, HP-UX, and Linux operating systems. This collection of systems, hardware, and applications requires constant monitoring to detect potential device failures or system bottlenecks before they can generate customer complaints or service failures. "Tivoli Netcool service monitors our Internet services, our mobile radio networks, and most importantly, provides round-the-clock management of our host and server devices. This ensures that important applications will never fail without being noticed," says Dr. Sabine Ringhofer, Senior Manager, Network Operations, T-Mobile Austria.[15]

Now that we have covered many of the basics of telecommunications, let's discuss the use of the Internet.

network operating system (NOS)
Systems software that controls the computer systems and devices on a network and allows them to communicate with each other.

network-management software
Software that enables a manager on a networked desktop to monitor the use of individual computers and shared hardware (such as printers), scan for viruses, and ensure compliance with software licenses.

USE AND FUNCTIONING OF THE INTERNET

The Internet is the world's largest computer network. Actually, the **Internet** is a collection of interconnected networks, all freely exchanging information. Research firms, colleges, and universities have long been part of the Internet, and businesses, high schools, elementary schools, and other organizations have joined it as well. Nobody knows exactly how big the Internet is because it is a collection of separately run, smaller computer networks. There is no single place where all the connections are registered. Figure 4.8 shows the staggering growth of the Internet, as measured by the number of Internet host sites or domain names. Domain names are discussed later in the chapter.

The Internet is truly international in scope, with users on every continent—including Antarctica. More than 215 million people in the United States (71.4 percent of the population) use the Internet. Although the United States has high Internet penetration among its population, it does not make up the majority of people online. Of all the people using the Internet, citizens of Asian countries make up 39 percent, Europeans 26 percent, and North America only 18 percent.[16] The Internet is expanding around the globe but at differing rates for each country. For example, most Internet usage in South Korea is through

Internet
A collection of interconnected networks, all freely exchanging information.

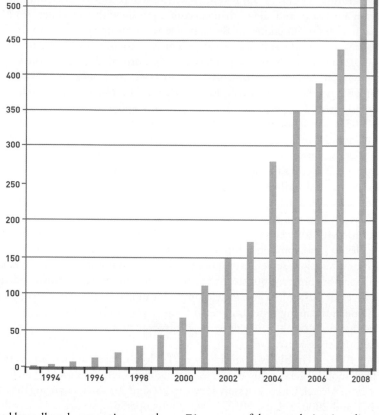

ARPANET

A project started by the U.S.
Department of Defense (DoD) in
1969 as both an experiment in
reliable networking and a means to
link DoD and military research
contractors, including many
universities doing military-funded
research.

Internet Protocol (IP)

A communication standard that
enables traffic to be routed from one
network to another as needed.

high-speed broadband connections, and over 71 percent of the population is online. In North Korea, however, Internet use and other civil liberties are restricted by the government. Several people and organizations are working to provide Internet access to developing countries.[17] More than 1.3 billion people use the Internet around the world, and if the rate of growth continues as shown in Figure 4.8, the number of users will surge to more than 2 billion in a few years.

The ancestor of the Internet was the **ARPANET**, a project started by the U.S. Department of Defense (DoD) in 1969. The ARPANET was both an experiment in reliable networking and a means to link DoD and military research contractors, including many universities doing military-funded research. (*ARPA* stands for the Advanced Research Projects Agency, the branch of the DoD in charge of awarding grant money. The agency is now known as DARPA—the added *D* is for *Defense*.) The ARPANET was highly successful, and every university in the country wanted to use it. This wildfire growth made it difficult to manage the ARPANET, particularly its large and rapidly growing number of university sites. So, the ARPANET was broken into two networks: MILNET, which included all military sites, and a new, smaller ARPANET, which included all the nonmilitary sites. The two networks remained connected, however, through use of the **Internet Protocol** (IP), which enables traffic to be routed from one network to another as needed. All the networks connected to the Internet use IP, so they all can exchange messages.

Today, people, universities, and companies are attempting to make the Internet faster and easier to use. To speed Internet access, a group of corporations and universities called the University Corporation for Advanced Internet Development (UCAID) is working on a faster, new Internet. Called Internet2 (I2), Next Generation Internet (NGI), or Abilene, depending on the universities or corporations involved, the new Internet offers the potential of faster Internet speeds, up to 2 Gbps per second or more.[18] An offshoot of Internet2 that some call Internet3, which is officially named the *National LambdaRail* (NLR), is a cross-country, high-speed (10 Gbps), fiber-optic network dedicated to research in high-speed networking applications.[19] The NLR provides a unique national networking infrastructure to advance networking research and next-generation network-based applications in science, engineering, and medicine. This new high-speed fiber-optic network will support the ever-increasing need of scientists to gather, transfer, and analyze massive amounts of scientific data.

How the Internet Works

The Internet transmits data from one computer (called a *host*) to another (see Figure 4.9). If the receiving computer is on a network to which the first computer is directly connected, it can send the message directly. If the receiving and sending computers are not directly connected to the same network, the sending computer relays the message to another computer that can forward it. The message might be sent through a router to reach the forwarding computer. The forwarding host, which presumably is attached to at least one other network, delivers the message directly if it can or passes it to another forwarding host. A message can pass through a dozen or more forwarders on its way from one part of the Internet to another.

Figure 4.9

Routing Messages over the Internet

The various networks that are linked to form the Internet work much the same way—they pass data around in chunks called *packets*, each of which carries the addresses of its sender and its receiver. Recall that the set of conventions used to pass packets from one host to another is IP. Many other protocols are used in connection with IP. The best known is the **Transmission Control Protocol (TCP)**. Many people use TCP/IP as an abbreviation for the combination of TCP and IP used by most Internet applications. After a network following these standards links to a **backbone**—one of the Internet's high-speed, long-distance communications links—it becomes part of the worldwide Internet community.

Each computer on the Internet has an assigned address called its **Uniform Resource Locator**, or **URL**, to identify it to other hosts. The URL gives those who provide information over the Internet a standard way to designate where Internet elements such as servers, documents, and newsgroups can be found. Consider the URL for Course Technology, *http://www.course.com*.

The "http" specifies the access method and tells your software to access a file using the Hypertext Transport Protocol. This is the primary method for interacting with the Internet. In many cases, you don't need to include http:// in a URL because it is the default protocol. Thus, *http://www.course.com* can be abbreviated as *www.course.com*.

The "www" part of the address signifies that the address is associated with the World Wide Web service, discussed later. The "course.com" part of the address is the domain name that identifies the Internet host site. Domain names must adhere to strict rules. They always have at least two parts, with each part separated by a dot (period). For some Internet addresses, the far right part of the domain name is the country code (such as au for Australia, ca for Canada, dk for Denmark, fr for France, and jp for Japan). Many Internet addresses have a

Transmission Control Protocol (TCP)

The widely used Transport-layer protocol that most Internet applications use with IP.

backbone

One of the Internet's high-speed, long-distance communications links.

Uniform Resource Locator (URL)

An assigned address on the Internet for each computer.

code denoting affiliation categories. (Table 4.5 contains a few popular categories.) The far left part of the domain name identifies the host network or host provider, which might be the name of a university or business.

Table 4.5

U.S. Top-Level Domain Affiliations

Affiliation ID	Affiliation
com	Business sites
edu	Educational sites
gov	Government sites
net	Networking sites
org	Nonprofit organization sites

Note that some other countries outside the United States use different top-level domain affiliations from the ones described in Table 4.5.

Millions of domain names have been registered. Some people, called *cyber-squatters*, have registered domain names in the hope of selling the names to corporations or people. The domain name Business.com, for example, sold for $7.5 million. In one case, a federal judge ordered the former owner of one Web site to pay the person who originally registered the domain name $40 million in compensatory damages and an additional $25 million in punitive damages. But some companies are fighting back, suing people who register domain names only to sell them to companies. Today, the Internet Corporation for Assigned Names and Numbers (ICANN) has the authority to resolve domain-name disputes.

Accessing the Internet

Although you can connect to the Internet in numerous ways (see Figure 4.10), Internet access is not distributed evenly throughout the world. Which access method you choose is determined by the size and capability of your organization or system.

Figure 4.10

Several Ways to Access the Internet

Users can access the Internet in several ways, including using a LAN server, dialing into a server using the telephone lines, using a high-speed service, or accessing the Internet over a wireless network.

Connect via LAN Server

This approach is used by businesses and organizations that manage a local area network (LAN). By connecting a server on the LAN to the Internet using a router, all users on the LAN are provided access to the Internet. Business LAN servers are typically connected to the Internet at very fast data rates, sometimes in the hundreds of Mbps. In addition, you can share the higher cost of this service among several dozen LAN users to allow a reasonable cost per user.

Connect via Dial-Up

Connecting to the Internet through a dial-up connection requires a modem that allows the computer to use standard phone lines. The modem then contacts a server managed by the Internet service provider (ISP). Dial-up connections use TCP/IP protocol software plus Serial Line Internet Protocol (SLIP) or Point-to-Point Protocol (PPP) software. SLIP and PPP are two communications protocols that transmit packets over telephone lines, allowing dial-up access to the Internet. After the connection is made, you are on the Internet and can access any of its resources. Dial-up is considered the slowest of connections because it is restricted by the 56 Kbps limitation of traditional phone line service. A dial-up connection also ties up the phone line so that it is unavailable for voice calls.

Connect via an Online Service

Several "high-speed" Internet services are available for home and business. They include cable modem connections from cable television companies, DSL connections from phone companies, and satellite connections from satellite television companies. High-speed services provide data transfer rates between 1 and 7 Mbps. Unlike dial-up, high-speed services provide "always connected" service that does not tie up the phone line.

Connect Wirelessly

In addition to connecting to the Internet through wired systems such as phone lines and television cables, wireless Internet access is very popular. Thousands of public Wi-Fi services are available in coffee shops, airports, hotels, and elsewhere, where Internet access is provided free, for an hourly rate, or for a monthly subscription fee. Wi-Fi is even making its way into aircraft, allowing business travelers to be productive during air travel by accessing e-mail and corporate networks.[20]

Cell phone carriers also provide Internet access for handsets or notebooks equipped with connect cards. New 3G mobile phone services rival wired high-speed connections enjoyed at home and work. Sprint, Verizon, AT&T, and other popular carriers are working to bring 4G service to subscribers soon. 4G cell phone service will compete strongly against today's wired services. Wireless devices also require specific protocols and approaches to connect. For example, *wireless application protocol (WAP)* is used to connect cell phones and other devices to the Internet.

Apple's popular iPhone[21] is a combination mobile phone, widescreen iPod, and Internet access device to support e-mail and Web browsing. The iPhone can connect to the Internet either via Wi-Fi or AT&T's Edge data network (considered to be a 2.5G network, not quite as fast as a 3G network). When Apple introduced the iPhone, one of its slogans was the "Internet in your pocket." The iPhone serves to prove the popularity of, and potential for, Internet services over a handset.[22] Intel picked up on Apple's slogan and added it to its own marketing campaign for its processor called the Atom, which is designed to bring the Internet to more mobile devices.[23]

Internet Service Providers

An **Internet service provider (ISP)** is any company that provides Internet access to people and organizations. Some ISPs such as America Online (AOL) or Microsoft Network (MSN) offer extended information services through software installed on the subscriber's PC. Many others simply offer a connection to the Internet that subscribers use with Internet software to access various services. Thousands of organizations serve as ISPs, ranging from universities making unused communications line capacity available to students and faculty to major

Internet service provider (ISP)
Any company that provides Internet access to people or organizations.

Apple iPhone

The iPhone is a combination mobile phone, widescreen iPod, and Internet access device.

(Source: Courtesy of Apple Computer.)

communications giants such as AT&T and Verizon. To use this type of connection, you must have an account with the service provider and software that allows a direct link via TCP/IP.

For dial-up connections, ISPs charge a monthly fee that can range from $10 to $30 for unlimited Internet access. The fee normally includes e-mail. Many ISPs and online services offer broadband Internet access through DSLs, cable or satellite transmission for $30 to $60 per month depending on the speed of access. Some businesses and universities use very fast T1 or T3 lines to connect to the Internet. These are special high-speed communications links capable of sending and receiving data at up to 1.536 Mbps for T1 and 4.736 Mbps for a T3. Table 4.6 compares the speed of modem, DSL, cable, and T1 Internet connections to perform basic tasks.

Table 4.6

Time to Perform Basic Tasks with Various Internet Connections

Task	Modem (56 Kbps)	T1 (1.4 Mbps)	DSL (3 Mbps)	Cable (7 Mbps)	T3 (44 Mbps)
Send 20-page term paper—500 KB	9 seconds	.36 seconds	.17 seconds	.07 seconds	.01 seconds
Send a four-minute song as an MP3 file—4.5 MB	80 seconds	3.2 seconds	1.5 seconds	.64 seconds	.1 seconds
Send a full-length motion picture as a compressed file—1.4 GB	About 7 hours	16 minutes	7.8 minutes	3.3 minutes	.53 seconds

Many factors can affect the actual speed of sending and receiving data on the Internet as discussed in the Ethical and Social Issues special feature.

Comcast, Packet Shaping and Net Neutrality

In 2008, the U.S. Federal Communications Commission (FCC) required the telecommunications and cable TV giant Comcast to testify about its practice of packet shaping. Packet shaping is a technique that some Internet service providers (ISPs) use to control the volume of network traffic to optimize or guarantee performance. It involves filtering Internet traffic, which travels the Internet in data packets, and slowing down some types of packets in order to speed up others.

Some Comcast customers noticed that when they attempted to upload files on Bit Torrent, a popular file-sharing utility, the transfers were extremely slow and often stopped all together. Reporters with the Associated Press ran some nationwide tests and determined that Comcast was indeed filtering Internet packets, determining which packets were uploading files with Bit Torrent and then shutting down those packets. When the news was released, it made front page headlines, and many Internet users were enraged.

Why were they enraged? The people who designed the Internet and Web intended it to be run with neutrality. Network neutrality refers to a principle applied to Internet services whereby all data is delivered to all users with equal priority. Most importantly, many believe that managers of Internet traffic do not have the right to examine what is in packets. To do so would be an invasion of privacy. A process called "deep packet inspection" allows automated systems to analyze the contents of every packet flowing through the ISP and take actions based on a rule. For example, packets with the words "bomb instructions" might be detained and the owner tracked down. But would such an action be in accordance with the privacy guaranteed to U.S. citizens in the Constitution?

Some organizations do not support net neutrality. Comcast claims that 80 percent of its bandwidth is taken up by 10 percent of its users involved in illegal file sharing, with many of the files being huge video files. If this were true, and if it were affecting the overall performance of the Internet, they might have a case for packet shaping.

Law enforcement would also like the ability to filter packets to find criminal activity. FBI director Robert Mueller has indicated that he would like the House of Representatives to grant permission to filter all Internet traffic to detect illegal activities. The Recording Industry Association of America and the Motion Picture Association of America would like ISPs to filter out packets that are involved in illegal file sharing and copyright violation. However, the ISPs are not eager to become the Internet's traffic cops. When false arrests are made, as they inevitably would be, the ISP would be liable.

After hearing Comcast's case and the voices of those supporting net neutrality, the FCC stated that it plans to regulate how and when ISPs can practice packet shaping. Comcast has promised to change its packet-shaping practices. Net neutrality is likely to be a hot topic in federal courts in coming years.

Discussion Questions

1. What is packet shaping and do ISPs use it?
2. What are the principles of network neutrality?

Critical Thinking Questions

1. Do you favor network neutrality or packet shaping? Why?
2. Is it necessary for the U.S. government to intervene and regulate ISPs? What laws would you like to see enacted?

Sources: Svensson, Peter, "Comcast blocks some Internet traffic," Associated Press/MSNBC, *www.msnbc.msn.com/id/21376597*; Kumar, Vishesh, "Comcast, BitTorrent to Work Together on Network Traffic," *The Wall Street Journal*, March 27, 2008, *http://online.wsj.com/article/SB120658178504567453.html*; Stokes, Jon, "FBI wants to move hunt for criminals into Internet backbone," *Ars Technica*, April 24, 2008, *http://arstechnica.com/news.ars/post/20080424-fbi-wants-to-move-hunt-for-criminals-into-internet-backbone.html*.

THE WORLD WIDE WEB

World Wide Web

A collection of tens of millions of server computers that work together as one in an Internet service using hyperlink technology to provide information to billions of users.

Web browser

Web client software, like Internet Explorer, Firefox, and Safari, used to view Web pages.

The World Wide Web was developed by Tim Berners-Lee at CERN, the European Organization for Nuclear Research in Geneva. He originally conceived of it as an internal document-management system. This server can be located at *http://public.web.cern.ch/public*. From this modest beginning, the **World Wide Web** has grown to a collection of tens of millions of server computers that work together as one in an Internet service using hyperlink technology to provide information to over a billion users. These computers, called *Web servers*, are scattered all over the world and contain every imaginable type of data. Web users use hyperlinks, highlighted text or graphics in a Web document, that, when clicked, open a new Web page or section of the same page containing related content. Thanks to the high-speed Internet circuits connecting them and some clever cross-indexing software, users can jump from one Web computer to another effortlessly—creating the illusion of using one big computer. Because of the vast amount of information available on the Web and the wide variety of media, the Web has become the most popular means of information access in the world today.

In short, the Web is a hyperlink-based system that uses the client/server model. It organizes Internet resources throughout the world into a series of linked files, called pages, accessed and viewed using Web client software, called a **Web browser**. Internet Explorer, Firefox, and Safari are three popular Web browsers. See Figure 4.11. A collection of pages on one particular topic, accessed under one Web domain, is called a Web site. The Web was originally designed to support formatted text and pictures on a page. It has evolved to support many more types of information and communication including user interactivity, animation, and video. Web *plug-ins* help provide additional features to standard Web sites. Adobe Flash and Real Player are examples of Web plug-ins.

Figure 4.11

Mozilla Firefox

Web browsers such as Firefox let you access Internet resources throughout the world using a series of linked Web pages.

A *Web portal* is an entry point or doorway to the Internet. Web portals include AOL, MSN, Google, Yahoo!, and others. For example, some people use Yahoo.com as their Web portal, which means they have set Yahoo! as their starting point in their browsers. When they enter the Internet, the Yahoo! Web site appears. Web portals may be customizable and allow users to choose from a variety of widgets—small useful applications and services, to add to the page.[24] Web browser settings use the term *home page* to refer to your starting point. This setting can apply to any Web page you prefer. A *corporate Web portal* refers to the company's Internet site, which is a gateway or entry point to corporate data and resources.

Hypertext Markup Language (HTML) is the standard page description language for Web pages. One way to think about HTML is as a set of highlighter pens that you use to mark up plain text to make it a Web page—red for the headings, yellow for bold, and so on. The **HTML tags** let the browser know how to format the text: as a heading, list, or body text, for example. Users "mark up" a page by placing HTML tags before and after a word or words. For example, to turn a sentence into a heading, you place the <h1> tag at the start of the sentence. At the end of the sentence, you place the closing tag </h1>. When you view this page in your browser, the sentence will be displayed as a heading. HTML also provides tags to import objects stored in files, such as photos, audio, and movies, into a Web page. In short, a Web page is made up of three components: text, tags, and references to files. The text is your message, the tags are codes that mark the way words will be displayed, and the references to files insert photos and media into the Web page at specific locations. All HTML tags are enclosed in a set of angle brackets (< and >), such as <h2>. The closing tag has a forward slash in it, such as </b> for closing bold. Consider the following text and tags:

```
<h1 align="center">Principles of Information Systems</h1>
```

This HTML code centers Principles of Information Systems as a major, or level 1, heading. The "h1" in the HTML code indicates a first-level heading. On some browsers, the heading might be 14-point type with a Times Roman font. On other browsers, it might be a larger 18-point type in a different font. Figure 4.12 shows a simple document and its corresponding HTML tags. Notice the <html> tag at the top indicating the beginning of the HTML code. The <title> indicates the beginning of the title: "Course Technology—Leading the Way in IT Publishing." The </title> tag indicates the end of the title.

Hypertext Markup Language (HTML)
The standard page description language for Web pages.

HTML tags
Codes that let the Web browser know how to format text—as a heading, as a list, or as body text—and whether images, sound, and other elements should be inserted.

Figure 4.12

Sample Hypertext Markup Language

The window on the left is a Web document, and the window on the right shows the corresponding HTML tags.

Some newer Web standards are gaining in popularity, including Extensible Markup Language (XML), Extensible Hypertext Markup Language (XHTML), Cascading Style Sheets (CSS), Dynamic HTML (DHMTL), and Wireless Markup Language (WML), which can display Web pages on small screens such as smartphones and PDAs. XHTML is a combination of XML and HTML that has been approved by the World Wide Web Consortium (W3C).

Extensible Markup Language (XML) is a markup language for Web documents containing structured information, including words and pictures. XML does not have a predefined tag set. With HTML, for example, the <h1> tag always means a first-level heading. The content and formatting are contained in the same HTML document. XML Web

Extensible Markup Language (XML)
The markup language for Web documents containing structured information, including words, pictures, and other elements.

documents contain the content of a Web page. The formatting of the content is contained in a style sheet. A few typical instructions in XML follow:

    <chapter>Hardware
    <topic>Input Devices
    <topic>Processing and Storage Devices
    <topic>Output Devices

A **Cascading Style Sheet (CSS)** is a file or portion of an HTML file that defines the visual appearance of content in a Web page. Using CSS is convenient because you only need to define the technical details of the page's appearance once, rather than in each HTML tag. For example the visual appearance of the preceding XML content may be contained in the following style sheet. This style sheet specifies that the chapter title "Hardware" is displayed on the Web page in a large Arial font (18 points). "Hardware" will also appear in bold blue text. The "Input Devices" title will appear in a smaller Arial font (12 points) and italic red text.

    chapter: (font-size: 18pt; color: blue; font-weight: bold; display: block; font-family: Arial; margin-top: 10pt; margin-left: 5pt)
    topic: (font-size: 12pt; color: red; font-style: italic; display: block; font-family: Arial; margin-left: 12pt)

Many new Web sites being developed use CSS to define the visual design and layout of Web pages, XML to define the content, and XHTML to join the content (XML) with the design (CSS). See Figure 4.13. This modular approach to Web design allows you to change the visual design without affecting the content, or to change the content without affecting the visual design.

Web 2.0 and the Social Web

Over the past few years, the Web has been evolving from a one-directional resource where users obtain information to a two-directional resource where users obtain and contribute information. Consider Web sites such as YouTube, Wikipedia, and MySpace as examples. The Web has also grown in power to support full-blown software applications, such as Google Docs, and is becoming a computing platform on its own. These two major trends in how the Web is used and perceived have created dramatic changes on the Web, so that the new form of the Web has earned the title of **Web 2.0**.[25]

The original Web, now referred to as Web 1.0, provided a platform for technology-savvy developers and the businesses and organizations that hired them to publish information for the general public to view. The introduction of user-generated content supported by Wikipedia, blogging, and podcasting made it clear that those using the Web were also interested in contributing to its content. This led to the development of Web sites with the sole purpose of supporting user-generated content and user feedback.

Web sites such as YouTube and Flickr allow users to share video and photos with other people, groups, and the world. With social networking Web sites such as Facebook and MySpace, users can post information about their interests and find like-minded people. Using microblogging sites such as Twitter and Jaiku, people can post thoughts and ideas throughout the day for friends to read. Social bookmarking sites such as Digg and del.icio.us allow users to pool their votes to determine what online news stories and Web pages are most interesting each moment of the day. Similarly, Epinions and many retail Web sites allow consumers to voice their opinions about products. All of these popular Web sites serve as examples of how the Web has transformed to become the town square where people share information, ideas, and opinions, meet with friends, and make new acquaintances.

The introduction of powerful Web-delivered applications such as Google Docs, Adobe Photoshop Express, Xcerion Web-based OS, and Microsoft Maps has elevated the Web from an online library to a platform for computing.[26] Many of the computer activities traditionally provided through software installed on a PC can now be carried out using rich Internet applications (RIAs) in a Web browser without installing any software. A **rich Internet application** is software that has the functionality and complexity of traditional application software, but runs in a Web browser and does not require local installation. RIAs are the result of continuously improving programming languages and platforms designed for the Web.

Cascading Style Sheet (CSS)
A file or portion of an HTML file that defines the visual appearance of content in a Web page.

Web 2.0
The Web as a computing platform that supports software applications and the sharing of information between users.

rich Internet application
Software that has the functionality and complexity of traditional application software, but does not require local installation and runs in a Web browser.

Figure 4.13

XML, CSS, and XHTML

Today's Web sites are created using XML to define content, CSS to define the visual style, and XHTML to put it all together.

Web Programming Languages

Several programming languages are key to the Web. **Java**, for example, is an object-oriented programming language from Sun Microsystems based on the C++ programming language, which allows small programs called *applets* to be embedded within an HTML document. When the user clicks the appropriate part of an HTML page to retrieve an applet from a Web server, the applet is downloaded onto the client workstation, where it begins executing. Unlike other programs, Java software can run on any type of computer. Programmers use Java to make Web pages come alive, adding splashy graphics, animation, and real-time updates.

In addition to Java, companies use a variety of other programming languages and tools to develop Web sites. JavaScript, VBScript, and ActiveX (used with Internet Explorer) are Internet languages used to develop Web pages and perform important functions, such as accepting user input. *Hypertext Preprocessor*, or *PHP*, is an open-source programming language. PHP code or instructions can be embedded directly into HTML code. Unlike some other Internet languages, PHP can run on a Web server, with the results being transferred to a client computer. PHP can be used on a variety of operating systems, including Microsoft Windows, Macintosh OS X, HP-UX, and others. It can also be used with a variety of database management systems, such as DB2, Oracle, Informix, MySQL, and many others. These characteristics—running on different operating systems and database management systems, and being an open-source language—make PHP popular with many Web developers.

Java

An object-oriented programming language from Sun Microsystems based on C++ that allows small programs (applets) to be embedded within an HTML document.

Developing Web Content

The art of Web design involves working within the technical limitations of the Web and using a set of tools to make appealing designs. A number of products make developing and maintaining Web content easier. Microsoft, for example, has introduced a development and Web services platform called .NET. The .NET platform allows developers to use different programming languages to create and run programs, including those for the Web. The .NET platform also includes a rich library of programming code to help build XML Web applications. Bubbler is another example of a Web design and building tool. Bubbler helps you obtain a domain name, develop attractive Web pages by dragging and dropping features and options, host your Web site, and maintain it. Other Web publishing packages include Homestead QuickSites and JobSpot. See Figure 4.14.

Figure 4.14

Developing a Web Page Using Microsoft Visual Studio

The window on the left shows the Web page being developed using Microsoft Visual Studio, part of the .NET Web development platform. The window on the right shows the Web page in a browser.

After you create Web pages, your next step is to place, or publish, the content on a Web server. Popular publishing options include using ISPs, free sites, and Web hosting. Web hosting services provide space on their Web servers for people and businesses that don't have the financial resources, time, or skills to host their own Web sites. A Web host can charge $15 or more per month, depending on services. Some Web hosting sites include domain name registration, Web authoring software, and activity reporting and monitoring of the Web site. Some ISPs also provide limited Web space, typically 1 to 6 MB, as part of their monthly fee. If more disk space is needed, additional fees are charged. Free sites offer limited space for a Web site. In return, free sites often require the user to view advertising or agree to other terms and conditions.

Some Web developers are creating programs and procedures to combine two or more Web applications into a new service, called a *mash-up*. A mash-up is named for the process of mixing two or more hip-hop songs into one song. A Web site that provides crime information, for example, can be mashed up with a mapping Web site to produce a Web site with crime information placed on top of a map of a metropolitan area. People are becoming very creative in how they mash-up several Web sites into new ones. Google and Yahoo both provide developers with tools for creating mash-ups. Google's is called the Mashup Editor (*editor.googlemashups.com*), and Yahoo's is called Pipes (*pipes.yahoo.com*).[27]

Web Services

Web services consist of standards and tools that streamline and simplify communication among Web sites, promising to revolutionize the way we develop and use the Web for business and personal purposes. Internet companies, including Amazon, eBay, and Google, are now using Web services. Amazon, for example, has developed Amazon Web Services (AWS) to make the contents of its huge online catalog available to other Web sites or software applications. Mitsubishi Motors of North America uses Web services to link about 700 automotive dealers on the Internet.

The key to Web services is XML. Just as HTML was developed as a standard for formatting Web content into Web pages, XML is used within a Web page to describe and transfer data between Web service applications.

Web services
Standards and tools that streamline and simplify communication among Web sites for business and personal purposes.

INTERNET AND WEB APPLICATIONS

Search Engines and Web Research

A **search engine** is a valuable tool that enables you to find information on the Web by specifying words that are key to a topic of interest—known as keywords. You can also use operators such as OR and NOT for more precise search results.[28]

Search engines make money from companies that place advertisements with the search results. Search companies learn about visitors' interests by the topics for which they search, and can target ads to each user based on those interests. Google, Microsoft, and Yahoo! are leaders in the search-engine business.

Search engines scour the Web with bots (automated programs) called spiders that follow all Web links in an attempt to catalog every Web page by topic. The process is called Web crawling, and due to the ever-changing nature of the Web, it is a job that never ends. Google maintains over four billion indexed Web pages in a database on 30 clusters of up to 2,000 computers, each totaling over 30 petabytes of data.

After building the search database, the next challenge facing a search engine is to determine which of the hundreds or thousands of Web pages associated with a particular keyword are most useful. The method of ranking Web pages from most relevant to least differs from search engine to search engine. Google uses a popularity contest approach. Web pages that are referenced from other Web pages are ranked higher than those that are not. Each reference is considered a vote for the referenced page. The more votes a Web page gets, the higher its rank. References from higher-ranked pages weigh more heavily than those from lower-ranked pages.

Search engines have become important to businesses as a tool to drive visitors to the business' Web site. Many businesses invest in search engine optimization (SEO)—a process for driving traffic to a Web site, by using techniques that improve the site's ranking in search results. SEO is based on the understanding that Web page links listed on the first page of search results, as high on the list as possible, have a far greater chance of being clicked. SEO professionals study the algorithms employed by search engines, altering the Web page contents and other variables to improve the page's chance of being ranked number one.[29]

Wikipedia, an encyclopedia with 1.9 million English-language entries created and edited by tens of thousands of people (see Figure 4.15), is another example of a Web site that can be used to research information.[30] In Hawaiian, wiki means quick, so a wikipedia provides quick access to information. The Web site is both open source and open editing, which means that people can add or edit entries in the encyclopedia at any time. Because thousands of people are monitoring Wikipedia, the Web-based encyclopedia is self-regulating. Incorrect, outdated, or offensive material is usually removed, although people with an axe to grind have distorted information on Wikipedia intentionally.

search engine
A valuable tool that enables you to find information on the Web by specifying words that are key to a topic of interest, known as keywords.

Figure 4.15

Wikipedia

Wikipedia captures the knowledge of tens of thousands of experts and those who think they are experts.

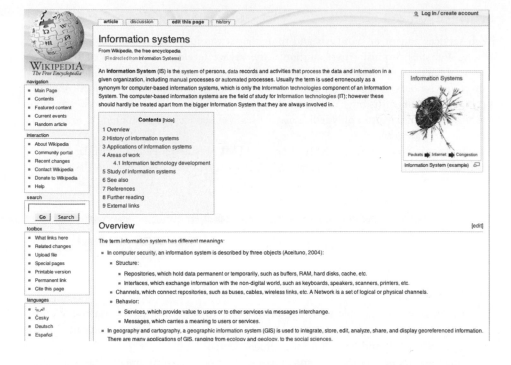

E-Mail, Instant Messaging, and Video Chat

E-mail is no longer limited to simple text messages. Depending on your hardware and software and the hardware and software of your recipient, you can embed sound and images in your message and attach files that contain text documents, spreadsheets, graphs, or executable programs. You can send e-mail messages to anyone in the world if you know that person's e-mail address.

For large organizations whose operations span a country or the world, e-mail allows people to work around the time zone changes. Some users of e-mail estimate that they eliminate two hours of verbal communications for every hour of e-mail use. But the person at the other end still must check the mailbox to receive messages.

Some companies use bulk e-mail to send legitimate and important information to sales representatives, customers, and suppliers around the world. With its popularity and ease of use, however, some people feel they are drowning in too much e-mail. Over a trillion e-mail messages are sent from businesses in North America each year—up from 40 billion e-mail messages in 1995. Users are taking a number of steps to cope with and reduce the mountain of e-mail. Some companies have banned the use of copying others on e-mails unless it is critical. Some e-mail services scan for possible junk or bulk mail, called *spam*, and delete it or place it in a separate file. Software products can help users sort and answer large amounts of e-mail. This software can recognize key words and phrases and respond to them.

Instant messaging is online, real-time communication between two or more people who are connected to the Internet. With instant messaging, two or more windows or panes open, with each one displaying what a person is typing. Because the typing is displayed in real-time, instant messaging is like talking to someone using the keyboard. See Figure 4.16.

Many companies offer instant messaging, including America Online, Yahoo!, and Microsoft. America Online is one of the leaders in instant messaging, with about 40 million users of its Instant Messenger and about 50 million people using its client program ICQ. In addition to being able to type messages on a keyboard and have the information instantly displayed on the other person's screen, some instant messaging programs are allowing voice communication or connection to cell phones. A wireless service provider announced that it has developed a technology that can detect when a person's cell phone is turned on. With this technology, someone on the Internet can use instant messaging to communicate with someone on a cell phone anywhere in the world.

instant messaging
A method that allows two or more people to communicate online using the Internet.

As more people are connecting to the Internet over broadband connections, increasing amounts of users are turning to video chat. Services such as Apple iChat and Skype provide computer-to-computer video chat so users can speak to each other face-to-face. Some video chat services support conference calling as well.

Career Information and Job Searching

The Internet is an excellent source of job-related information. People looking for their first job or seeking information about new job opportunities can find a wealth of information. Search engines can be a good starting point for searching for specific companies or industries. You can use a directory on Yahoo's home page, for example, to explore industries and careers. Most medium and large companies have Internet sites that list open positions, salaries, benefits, and people to contact for further information. The IBM Web site, *www.ibm.com*, has a link to "Jobs at IBM." When you click this link, you can find information on jobs with IBM around the world. Some Internet sites specialize in certain careers or industries. The site *www.directmarketingcareers.com* lists direct marketing jobs and careers. Some sites can help you develop a résumé and find a good job. They can also help you develop an effective cover letter for a résumé, prepare for a job interview, negotiate an employment contract, and more. In addition, several Internet sites specialize in helping you find job information and even apply for jobs online, including *www.monster.com*, *www.hotjobs.com*, and *www.careerbuilder.com*. You must be careful when applying for jobs online, however. Some bogus companies or Web sites will steal your identity by asking for personal information. People eager to get a job often give their Social Security number, birth date, and other personal information. The result can be no job, large bills on your credit card, and ruined credit.

Telnet, SSH, and FTP

Telnet is a network protocol that enables users to log on to networks remotely over the Internet. Telnet software uses a command-line interface that allows the user to work on a remote server directly. Since Telnet is not secured with encryption, most users are switching to *secure shell (SSH)*, which provides Telnet functionality through a more secure connection.

File Transfer Protocol (FTP) is a protocol that supports file transfers between a host and a remote computer. Using FTP, users can copy files from one computer to another. For example, companies use it to transfer vast amounts of business transactional data to the computers of its customers and suppliers. You can also use FTP to gain access to a wealth of free software on the Internet and to upload or download content to a Web site. The authors and editors of this book used an FTP site provided by the publisher, Course Technology, to share and transfer important files during the publication process. Chapter files and artwork, for example, were uploaded to a Course Technology Web site and downloaded by authors

File Transfer Protocol (FTP)
A protocol that describes a file transfer process between a host and a remote computer and allows users to copy files from one computer to another.

Several Internet sites specialize in helping people get job information and even apply for jobs online.

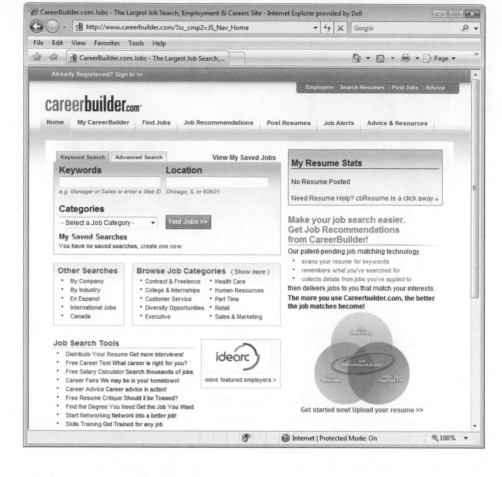

and editors to review. Like Telnet, FTP connections are not encrypted, and are therefore not secure. Many users are switching to secure FTP (SFTP) for more secure file transfers.

Web Log (Blog), Video Log (Vlog), and Podcasting

A **Web log**, typically called a **blog**, is a Web site that people can create and use to write about their observations, experiences, and opinions on a wide range of topics.[31] The community of blogs and bloggers is often called the *blogosphere*. A *blogger* is a person who creates a blog, while *blogging* refers to the process of placing entries on a blog site. A blog is like a journal. When people post information to a blog, it is placed at the top of the blog page. Blogs can include links to external information and an area for comments submitted by visitors. Video content can also be placed on the Internet using the same approach as a blog. This is often called a *video log* or *vlog*. Blogs are easy to post, but they can cause problems when people tell or share too much. People have been fired for blogging about work, and the daughter of a politician embarrassed her father when she made personal confessions on a blog.

Blog sites, such as *www.blogger.com* and *www.blogcatalog.com*, can include information and tools to help people create and use Web logs. To set up a blog, you can go to the Web site of a blog service provider, such as *www.livejournal.com*, create a username and password, select a theme, choose a URL, follow any other instructions, and start making your first entry. Blog search engines include Technorati and Blogdigger. You can also use Google to locate a blog.

A corporate blog can be useful for communicating with customers, partners, and employees. However, companies and their employees need to be cautious about the legal risks of blogging.[32] Blogging can expose a corporation and its employees to charges of defamation, copyright and trademark infringement, invasion of privacy, and revealing corporate secrets.

A *podcast* is an audio broadcast over the Internet. The name podcast comes from the word *iPod*, Apple's portable music player, and the word *broadcasting*. A podcast is an audio blog,

Web log (blog)
A Web site that people can create and use to write about their observations, experiences, and feelings on a wide range of topics.

like a personal radio station on the Internet, and extends blogging by adding audio messages. Using PCs, recording software, and microphones, you can record audio messages and place them on the Internet. You can then listen to the podcasts on your PC or download the audio material to a digital audio player, such as Apple's iPod. You can also use podcasting to listen to TV programs, your favorite radio personalities, music, and messages from your friends and family at any time and place. Finding good podcasts, however, can be challenging. Apple's iTunes provides free access to tens of thousands of podcasts sorted by topic and searchable by keyword. See Figure 4.17. After you find a podcast, you can download it to a PC (Windows or Mac) and to an MP3 music player such as the iPod for future listening.

Figure 4.17

iTunes Podcasts

iTunes provides free access to tens of thousands of podcasts.

Many blogs, vlogs, and podcasts offer automatic updates to a PC using a technology called Really Simple Syndication (RSS). RSS is a collection of Web technologies that allow users to subscribe to Web content that is frequently updated. With RSS, you can receive a blog update without actually visiting the blog Web site. You can also use RSS to receive other updates on the Internet from news Web sites and podcasts. Software used to subscribe to RSS feeds is called *aggregator software*. Google Reader is a popular aggregator for subscribing to blogs.

Chat Rooms

A **chat room** is a facility that enables two or more people to engage in interactive "conversations" over the Internet. When you participate in a chat room, dozens of people might be participating from around the world. Multiperson chats are usually organized around specific topics, and participants often adopt nicknames to maintain anonymity. One form of chat room, Internet Relay Chat (IRC), requires participants to type their conversation rather than speak. Voice chat is also an option, but you must have a microphone, sound card, speakers, a fast modem or broadband, and voice-chat software compatible with the other participants'. Most of the functionality of chat is available in instant messaging software.

chat room
A facility that enables two or more people to engage in interactive "conversations" over the Internet.

Internet Phone and Videoconferencing Services

Internet phone service enables you to communicate with others around the world. This service is relatively inexpensive and can make sense for international calls. With some services, you can use the Internet to call someone who is using a standard phone. You can also keep your phone number when you move to another location. Cost is often a big factor for those using Internet phones—a call can be as low as 1 cent per minute for calls within the United States. Low rates are also available for calling outside the United States. In addition, voice mail and fax capabilities are available. Some cable TV companies, for example, are offering cable TV, phone service, and caller ID for under $40 a month. Skype offers free and low-priced Internet phone and video phone service from any Internet connected computer.

Internet videoconferencing, which supports both voice and visual communications, is another important Internet application. Microsoft's NetMeeting, a utility within Windows XP, is an inexpensive and easy way for people to meet and communicate on the Web. Windows Vista offers Windows Meeting Space to provide the same service. The Internet can also be used to broadcast group meetings, such as sales seminars, using presentation software and videoconferencing equipment. These Internet presentations are often called Webcasts or Webinars. WebEx and GoToMeeting are two popular Web conferencing tools. The ideal video product will support multipoint conferencing, in which users appear simultaneously on multiple screens. Hewlett-Packard (HP) has produced such a system called Halo (see Figure 4.18). When using Halo, it appears as though you are speaking with a number of people across a table, though those people may actually be located around the world.

Figure 4.18

Halo Collaboration Meeting Room

HP's Halo telepresence system allows people at various locations to meet as though they were gathered around a table.

Social Networks

Social networking Web sites provide Web-based tools for users to share information about themselves with people on the Web and to find, meet, and converse with other members.

The most popular social networking sites are MySpace and Facebook. Both sites provide members with their own personal Web page and allow members to post photos and information about themselves (see Figure 4.19). Social networking sites allow members to send messages to each other and post comments on each other's pages. Members accumulate friends through an invitation process. Special interest groups can be created, and joined as well. Tools are provided to search for others with similar interests.

Figure 4.19

Social Networking Web Sites

Facebook is a social networking site that provides members with Web pages to post photos and information about themselves.

The power of social networks is being harnessed for business purposes as well.[33] Many businesses are using the information posted in member profiles to find potential clients. Linked In is a social network that allows professionals to find others who work in the same field, applying social networking techniques for business networking.

Another social networking site called Twitter (*www.twitter.com*) allows members to report on what they are doing throughout the day. Referred to as a microblogging service, Twitter allows users to send short text updates (up to 140 characters long) from cell phones or the Internet to their Twitter page to let others know what they are doing.[34] Twitter updates can be forwarded to MySpace or Facebook Web sites.

Media Sharing

Media-sharing Web sites such as YouTube for video sharing and Flickr for photo sharing provide methods for members to store and share digital media files on the Web. YouTube allows members to post homemade video content in categories such as comedy, entertainment, film and animation, how-to, news, people, pets, sports, and travel. As mentioned earlier, Flickr allows its members to upload photos to their own personal online photo album and choose photos to share with the community.

What makes these media-sharing sites part of Web 2.0 is their focus on community. Both Flickr and YouTube provide ways for members to comment on the media. YouTube allows visitors to e-mail favorite video clips to friends. Both sites provide methods for visitors to view the most popular media or search on a particular topic.

Content Streaming

Content streaming is a method for transferring multimedia files, radio broadcasts, and other content over the Internet so that the data stream of voice and pictures plays more or less continuously, without a break, or with very few of them. It also enables users to browse large files in real time. For example, rather than wait for an entire 5 MB video clip to download before they can play it, users can begin viewing a streamed video as it is being received. Content streaming works best when the transmission of a file can keep up with the playback of the file.

content streaming
A method for transferring multimedia files over the Internet so that the data stream of voice and pictures plays more or less continuously without a break, or very few of them; enables users to browse large files in real time.

Shopping on the Web

Shopping on the Web for books, clothes, cars, medications, and even medical advice can be convenient, easy, and cost effective. Amazon.com, for example, sells short stories by popular authors for $.49 per story. The service, called Amazon Shorts, has stories that vary in length from 2,000 to 10,000 words by authors such as Danielle Steel, Terry Brooks, and others. The company also sells traditional books and other consumer products. To add to their other conveniences, many Web sites offer free shipping and pickup for returned items that don't fit or otherwise meet a customer's needs.

Increasingly, people are using bots to help them search for information or shop on the Internet. A **bot**, also called an *intelligent agent*, is a software tool that searches the Web for information, products, or prices. A bot, short for *robot*, can find the best prices or features from multiple Web sites. Shopping.com uses bots to identify the best prices on merchandise.

Web Auctions

A **Web auction** is a way to connect buyers and sellers. Web auction sites are a place where businesses are growing their markets or reaching customers for a low cost per transaction. Web auctions are transforming the customer-supplier relationship.

One of the most popular auction sites is eBay, which often has millions of auctions occurring at the same time. The eBay site is easy to use, and includes thousands of products and services in many categories. eBay remains a good way to get rid of things you don't need or find bargains on things you do need. eBay drop-off stores allow people who are inexperienced with Internet auctions or too busy to develop their own listings to sell items on the popular Web site. In addition to eBay, you can find a number of other auction sites on the Web. Traditional companies are even starting their own auction sites.

Although auction Web sites are excellent for matching buyers and sellers, they can present problems. Auction sites cannot always determine whether the people and companies listing products and services are legitimate. In addition, some Web sites have had illegal or questionable items offered. Many Web sites have an aggressive fraud investigation system to prevent and help prosecute fraudulent use of their sites. Even with these potential problems, the use of Web auction sites is expected to continue to grow rapidly.

bot
A software tool that searches the Web for information such as products and prices.

Web auction
An Internet site that matches buyers and sellers.

eBay is a popular auction Web site.

Music, Radio, Video, and TV on the Internet

Music, radio, and video are hot growth areas on the Internet. Audio and video programs can be played on the Internet or files can be downloaded for later use. Using music players and music formats such as MP3, you can download music from the Internet and listen to it anywhere using small portable music players. A subscriber to a music service such as Napster or Rhapsody can download an unlimited number of songs from the site, as long as they pay the annual subscription service fee. Pay-per-song services such as those offered by Apple iTunes and Amazon allow users to purchase songs for around $0.99 per song.[35] A music service called SpiralFrog supports free music downloads for users willing to provide marketing information and watch ads.[36]

Radio broadcasts are now available on the Internet. Entire audio books can also be downloaded for later listening using services such as Audible.com. Audible provides a subscription service that allows users to download one or more books per month that can then be played using PC media software and portable digital music players.

Video and TV are also becoming available on the Internet. One way to put TV programming on the Internet is to use the Internet Protocol Television (IPTV) protocol. With the potential of offering an almost unlimited number of programs, IPTV can serve a vast array of programming on specialty areas, such as yoga, vegetarian food, unusual sporting events, and news from a city or region of a state.[37]

Internet TV service Joost offers 20,000 free TV shows over 400 channels.

A number of innovative devices let you record TV programs and view them at any time and place.[38] A California company called Sling Media (*www.slingmedia.com*), for example, offers a device that can broadcast any TV program coming into your home to a broadband Internet-connected PC.[39] Once on the Internet, you can watch the TV program at any time and place that has broadband Internet service. The device, called a Slingbox, costs about $250 and doesn't require monthly service fees.

Office on the Web

Having an Internet office with access to files and information can be critical for people who travel frequently or work at home. An Internet office is a Web site that contains files, phone numbers, e-mail addresses, an appointment calendar, and more. Using a standard Web browser, you can access important business information. An Internet office allows your desktop computer, phone books, appointment schedulers, and other important information to be with you wherever you are.

Many services and software products offer remote access to your files and programs over the Internet on the Web. As mentioned earlier, Microsoft and Google both support online document storage and sharing. Both companies and others also provide ways to access contact lists and calendar software and data online. For example, 37signals.com provides online project management, contact management, calendar, and group chat applications. Microsoft SharePoint provides businesses with collaborative workspaces and social computing tools to allow people at different locations to work on projects together. Over time, increasing amounts of software, services, and storage will be available online through your Web browser.

Other Internet Services and Applications

Other Internet services are constantly emerging. A vast amount of information is available over the Internet from libraries. Many articles that served as the basis of the sidebars, cases, and examples used throughout this book were obtained from university libraries online. Movies can be ordered and even delivered over the Internet. The Internet can provide critical information during times of disaster or terrorism. During a medical emergency, critical medical information can be transmitted over the Internet. People wanting to consolidate their credit card debt or to obtain lower payments on their existing home mortgages have turned to sites such as Quicken Loan, E-Loan, and LendingTree for help.

The Internet can also be used to translate words, sentences, or complete documents from one language into another. For example, Babel Fish Translation (*www.world.altavista.com*) and Free Translation (*www.freetranslation.com*) can translate a block of text from one language into another. Some search engines can translate Web pages and allow you to search for Web sites published in certain languages or countries. Clicking Language Tools on the home page of Google, for example, provides these capabilities.

The Internet also facilitates distance learning, which has dramatically increased in the last several years. Many colleges and universities now allow students to take courses without visiting campus. In fact, you might be taking this course online. MIT is offering all of its 1,800 courses free online.[40] Businesses are also taking advantage of distance learning through the Internet. Video cameras can be attached to computers and connected to the Internet. These Internet cameras can be used to conduct job interviews, hold group meetings with people around the world, monitor young children at daycare centers, check rental properties and second homes from a distance, and more.

INTRANETS AND EXTRANETS

intranet
An internal corporate network built using Internet and World Wide Web standards and products; used by employees to gain access to corporate information.

An **intranet** is an internal corporate network built using Internet and World Wide Web standards and products. Employees of an organization use it to gain access to corporate information. After getting their feet wet with public Web sites that promote company products and services, corporations are seizing the Web as a swift way to streamline—even transform—their organizations. These private networks use the infrastructure and standards of the Internet and the World Wide Web. Using an intranet offers one considerable advantage: Many people are already familiar with Internet technology, so they need little training to make effective use of their corporate intranet.

An intranet is an inexpensive yet powerful alternative to other forms of internal communication, including conventional computer setups. One of an intranet's most obvious virtues is its ability to reduce the need for paper. Because Web browsers run on any type of computer, the same electronic information can be viewed by any employee. That means that all sorts of documents (such as internal phone books, procedure manuals, training manuals, and requisition forms) can be inexpensively converted to electronic form on the Web and be constantly updated. An intranet provides employees with an easy and intuitive approach to accessing information that was previously difficult to obtain. For example, it is an ideal solution to providing information to a mobile sales force that needs access to rapidly changing information.

A growing number of companies offer limited network access to selected customers and suppliers. Such networks are referred to as extranets, which connect people who are external to the company. An **extranet** is a network that links selected resources of the intranet of a company with its customers, suppliers, or other business partners. Again, an extranet is built around Web technologies.

Security and performance concerns are different for an extranet than for a Web site or network-based intranet. User authentication and privacy are critical on an extranet so that information is protected. Obviously, the network must perform well to provide quick response to customers and suppliers. Table 4.7 summarizes the differences between users of the Internet, intranets, and extranets.

extranet
A network based on Web technologies that links selected resources of a company's intranet with its customers, suppliers, or other business partners.

Type	Users	Need User ID and Password?
Internet	Anyone	No
Intranet	Employees	Yes
Extranet	Business partners	Yes

Table 4.7

Summary of Internet, Intranet, and Extranet Users

Secure intranet and extranet access applications usually require the use of a virtual private network (VPN). A **virtual private network (VPN)** is a secure connection between two points on the Internet. VPNs transfer information by encapsulating traffic in IP packets and sending the packets over the Internet, a practice called **tunneling**. Most VPNs are built and run by ISPs. Companies that use a VPN from an ISP have essentially outsourced their networks to save money on wide area network equipment and personnel.

virtual private network (VPN)
A secure connection between two points on the Internet.

tunneling
The process by which VPNs transfer information by encapsulating traffic in IP packets over the Internet.

NET ISSUES

The topics raised in this chapter apply not only to the Internet and intranets but also to LANs, private WANS, and every type of network. Control, access, hardware, and security problems affect all networks, so you should be familiar with the following issues:

- **Management issues.** Although the Internet is a huge, global network, it is managed at the local level; no centralized governing body controls the Internet. Preventing attacks is always an important management issue. Increasingly, states are proposing legislation to help collect sales tax from Internet sales.

- **Service and speed issues.** The growth in Internet traffic continues to be significant. Traffic volume on company intranets is growing even faster than the Internet. Companies setting up an Internet or intranet Web site often underestimate the amount of computing power and communications capacity they need to serve all the "hits" (requests for pages) they get from Web cruisers.

- **Privacy, fraud, security, and unauthorized Internet sites.** As use of the Internet grows, privacy, fraud, and security issues become even more important. People and companies are reluctant to embrace the Internet unless these issues are successfully addressed. Unauthorized and unwanted Internet sites are also problems some companies face. A competitor or an unhappy employee can create an Internet site with an address similar to a company's. When someone searches for information about the company, he might find an unauthorized site instead. While the business use of the Web has soared, online scams have put the brakes on some Internet commerce. Many Internet users have cut back on their Internet shopping and banking because of potential Internet scams and concerns about privacy and identity theft. In a business setting, the Web can also be a distraction to doing productive work. Although many businesses block certain Web sites at work, others monitor Internet usage. Workers have been fired for inappropriate or personal use of the Internet while on the job.

SUMMARY

Principle

A telecommunications system has many fundamental components that must be carefully selected and work together effectively to enable people to meet personal and organization objectives.

Telecommunications refers to the electronic transmission of signals for communications, including telephone, radio, and television. Telecommunications is creating profound changes in business because it removes the barriers of time and distance.

The elements of a telecommunications system include the sending and receiving devices, modems, the transmission media, and the message. The sending unit transmits a signal to a modem, which performs a number of functions such as converting the signal into a different form or from one type to another. The modem then sends the signal through a medium, which is anything that carries an electronic signal and serves as an interface between a sending device and a receiving device. The signal is received by another modem that is connected to the receiving computer. The process can then be reversed, and another message can pass from the receiving unit to the original sending unit. A communications channel is the transmission medium that carries a message from the source to its receivers.

The telecommunications media that physically connect data communications devices can be divided into two broad categories: guided transmission media, in which communications signals are guided along a solid medium, and wireless media, in which the communications signal is sent over airwaves. Guided transmission media include twisted-pair wire cable, coaxial cable, fiber-optic cable, and broadband over power lines. Wireless telecommunications involves the broadcast of communications in one of three frequency ranges: microwave, radio, and infrared.

Wireless communications options include near field communications, Bluetooth, ultra wideband, Wi-Fi, wireless mesh, 3G, 4G and WiMAX.

Telecommunications uses various devices including modems, multiplexers, PBXs, front end processors, switches, bridge, routers, and gateways.

Telecommunications carriers offer a wide array of phone and dialing services including digital subscriber line and Voice over Internet.

The effective use of networks can turn a company into an agile, powerful, and creative organization, giving it a long-term competitive advantage. Networks let users share hardware, programs, and databases across the organization. They can transmit and receive information to improve organizational effectiveness and efficiency. They enable geographically separated workgroups to share documents and opinions, which fosters teamwork, innovative ideas, and new business strategies.

The physical distance between nodes on the network and the communications and services provided by the network determines whether it is called a personal area network (PAN), local area network (LAN), metropolitan area network (MAN), or a wide area network (WAN).

The electronic flow of data across international and global boundaries is often called transborder data flow.

When an organization needs to use two or more computer systems, it can follow one of three basic data processing strategies: centralized, decentralized, or distributed.

A client/server system is a network that connects a user's computer (a client) to one or more server computers (servers). A client is often a PC that requests services from the server, shares processing tasks with the server, and displays the results.

A communications protocol is a set of rules that govern the exchange of information over a communications channel. There are myriad communications protocols including international, national, and industry standards.

A network operating system controls the computer systems and devices on a network, allowing them to communicate with one another. Network-management software enables a manager to monitor the use of individual computers and shared hardware, scan for viruses, and ensure compliance with software licenses.

Principle

The Internet and the Web provide a wide range of services, some of which are effective and practical for use today, others are still evolving, and still others will fade away from lack of use.

The Internet started with ARPANET, a project sponsored by the U.S. Department of Defense (DoD). Today, the Internet is the world's largest computer network. Actually, it is a collection of interconnected networks, all freely exchanging information. The Internet transmits data from one computer (called a host) to another. The set of conventions used to pass packets from one host to another is known as the Internet Protocol (IP). Many other protocols are used with IP. The best known is the Transmission Control Protocol (TCP). Each computer on the Internet has an assigned address to identify it from other hosts, called its Uniform Resource Locator (URL). There are several ways to connect to the Internet: via a LAN whose server is an Internet host, via SLIP or PPP, and via an online service that provides Internet access.

An Internet service provider is any company that provides access to the Internet. To use this type of connection, you must have an account with the service provider and software that allows a direct link via TCP/IP.

The Web is a collection of independently owned computers that work together as one. High-speed Internet circuits connect these computers, and cross-indexing software is employed to enable users to jump from one Web computer to another effortlessly. Because of its ability to handle multimedia objects and hypertext links between distributed objects, the Web is emerging as the most popular means of information access on the Internet today.

A Web site is like a magazine, with a cover page called a *home page* that has graphics, titles, and black and highlighted text. Web pages are loosely analogous to chapters in a book. Hypertext links are maintained using URLs, a standard way of coding the locations of the HTML (Hypertext Markup Language) documents. In addition to HTML, a number of newer Web standards are gaining in popularity, including Extensible Markup Language (XML), Extensible Hypertext Markup Language (XHTML), and Cascading Style Sheets (CSS).

With a Web browser on your PC, you can call up any Web document—no matter what kind of computer it is on. Because Web browsers run on any type of computer, the same electronic information can be viewed by any employee. That means that all sorts of documents can be converted to electronic form on the Web and constantly be updated. Internet Explorer, Firefox and Safari are examples of Web browsers.

Computers using Web server software store and manage documents built on the Web's HTML format. A search engine helps find information on the Internet. Popular search engines include Yahoo! and Google. A meta-search engine submits keywords to several individual search engines and returns the results.

Web 2.0 refers to the use of the Web as a computing platform that supports software applications and the sharing of information among users.

Web programming languages include Java, JavaScript, VBScript, ActiveX, and Hypertext Preprocessor.

Microsoft .NET is one example of a product that makes developing and maintaining Web content easier.

Web services consist of a collection of standards and tools that streamline and simplify communication among Web sites.

The Web offers a dizzying array of services including search engines to aid in Web research, E-mail, instant messaging, video chat, information and tools to aid in finding career information and job searching, Telenet, SSH, FTP, Blog, Vlog, Podcasting, chat rooms, Internet Phone and videoconferencing services, social networks, media sharing, content streaming, shopping on the Web, Web auctions, music, radio, video and TV on the Internet, and office on the Web.

Principle

Because use of the Internet and the World Wide Web is becoming universal in the business environment, management, service and speed, privacy, and security issues must continually be addressed and resolved.

An intranet is an internal corporate network built using Internet and World Wide Web standards and products. It is used by the employees of an organization to gain access to corporate information.

An extranet is a network that links selected resources of the intranet of a company with its customers, suppliers, or other business partners. It is also built around Web technologies. Security and performance concerns are different for an extranet than for a Web site or network-based intranet. User authentication and privacy are critical on an extranet. Obviously, performance must be good to provide quick response to customers and suppliers.

Management issues and service and speed affect all networks. No centralized governing body controls the Internet. Also, because the amount of Internet traffic is so large, service bottlenecks often occur. Privacy, fraud, and security issues must continually be addressed and resolved.

CHAPTER 4: SELF-ASSESSMENT TEST

A telecommunications system has many fundamental components that must be carefully selected and work together effectively to enable people to meet personal and organization objectives.

1. _____ is a relative term that generally means a telecommunications system that can exchange data very quickly.

2. _____ involves the broadcast of communications in one of three frequency ranges: microwave, radio or infrared.

3. A telecommunications service that delivers high-speed Internet access to homes and small businesses over existing phone lines is called _____.
   a. BPL
   b. DSL
   c. Wi-Fi
   d. Ethernet

4. A(n) _____ is a network that can connect technology devices within a range of 33 feet or so.

The Internet and the Web provide a wide range of services, some of which are effective and practical for use today, others are still evolving, and still others will fade away from lack of use.

5. On the Internet, _____ enables traffic to flow from one network to another.
    a. Internet Protocol
    b. ARPANET
    c. Uniform Resource Locator
    d. LAN server

6. _____ is a company that provides people and organizations with access to the Internet.

7. Firefox, Safari and Internet Explorer are examples of _____.
    a. Web browsers
    b. chat rooms
    c. search engines
    d. Web programming languages

8. _____ can be used to route phone calls over networks and the Internet.

9. The standard page description language for Web pages is _____.
    a. Home Page Language
    b. Hypermedia Language
    c. Java
    d. Hypertext Markup Language (HTML)

Because use of the Internet and World Wide Web is becoming universal in the business environment, management, service and speed, privacy, and security issues must continually be addressed and resolved.

10. A(n) _____ is a network based on Web technology that links customers, suppliers, and others to the company.

11. Privacy, fraud, and security issues on the Internet have all but been eliminated through the diligent efforts of ISPs and software providers. True or False?

CHAPTER 4: SELF-ASSESSMENT TEST ANSWERS

(1) broadband communications (2) wireless telecommunications (3) b (4) personal area network (5) a (6) Internet service provider (ISP) (7) a (8) Voice over IP (VoIP) (9) d (10) extranet (11) False

REVIEW QUESTIONS

1. Describe the elements and steps involved in the telecommunications process.
2. What is a telecommunications protocol?
3. What is a VoIP? What capabilities does it provide?
4. What is mesh networking? What are some of its advantages?
5. What is a metropolitan area network?
6. Identify the features of the Internet that make it unlikely to stop working from a single point of failure. Why do you think the Internet has such a high degree of redundancy?
7. Explain the naming conventions used to identify Internet host computers.
8. Briefly describe four different ways to connect to the Internet. What are the advantages and disadvantages of each approach?
9. What is an Internet service provider? What services do they provide?
10. What is a podcast?
11. What is the ARPANET?
12. What is content streaming?
13. What is instant messaging?
14. What is the Web? Is it another network like the Internet or a service that runs on the Internet?
15. What is a URL and how is it used?
16. What is HTML and how is it used?
17. What is a Web browser? How is it different from a Web search engine?
18. What is an intranet? Provide three examples of the use of an intranet.
19. What is an extranet? How is it different from an intranet?
20. Identify three important Internet issues.

DISCUSSION QUESTIONS

1. Why is an organization that employs centralized processing likely to have a different management decision-making philosophy than an organization that employs distributed processing?

2. What are the pros and cons of distributed processing versus centralized processing for a large retail chain?

3. Identify at least three wireless telecommunications protocols. Why do you think that there are so many protocols? Will the number of protocols increase or shrink over time?

4. Instant messaging is being widely used today. Describe how this technology could be used in a business setting. Are there any drawbacks or limitations to using instant messaging in a business setting?

5. Your company is about to develop a new Web site. Describe how you could use Web services for your site.

6. Describe how a company could use a blog and podcasting.

7. Why is it important to have an organization that manages IP addresses and domain names?

8. Briefly describe some of the tools that can be used to build a Web page.

9. One of the key issues associated with the development of a Web site is getting people to visit it. If you were developing a Web site, how would you inform others about it and make it interesting enough that they would return and also tell others about it?

10. Getting music, radio, and video programs from the Internet is easier than in the past, but some companies are still worried that people will illegally obtain copies of this programming without paying the artists and producers royalties. If you were an artist or producer, what would you do?

11. How could you use the Internet if you were a traveling salesperson?

12. Briefly summarize the differences in how the Internet, a company intranet, and an extranet are accessed and used.

PROBLEM-SOLVING EXERCISES

1. You have been hired as a telecommunications consultant to help an organization assess the benefits and potential cost savings associated with replacing an existing wired LAN with a wireless Wi-Fi network for an organization of 450 people located in a three-story building. You have determined that the cost to remove the current LAN and replace it with a new Wi-Fi network is $150,000. The cost of moving an existing LAN jack or installing a new one for the old LAN was $125 per change. The number of moves and installs averaged 25 per month. Develop a spreadsheet to analyze the costs and savings over a five-year period. Write

a recommendation to management based on your findings and any other factors that might support or not support the installation of a Wi-Fi network.

2. Do research on the Web to find several popular Web auction sites. After researching these sites, use a word processor to write a report on the advantages and potential problems of using a Web auction site to purchase a product or service. Also discuss the advantages and potential problems of selling a product or service on a Web auction site. How could you prevent scams on an auction Web site?

TEAM ACTIVITIES

1. With a group of your classmates, develop a proposal to install videoconferencing equipment in one of your school's classrooms so that students can view lectures at a distant videoconferencing facility. What sort of equipment is required, who provides this equipment, and what does it cost to install and operate?

2. Have each team member use a different search engine to find information about podcasting. Meet as a team and decide which search engine was the best for this task. Write a brief report to your instructor summarizing your findings.

WEB EXERCISES

1. Do research on the Web on the current status of Wi-Fi versus WiMAX. Which communications protocol is in widest use? Why? Write a short report on what you found.

2. Your group will use Web 2.0 sites to organize a social gathering. First choose a group name based on what type of social event you are planning. This could be an actual event

that group members will attend such as "Pizza Extravaganza." Use Facebook to create a group page and use it to communicate with group members. Use the group page to establish who will be the group leader. Each member should use Google Calendar to post his or her activities for the week the event is to take place. Share your calendars with everyone in the group. The group leader should examine everyone's calendar to determine a date and time when everyone is available for the event. Create the event and invite the other group members using Google Calendar and Gmail.

The leader should create a document using Google Docs that lists details of the event—the title, purpose, activities on the agenda, food that will be available, the responsibilities of those attending, etc. Share the document with group members. Group members should share their ideas by editing the document. The group leader should judiciously decide which edits to keep and which to reject. Present your instructor with information to join your Facebook group and to view your calendars and Google doc. Write a summary of your experiences with this exercise.

CAREER EXERCISES

1. One of the many online job-search companies includes Monster.com. Investigate one or more of these companies and research the positions available in the telecommunications industry, including the Internet. You might be asked to summarize your findings for your class in a written or verbal report.

2. Describe how the Internet can be used on the job for two careers that interest you.

CASE STUDIES

Case One
Del Monte Provides Secure Connections for Telecommuters

San Francisco-based Del Monte Foods is one of America's largest and most well-known food companies and the second largest pet foods company. It generated approximately $3.4 billion in net sales in 2007 through its numerous brands, which include Del Monte, StarKist, Contadina, Milkbone, 9Lives, Meow Mix, and Nature's Recipe.

Del Monte depends on telecommunications networks to supply its 7,800 full-time employees with access to information systems such as the corporate enterprise resource planning (ERP), data warehouse, and customer relationship management (CRM) applications. An increasing number of Del Monte employees work from home offices or remote sales offices. Del Monte needed a system that employees could use to access the corporate network so that they could work as effectively as employees in the corporate offices.

The challenge with providing access to corporate networks to people outside the network is security. By opening connections over the Internet, a company makes its network more vulnerable to hackers. Del Monte wanted to provide access to its corporate data and services to employees outside the office and to select business partners and other third parties without putting its network at risk. Del Monte needed a secure intranet and extranet.

Del Monte worked with telecommunications professionals to set up a secure Web site that employees and partners could access from any Internet connection. The Web site uses VPN authentication and Cisco's Secure Access Control Server to keep hackers out and allow authorized users in. Once logged on, the user can access only the portions of the network and data that they have been authorized to access. For example, a sales representative may need access to Del Monte's data warehouse to track an order, while an accountant may be provided with access to the ERP.

Del Monte supplies an even more secure connection for employees that work from home. These employees are given a Cisco ASA 5500 Series Adaptive Security Appliance, a network device that provides a firewall and intrusion prevention system (IPS) to keep hackers out, and virtual private networking (VPN) to encrypt and safeguard data flowing over the network. Telecommuters connect their PCs and telephones to the device to enjoy the same quality of network service as those in corporate headquarters.

The Cisco device allows professionals to receive business phone calls at home, while referencing data acquired from the corporate information systems. Del Monte calls it an "office in a box." Telecommuters can even use the system to attend meetings through videoconferencing software.

Using Del Monte's secure extranet and secure home office system, employees at home offices and remote locations can communicate with employees anywhere to collaborate more effectively over one integrated network for voice, video, and data. The system saves Del Monte money by allowing the company to remove expensive T1 lines from remote offices and replace them with high-speed Internet connections. The system is also easy to manage and expand. Del Monte could easily add new security features as they are needed.

Network engineers at Del Monte are currently working on using the service for disaster recovery. If Del Monte's corporate offices were to experience a fire, earthquake, or another natural disaster, the company could continue operations by using the extranet to allow all employees to access network resources from home.

Discussion Questions

1. What is a primary concern of making a private network available to employees who are outside the office? Why?
2. What technologies did Del Monte employ to address this primary concern?

Critical Thinking Questions

1. What benefits do Del Monte and its employees enjoy by providing extranet access to the Del Monte network?
2. If you could choose whether to work at home or in a corporate office, which would you choose and why?

Sources: Cisco Staff, "Food Manufacturer Extends its Workplace with Secure Remote Access," Cisco Success Story, *www.cisco.com/en/US/prod/collateral/ vpndevc/ps6032/ps6094/ps6120/case_study_c36-464676_v1.pdf*, accessed April 28, 2008; Del Monte Web site, *www.delmonte.com*, accessed April 28, 2008.

Case Two
Procter & Gamble Implement Enterprise 2.0

Procter & Gamble (P&G) owns a large portfolio of familiar brands such as Pampers, Tide, Bounty, Pringles, Charmin, and Crest. P&G operates in more than 80 countries worldwide, with net sales increasing continuously over the past ten years to over $76 billion in 2007.

Procter & Gamble's CEO, A.G. Lafley, believes in communication and collaboration. He is pushing P&G IT Innovation Manager Joe Schueller to find more effective and innovative ways for P&G's 138,000 employees to collaborate online. Naturally, Schueller looked immediately to Web 2.0 technologies for ideas. When applied to an enterprise, Web 2.0 technologies are referred to as Enterprise 2.0.

Schueller is not a fan of e-mail. He sees it as a barrier to employees' use of more effective means of communication. Replying to all recipients of a message ends up wasting the time of people who do not need to receive, read, and respond to the message. Instead Schueller has equipped P&G employees with easy access to a corporate blog. For some types of group communications, Schueller finds blogs the ideal tool. Information is not forced on people. Those interested can follow the blog and post comments to add to the dialog.

Schueller is harnessing the power of the wiki as a content and knowledge management system. Members of the organization who have valuable knowledge about P&G topics can post articles and advice. That helps corporate knowledge stay within the company, even when knowledgeable employees leave.

P&G banked on Microsoft products to provide most of its Enterprise 2.0 functionality. Microsoft Live Communications Server provides instant messaging, unified communications, and presence—the ability to access communications services from any location. Live Meeting provides Web conferencing, and SharePoint provides a platform for content management and collaboration. Roughly 80,000 P&G employees use corporate instant messaging tools.

Besides using Microsoft products, P&G also uses software and tools from other vendors for its Enterprise 2.0 investments. For example, P&G uses a product from Connectbeam that works with Google search tools to allow employees to share bookmarks and tag articles, pages, and documents with descriptive words to make information easier to find. P&G has launched a corporate social networking site so that employees can let others know who they are and in which areas of corporate activities they are involved. The goal is to encourage employees to easily find others with expert knowledge. All of these Enterprise 2.0 applications are accessed through a unified portal that also includes RSS feeds of business news.

P&G is serving as inspiration to other companies who are developing an interest in Enterprise 2.0. Information systems departments see Web 2.0 technologies as a chance to provide real value to the organization. Bank of America, Boeing, the CIA, FedEx, Morgan Stanley, and Pfizer are examining Schueller's example. Motorola has also invested in Enterprise 2.0, with an intranet that includes 4,400 blogs and 4,200 wiki pages.

Discussion Questions

1. What qualities of Web 2.0 applications are appealing for enterprise use?
2. Why might a company not want to use Web 2.0 applications?

Critical Thinking Questions

1. How can each of the five Enterprise 2.0 applications used by P&G help its employees be more effective and efficient?
2. Compare and contrast e-mail, IM, and blogs as tools for effective communications.

Sources: Hoover, Nicholas, "Beyond E-Mail," *Information Week*, June 25, 2007, pages 29-30; Procter and Gamble Corporate Web site, *www.pg.com/en_US/ index.jhtml*, accessed May 7, 2008.

Questions for Web Case

See the Web site for this book to read about the Whitmann Price Consulting case for this chapter. Following are questions concerning this Web case.

Whitmann Price Consulting: Telecommunications, the Internet, Intranets, and Extranets

Discussion Questions

1. What role does bandwidth play in the successful delivery of the Advanced Mobile Communications and Information System?
2. When does functionality transform the standard BlackBerry device into an Advanced Mobile Communications and Information System?

Critical Thinking Questions

1. Describe three telecommunications and network technologies used to connect the BlackBerry with other devices.
2. At this stage in the process, what actions might Sandra and Josh take to reduce the overall costs of the Advanced Mobile Communications and Information System?

NOTES

1 Nystedt, Dan, "Wal-Mart Eyes $287 Million Benefit from RFID," *Network World*, October 12, 2007.
2 Reese, Brad, "Procter & Gamble CIO Filippo Passerini Deploys Expensive Rollout of Cisco Telepresence System," *Network World*, December 30, 2007.
3 Desmond, Paul, "Building Virtual Worlds at Boeing," *Network World*, October 29, 2007.
4 Hoover, Nicholas, J., "TSA to Test Broadband Over Powerline Technology," *Information Week*, January 6, 2007.
5 Brown, Damon, "Letting Go of Your Landline, *Technology, Inc.*, January 2007.
6 Malykhina, Elena, "Deployments of Contactless Payment Systems Slower than Expected," *InformationWeek*, January 10, 2008.
7 Griffith, Eric, "In 2008, UWB Takes the World," *Ultrawideband Planet.com*, June 18, 2007.
8 Lawson, Stephen, "Year End—New Mobile Approaches Got a Reality Check," *Network World*, December 18, 2007.
9 "Pine Cellular and Choctaw Electric Cooperative Will Use Nortel 4G WiMAX to Bring Broadband to Areas of Southeastern Oklahoma," *Fierce Wireless*, September 27, 2007.
10 Gohring, Nancy, "CES—Sprint Insists Its WiMAX Network Is on Track," *Network World*, January 8, 2008.
11 Shifrin, Tash, "Merrill Lynch Picks VoIP-Powered Trading Systems, *Computerworld*, November 15, 2007.
12 Flaccus, Gillian, "Disneyland to Open House of the Future 2.0," *Tech News World*, February 13, 2008.
13 "Case Study: Miami-Dade Police Department," *www.nortel.com*, accessed February 15, 2008.
14 Hamblen, Matt, "Jenny Craig Upgrades WAN to 500 Sites," *Computerworld*, February 15, 2007.
15 "T-Mobile Austria Looks to IBM Tivoli Netcool Software for End-to-End Visibility of Its Wireless Networks," March 27, 2007, *www-01.ibm.com/software/success*.
16 Internet Usage World Stats Web site, *www.internetworldstats.com*, accessed April 1, 2008.
17 Zhao, Michael, "60-Mile Wi-Fi," *Forbes*, April 9, 2007, pages 76-78.
18 Internet2 Web site, *www.internet2.edu*, accessed April 1, 2008.
19 National LambdaRail Web site, *www.nlr.net*, accessed April 1, 2008.
20 Swibel, Matthew, "Fly the Connected Skies," *Forbes*, November 26, 2007, page 56.
21 Gardner, David W., "Google's Android Builds on Past Features, Hints at Future Enhancements," *InformationWeek*, November 5, 2007.
22 Worthen, Ben, "Web Surfing on iPhone Erases Doubts Of Mobile Devices' Future Online Role," *Wall Street Journal*, December 11, 2007, Business Technology section, page 84.
23 Case, Loyd, "Intel Launches Low-Power Atom Processor," PC Magazine, April 2, 2008, *www.pcmag.com/article2/0,2817,2280892,00.asp*.
24 Mossberg, Walter, "Desktop Modules Help To Personalize Data, Cut Through Clutter," *The Wall Street Journal*, February 1, 2007, Marketplace Section, Page B1.
25 Neville, Jeffrey, "Web 2.0's Wild Blue Yonder," *Information Week*, January 8, 2007, Page 45.
26 Larkin, Erik, "Back Up Your Files Online Without Even Trying," *PC World*, May 2007, page 112.
27 Claburn, Thomas, "Mashups Made Easy," *Information Week*, February 12, 2007, Page 14.
28 Spanbauer, Scott, "Advanced Google: Search Faster, Find More," *PC World*, February 2008, page 128-130.
29 Delaney, Kevin, "How Search-Engine Rules Cause Sites to Go Missing," *The Wall Street Journal*, Page B1.
30 Gomes, Lee, "Forget the Articles, Best Wikipedia Read Is Its Discussions," *The Wall Street Journal*, August 15, 2007, Marketplace Section, page B1.
31 Lyons, Daniel, "Easy Blogging," *Forbes*, April 9, 2007, pages 56-57.
32 Savell, Lawrence, "Blogger Beware!," *Computerworld*, September 24, 2007, pages 32-34.
33 Enrich, David, "Turning an Online Community Into a Business," *The Wall Street Journal*, Page B8.
34 Gomes, Lee, "Web Is Now So Filled With Idle Chat, It's Almost Like Phoning," *The Wall Street Journal*, July 11, 2007, Page B1.
35 Smith, Ethan and Vara, Vauhini, "Music Service from Amazon Takes on iTunes," *The Wall Street Journal*, May 17, 2007, Personal Journal section, page D1.
36 Coyle, Jake, "Site allows free music downloads," *Rocky Mountain News*, September 24, 2007, Business Section page 4.
37 Yuan, Li, "Cellphone Video Gets On the Beam," *The Wall Street Journal*, January 4, 2007, Technology Journal Section, Page B3.
38 Grant, Peter, "Find It on the Web, Watch It on TV," *The Wall Street Journal*, January 3, 2007, Marketplace Section, Page B1.
39 Cassavoy, Liane, "Slingbox Keeps You in Touch With Your TV," *PCWorld*, January 2007, page 78.
40 Reuters Staff, "MIT to offer its courses free online by year end," Reuters, March 9, 2007, *www.reuters.com/article/domesticNews/idUSN0927676520070310*.

Business Information Systems

CHAPTER
· 5 ·

Electronic and Mobile Commerce and Enterprise Systems

PRINCIPLES	LEARNING OBJECTIVES
▪ Electronic commerce and mobile commerce are evolving, providing new ways of conducting business that present both opportunities for improvement and potential problems.	▪ Describe the current status of various forms of e-commerce, including B2B, B2C, C2C, and m-commerce. ▪ Identify several e-commerce and m-commerce applications. ▪ Identify several advantages associated with the use of e-commerce and m-commerce.
▪ E-commerce and m-commerce require the careful planning and integration of a number of technology infrastructure components.	▪ Identify the key components of technology infrastructure that must be in place for e-commerce and m-commerce to work. ▪ Discuss the key features of the electronic payment systems needed to support e-commerce and m-commerce.
▪ An organization must have information systems that support the routine, day-to-day activities that occur in the normal course of business and help a company add value to its products and services.	▪ Identify the basic activities and business objectives common to all transaction processing systems. ▪ Identify key control and management issues associated with transaction processing systems.
▪ A company that implements an enterprise resource planning system is creating a highly integrated set of systems, which can lead to many business benefits.	▪ Discuss the advantages and disadvantages associated with the implementation of an enterprise resource planning system. ▪ Identify the challenges multinational corporations must face in planning, building, and operating their TPSs.

Information Systems in the Global Economy
Maporama, France

Maporama Gains Ground Through New Enterprise System

Based in Paris, France, Maporama is a world leader in developing location-based services for businesses. Maporama develops GPS-based systems that allow businesses to track mobile work forces, sales outlets, and competition. Maporama's Web site states that its custom-designed products "empower existing mission-critical applications or processes within an organization and display results enterprise-wide, as needed, via connected devices (computers, mobile phones, PDAs, etc.)" The company claims 500 customers supporting 26 languages on five continents, and boasts the most complete pan-European map coverage.

Recently Maporama had a pressing need to gain tighter control over its global mobile workforce and associated customer information. The company's rapid growth to 500 customers—with 10,000 contacts to track—left its European and North American sales force with more information than it could manage. The company decided to invest in a customer relationship management system that could help manage its customer relations and synchronize sales information across the entire enterprise.

When the company was smaller, Maporama used Microsoft Outlook and Exchange to manage customer information. Over time, Maporama outgrew that system and could no longer easily manage customer accounts. It became difficult to synchronize the information between departments, which resulted in disconnected islands of information. The company needed a system that would collect all information into a central database, which could be leveraged to improve sales and customer service.

Initially, Maporama investigated systems that they could implement themselves through an internal server. The company decided that such a system would be too costly and a burden to maintain. When Maporama discovered a hosted system accessed over the Internet, the company knew that it was the perfect solution. Software as a Service (SaaS) allowed Maporama to use a full-service CRM system for a monthly fee delivered to any Internet-connected device around the world.

Maporama's new CRM system was set up and ready to go in 15 days. (It would have taken over a year for Maporama to set up such a system in-house.) Maporama's sales, marketing, and support teams use the system at headquarters and worldwide offices. According to Dominique Grillet, CEO of Maporama, the new system allows staff and management to find answers to questions such as: "Who should a salesperson call on today?", "Which products have customers bought in the past, and when?", "What was the last contact with a customer and by whom?", "Has a sister company bought our products in another country?", "How effective are our telesales and marketing campaign?", "What are our sales forecasts?", "In what industry sectors are we winning the most business?"

With the new enterprise system, Maporama can quickly respond to customers' needs and boost revenue opportunities. The company can also maximize the productivity of its sales force as individuals and a team, and implement more effective marketing campaigns for less money by improving customer targeting.

As you read this chapter, consider the following:

- What advantages do e-commerce and m-commerce offer sellers and vendors over traditional shopping venues?
- What types of information systems are critical to the success of a business and how are the systems related to one another?

Why Learn About Electronic and Mobile Commerce and Enterprise Systems?

Electronic and mobile commerce and enterprise systems have transformed many areas of our lives and careers. One fundamental change has been the manner in which companies interact with their suppliers, customers, government agencies, and other business partners. As a result, most organizations today have or are considering setting up business on the Internet and implementing integrated enterprise systems. To be successful, all members of the organization need to participate in that effort. As a sales or marketing manager, you will be expected to help define your firm's e-commerce business model. Customer service employees can expect to use enterprise systems to provide improved customer service. As a human resource or public relations manager, you will likely be asked to provide content for a Web site directed to potential employees and investors. Analysts in finance need to know how to set up and use enterprise systems to capture and report the data needed to manage and control the firm's operations. Clearly, as an employee in today's organization, you must understand the potential role of e-commerce and enterprise systems, how to capitalize on their many opportunities, and how to avoid their pitfalls. The emergence of m-commerce adds an exciting new dimension to these opportunities and challenges. This chapter begins by providing a brief overview of the dynamic world of e-commerce and defines its various components.

AN INTRODUCTION TO ELECTRONIC COMMERCE

electronic commerce
Conducting business activities (e.g., distribution, buying, selling, marketing, and servicing of products or services) electronically over computer networks such as the Internet, extranets, and corporate networks.

Electronic commerce is the conducting of business activities (e.g., distribution, buying, selling, marketing, and servicing of products or services) electronically over computer networks such as the Internet, extranets, and corporate networks. Business activities that are strong candidates for conversion to e-commerce are paper based, time-consuming, and inconvenient for customers. Thus, some of the first business processes that companies converted to an e-commerce model were those related to buying and selling. For example, after Cisco Systems, the maker of Internet routers and other telecommunications equipment, put its procurement operation online in 1998, the company reported that it halved cycle times and saved an additional $170 million in material and labor costs. Similarly, Charles Schwab & Co. slashed transaction costs by more than half by shifting brokerage transactions from traditional channels such as retail and phone centers to the Internet.

Business-to-Business (B2B) E-Commerce

business-to-business (B2B) e-commerce
A subset of e-commerce where all the participants are organizations.

Business-to-business (B2B) e-commerce is a subset of e-commerce where all the participants are organizations. B2B e-commerce is a useful tool for connecting business partners in a virtual supply chain to cut resupply times and reduce costs. Although the business-to-consumer market grabs more of the news headlines, the B2B market is considerably larger and is growing more rapidly. As early as 2003, over 80 percent of U.S. companies had already experimented with some form of B2B online procurement.[1]

Covisint operates a Web portal that supports B2B by performing data translations and code conversions to enable auto makers and parts suppliers to collaborate on orders, scheduling, shipping, and other manufacturing-related tasks. Covisint is expanding its data translation and collaboration services into the healthcare industry to enable sharing of patient care data among healthcare providers and insurance companies.[2]

Business-to-Consumer (B2C) E-Commerce

business-to-consumer (B2C) e-commerce
A form of e-commerce in which customers deal directly with an organization and avoid intermediaries.

Early **business-to-consumer (B2C) e-commerce** pioneers competed with the traditional "brick-and-mortar" retailers in an industry selling their products directly to consumers. For example, in 1995, upstart Amazon.com challenged well-established booksellers Waldenbooks and Barnes and Noble. Although Amazon did not become profitable until 2003, the firm has grown from selling only books on a U.S.-based Web site to selling a wide variety of products (including apparel, CDs, DVDs, home and garden supplies, and consumer electronic devices) from international Web sites in Canada, China, France, Germany, Japan, and the United Kingdom. Although it is estimated that B2C e-commerce represents only about 3.4 percent of retail sales in the U.S., the rate of growth of online purchases is three times faster than the growth in total retail sales.[3] One reason for the rapid growth is that shoppers find many goods and services are cheaper when purchased via the Web, including stocks, books, newspapers, airline tickets, and hotel rooms. They can also compare information about automobiles, cruises, loans, insurance, and home prices to find better values.

More than just a tool for placing orders, the Internet is an extremely useful way to compare prices, features, and value. Internet shoppers can, for example, unleash shopping bots or access sites such as eBay's Shopping.com, Google's Froogle, Shopzilla, PriceGrabber, Yahoo! Shopping, or Excite to browse the Internet and obtain lists of items, prices, and merchants. Yahoo! is adding what it calls "social commerce" to its Web site by creating a new section of Yahoo! where users can go to see only those products that have been reviewed and listed by other shoppers.

By using B2C e-commerce to sell directly to consumers, producers or providers of consumer products can eliminate the middlemen, or intermediaries, between them and the consumer. In many cases, this squeezes costs and inefficiencies out of the supply chain and can lead to higher profits and lower prices. The elimination of intermediate organizations between the producer and the consumer is called *disintermediation*.

Dell is an example of a manufacturer that has successfully embraced this model to achieve a strong competitive advantage. People can specify their own unique computer online, and Dell assembles the components and ships the computer directly to the consumer within five days. Dell does not inventory computers and does not sell through intermediate resellers or distributors. The savings are used to increase Dell profits and reduce consumer prices.

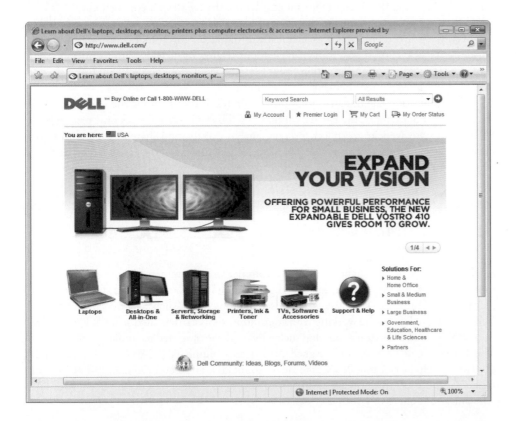

Dell sells its products through the Dell.com Web site.

Many retailers have elected to increase their sales by adding a Web site component to their operations. For example, American Eagle Outfitters launched a B2C Web site for Martin + OSA, its brand targeting 28- to 40-year old men and women. Says Laura Dubin-Wander, president of Martin + Osa: "We're excited to introduce Martin + Osa as a global brand through our e-commerce Web site. Free shipping and returns, along with unique shopping tools, give customers a world-class online shopping experience that's both frictionless and fun."[4]

Consumer-to-Consumer (C2C) E-Commerce

consumer-to-consumer (C2C) e-commerce
A subset of e-commerce that involves consumers selling directly to other consumers.

Consumer-to-consumer (C2C) e-commerce is a subset of e-commerce that involves consumers selling directly to other consumers. eBay is an example of a C2C e-commerce site; customers buy and sell items directly to each other through the site. Founded in 1995, eBay has become one of the most popular Web sites in the world; in 2007, 2.3 billion items were listed for sale and 276 million registered users bought and sold items valued at more than $57 billion.[5]

The Web offers many C2C sites, with some of the more popular being Bidzcom, Craigslist, eBid, ePier, Ibidfree, Ubid, and QXL. The growth of C2C is responsible for reducing the use of the classified pages of a newspaper to advertise and sell personal items. Many people make a living out of selling items on auction Web sites.

e-Government

e-Government
The use of information and communications technology to simplify the sharing of information, speed formerly paper-based processes, and improve the relationship between citizen and government.

e-Government is the use of information and communications technology to simplify the sharing of information, speed formerly paper-based processes, and improve the relationship between citizens and government. Government-to-consumer (G2C), government-to-business (G2B), and government-to-government (G2G) are all forms of e-Government, each with different applications.

Citizens can use G2C applications to submit their state and federal tax returns online, renew auto licenses, apply for student loans, and make campaign contributions. Information about the 2008 economic stimulus payments that were sent to over 130 million taxpayers was available on the IRS Web site for months before the rebates were mailed out.

G2B applications support the purchase of materials and services from private industry by government procurement offices, enable firms to bid on government contracts, and help businesses receive current government regulations related to their operations. Business.gov allows businesses to access information about laws and regulations and relevant forms needed to comply with federal requirements for their business.

G2G applications are designed to improve communications among the various levels of government. For example, the E-Vital initiative establishes common electronic processes for federal and state agencies to collect, process, analyze, verify, and share death record information. Geospatial One-Stop's Web portal, GeoData.gov, makes it easier, faster, and less expensive to find, share, and access geospatial information for all levels of government.

MOBILE COMMERCE

As discussed briefly in Chapter 1, mobile commerce (m-commerce) relies on the use of mobile, wireless devices, such as personal digital assistants, cell phones, and smartphones, to place orders and conduct business. Handset manufacturers such as Ericsson, Motorola, Nokia, and Qualcomm are working with communications carriers such as AT&T, Cingular, Sprint/Nextel, and Verizon to develop such wireless devices, related technology, and services. The Internet Corporation for Assigned Names and Numbers (ICANN) created a .mobi domain to help attract mobile users to the Web. mTID Top Level Domain Ltd of Dublin, Ireland administers this domain and helps to ensure that the .mobi destinations work fast, efficiently, and effectively with user handsets.

Mobile Commerce in Perspective

The market for m-commerce in North America is maturing much later than in Western Europe and Japan for several reasons. In North America, responsibility for network infrastructure is fragmented among many providers, consumer payments are usually made by credit card, and many Americans are unfamiliar with mobile data services. In most Western European countries, communicating via wireless devices is common, and consumers are much more willing to use m-commerce. Japanese consumers are generally enthusiastic about new technology and are much more likely to use mobile technologies for making purchases.

M-commerce spending in the U.S. is expected to exceed $500 million in 2008 and grow to almost $2 billion by 2010 according to Juniper Research. For perspective, U.S. e-commerce exceeded $100 billion in 2006.[6]

It is estimated that 40 percent of U.S. companies with annual revenue exceeding $50 million have established mobile Web sites.[7] The number of mobile Web sites is expected to grow because of advances in wireless broadband technologies, the development of new and useful applications, and the availability of less costly but more powerful handsets. For example, Yahoo's oneSearch 2.0 mobile search service includes a predictive text-search completion capability as well as voice recognition technology that adapts to a user's vocal patterns.[8] However, the relative clumsiness of mobile browsers and security concerns must be overcome to ensure rapid m-commerce growth.[9]

When it comes to mobile Web sites and mobile Web browsing capabilities, "just because you build it, doesn't mean they'll come," says Nikki Baird, managing partner at Retail Systems Research LLC. "You have to make consumers aware. It's all about getting people to try something new in the hope they'll come back for more."[10]

ELECTRONIC AND MOBILE COMMERCE APPLICATIONS

E-commerce and m-commerce are being used in innovative and exciting ways. This section examines a few of the many B2B, B2C, C2C, and m-commerce applications in the retail and wholesale, manufacturing, marketing, investment and finance, real estate services, and auction arenas.

Retail and Wholesale

E-commerce is being used extensively in retailing and wholesaling. **Electronic retailing**, sometimes called *e-tailing*, is the direct sale of products or services by businesses to consumers through electronic storefronts, which are typically designed around the familiar electronic catalog and shopping cart model. Companies such as Office Depot, Wal-Mart, and many others have used the same model to sell wholesale goods to employees of corporations. Tens of thousands of electronic retail Web sites sell everything from soup to nuts.

Cybermalls are another means to support retail shopping. A **cybermall** is a single Web site that offers many products and services at one Internet location—similar to a regular shopping mall. An Internet cybermall pulls multiple buyers and sellers into one virtual place, easily reachable through a Web browser.

Sears, the company that pioneered the use of the mail-order catalog back in the 1890s, is making a major investment in B2C e-commerce employing more than 100 technology workers to improve its online sales. It ranks as the second largest mass merchant retailer online with recent sales of $2.6 billion. With the number of unique visitors per month growing at over 20 percent, Sears is the second fastest growing site among mass retailers. Some industry experts believe that Sears.com may turn into a cybermall that sells all kinds of products and competes with companies such as Amazon.com.[11]

A key sector of wholesale e-commerce is spending on manufacturing, repair, and operations (MRO) goods and services—from simple office supplies to mission-critical equipment, such as the motors, pumps, compressors, and instruments that keep manufacturing facilities running smoothly. MRO purchases often approach 40 percent of a manufacturing company's total revenues, but the purchasing system can be haphazard, without automated controls. In addition to these external purchase costs, companies face significant internal costs resulting from outdated and cumbersome MRO management processes. For example, studies show that a high percentage of manufacturing downtime is often caused by not having the

electronic retailing (e-tailing)
The direct sale from business to consumer through electronic storefronts, typically designed around an electronic catalog and shopping cart model.

cybermall
A single Web site that offers many products and services at one Internet location.

right part at the right time in the right place. The result is lost productivity and capacity. E-commerce software for plant operations provides powerful comparative searching capabilities to enable managers to identify functionally equivalent items, helping them spot opportunities to combine purchases for cost savings. Comparing various suppliers, coupled with consolidating more spending with fewer suppliers, leads to decreased costs. In addition, automated workflows are typically based on industry best practices, which can streamline processes.

Manufacturing

electronic exchange
An electronic forum where manufacturers, suppliers, and competitors buy and sell goods, trade market information, and run back-office operations.

One approach taken by many manufacturers to raise profitability and improve customer service is to move their supply chain operations onto the Internet. Here they can form an **electronic exchange** to join with competitors and suppliers alike, using computers and Web sites to buy and sell goods, trade market information, and run back-office operations, such as inventory control, as shown in Figure 5.1. With such an exchange, the business center is not a physical building but a network-based location where business interactions occur. This approach has greatly speeded up the movement of raw materials and finished products among all members of the business community, thus reducing the amount of inventory that must be maintained. It has also led to a much more competitive marketplace and lower prices. Private exchanges are owned and operated by a single company. The owner uses the exchange to trade exclusively with established business partners. Public exchanges are owned and operated by industry groups. They provide services and a common technology platform to their members and are open, usually for a fee, to any company that wants to use them.

Figure 5.1

Model of an Electronic Exchange

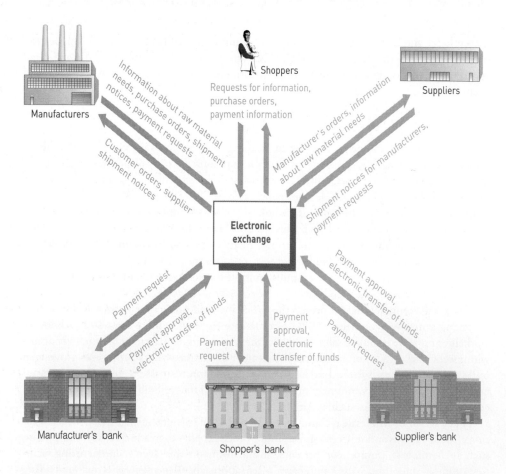

The Detroit Trading Exchange lets auto dealers and others bid to buy more than 300,000 sales leads generated from consumers who visit a host of auto-related Web sites. The sales leads can be sorted by zip code, financial factors, and other parameters so buyers tailor the sales leads they receive.[12]

Several strategic and competitive issues are associated with the use of exchanges. Many companies distrust their corporate rivals and fear they might lose trade secrets through participation in such exchanges. Suppliers worry that the online marketplaces and their auctions will drive down the prices of goods and favor buyers. Suppliers also can spend a great deal of money in the setup to participate in multiple exchanges. For example, more than a dozen new exchanges have appeared in the oil industry, and the printing industry is up to more than 20 online marketplaces. Until a clear winner emerges in particular industries, suppliers are more or less forced to sign on to several or all of them. Yet another issue is potential government scrutiny of exchange participants—when competitors get together to share information, it raises questions of collusion or antitrust behavior.

Many companies that already use the Internet for their private exchanges have no desire to share their expertise with competitors. At Wal-Mart, the world's number-one retail chain, executives turned down several invitations to join exchanges in the retail and consumer goods industries. Wal-Mart is pleased with its in-house exchange, Retail Link, which connects the company to 7,000 worldwide suppliers that sell everything from toothpaste to furniture.

Marketing

The nature of the Web allows firms to gather much more information about customer behavior and preferences than they could using other marketing approaches. Marketing organizations can measure many online activities as customers and potential customers gather information and make their purchase decisions. Analysis of this data is complicated because of the Web's interactivity and because each visitor voluntarily provides or refuses to provide personal data such as name, address, e-mail address, telephone number, and demographic data. Internet advertisers use the data they gather to identify specific portions of their markets and target them with tailored advertising messages. This practice, called **market segmentation**, divides the pool of potential customers into subgroups, which are usually defined in terms of demographic characteristics, such as age, gender, marital status, income level, and geographic location.

market segmentation
The identification of specific markets to target them with advertising messages.

comScore Networks is a global information provider to large companies seeking information on consumer behavior to boost their marketing, sales, and trading strategies.

technology-enabled relationship management
Occurs when a firm obtains detailed information about a customer's behavior, preferences, needs, and buying patterns and uses that information to set prices, negotiate terms, tailor promotions, add product features, and otherwise customize its entire relationship with that customer.

Technology-enabled relationship management is a new twist on establishing direct customer relationships made possible when firms promote and sell on the Web. **Technology-enabled relationship management** occurs when a firm obtains detailed information about a customer's behavior, preferences, needs, and buying patterns and uses that information to set prices, negotiate terms, tailor promotions, add product features, and otherwise customize its entire relationship with that customer.

Cliff Conneighton, senior vice president of e-commerce platform provider Art Technology Group (ATG), says: "The secret to improved sales on the Web is to deliver the right offer to someone at the right time. [You have] to know something about who you're selling to, and try to show them the goods and the offer that's more relevant." American Eagle Outfitters, an ATG client, followed this advice and doubled the revenue generated at its Web site in only one year.[13]

Investment and Finance

The Internet has revolutionized the world of investment and finance. Perhaps the changes have been so significant because this industry had so many built-in inefficiencies and so much opportunity for improvement.

The brokerage business adapted to the Internet faster than any other arm of finance. The allure of online trading that enables investors to do quick, thorough research and then buy shares in any company in a few seconds and at a fraction of the cost of a full-commission firm has brought many investors to the Web. Online brokerage firms have consolidated, with Ameritrade acquiring TD Waterhouse and E-Trade acquiring Harrisdirect and the online brokerage services of JP Morgan. In spite of the wealth of information available online, the average consumer buys stocks based on a tip or a recommendation rather than as the result of research and analysis. It is the more sophisticated investor that really takes advantage of the data and tools available on the Internet.

E-Trade is an online brokerage site that offers information, tools, and account-management services for investors.

Online banking customers can check balances of their savings, checking, and loan accounts; transfer money among accounts; and pay their bills. These customers enjoy the convenience of not writing checks by hand, tracking their current balances, and reducing expenditures on envelopes and stamps. In addition, the PayItGreen Alliance reports that paying bills online is good for the environment. By its estimates, the average household makes seven paper bill payments per month. If just ten percent of the United States population converted to online bill payment, the environmental savings would total more than 75 million pounds of paper, nearly one million trees and two million pounds of greenhouse gases.[14]

All of the major banks in the United States and many of the smaller banks enable their customers to pay bills online; many support bill payment via cell phone or other wireless device. Banks are eager to gain more customers who pay bills online because such customers tend to stay with the bank longer, have higher cash balances, and use more of the bank's products. To encourage the use of this service, many banks have eliminated all fees associated with online bill payment.

The next advance in online bill paying is **electronic bill presentment**, which eliminates all paper, down to the bill itself. With this process, the vendor posts an image of your statement on the Internet and alerts you by e-mail that your bill has arrived. You then direct your bank to pay it. ePower is an electronic bill presentment and payment solution provider that enables utility firms to provide interactive financial statements to their customers via e-mail and on the Internet at *www.payabill.com*.[15]

electronic bill presentment
A method of billing whereby a vendor posts an image of your statement on the Internet and alerts you by e-mail that your bill has arrived.

INFORMATION SYSTEMS @ WORK

MoneyAisle.com Puts Customers in Charge

E-commerce has shaken up traditional forms of commerce, and in some cases turned them completely upside down. Consider the consumer's ability to comparison shop online, sampling prices from businesses of all sizes located around the world. Consider how easy it is to find rare items, out-of-print books, and collectables. Online auction houses such as eBay have created an entirely new kind of marketplace.

While e-commerce has dramatically affected retail sales, other types of transactions have remained relatively stable. For example, consider choosing a bank for financial services. In the days before the popularity of the Internet, if you wanted to invest in a high-yield savings account or a certificate of deposit (CD), you would visit a number of local banks and find the best interest rate for the amount of money you planned to invest. The banks simply advertised their rates, and the customer did the work of collecting the data and making the decision based on value and bank reputation.

The Web has simplified this task, making thousands of bank quotes available online, but still the process is time consuming, and when you're done, it's hard to know if you've really found the best deal. Web sites such as LendingTree.com and Bankrate.com aggregate quotes from numerous banks, reducing the customer's research time, but the rates are still inflexible and the banks are in control.

MoneyAisle.com is working to change these factors by providing a service that puts the consumer in charge. At MoneyAisle.com, more than 100 reputable banks compete for your business. Unlike other services that merely give the impression of competition between banks for your business, at MoneyAisle.com the banks actually work to outbid each other for your business in a live, real-time auction. Customers use the form at MoneyAisle.com to provide the amount they want to invest and their state of residence, and then click the Start Auction button. After a few minutes, the customer watches banks bid in real time, round after round, until all but one bank has dropped out, offering the highest amount of interest.

MoneyAisle.com chief executive, Mukesh Chatter, thought of the idea for the business after noticing that prices for high-definition TVs varied significantly depending on the vendor. He saw similar variations in pricing elsewhere as well, including in banks. Chatter worked with partners to develop the algorithms to allow banks to place their bids for investor dollars, which is how MoneyAisle works. The site now earns revenue from charging participating banks a small fee. This provides a benefit to investors by finding them the best return on investment with the lowest amount of effort. It also provides benefits to smaller banks with less advertising capital. It is ordinarily difficult for smaller banks to compete with big banks with big advertising budgets. MoneyAisle.com levels the playing field giving banks of all sizes an equal opportunity.

The service offered by MoneyAisle.com meets the needs of smaller banks looking to increase business through online tools. The challenge for MoneyAisle.com will be to generate enough traffic to let banks know that using the service is worth the effort. So far, the idea seems to be catching on. In its first week of business, MoneyAisle.com was used as a tool for investing over $1 million at small and mid-size banks.

Discussion Questions

1. How does MoneyAisle.com turn the process of investing upside down?
2. Who benefits from the service offered by MoneyAisle.com? Who is negatively affected?

Critical Thinking Questions

1. For what other types of products might reverse bidding be useful? What makes a product a good subject for reverse bidding?
2. How does reverse bidding impact the way that a bank operates and its budget and profit margins?

SOURCES: Rosencrance, Linda, "MoneyAisle launches 'reverse' consumer auction Web site for banks," *Computerworld*, June 9, 2008, *www.computerworld.com/action/article.do? command=viewArticleBasic&taxonomyNam e=internet_business&articleId=9094758&taxonomyId=71&intsrc=kc_top;* MoneyAisle.com Web site, *https://www.moneyaisle.com*, accessed June 21, 2008; Rosencrance, Linda, "$1M deposited in banks via MoneyAisle in first week," *Computerworld*, June 17, 2008, *www.computerworld.com/action/ article.do?command=viewArticleBasic&articleId=9099 458*; Johnson, Carolyn, "Website lets banks bid for customers," Boston Globe, June 9, 2008, *www.boston.com/business/technology/articles/2008/06/09/ website_lets_banks_bid_for_customers*.

Online Real Estate Services

Cyberhomes, FHA Anonymous, Lonexa, Realtor.com, Redfin, Terabitz, Trulia, and Zillow are just a few of the hundreds of Web sites that offer interesting services for those looking to buy a home. Many of the sites offer the capability to search the U.S. for homes based on geographic location, price range, number of bedrooms or bathrooms, and special features such as a pool or hot tub.

Online real estate service Zillow has set up a large number of Web sites based on specific communities and enables users to exchange data such as demographics and the crime rate in neighborhoods. This is data that for various legal and ethical reasons, real estate agents can't freely discuss.[16]

Redfin is an online real estate company that provides both online real estate search capabilities and access to live agents. The firm employs its agents so it can better manage customer service—unlike traditional real estate firms that license their name to independent agents. Redfin pays bonuses to agents when they receive high customer satisfaction ratings. It claims to reimburse home buyers roughly two-thirds of their real estate fees immediately upon closing, thus reducing the purchase price by many thousands of dollars.[17]

From the customer's viewpoint, an important service is the ability to receive competitive quotes from lenders without giving out personally identifying information that makes them a target of aggressive loan officers. Consumers can anonymously request loan quotes through several Web sites including FHA Anonymous, Lonexa, and Zillow.

Auctions

eBay has become synonymous with online auctions for both private sellers and small companies. Other popular online auction Web sites include Craigslist, uBid, Auctions, Onsale, WeBidz, and many others. The most frequent complaints about online auctions are increases in fees and problems with unscrupulous buyers. Auction sites are used by criminals to unload stolen, diverted, and counterfeit products. Law enforcement organizations regularly monitor such Web sites to capture criminals and recover stolen goods. Another frequent problem with online auctions is inaccurate or incomplete representation of the item for sale. Descriptions may omit important aspects or photos may not be clear enough to show the item's features.

There are two common types of online auctions. In an English auction, the initial price starts low and is bid up by successive bidders. In a reverse auction, sellers compete to obtain business by submitting successively lower prices for their goods or services. Reverse auctions are frequently used in B2B procurement.

Blair Corporation is a multi-channel direct marketer of fashions for men, women, and homes. The firm worked with eDynaQuote to conduct its first reverse auction and ensure broad supplier participation. Blair achieved significant cost savings on its first reverse auction for $1 million in packaging supplies.[18]

Anywhere, Anytime Applications of Mobile Commerce

Because m-commerce devices usually have a single user, they are ideal for accessing personal information and receiving targeted messages for a particular consumer. Through m-commerce, companies can reach individual consumers to establish one-to-one marketing relationships and communicate whenever it is convenient—in short, anytime and anywhere. Following are just a few examples of potential m-commerce applications:

Mobile Banking

With mobile banking, consumers can manage their finances from anywhere without driving to their bank or credit union or booting up their computer. Consumers can use mobile banking to access multiple banks, accounts, and financial services to:

- View account balances (checking, savings, Money Market, and credit cards)
- Transfer funds between accounts
- View and pay bills
- Review a history of account transactions

Such capability allows a consumer to check their credit card balance before making a major purchase and avoid having the credit provider rejecting the purchase. They can also transfer funds from savings to checking accounts to avoid an overdraft.

To begin using mobile banking with their wireless phones, consumers must visit their bank's online banking site and enroll in mobile banking. They then download the mobile application to their phone. As a security measure, the mobile banking user must enter their personal PIN to unlock the application each time they use it.

Mobile Price Comparison

A growing number of companies are employing a strategy that encourages shoppers to do Web-based price comparisons while they are in the stores. The idea is to drive the shopper who is ready to make a purchase from one retailer to another based on price and product comparisons. Web sites, like Google Maps, can be used to locate stores, restaurants, gas stations, and other retailers while you are on the move.

ShopLocal offers product location and comparison on mobile devices via a service called Where from mobile technology vendor uLocate. Shoppers can download the Where application using a text message from uLocate. The application works with GPS-enabled phones and provides comparison shoppers with product, price, and retailer information including step-by-step directions to the selected retailer's store. The Where service is available for $2.99 per month with users who have wireless phone plans with Alltel, Boost, or Sprint Nextel.[19]

Mobile Advertising

While some 58 million U.S. wireless subscribers viewed an ad on their cell phones in February 2008, many advertisers are not yet convinced that mobile advertising is effective and are taking a wait-and-see approach.[20]

Traditional Web sites designed for access by users with personal computers place cookies on your computer to track your browsing behavior and pass the data on to advertisers and ad-placement networks. However, the wireless industry service providers block cookies before they get to the cell phone out of concern that the cookies could provide access to their networks for computer viruses. They also fear that the cookies might cause a dramatic increase in the volume of data traffic as the cookies report back to the advertisers and ad-placement networks. The increase in volume could be enough to choke the network and seriously degrade performance. Thus advertisers are frustrated in their attempt to gather data to measure the number of views or effectiveness of mobile ads.

Mobile Coupons

About two percent of advertisers surveyed by Jupiter Research are using mobile coupons.[21] The Clorox Company, Del Monte Corporation, General Mills, Kimberly-Clark, and Procter & Gamble are collaborating with grocery retailer Kroger to test how consumers will react to using mobile coupons. Users in the test must first download a mobile marketing application to their cell phones so that coupons from the companies can be loaded onto their cell phones. While in a Kroger store, a shopper can choose an item, select the appropriate coupon from the cell phone, and have the coupon information sent to Kroger's in-store computer, which identifies the shopper by her loyalty card. At checkout, the coupon discount is applied when the loyalty card is scanned.[22]

As with any new technology, m-commerce will succeed only if it provides users with real benefits. Companies involved in m-commerce must think through their strategies carefully and ensure that they provide services that truly meet customers' needs.

Advantages of Electronic and Mobile Commerce

Conversion to an e-commerce or m-commerce system enables organizations to reduce the cost of doing business, speed the flow of goods and information, increase the accuracy of order processing and order fulfillment, and improve the level of customer service. These advantages are summarized in Table 5.1.

Table 5.1

Advantages of Electronic and Mobile Commerce

Advantages	Explanation
Provides global reach	Allows manufacturers to buy at a low cost worldwide and offers enterprises the chance to sell to a global market right from the very start-up of their business.
Reduces costs	Eliminates time-consuming and labor-intensive steps throughout the order and delivery process so that more sales can be completed in the same period and with increased accuracy.
Speeds flow of goods and information	The flow of information is accelerated because of the established electronic connections and communications processes.
Increased accuracy	Enables buyers to enter their own product specifications and order information directly so that human data-entry error is eliminated.
Improves customer service	Increased and more detailed information about delivery dates and current status increases customer loyalty.

Now that we've examined several e-commerce and m-commerce applications, let's look at the key components of technology infrastructure that must be in place to make this all work.

TECHNOLOGY INFRASTRUCTURE REQUIRED TO SUPPORT E-COMMERCE AND M-COMMERCE

Successful implementation of e-business requires significant changes to existing business processes and substantial investment in IS technology. These technology components must be chosen carefully and integrated to support a large volume of transactions with customers, suppliers, and other business partners worldwide. Online consumers complain that poor Web site performance (e.g., slow response time, inadequate customer support, and lost orders) drives them to abandon some e-commerce sites in favor of those with better, more reliable performance. This section provides a brief overview of the key technology infrastructure components (see Figure 5.2).

Figure 5.2

Key Technology Infrastructure Components

Hardware

A Web server hardware platform complete with the appropriate software is a key e-commerce infrastructure ingredient. The amount of storage capacity and computing power required of the Web server depends primarily on two things: the software that must run on the server and the volume of e-commerce transactions that must be processed. Although IS staff can sometimes define the software to be used, they can only estimate how much traffic the site will generate. As a result, the most successful e-commerce solutions are designed to be highly scalable so that they can be upgraded to meet unexpected user traffic.

A key decision facing new e-commerce companies is whether to host their own Web site or to let someone else do it. Many companies decide that using a third-party Web service provider is the best way to meet initial e-commerce needs. The third-party company rents space on its computer system and provides a high-speed connection to the Internet, which minimizes the initial out-of-pocket costs for e-commerce start-up. The third party can also provide personnel trained to operate, troubleshoot, and manage the Web server. Of course, many companies decide to take full responsibility for acquiring, operating, and supporting the Web server hardware and software themselves, but this approach requires considerable up-front capital and a set of skilled and trained workers. No matter which approach a company takes, it must have adequate hardware backup to avoid a major business disruption in case of a failure of the primary Web server.

Web Server Software

In addition to the Web server operating system, each e-commerce Web site must have Web server software to perform fundamental services, including security and identification, retrieval and sending of Web pages, Web site tracking, Web site development, and Web page development. The two most popular Web server software packages are Apache HTTP Server and Microsoft Internet Information Services.

Security and Identification

Security and identification services are essential for intranet Web servers to identify and verify employees accessing the system from the Internet. Access controls provide or deny access to files based on the username or URL. Web servers support encryption processes for transmitting private information securely over the public Internet.

Web Site Development

Web site development tools include features such as an HTML/visual Web page editor (e.g., Microsoft Expression Web, Adobe Dreamweaver, NetStudio Easy Web Graphics, and Soft-Quad HoTMetaL Pro), software development kits that include sample code and code development instructions for languages such as Java or Visual Basic, and Web page upload support to move Web pages from a development PC to the Web site. The tools bundled with the Web server software depends on which Web server software you select.

Web Page Construction

Web page construction software uses HTML editors and extensions to produce Web pages—either static or dynamic. **Static Web pages** always contain the same information—for example, a page that provides text about the history of the company or a photo of corporate headquarters. **Dynamic Web pages** contain variable information and are built to respond to a specific Web site visitor's request. For example, if a Web site visitor inquires about the availability of a certain product by entering a product identification number, the Web server searches the product inventory database and generates a dynamic Web page based on the current product information it found, thus fulfilling the visitor's request. This same request by another visitor later in the day might yield different results due to ongoing changes in product inventory. A server that handles dynamic content must be able to access information from a variety of databases. The use of open database connectivity enables the Web server to assemble information from different database management systems, such as SQL Server, Oracle, and Informix.

Web site development tools
Tools used to develop a Web site, including HTML or visual Web page editor, software development kits, and Web page upload support.

Web page construction software
Software that uses Web editors and extensions to produce both static and dynamic Web pages.

static Web pages
Web pages that always contain the same information.

dynamic Web pages
Web pages containing variable information that are built to respond to a specific Web visitor's request.

E-Commerce Software

After you have located or built a host server, including the hardware, operating system, and Web server software, you can begin to investigate and install e-commerce software. E-commerce software must support five core tasks: catalog management, product configuration, shopping cart facilities, e-commerce transaction processing, and Web traffic data analysis.

The specific e-commerce software you choose to purchase or install depends on whether you are setting up for B2B or B2C transactions. For example, B2B transactions do not include sales tax calculations if they involve items purchased for resale, and software to support B2B must incorporate electronic data transfers between business partners, such as purchase orders, shipping notices, and invoices. B2C software, on the other hand, must handle the complication of accounting for sales tax based on the current state laws and rules. However, it does not need to support negotiation between buyer and seller.

Catalog Management

Any company that offers a wide range of products requires a real-time interactive catalog to deliver customized content to a user's screen. *Catalog management software* combines different product data formats into a standard format for uniform viewing, aggregating, and integrating catalog data. It also provides a central repository for easy access, retrieval, and updating of pricing and availability changes. The data required to support large catalogs is almost always stored in a database on a computer that is separate from, but accessible to, the e-commerce server machine.

Product Configuration

Customers need help when an item they are purchasing has many components and options. *Product configuration software* tools were originally developed in the 1980s to assist B2B salespeople to match their company's products to customer needs. Buyers use the new Web-based product configuration software to build the product they need online with little or no help from salespeople. For example, Dell customers use product configuration software to build the computer that meets their needs. Such software is also used in the service arena to help people decide what sort of consumer loan or insurance is best for them.

Shopping Cart

Today many e-commerce sites use an *electronic shopping cart* to track the items selected for purchase, allowing shoppers to view what is in their cart, add new items to it, or remove items from it, as shown in Figure 5.3. To order an item, shoppers simply click an item. All the details about it—including its price, product number, and other identifying information—are stored automatically. If shoppers later decide to remove one or more items from the cart, they can view the cart's contents and remove any unwanted items. When shoppers are ready to pay for the items, they click a button (usually labeled "proceed to checkout") and begin a purchase transaction. Clicking the Checkout button opens another window that usually asks shoppers to fill out billing, shipping, and payment method information and to confirm the order.

Web Services

Web services are software modules supporting specific business processes that users can interact with over a network (such as the Internet) as necessary. Web services can combine software and services from different companies to provide an integrated way to communicate. For example, an organization could use a supplier provided Web service to streamline the payment of vendor invoices. The Web service could be developed so that when the user moves the mouse over a purchase order number in an e-mail from the supplier, the amount of funds remaining in the purchase order are displayed. The user can then approve payment by clicking a button or link.

Software manufacturers are scrambling to meet customer demands by offering software applications for use over the Web as services supported by advertising or subscription fees. SAP, for example, offers more than 500 components that run as Web services to support

Web services
Software modules supporting specific business processes that users can interact with over a network (such as the Internet) on an as-needed basis.

Figure 5.3

Electronic Shopping Cart

An electronic shopping cart (or bag) allows online shoppers to view their selections and add or remove items.

business functions such as finance, human resources, logistics, manufacturing, procurement, and product development. Dun & Bradstreet provides an address verification service called GlobalAccess that checks and completes the addresses of prospects or customers to ensure the accuracy and completeness of this key information. Oanda.com offers a currency exchange rate service that downloads the most current set of rates to support the running of accounting processes such as consolidation reporting that require the translation of multicurrency transactions into a single corporate currency. UPS provides a shipment tracking service for determining the cost, current location, and the receiving party of a specific package to enhance the order fulfillment process and provide shippers with greater visibility into the shipping process.[23]

Technology Needed for Mobile Commerce

For m-commerce to work effectively, the interface between the wireless device and its user must improve to the point that it is nearly as easy to purchase an item on a wireless device as it is to purchase it on a PC. In addition, network speed must improve so that users do not become frustrated. Security is also a major concern, particularly in two areas: the security of the transmission itself and the trust that the transaction is being made with the intended party. Encryption can provide secure transmission. Digital certificates, discussed later in this chapter, can ensure that transactions are made between the intended parties.

The handheld devices used for m-commerce have several limitations that complicate their use. Their screens are small, perhaps no more than a few square inches, and might be able to display only a few lines of text. Their input capabilities are limited to a few buttons, so entering data can be tedious and error prone. They also have less processing power and less bandwidth than desktop computers, which are usually hardwired to a high-speed LAN. They also operate on limited-life batteries. For these reasons, it is currently impossible to directly access many Web sites with a handheld device. Web developers must rewrite Web applications so that users with handheld devices can access them.

To address the limitations of wireless devices, the industry has undertaken a standardization effort for their Internet communications. The Wireless Application Protocol (WAP) is a standard set of specifications for Internet applications that run on handheld, wireless devices. It effectively serves as a Web browser for such devices. WAP is a key underlying technology of m-commerce that is supported by an entire industry association of over 200 vendors of wireless devices, services, and tools. In the future, devices and service systems based on WAP and its derivatives (including WAP 2.0 and Wireless Internet Protocol) will be able to interoperate. Japan's largest wireless network provider, DoCoMo, developed a competing standard called the i-mode system.

For equipment and service providers, the existence of competing standards makes it much more difficult to meet the needs of their customers. In many cases, the providers must develop their services or products based on one standard and forfeit the market for customers who elect to adopt the competing standard. Of course, multiple standards also create problems for customers who must make a decision on which set of services and equipment to adopt. Early adopters may find to their dismay they have chosen a standard that falls out of favor.

WAP uses the Wireless Markup Language (WML), which is designed for effectively displaying information on small devices. A user with a WAP-complaint device uses the built-in microbrowser to make a WML request. The request is forwarded to a special WAP gateway to fetch the information from the appropriate Internet server. If the information is already in WML format, it can be passed from the Internet server through the gateway directly to the user's device. If the information is in HTML format, the gateway translates the HTML content into WML so it can be displayed on the user's device.

Electronic Payment Systems

Electronic payment systems are a key component of the e-commerce infrastructure. Current e-commerce technology relies on user identification and encryption to safeguard business transactions. Actual payments are made in a variety of ways, including electronic cash, electronic wallets, and smart, credit, charge, and debit cards. Web sites that accept multiple payment types convert more visitors to purchasing customers than merchants who offer only a single payment method.

Authentication technologies are used by many organizations to confirm the identity of a user requesting access to information or assets. A **digital certificate** is an attachment to an e-mail message or data embedded in a Web site that verifies the identity of a sender or Web site. A **certificate authority (CA)** is a trusted third-party organization or company that issues digital certificates. The CA is responsible for guaranteeing that the people or organizations granted these unique certificates are, in fact, who they claim to be. Digital certificates thus create a trust chain throughout the transaction, verifying both purchaser and supplier identities.

digital certificate
An attachment to an e-mail message or data embedded in a Web site that verifies the identity of a sender or Web site.

certificate authority (CA)
A trusted third-party organization or company that issues digital certificates.

Secure Sockets Layer

All online shoppers fear the theft of credit card numbers and banking information. To help prevent this type of identity theft, the **Secure Sockets Layer** (SSL) communications protocol is used to secure sensitive data. The SSL communications protocol includes a handshake stage, which authenticates the server (and the client, if needed), determines the encryption and hashing algorithms to be used, and exchanges encryption keys. Following the handshake stage, data might be transferred. The data is always encrypted, ensuring that your transactions are not subject to interception or "sniffing" by a third party. Although SSL handles the encryption part of a secure e-commerce transaction, a digital certificate is necessary to provide server identification.

Secure Sockets Layer (SSL)
A communications protocol is used to secure sensitive data during e-commerce.

Electronic Cash

Electronic cash is an amount of money that is computerized, stored, and used as cash for e-commerce transactions. Typically, consumers must open an account with an electronic cash service provider by providing identification information. When the consumers want to withdraw electronic cash to make a purchase, they access the service provider via the Internet and present proof of identity—a digital certificate issued by a certification authority or a

electronic cash
An amount of money that is computerized, stored, and used as cash for e-commerce transactions.

username and password. After verifying a consumer's identity, the system debits the consumer's account and credits the seller's account with the amount of the purchase. PayPal, BillMeLater, MoneyZap, and TeleCheck are four popular forms of electronic cash.

The PayPal service of eBay enables any person or business with an e-mail address to securely, easily, and quickly send and receive payments online. To send money, you enter the recipient's e-mail address and the amount you want to send. You can pay with a credit card, debit card, or funds from a checking account. The recipient gets an e-mail that says, "You've Got Cash!" Recipients can then collect their money by clicking a link in the e-mail that takes them to *www.paypal.com*. To receive the money, the user also must have a credit card or checking account to accept fund transfers. To request money for an auction, invoice a customer, or send a personal bill, you enter the recipient's e-mail address and the amount you are requesting. The recipient gets an e-mail and instructions on how to pay you using PayPal. PayPal serves more than 60 million active accounts worldwide. It is available in 190 markets and processes payments in 17 currencies around the world.[24]

Credit, Charge, Debit, and Smart Cards

Many online shoppers use credit and charge cards for most of their Internet purchases. A credit card, such as Visa or MasterCard, has a preset spending limit based on the user's credit history, and each month the user can pay part or all of the amount owed. Interest is charged on the unpaid amount. A charge card, such as American Express, carries no preset spending limit, and the entire amount charged to the card is due at the end of the billing period. Charge cards do not involve lines of credit and do not accumulate interest charges. American Express became the first company to offer disposable credit card numbers in 2000. Other banks, such as Citibank, protect the consumer by providing a unique number for each transaction. Debit cards look like credit cards or automated teller machine (ATM) cards, but they operate like cash or a personal check. Credit, charge, and debit cards currently store limited information about you on a magnetic strip. This information is read each time the card is swiped to make a purchase. All credit card customers are protected by law from paying more than $50 for fraudulent transactions.

smart card
A credit card–sized device with an embedded microchip to provide electronic memory and processing capability.

The **smart card** is a credit card–sized device with an embedded microchip to provide electronic memory and processing capability. Smart cards can be used for a variety of purposes, including storing a user's financial facts, health insurance data, credit card numbers, and network identification codes and passwords. They can also store monetary values for spending.

Smart cards are better protected from misuse than conventional credit, charge, and debit cards because the smart-card information is encrypted. Conventional credit, charge, and debit cards clearly show your account number on the face of the card. The card number, along with a forged signature, is all that a thief needs to purchase items and charge them against your card. A smart card makes credit theft practically impossible because a key to unlock the encrypted information is required, and there is no external number that a thief can identify and no physical signature a thief can forge.

The smart card connects to a reader with direct physical contact or via remote contactless radio frequency interface. Smart cards have been around for over a decade and are widely used in Europe, Australia, and Japan. United Kingdom credit card giant Barclaycard is conducting a pilot test of contactless retail and transit payment using mobile phones that support near field communications (NFC).[25] Smart card use has not caught on in the United States because there are few smart-card readers to record payments and U.S. banking regulations have slowed smart-card marketing and acceptance as well. Table 5.2 compares various types of payment systems.

Payments Using Cell Phones

The retail and banking industries are keenly interested in using a cell phone like a credit card by waving the end of the phone near a scanner device to pay for purchases. Some people believe that mobile device-based transactions will exceed card-based transactions.

U.S. Bank began testing the concept of a credit card "buried" inside a cell phone and the use of no contact scanners. When two NFC devices (the scanner and cell phone) come within

Table 5.2

Comparison of Payment
Systems

Payment System	Description	Advantages	Disadvantages
Credit card	Carries preset spending limit based on the user's credit history.	Each month the user can pay part or all of the amount owed.	Unpaid balance accumulates interest charges—often at a high rate of interest.
Charge card	Looks like a credit card but carries no preset spending limit.	Charge cards do not involve lines of credit and do not accumulate interest charges.	The entire amount charged to the card is due at the end of the billing period.
Debit card	Look like a credit cards or automated teller machine (ATM) cards.	Operates like cash or a personal check	Money is immediately deducted from user's account balance.
Smart card	Credit card device with embedded microchip capable of storing facts about card holder	Better protected from misuse than conventional credit, charge, and debit cards because the smart-card information is encrypted	Not widely used in the U.S.

about 3 inches of each other, they can exchange data using radio signals including encrypted credit card account numbers. The U.S. Bank pilot supports the use of only one credit card; however, if successful, banks and wireless service providers may allow customers to load their "tap and go" phone with multiple credit cards or merchant reward cards. Says Dominic Venturo, the bank vice president helping to manage the U.S. Bank pilot test: "Anytime you can combine the phone, which most of us have in our wallet, with the bank payment card many of us carry in our wallets, into a single system, you've created a simpler and easier way for your customers to manage their lives."[26] Japan, Australia, and Korea are also experimenting with "tap and go" phones.

AN OVERVIEW OF ENTERPRISE SYSTEMS: TRANSACTION PROCESSING SYSTEMS AND ENTERPRISE RESOURCE PLANNING

An **enterprise system** is central to an organization and ensures information can be shared across all business functions and all levels of management to support the running and managing of a business. Enterprise systems employ a database of key operational and planning data that can be shared by all. This eliminates the problems of lack of information and inconsistent information caused by multiple transaction processing systems that support only one business function or one department in an organization. Examples of enterprise systems include enterprise resource planning systems that support supply-chain processes, such as order processing, inventory management, and purchasing and customer relationship management systems that support sales, marketing, and customer service-related processes.

Businesses rely on such systems to perform many of their daily activities in areas such as product supply, distribution, sales, marketing, human resources, manufacturing, accounting, and taxation so that work is performed quickly, while avoiding waste and mistakes. Without such systems, recording and processing business transactions would consume huge amounts of an organization's resources. This collection of processed transactions also forms a storehouse of data invaluable to decision making. The ultimate goal is to satisfy customers and provide a competitive advantage by reducing costs and improving service.

Every organization has many *transaction processing systems*, which capture and process the detailed data necessary to update records about the fundamental business operations of the organization. These systems include order entry, inventory control, payroll, accounts

enterprise system
A system central to the organization that ensures information can be shared across all business functions and all levels of management to support the running and managing of a business.

payable, accounts receivable, and the general ledger, to name just a few. The input to these systems includes basic business transactions, such as customer orders, purchase orders, receipts, time cards, invoices, and customer payments. The processing activities include data collection, data editing, data correction, data manipulation, data storage, and document production. The result of processing business transactions is that the organization's records are updated to reflect the status of the operation at the time of the last processed transaction.

A TPS also provides employees involved in other business processes—via management information system/decision support system (MIS/DSS) and the special-purpose information systems—with data to help them achieve their goals. (MIS/DSS systems are discussed in Chapter 6.) A transaction processing system serves as the foundation for these other systems (see Figure 5.4).

Figure 5.4

TPS, MIS/DSS, and Special Information Systems in Perspective

Transaction processing systems support routine operations associated with customer ordering and billing, employee payroll, purchasing, and accounts payable. The amount of support for decision making that a TPS directly provides managers and workers is low.

TPSs work with a large amount of input and output data and use this data to update the official records of the company about such things as orders, sales, and customers. As systems move from transaction processing to management information/decision support and special-purpose information systems, they involve less routine, more decision support, less input and output, and more sophisticated and complex analysis. These higher-level systems require the basic business transaction data captured by the TPS.

Because TPSs often perform activities related to customer contacts—such as order processing and invoicing—these information systems play a critical role in providing value to the customer. For example, by capturing and tracking the movement of each package, shippers such as Federal Express and United Parcel Service (UPS) can provide timely and accurate data on the exact location of a package. Shippers and receivers can access an online database and, by providing the airbill number of a package, find the package's current location. If the package has been delivered, they can see who signed for it (a service that is especially useful in large companies where packages can become "lost" in internal distribution systems and mailrooms). Such a system provides the basis for added value through improved customer service.

Traditional Transaction Processing Methods and Objectives

batch processing system
A form of data processing where business transactions are accumulated over a period of time and prepared for processing as a single unit or batch.

With **batch processing systems**, business transactions are accumulated over a period of time and prepared for processing as a single unit or batch (see Figure 5.5a). Transactions are accumulated for the length of time needed to meet the needs of the users of that system. For example, it might be important to process invoices and customer payments for the accounts receivable system daily. On the other hand, the payroll system might receive time cards and process them biweekly to create checks, update employee earnings records, and distribute labor costs. The essential characteristic of a batch processing system is that there is some delay between an event and the eventual processing of the related transaction to update the organization's records.

With **online transaction processing (OLTP)**, each transaction is processed immediately, without the delay of accumulating transactions into a batch (see Figure 5.5b). Consequently, at any time, the data in an online system reflects the current status. This type of processing is essential for businesses that require access to current data such as airlines, ticket agencies, and stock investment firms. Many companies find that OLTP helps them provide faster, more efficient service—one way to add value to their activities in the eyes of the customer. Increasingly, companies are using the Internet to capture and process transaction data such as customer orders and shipping information from e-commerce applications.

online transaction processing (OLTP)

A form of data processing where each transaction is processed immediately, without the delay of accumulating transactions into a batch.

Figure 5.5

Batch Versus Online Transaction Processing

(a) Batch processing inputs and processes data in groups. (b) In online processing, transactions are completed as they occur.

Data entry of accumulated transactions → Input (batched) → **(a) Batch Processing** → Output

Terminal

Terminal

Immediate processing of each transaction

Central computer (processing) → Output

Terminal

Terminal

Terminal

(b) Online Transaction Processing

Although the technology is advanced enough, TPS applications do not always run using online processing. For many applications, batch processing is more appropriate and cost effective. Payroll transactions and billing are typically done via batch processing. Specific goals of the organization define the method of transaction processing best suited for the various applications of the company.

Figure 5.6 shows the flow of key pieces of information from one TPS to another for a typical manufacturing organization. TPSs can be designed so that the flow of information from one system to another is automatic and requires no manual intervention or reentering of data. Such a set of systems is called an *integrated information system*. Many organizations have limited or no integration among its TPSs. In this case, data input to one TPS must be printed out and manually reentered into other systems. Of course, this increases the amount of effort required and introduces the likelihood of processing delays and errors.

Figure 5.6

Integration of a Firm's TPSs

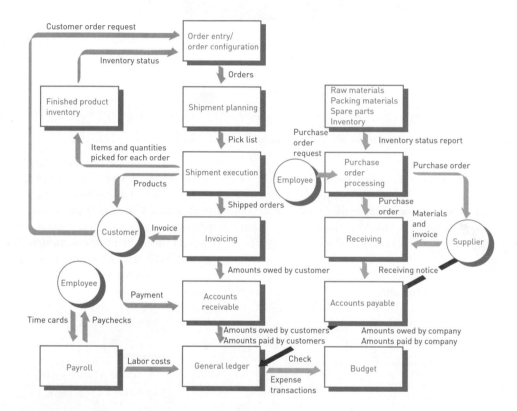

Because of the importance of transaction processing, organizations expect their TPSs to accomplish a number of specific objectives including:

- Process data generated by and about transactions.
- Maintain a high degree of accuracy and integrity.
- Avoid processing fraudulent transactions.
- Produce timely user responses and reports.
- Increase labor efficiency.
- Help improve customer service and/or loyalty.

Depending on the specific nature and goals of the organization, any of these objectives might be more important than others. By meeting these objectives, TPSs can support corporate goals such as reducing costs; increasing productivity, quality, and customer satisfaction; and running more efficient and effective operations. For example, overnight delivery companies such as FedEx expect their TPSs to increase customer service. These systems can locate a client's package at any time—from initial pickup to final delivery. This improved customer information allows companies to produce timely information and be more responsive to customer needs and queries.

TRANSACTION PROCESSING ACTIVITIES

Along with having common characteristics, all TPSs perform a common set of basic data-processing activities. TPSs capture and process data that describes fundamental business transactions. This data is used to update databases and to produce a variety of reports people both within and outside the enterprise use. The business data goes through a **transaction processing cycle** that includes data collection, data editing, data correction, data manipulation, data storage, and document production (see Figure 5.7).

transaction processing cycle
The process of data collection, data editing, data correction, data manipulation, data storage, and document production.

Figure 5.7

Data-Processing Activities Common to Transaction Processing Systems

Data Collection

Capturing and gathering all data necessary to complete the processing of transactions is called **data collection**. In some cases it can be done manually, such as by collecting handwritten sales orders or changes to inventory. In other cases, data collection is automated via special input devices such as scanners, point-of-sale devices, and terminals.

Data collection begins with a transaction (e.g., taking a customer order) and results in data that serves as input to the TPS. Data should be captured at its source and recorded accurately in a timely fashion, with minimal manual effort, and in an electronic or digital form that can be directly entered into the computer. This approach is called *source data automation*. An example of source data automation is an automated device at a retail store that speeds the checkout process—either UPC codes read by a scanner or RFID signals picked up when the items approach the checkout stand. Using both UPC bar codes and RFID tags is quicker and more accurate than having a clerk enter codes manually at the cash register. The product ID for each item is determined automatically, and its price retrieved from the item database. The point-of-sale TPS uses the price data to determine the customer's bill. The store's inventory and purchase databases record the number of units of an item

data collection
Capturing and gathering all data necessary to complete the processing of transactions.

purchased, the date, the time, and the price. The inventory database generates a management report notifying the store manager to reorder items that have fallen below the reorder quantity. The detailed purchases database can be used by the store or sold to marketing research firms or manufacturers for detailed sales analysis (see Figure 5.8).

Figure 5.8

Point-of-Sale Transaction Processing System

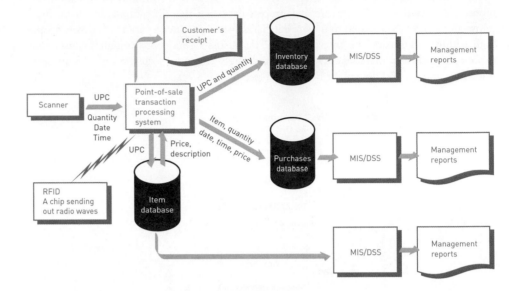

The purchase of items at the checkout stand updates a store's inventory database and its database of purchases.

Many grocery stores combine point-of-sale scanners and coupon printers. The systems are programmed so that each time a specific product—for example, a box of cereal— crosses a checkout scanner, an appropriate coupon—perhaps a milk coupon—is printed. Companies can pay to be promoted through the system, which is then reprogrammed to print those companies' coupons if the customer buys a competitive brand. These TPSs help grocery stores increase profits by improving their repeat sales and bringing in revenue from other businesses.

Data Editing

data editing
The process of checking data for validity and completeness.

An important step in processing transaction data is to perform **data editing** for validity and completeness to detect any problems. For example, quantity and cost data must be numeric and names must be alphabetic; otherwise, the data is not valid. Often, the codes associated with an individual transaction are edited against a database containing valid codes. If any code entered (or scanned) is not present in the database, the transaction is rejected.

Data Correction

data correction
The process of reentering data that was not typed or scanned properly.

It is not enough simply to reject invalid data. The system should also provide error messages that alert those responsible for editing the data. Error messages must specify the problem so proper corrections can be made. A **data correction** involves reentering data that was not typed or scanned properly. For example, a scanned UPC code must match a code in a master table of valid UPCs. If the code is misread or does not exist in the table, the checkout clerk is given an instruction to rescan the item or type the information manually.

Data Manipulation

data manipulation
The process of performing calculations and other data transformations related to business transactions.

Another major activity of a TPS is **data manipulation**, the process of performing calculations and other data transformations related to business transactions. Data manipulation can include classifying data, sorting data into categories, performing calculations, summarizing results, and storing data in the organization's database for further processing. In a payroll TPS, for example, data manipulation includes multiplying an employee's hours worked by the hourly pay rate. Overtime pay, federal and state tax withholdings, and deductions are also calculated.

Data Storage

Data storage involves updating one or more databases with new transactions. After being updated, this data can be further processed and manipulated by other systems so that it is available for management reporting and decision making. Thus, although transaction databases can be considered a by-product of transaction processing, they have a pronounced effect on nearly all other information systems and decision-making processes in an organization.

data storage
The process of updating one or more databases with new transactions.

Document Production and Reports

Document production involves generating output records, documents, and reports. These can be hard-copy paper reports or displays on computer screens (sometimes referred to as *soft copy*). Printed paychecks, for example, are hard-copy documents produced by a payroll TPS, while an outstanding balance report for invoices might be a soft-copy report displayed by an accounts receivable TPS. Often, results from one TPS flow downstream to become input to other systems, which might use the results of updating the inventory database to create the stock exception report (a type of management report) of items whose inventory level is below the reorder point.

document production
The process of generating output records and reports.

In addition to major documents such as checks and invoices, most TPSs provide other useful management information and decision support, such as printed or on-screen reports that help managers and employees perform various activities. A report showing current inventory is one example; another might be a document listing items ordered from a supplier to help a receiving clerk check the order for completeness when it arrives. A TPS can also produce reports required by local, state, and federal agencies, such as statements of tax withholding and quarterly income statements.

CONTROL AND MANAGEMENT ISSUES

Transaction processing systems process the fundamental business transactions that are the lifeblood of the firm's operation. They capture facts about basic business operations of the organization—facts without which orders cannot be shipped, customers cannot be invoiced, and employees and suppliers cannot be paid. In addition, the data captured by TPSs flows downstream to other systems in the organization where it is used to support analysis and decision making. TPSs are so critical to the operation of most firms that many business activities would come to a halt if the supporting TPSs failed. Because firms must ensure the reliable operation of their TPSs, they must also engage in disaster recovery planning and TPS audits.

Disaster Recovery Plan

Unfortunately, recent history reminds us of the need to be prepared in the event of a natural or man-made accident or disaster. The **disaster recovery plan (DRP)** is a firm's plan to recover data, technology, and tools that support critical information systems and necessary information systems components such as the network, databases, hardware, software, and operating systems.

disaster recovery plan (DRP)
A formal plan describing the actions that must be taken to restore computer operations and services in the event of a disaster.

Those TPSs that directly affect the cash flow of the firm (such as order processing, accounts receivable, accounts payable, and payroll) are typically identified as critical business information systems. A lengthy disruption in the operation of any of those systems could create a serious cash flow problem for the firm and potentially put it out of business. Companies vary widely in the thoroughness and effectiveness of their disaster recovery planning, and, as a result, some have a harder time resuming business than others.

Fires, hurricanes, floods, earthquakes, and tornados are the most dramatic causes of business disasters. TiVo operates with 600 employees and 700 servers that store more than 100 TB of data. Its headquarters is in a part of the country where there are occasional earthquakes and other natural disasters. As a result, the firm set up its disaster recovery site in Las Vegas, a location relatively free of natural disasters and with an infrastructure proven reliable for operating the city's many casinos.[27] If a disaster strikes its headquarters, computer operations will be relocated to this site until normal operations can be restored.

Companies such as Iron Mountain provide a secure, off-site environment for records storage. In the event of a disaster, vital data can be recovered.

(Source: Geostock/Getty Images.)

JetBlue—Trial by Fire and Ice

Today's global society depends on air travel for business and pleasure more than ever before. Services offered by airlines have come under scrutiny due to incidents that point to unreliability. Some of these incidents are caused by the inefficient use of information systems.

On the list of the major corporate disaster recovery challenges of recent times, JetBlue and the St. Valentine's Day ice storm appears in the top ten. JetBlue has built a reputation as an airline that caters to the needs of its customers. Plush leather seats, expanded leg room, complimentary beverages and snacks, snooze kits, seat-back displays providing 36 channels of entertainment, satellite radio, first-run movies, wireless Internet, and smiling crew members, all at affordable rates, are amenities rare at traditional airlines. On February 14, 2007, the honeymoon seemed to be over for JetBlue and its customers.

Weather forecasters predicted an ice storm would hit the east coast on Valentine's Day. While it was unclear how much it might affect air traffic, most airlines took precautionary measures, canceling dozens of flights. In its efforts to please passengers, JetBlue gambled and waited it out until it was too late. Rather than improving, conditions only worsened over the course of the day, leaving hundreds of JetBlue passengers stranded in planes on tarmacs at JFK International airport in New York and other major airports including Washington, DC, and Newark, New Jersey—some for as long as 11 hours. Around 3:00 pm, JetBlue gave up hope and called in buses to rescue the passengers from the planes. By then the damage was done.

Thousands of passengers waiting in airline terminals were hoping to complete their trips despite the storms. More passengers were arriving at airports unaware of delays and cancellations. Still other passengers were returning to the terminals by bus from stranded aircraft. JetBlue wound up with thousands of irate passengers at their counters and no flights departing or arriving on the east coast. Chief executive officer David Neeleman admitted to doing a horrible job. "We got ourselves into a situation where we were doing rolling cancellations instead of a massive cancellation. Communications broke down, we weren't able to reach out to passengers and they continued to arrive at the airports... it had a cascading effect."

Charles "Duffy" Mees won't ever forget that day. Duffy Mees was vice president and CIO of JetBlue Airways at the time. He came to JetBlue a few months prior to the disaster with years of experience in the airline industry. During his first few months, he oversaw the completion of an enterprise resource planning (ERP) installation at JetBlue. However, his experience did not prepare him for handling the Valentine's Day crisis.

The impact of the storm on JetBlue's information systems lasted a week. During the days that followed, many systems were pushed beyond their limits. Massive flight cancellations and rescheduling placed an unprecedented amount of traffic on JetBlue's reservation systems. Since JetBlue did not support rebooking flights online or at airport kiosks, customers had only one option for rebooking: call the JetBlue reservation office. JetBlue's Salt Lake City reservation agents were flooded with calls from angry passengers. Limitations in the system allowed only up to 650 agents to work at a time, plenty for normal days, but not for an exceptional demand. Many customers were stranded on hold waiting to rebook flights. Mees worked with their software provider to boost the limit to 950, which helped to open the bottleneck. Still, it was days before many passengers could get through to an agent.

Meanwhile, customer luggage was piling up in huge mounds at airports. JetBlue had no computerized system in place for tracking bags. The company had placed that system on the back burner while concentrating on its new ERP system. JetBlue had to haul mountains of luggage to off-site locations, where extra workers were employed to sort and identify bags. An information system was developed on the fly to scan bag tags and identify the owners from passenger records.

Besides reservations and baggage problems, managers faced outages and failure from important systems that control core operations. SkySolver software, which operations planners use to redeploy planes and crews, could not transfer new schedules to the primary flight scheduling systems. Programmers from the vendors attacked the problem and solved it within hours, but the delay caused more havoc. JetBlue was caught in a tailspin of system failures triggered by too much information all at once.

Mees and his crew spent three days and nights working to bring JetBlue systems back online. They pushed systems to their limits and created databases, tools, and applications on the fly in their efforts to find solutions.

During the crisis, JetBlue management learned many lessons and discovered many solutions, including preventative measures. A new system has now been implemented that allows passengers to rebook canceled flights online. Computer terminals have been installed at airports to allow passengers to rebook onsite. Software allows double the number of booking agents to respond in emergencies. A lost-bag system has been installed to track luggage—which is particularly valuable when flights are cancelled. A new system has been implemented that notifies passengers by e-mail, phone, or Web when flights are cancelled or changed.

Most significantly, the crisis motivated JetBlue to create a customer bill of rights offering compensation to customers whose flights have been cancelled or are left sitting too long on planes.

The cost to JetBlue for the Valentine's Day disaster has been estimated at around $30 million. What about the cost to JetBlue's reputation? After JetBlue offered many apologies and fired several top-level executives, it appears that JetBlue is still loved by its customers. J.D. Power and Associates 2007 airline satisfaction survey ranked JetBlue Number 1 by far for the third year in a row. In this case, good intentions seem to have won out over poor management.

Discussion Questions

1. What could JetBlue have done to prevent the Valentine's Day disaster in terms of information systems and management decisions?
2. What information systems does JetBlue use to manage air travel?

Critical Thinking Questions

1. How do you think JetBlue's disaster affected the airline industry from the airlines' perspective and the traveler's perspective?
2. JetBlue offers many amenities to its customers that other airlines have discontinued in order to cut costs. What are the benefits and dangers of JetBlue's approach, and how does this incident illustrate the dangers?

Sources: Duvall, Mel, "What Really Happened At JetBlue," *CIO Insight*, April 5, 2007, *www.cioinsight.com/c/a/Past-News/What-Really-Happened-At-JetBlue/1*; Ho, David, "Fans stand behind JetBlue," *Atlanta Journal-Constitution*, June 10, 2008, *www.ajc.com/business/content/business/stories/2008/06/10/jet_blue.html*; JetBlue Web site, *www.jetblue.com*, accessed June 29, 2008.

Transaction Processing System Audit

transaction processing system audit

A check of a firm's TPS systems to prevent accounting irregularities and/or loss of data privacy.

The Sarbanes-Oxley Act, enacted as a result of several major accounting scandals, requires public companies to implement procedures to ensure their audit committees can document financial data, validate earnings reports, and verify the accuracy of information. The Financial Services Modernization Act (Gramm-Leach-Bliley) requires systems security for financial service providers, including specific standards to protect customer privacy. The Health Insurance Portability and Accountability Act (HIPAA) defines regulations covering healthcare providers to ensure that their patient data is adequately protected. Many organizations conduct ongoing **transaction processing system audits** to prevent the kind of accounting irregularities or loss of data privacy that can put their firm in violation of these acts and erase investor confidence. The audit can be performed by the firm's own internal audit group or an outside auditor might be hired to provide a higher degree of objectivity. A transaction processing system audit attempts to answer four basic questions:

- Does the system meet the business need for which it was implemented?
- What procedures and controls have been established?
- Are these procedures and controls being used properly?
- Are the information systems and procedures producing accurate and honest reports?

A typical audit also examines the distribution of output documents and reports, determines if only appropriate people can execute key system functions (e.g., approve the payment of an invoice), assesses the training and education associated with existing and new systems, and determines the effort required to perform various tasks and to resolve problems in the system. General areas of improvement are also identified and reported during the audit.

TRADITIONAL TRANSACTION PROCESSING APPLICATIONS

A TPS typically includes the following types of systems:

- **Order processing systems.** Running these systems efficiently and reliably is so critical that the order processing systems are sometimes referred to as the "lifeblood of the organization." The processing flow begins with the receipt of a customer order. The finished product inventory is checked to see if sufficient inventory is on hand to fill the order. If sufficient inventory is available, the customer shipment is planned to meet the customer's desired receipt date. A product pick list is printed at the warehouse from which the order is to be filled on the day the order is planned to be shipped. At the warehouse, workers gather the items needed to fill the order, and enter the item identifier and quantity for each item to update the finished product inventory. When the order is complete and sent on its way, a customer invoice is created with a copy included in the customer shipment.

- **Accounting systems.** The accounting systems must track the flow of data related to all the cash flows that affect the organization. As mentioned earlier, the order processing system generates an invoice for customer orders to include with the shipment. This information is also sent to the accounts receivable system to update the customer's account. When the customer pays the invoice, the payment information is also used to update the customer's account. The necessary accounting transactions are sent to the general ledger system to keep track of amounts owed and amounts paid. Similarly, as the purchasing systems generate purchase orders and those items are received, information is sent to the accounts payable system to manage the amounts owed by the company. Data about amounts owed and paid by customers to the company and from the company to vendors and others are sent to the general ledger system that records and reports all financial transactions for the company.

- **Purchasing systems.** The traditional transaction processing systems that support the purchasing business function include inventory control, purchase order processing, receiving, and accounts payable. Employees place purchase order requests in response to shortages identified in inventory control reports. Purchase order information flows to the receiving system and accounts payable systems. A record of receipt is created upon receipt of the items ordered. When the invoice arrives from the supplier, it is matched to the original order and the receiving report and a check is generated if all data is complete and consistent.

TRANSACTION PROCESSING SYSTEMS FOR SMALL AND MEDIUM SIZE ENTERPRISES (SMES)

Many software packages provide integrated transaction processing system solutions for small and medium size enterprises (SMEs) where small is an enterprise with less than 50 employees and medium is one with fewer than 250 employees. These systems are typically easy to install, easy to operate, and have a low total cost of ownership with an initial cost of a few hundred to a few thousand dollars. Such solutions are highly attractive to firms that have outgrown their current software but cannot afford a complex, high-end integrated system solution. Table 5.3 presents some of the dozens of such software solutions available.

Vendor	Software	Type of TPS Offered	Target Customers
AccuFund	AccuFund	Financial reporting and accounting	Non-profit, municipal and government organizations
OpenPro	OpenPro	Complete ERP solution including financials, supply chain management, e-commerce, customer relationship management, and retail POS system	Manufacturers, distributors, and retailers
Intuit	QuickBooks	Financial reporting and accounting	Manufacturers, professional services, contractors, nonprofits, and retailers
Sage	Timberline	Financial reporting, accounting, and operations	Contractors, real estate developers, and residential builders
Redwing	TurningPoint	Financial reporting and accounting	Professional services, banks, and retailers

Table 5.3

Sample of Integrated TPS Solutions for SMEs

ENTERPRISE RESOURCE PLANNING, SUPPLY CHAIN MANAGEMENT, AND CUSTOMER RELATIONSHIP MANAGEMENT

As previously defined, enterprise resource planning (ERP) is a set of integrated programs that manage a company's vital business operations for an entire multisite, global organization. Recall that a business process is a set of coordinated and related activities that takes one or more kinds of input and creates an output of value to the customer of that process. The customer might be a traditional external business customer who buys goods or services from the firm. An example of such a process is a sales order, which takes customer input and generates an order. The customer of a business process might also be an internal customer such as a worker in another department of the firm. For example, the shipment process creates the necessary internal documents needed by workers in the warehouse and shipping functions to pick, pack, and ship orders. At the core of the ERP system is a database that is shared by all users so that all business functions have access to current and consistent data for operational decision making and planning as shown in Figure 5.9.

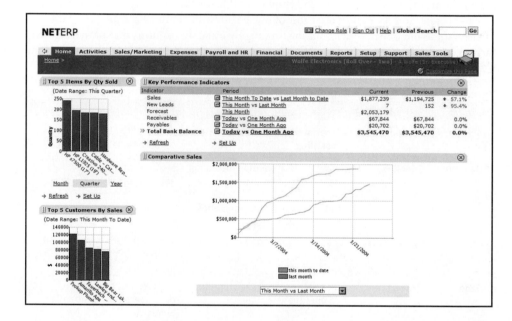

An Overview of Enterprise Resource Planning

ERP systems evolved from materials requirement planning systems (MRP) developed in the 1970s. These systems tied together the production planning, inventory control, and purchasing business functions for manufacturing organizations. During the late 1980s and early 1990s, many organizations recognized that their legacy transaction processing systems lacked the integration needed to coordinate activities and share valuable information across all the business functions of the firm. As a result, costs were higher and customer service poorer than desired. The impending year 2000 (Y2K) problem that people expected to cause date-related processing to operate incorrectly after January 1, 2000 provided further impetus for organizations all over the world to review, modify, and upgrade their computer systems. Many firms used the Y2K issue to justify scrapping large parts of their existing information systems and converting to new ERP systems. Large organizations, members of the Fortune 1000, were the first to take on the challenge of implementing ERP. As they did, they uncovered many advantages as well as some disadvantages summarized in the following sections.

Advantages of ERP

Increased global competition, new needs of executives for control over the total cost and product flow through their enterprises, and ever-more-numerous customer interactions drive the demand for enterprise-wide access to real-time information. ERP offers integrated software from a single vendor to help meet those needs. The primary benefits of implementing ERP include improved access to data for operational decision making, elimination of inefficient or outdated systems, improvement of work processes, and technology standardization. ERP vendors have also developed specialized systems for specific applications and market segments.

Improved Access to Data for Operational Decision Making

ERP systems operate via an integrated database, using one set of data to support all business functions. The systems can support decisions on optimal sourcing or cost accounting, for instance, for the entire enterprise or business units from the start, rather than gathering data from multiple business functions and then trying to coordinate that information manually or reconciling data with another application. The result is an organization that looks seamless, not only to the outside world but also to the decision makers who are deploying resources within the organization. The data is integrated to facilitate operational decision making and allows companies to provide greater customer service and support, strengthen customer and supplier relationships, and generate new business opportunities.

Elimination of Costly, Inflexible Legacy Systems

Adoption of an ERP system enables an organization to eliminate dozens or even hundreds of separate systems and replace them with a single, integrated set of applications for the entire enterprise. In many cases, these systems are decades old, the original developers are long gone, and the systems are poorly documented. As a result, the systems are extremely difficult to fix when they break, and adapting them to meet new business needs takes too long. They become an anchor around the organization that keeps it from moving ahead and remaining competitive. An ERP system helps match the capabilities of an organization's information systems to its business needs—even as these needs evolve.

Improvement of Work Processes

Competition requires companies to structure their business processes to be as effective and customer oriented as possible. ERP vendors do considerable research to define the best business processes. They gather requirements of leading companies within the same industry and combine them with research findings from research institutions and consultants. The individual application modules included in the ERP system are then designed to support these **best practices**, the most efficient and effective ways to complete a business process. Thus, implementation of an ERP system ensures good work processes based on best practices. For example, for managing customer payments, the ERP system's finance module can be configured to reflect the most efficient practices of leading companies in an industry. This increased efficiency ensures that everyday business operations follow the optimal chain of activities, with all users supplied the information and tools they need to complete each step.

best practices
The most efficient and effective ways to complete a business process.

Upgrade of Technology Infrastructure

When implementing an ERP system, an organization has an opportunity to upgrade the information technology (hardware, operating systems, databases, etc.) that it uses. While centralizing and formalizing these decisions, the organization can eliminate the hodgepodge of multiple hardware platforms, operating systems, and databases it is currently using—most likely from a variety of vendors. Standardizing on fewer technologies and vendors reduces ongoing maintenance and support costs as well as the training load for those who must support the infrastructure.

Disadvantages of ERP Systems

Unfortunately, implementing ERP systems can be difficult and can disrupt current business practices. Some of the major disadvantages of ERP systems are the expense and time required for implementation, the difficulty in implementing the many business process changes that accompany the ERP system, the problems with integrating the ERP system with other systems, the risks associated with making a major commitment to a single vendor, and the risk of implementation failure.

Expense and Time in Implementation

Getting the full benefits of ERP takes time and money. Although ERP offers many strategic advantages by streamlining a company's TPSs, large firms typically need three to five years and spend tens of millions of dollars to implement a successful ERP system.

Difficulty Implementing Change

In some cases, a company has to radically change how it operates to conform to the ERP's work processes—its best practices. These changes can be so drastic to long-time employees that they retire or quit rather than go through the change. This exodus can leave a firm short of experienced workers. Sometimes, the best practices simply are not appropriate for the firm and cause great work disruptions.

Difficulty Integrating with Other Systems

Most companies have other systems that must be integrated with the ERP system, such as financial analysis programs, e-commerce operations, and other applications. Many companies have experienced difficulties making these other systems operate with their ERP system. Other companies need additional software to create these links.

Risks in Using One Vendor

The high cost to switch to another vendor's ERP system makes it extremely unlikely that a firm will do so. After a company has adopted an ERP system, the vendor has less incentive to listen and respond to customer concerns. The high cost to switch also increases risk—in the event the ERP vendor allows its product to become outdated or goes out of business. Selecting an ERP system involves not only choosing the best software product but also the right long-term business partner. It was unsettling for many companies that had implemented PeopleSoft, J.D. Edwards, or Siebel Systems enterprise software when these firms were acquired by Oracle.

Risk of Implementation Failure

Implementing an ERP system for a large organization is extremely challenging and requires tremendous amounts of resources, the best IS and businesspeople, and plenty of management support. Unfortunately, large ERP installations occasionally fail, and problems with an ERP implementation can require expensive solutions.

The following list provides tips for avoiding many common causes for failed ERP implementations:

- Assign a full-time executive to manage the project.
- Appoint an experienced, independent resource to provide project oversight and to verify and validate system performance.
- Allow sufficient time for transition from the old way of doing things to the new system and new processes.
- Plan to spend a lot of time and money training people; many project managers recommend that $10,000–$20,000 per employee be budgeted for training of personnel.
- Define metrics to assess project progress and to identify project-related risks.
- Keep the scope of the project well defined and contained to essential business processes.
- Be wary of modifying the ERP software to conform to your firm's business practices.

ERP for Small and Medium Size Enterprises (SMEs)

It is not only large *Fortune* 1000 companies that are successful in implementing ERP. SMEs (both for-profit and not-for-profit) can achieve real business benefits from their ERP efforts. Many of the SMEs elected to implement open source ERP systems. With open-source software, anyone can see and modify the source code to customize it to meet their needs. Such systems are much less costly to acquire and are relatively easy to modify to meet business needs. A wide range of organizations can perform the system development and maintenance. Table 5.4 lists some of the open source ERP systems geared for SMEs.

The following sections outline how an ERP system can support the various major business processes.

Vendor	ERP Solutions
Apache	Open For Business ERP
Compiere	Compiere Open Source ERP
Openbravo	Openbravo Open Source ERP
WebERP	WebERP

Table 5.4

Open Source ERP Systems

Production and Supply Chain Management

ERP systems follow a systematic process for developing a production plan that draws on the information available in the ERP system database.

The process starts with *sales forecasting* to develop an estimate of future customer demand. This initial forecast is at a fairly high level with estimates made by product group rather than by each individual product item. The sales forecast extends for months into the future. The sales forecast might be developed using an ERP software module or it might be produced by other means using specialized software and techniques. Many organizations are moving to a collaborative process with major customers to plan future inventory levels and production rather than relying on an internally generated sales forecast.

The *sales and operations plan* takes demand and current inventory levels into account and determines the specific product items that need to be produced and when to meet the forecast future demand. Production capacity and any seasonal variability in demand must also be considered. The result is a high-level production plan that balances market demand to production capacity.

Demand management refines the production plan by determining the amount of weekly or daily production needed to meet the demand for individual products. The output of the demand management process is the master production schedule which is a production plan for all finished goods.

Detailed scheduling uses the production plan defined by the demand management process to develop a detailed production schedule specifying details such as which item to produce first and when production should be switched from one item to another. A key decision is how long to make the production runs for each product. Longer production runs reduce the number of machine setups required, thus reducing production costs. Shorter production runs generate less finished product inventory and reduce inventory holding costs.

Materials requirement planning determines the amount and timing for placing raw material orders with suppliers. The types and amounts of raw materials required to support the planned production schedule are determined based on the existing raw material inventory and the bill of materials or BOM, a sort of "recipe" of ingredients needed to make each product item. The quantity of raw materials to order also depends on the lead time and lot sizing. Lead time is the time it takes from the time a purchase order is placed until the raw materials arrive at the production facility. Lot size has to do with discrete quantities that the supplier will ship and the amount that is economical for the producer to receive and/or store. For example, a supplier might ship a certain raw material in units of 80,000 pound rail cars. The producer might need 95,000 pounds of the raw material. A decision must be made to order one or two rail cars of the raw material.

Purchasing uses the information from materials requirement planning to place purchase orders for raw materials and transmit them to qualified suppliers. Typically, the release of these purchase orders is timed so that raw materials arrive just in time to be used in production and minimize warehouse and storage costs. Often, producers will allow suppliers to tap into data via an extranet that enables them to determine what raw materials the supplier needs thus minimizing the effort and lead time to place and fill purchase orders.

Production uses the detailed schedule to plan the details of running and staffing the production operation.

Customer Relationship Management and Sales Ordering

Customer Relationship Management

customer relationship management (CRM) system
A system that helps a company manage all aspects of customer encounters, including marketing and advertising, sales, customer service after the sale, and programs to retain loyal customers.

As discussed in Chapter 1, a **customer relationship management (CRM) system** helps a company manage all aspects of customer encounters, including marketing and advertising, sales, customer service after the sale, and programs to keep and retain loyal customers (see Figure 5.10). The goal of CRM is to understand and anticipate the needs of current and potential customers to increase customer retention and loyalty while optimizing the way that products and services are sold. CRM is used primarily by people in the sales, marketing, and service organizations to capture and view data about customers and improve communications. Businesses implementing CRM systems report business benefits such as improved customer satisfaction, increased customer retention, reduced operating costs, and the ability to meet customer demand.

Figure 5.10

Customer Relationship Management System

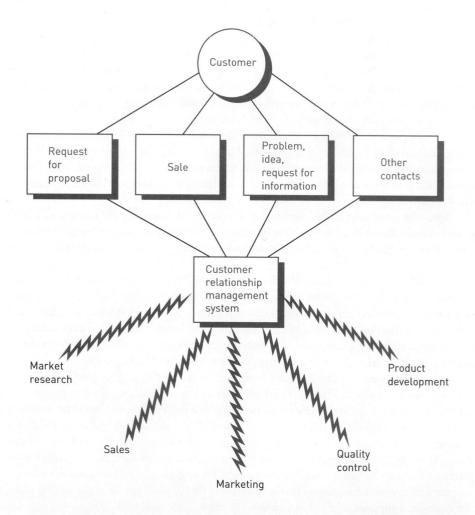

CRM software automates and integrates the functions of sales, marketing, and service in an organization. The objective is to capture data about every contact a company has with a customer through every channel and store it in the CRM system so the company can truly understand customer actions. CRM software helps an organization build a database about its customers that describes relationships in sufficient detail so that management, salespeople, customer service providers—and even customers—can access information to match customer needs with product plans and offerings, remind them of service requirements, and know what other products they have purchased. Figure 5.11 shows contact manager software from SAP that fills this CRM role.

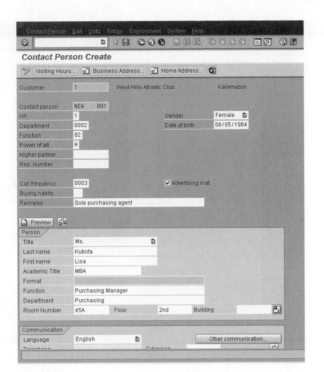

Figure 5.11

SAP Contact Manager

(Source: Copyright © by SAP AG.)

The focus of CRM involves much more than installing new software. Moving from a culture of simply selling products to placing the customer first is essential to a successful CRM deployment. Before any software is loaded onto a computer, a company must retrain employees. Who handles customer issues and when must be clearly defined, and computer systems need to be integrated so that all pertinent information is available immediately, whether a customer calls a sales representative or customer service representative. In addition to using stationary computers, most CRM systems can now be accessed via wireless devices.

Organizations choose to implement CRM for a variety of reason depending on their needs. American Eagle implemented a CRM system to improve marketers' ability to interact with customers via multiple channels including stores, the Web, mobile devices, and other means.[28] American of Martinsville, a contract furniture manufacturer, implemented CRM to automate its process for developing customer quotes and to improve customer communications.[29] Central Michigan University implemented a CRM system to improve its operational efficiency by reducing the elapsed time it takes to convert a request for information into an applicant and to retain those students through matriculation.[30] Kabel, a provider of cable TV, Internet, and telephone services, implemented CRM to improve its communications with customers and to provide support for an anticipated increase in customers.[31]

Sales Ordering

Sales ordering is the set of activities that must be performed to capture a customer sales order. A few of the essential steps include recording the items to be purchased, setting the sales price, recording the order quantity, determining the total cost of the order including delivery costs, and confirming the customer's available credit. The determination of the sales prices can become quite complicated and include quantity discounts, promotions, and incentives. After the total cost of the order is determined, it is necessary to check the customer's available credit to see if this order puts the customer over his credit limit. Figure 5.12 shows a sales order entry window in SAP business software.

Many small-to-midsize businesses are turning to ERP software to make it easier for their large customers to place orders with them. Car distributor smart USA is the only authorized distributor in the U.S. for a car called the smart fortwo. This rear-engine, two-passenger auto can achieve more than 50 miles per gallon. The firm implemented an ERP system with features to manage vehicle and parts ordering plus manage warranty claims from the dealer network to the manufacturer in Germany.[32]

Financial and Managerial Accounting

The general ledger is the main accounting record of a business. It is often divided into different categories including assets, liabilities, revenue, expenses, and equity. These categories in turn, are subdivided into subledgers to capture details such as cash, accounts payable, and accounts receivable. In an ERP system, input to the general ledger occurs simultaneously with the input of a business transaction to a specific module. Here are several examples of how this occurs:

- An order clerk records a sale and the ERP system automatically creates an accounts receivable entry indicating that a customer owes money for goods received.
- A buyer enters a purchase order and the ERP system automatically creates an accounts payable entry in the general ledger registering that the company has an obligation to pay for goods that will be received at some time in the future.
- A dock worker enters a receipt of purchased materials from a supplier and the ERP system automatically creates a general ledger entry to increase the value of inventory on hand.
- A production worker withdraws raw materials from inventory to support production and the ERP system generates a record to reduce the value of inventory on hand.

Thus, the ERP system captures transactions entered by workers in all functional areas of the business. The ERP system then creates the associated general ledger record to track the financial impact of the transaction. This set of records is an extremely valuable resource that companies can use to support financial accounting and managerial accounting.

Financial accounting consists of capturing and recording all the transactions that affect a company's financial state and then using these documented transactions to prepare financial statements to external decision makers, such as stockholders, suppliers, banks, and government agencies. These financial statements include the profit and loss statement, balance sheet, and cash flow statement. They must be prepared in strict accordance to rules and guidelines of agencies such as the Securities and Exchange Commission, Internal Revenue Service, and the Financial Accounting Standards Board. Data gathered for financial accounting can also form the basis for tax accounting because this involves external reporting of a firm's activities to the local, state, and federal tax agencies.

Managerial accounting involves the use of "both historical and estimated data in providing information that management uses in conducting daily operations, in planning future operations, and in developing overall business strategies."[33] Managerial accounting provides data to enable the firm's managers to assess the profitability of a given product line or specific product, identify underperforming sales regions, establish budgets, make profit forecasts, and measure the effectiveness of marketing campaigns.

All transactions that affect the financial state of the firm are captured and recorded in the database of the ERP system. This data is used in the financial accounting module of the ERP system to prepare the statements required by various constituencies. The data can also be used in the managerial accounting module of the ERP system along with various assumptions and forecasts to perform various analyses such as generating a forecasted profit and loss statement to assess the firm's future profitability.

Hosted Software Model for Enterprise Software

Many business application software vendors are pushing the use of the hosted software model for SMEs. The goal is to help customers acquire, use, and benefit from the new technology while avoiding much of the associated complexity and high start-up costs. SAP, Microsoft, NetSuite, Intacct, Oracle, BizAutomation.com, Salesforce.com, NetBooks, and Workday are among the software vendors who offer hosted versions of their ERP or CRM software at a cost of $50–$200 per month per user.[34]

This pay-as-you-go approach is appealing to SMEs because they can experiment with powerful software capabilities without making a major financial investment. Organizations can then dispose of the software without large investments if the software fails to provide value or otherwise misses expectations. Also, using the hosted software model means the small business firm does not need to employ a full-time IT person to maintain key business applications. The small business firm can expect additional savings from reduced hardware costs and costs associated with maintaining an appropriate computer environment (such as air conditioning, power, and an uninterruptible power supply).

Potential problems can occur if the hosted software vendor cannot provide a reliable operation environment that ensures both that the software is available when needed and that company sensitive data is safe from compromise. Car Toys Inc. experienced an outage of its hosted BI software that disrupted month-end reporting when its hosted software provider decided to move the firm's data to new hardware with no prior warning.[35] Table 5.5 lists the advantages and disadvantages of hosted software.

Advantages	Disadvantages
Decreased total cost of ownership	Potential availability and reliability issues
Faster system startup	Potential data security issues
Lower implementation risk	Potential problems integrating the hosted products of different vendors
Management of systems outsourced to experts	Savings anticipated from outsourcing may be offset by increased effort to manage vendor

Table 5.5

Advantages and Disadvantages of Hosted Software Model

INTERNATIONAL ISSUES ASSOCIATED WITH ENTERPRISE SYSTEMS

Enterprise systems must support businesses that interoperate with customers, suppliers, business partners, shareholders, and government agencies in multiple countries. Different languages and cultures, disparities in IS infrastructure, varying laws and customs rules, and multiple currencies are among the challenges that must be met by an enterprise system of a multinational company.

SUMMARY

Principle

Electronic commerce and mobile commerce are evolving, providing new ways of conducting business that present both opportunities for improvement and potential problems.

E-commerce is the conducting of business activities electronically over networks. Business-to-business (B2B) e-commerce allows manufacturers to buy at a low cost worldwide, and it offers enterprises the chance to sell to a global market. B2B e-commerce is currently the largest type of e-commerce. Business-to-consumer (B2C) e-commerce enables organizations to sell directly to consumers, eliminating intermediaries. In many cases, this squeezes costs and inefficiencies out of the supply chain and can lead to higher profits and lower prices for consumers. Consumer-to-consumer (C2C) e-commerce involves consumers selling directly to other consumers. Online auctions are the chief method by which C2C e-commerce is currently conducted. e-Government is the use of information and communications technology to simplify the sharing of information, speed formerly paper-based processes, and improve the relationship between citizens and government.

Mobile commerce is the use of wireless devices such as PDAs, cell phones, and smartphones to facilitate the sale of goods or services—anytime, anywhere. The market for m-commerce in North America is expected to mature much later than in Western Europe and Japan. Although some industry experts predict great growth in this arena, several hurdles must be overcome, including improving the ease of use of wireless devices, addressing the security of wireless transactions, and improving network speed.

Electronic retailing (e-tailing) is the direct sale from a business to consumers through electronic storefronts designed around an electronic catalog and shopping cart model.

A cybermall is a single Web site that offers many products and services at one Internet location.

Manufacturers are joining electronic exchanges, where they can work with competitors and suppliers to use computers and Web sites to buy and sell goods, trade market information, and run back-office operations such as inventory control. They are also using e-commerce to improve the efficiency of the selling process by moving customer queries about product availability and prices online.

The Web allows firms to gather much more information about customer behavior and preferences than they could using other marketing approaches. This new technology has greatly enhanced the practice of market segmentation and enabled companies to establish closer relationships with their customers. Detailed information about a customer's behavior, preferences, needs, and buying patterns allow companies to set prices, negotiate terms, tailor promotions, add product features, and otherwise customize a relationship with a customer.

The Internet has also revolutionized the world of investment and finance, especially online stock trading and online banking.

There are hundreds of Websites that offer interesting services for those looking to buy a home.

The Internet has also created many options for electronic auctions, where geographically dispersed buyers and sellers can come together.

M-commerce provides a unique opportunity to establish one-on-one marketing relationships and support communications anytime and anywhere. M-commerce transactions are being used in many application arenas including mobile banking, mobile price comparison, mobile advertising, and mobile coupons.

Businesses and people use e-commerce to reduce transaction costs, speed the flow of goods and information, improve the level of customer service, and enable the close coordination of actions among manufacturers, suppliers, and customers. E-commerce also enables consumers and companies to gain access to worldwide markets. E-commerce offers great promise for developing countries, helping them to enter the prosperous global marketplace, and hence helping to reduce the gap between rich and poor countries.

Principle

E-commerce and m-commerce require the careful planning and integration of a number of technology infrastructure components.

A number of infrastructure components must be chosen and integrated to support a large volume of transactions with customers, suppliers, and other business partners worldwide. These components include hardware, Web server software, security and identification services, Web site development tools, e-commerce software and Web services.

The Wireless Application Protocol (WAP) is a standard set of specifications to enable development of m-commerce software for wireless devices. WAP uses the Wireless Markup Language, which is designed for effectively displaying information on small devices. The development of the Wireless Application Protocol (WAP) and its derivatives addresses many m-commerce issues.

Electronic payment systems are a key component of the e-commerce infrastructure. A digital certificate is an attachment to an e-mail message or data embedded in a Web page that verifies the identity of a sender or a Web site. To help prevent the theft of credit card numbers and banking information, the Secure Sockets Layer (SSL) communications protocol is used to secure all sensitive data. There are several electronic cash alternatives that require the purchaser to open an account with an electronic cash service provider and to present proof of identity whenever payments are to be made. Payments can also be made by credit, charge, debit, and smart cards.

Principle

An organization must have information systems that support the routine, day-to-day activities that occur in the normal course of business and help a company add value to its products and services.

Transaction processing systems (TPSs) are at the heart of most information systems in businesses today. A TPS is an organized collection of people, procedures, software, databases, and devices used to capture fundamental data about events that affect the organization (transactions) and use that data to update the official records of the organization.

The methods of transaction processing systems include batch and online. Batch processing involves the collection of transactions into batches, which are entered into the system at regular intervals as a group. Online transaction processing (OLTP) allows transactions to be entered as they occur.

Organizations expect TPSs to accomplish a number of specific objectives, including processing data generated by and about transactions, maintaining a high degree of accuracy and information integrity, compiling accurate and timely reports and documents, increasing labor efficiency, helping provide increased and enhanced service, and building and maintaining customer loyalty. In some situations, an effective TPS can help an organization gain a competitive advantage.

All TPSs perform the following basic activities: data collection, which involves the capture of source data to complete a set of transactions; data editing, which checks for data validity and completeness; data correction, which involves providing feedback of a potential problem and enabling users to change the data; data manipulation, which is the performance of calculations, sorting, categorizing, summarizing, and storing data for further processing; data storage, which involves placing transaction data into one or more databases; and document production, which involves outputting records and reports.

Because of the importance of TPSs to ongoing operations, organizations must develop a disaster recovery plan that focuses on the actions that must be taken to restore computer operations and services in the event of a disaster. Although companies have known about the importance of disaster planning and recovery for decades, many do not adequately prepare.

Many organizations conduct ongoing TPS audits to prevent accounting irregularities or loss of data privacy that can violate federal acts and diminish investor confidence. The TPS audit attempts to answer four basic questions: (1) Does the system meet the business need for which it was implemented? (2) What procedures and controls have been established? (3) Are these procedures and controls being used properly? (4) Are the information systems and procedures producing accurate and honest reports?

Traditional TPS systems include the following types of systems: order processing, accounting and purchasing systems.

The traditional TPSs that support the purchasing function include inventory control, purchase order processing, accounts payable, and receiving.

Many software packages provide integrated transaction processing solutions for SMEs.

Principle

A company that implements an enterprise resource planning system is creating a highly integrated set of systems, which can lead to many business benefits.

Enterprise resource planning (ERP) is software that supports the efficient operation of business processes by integrating activities throughout a business, including sales, marketing, manufacturing, logistics, accounting, and staffing.

Implementation of an ERP system can provide many advantages, including providing access to data for operational decision making; elimination of costly, inflexible legacy systems; providing improved work processes; and creating the opportunity to upgrade technology infrastructure.

Some of the disadvantages associated with an ERP system are that they are time consuming, difficult, and expensive to implement.

Many SMEs have found open source ERP systems to be effective solutions to their transaction processing and management reporting needs.

The production and supply chain management process starts with sales forecasting to develop an estimate of future customer demand. This initial forecast is at a fairly high level with estimates made by product group rather than by each individual product item. The sales and operations plan takes demand and current inventory levels into account and determines the specific product items that need to be produced and when to meet the forecast future demand. Demand management refines the production plan by determining the amount of weekly or daily production needed to meet the demand for individual products. Detailed scheduling uses the production plan defined by the demand management process to develop a detailed production schedule specifying details such as which item to produce first and when production should be switched from one item to another. Materials requirement planning determines the amount and timing for placing raw material orders with suppliers. Purchasing uses the information from materials requirement planning to place purchase orders for raw materials and transmit them to qualified suppliers. Production uses the detailed schedule to plan the details of running and staffing the production operation.

Business application software vendors are experimenting with the hosted software model to see if the approach meets customer needs and is likely to generate significant revenue.

Numerous complications arise that multinational corporations must address in planning, building, and operating their TPSs. These challenges include dealing with different languages and cultures, disparities in IS infrastructure, varying laws and customs rules, and multiple currencies.

CHAPTER 5: SELF-ASSESSMENT TEST

Electronic commerce and mobile commerce are evolving, providing new ways of conducting business that present both opportunities for improvement and potential problems.

1. The market for m-commerce in North America is maturing much later than in Western Europe and Japan. True or False?

2. A form of e-commerce in which customers deal directly with an organization and avoids intermediaries is called _____.

3. An advancement in online bill payment that uses e-mail for the biller to post an image of your statement on the Internet so you can direct your bank to pay it is called _____.

E-commerce and m-commerce require the careful planning and integration of a number of technology infrastructure components.

4. Many companies use a third-party Web service to host their Web site. True or False?

5. An attachment to an e-mail message or data embedded in a Web site that verifies the identity of a sender or Web site is called a(n) _____.

An organization must have information systems that support the routine, day-to-day activities that occur in the normal course of business and help a company add value to its products and services.

6. Identify the missing TPS basic activity: data collection, data editing, data _____, data manipulation, data storage, and document production.

7. A form of TPS where business transactions are accumulated over a period of time and processed all at once is called _____.

8. Many organizations conduct ongoing transaction processing system _____ to prevent accounting irregularities or loss of data privacy that might violate federal acts.

A company that implements an enterprise resource planning system is creating a highly integrated set of systems, which can lead to many business benefits.

9. A firm's plan to recover data, technology, and tools that support critical information systems and necessary information system components such as the network, databases, hardware, software and operating systems is called a _____.

10. Because it is so critical to the operation of an organization, most companies can implement an ERP system without major difficulty. True or False?

11. Only large, multinational companies can justify the implementation of ERP systems. True or False?

CHAPTER 5: SELF-ASSESSMENT TEST ANSWERS

(1) True (2) B2C (3) electronic bill presentment (4) True (5) digital certificate (6) correction (7) batch processing (8) audit (9) disaster recovery plan (10) False (11) False

REVIEW QUESTIONS

1. Define the term e-Government. Identify three forms of e-Government and give an example of each.
2. Identify and briefly describe three limitations that complicate the use of handheld devices used for m-commerce.
3. What are Web services? Provide a brief example of how you might use a Web service in conducting e-commerce.
4. What benefits can a firm achieve by converting to an e-commerce supply chain system?

5. Briefly explain the differences among smart, credit, charge, and debit cards.
6. Identify the key elements of the technology infrastructure required to successfully implement e-commerce within an organization.
7. What is the Secure Sockets Layer and how does it support e-commerce?
8. What problems can arise when an organization's TPS systems are not integrated?
9. An ERP system follows a systematic process for developing a production plan that draws on the information available in the ERP system database. Outline this process and identify the software modules that are used to support it.

10. Identify four complications that multinational corporations must address in planning, building, and operating their ERP systems.
11. A transaction processing system audit attempts to answer what fundamental questions?
12. What is the role of a CRM system? What sort of business benefits can such a system produce?
13. Identify three characteristics of integrated processing software package solutions that make them attractive to SMEs.
14. What is the difference between managerial and financial accounting?

DISCUSSION QUESTIONS

1. Wal-Mart, the world's largest retail chain, has turned down several invitations to join exchanges in the retail and consumer goods industries. Is this good or bad for the overall U.S. economy? Why?
2. What do you think are the biggest barriers to wide-scale adoption of m-commerce by consumers? Who do you think is working on solutions to these problems and what might the solutions entail?
3. Identify and briefly describe three m-commerce applications you have used.
4. Discuss the use of e-commerce to improve spending on manufacturing, repair, and operations (MRO) of goods and services.
5. Identify three kinds of business organizations that would have difficulty in becoming a successful e-commerce organization.
6. Imagine that you are the new IS manager for a *Fortune* 1000 company. You have uncovered that there has been no thorough system processing audits conducted for over three years. Prepare a brief outline of a talk you will make

to senior company managers to convince them that substantial resources must be assigned to conducting such an audit.
7. Do you think that an SME would have less or more difficulty implementing an ERP system than a large, multinational corporation? Defend your position.
8. You are the key user of your firm's accounts receivable system and have been asked to assess the disaster recovery plan for this system. Outline the steps you would take to complete the assessment. Identify specific problems you would look for.
9. What sort of benefits should the suppliers and customers of a firm that has successfully implemented an ERP system see? What sort of issues might arise for suppliers and customers during an ERP implementation?
10. Many organizations are moving to a collaborative process with their major customers to get their input on planning future inventory levels and production rather than relying on an internally generated demand forecast. Explain how such a process might work. What issues and concerns might a customer have in entering into an agreement to do this?

PROBLEM-SOLVING EXERCISES

1. Develop a set of criteria you would use to evaluate various business-to-consumer Web sites based on factors such as ease of use, protection of consumer data, and security of payment process. Develop a simple spreadsheet containing these criteria. Evaluate five popular Web sites using the criteria you developed. What changes would you recommend to the Web developer of the site that scored lowest?
2. Research the growth of B2B and B2C e-commerce and retail sales for the period 2000 to present. Use the graphics

capability of your spreadsheet software to plot the growth of all three. Using current growth rates, predict the year that B2C e-commerce will exceed 10 percent of retail sales.
3. Use a spreadsheet program to develop a sales forecasting system for a new car dealership that can estimate monthly sales for each make and model based on historical sales data and various parameters. Suggestion: Assume that this month's sales will be the same as the sales for this month last year except for adjustments due to the cost of gas and

each make of car's miles per gallon. You can further refine the model to take into account change in interest rates for

new cars or other parameters you wish to include. Document the assumptions you make in building your model.

TEAM ACTIVITIES

1. Imagine that your team has been hired as consultants to provide recommendations to boost the traffic to a Web site that sells environmentally friendly household cleaning products. Identify as many ideas as possible for how you can increase traffic to this Web site. Next, rank your ideas from best to worst.

2. Your team members should interview several business managers at a firm that has implemented a CRM system. Try to define the scope and schedule for the overall project. Make a list of what they see as the primary benefits of the implementation. What were the biggest hurdles they had to overcome? Did the firm need to retrain its employees to place greater emphasis on putting the customer first?

WEB EXERCISES

1. Visit e-Bay or another online auction Web site and choose an item on which to bid. Before entering the bid process, research the site for information about any rules associated with bidding and how to bid effectively. Follow the suggested processes and record your results. Write a brief memo to your instructor summarizing your experience.

2. Do research on the Web and find a Web site that offers a demo of an ERP or CRM system. View the demo, perhaps more than once. Write a review of the software based on the demo. What are its strengths and weaknesses? What additional questions about the software do you have? E-mail your questions to the vendor and document their response to your questions.

CAREER EXERCISES

1. For your chosen career field, describe how you might use or be involved with e-commerce. If you have not chosen a career yet, answer this question for someone in marketing, finance, or human resources.

2. Initially thought to be cost-effective for only very large companies, CRM systems are now being implemented in small and midsized companies to reduce costs and improve service. A firm's operations and accounting personnel play

a dual role in the implementation of such a system: (1) They must ensure a good payback on the investment in CRM, and (2) they must also ensure that the system meets the needs of the operations and accounting organizations. Identify three or four tasks that the operations and accounting personnel need to perform to ensure that both these goals are met.

CASE STUDIES

Case One
Paying with Cell Phones in Canada

Companies in Canada are racing to see who can motivate Canadians to start paying at the register with their cell

phones. You are probably familiar with MasterCard's PayPass technology that allows people in many countries to make payments at checkouts by touching their credit card to a pad. The cards are equipped with a chip that supports near-field communication (NFC), passing credit card information to the receiver in the payment pad. Now vendors want to do away with plastic credit cards altogether by embedding NFC chips in cell phones and allowing customers to make payments by touching their cell phones to pads rather than credit cards.

In Canada, MasterCard and Visa are rolling out mobile wireless payment pilot programs this year. However, credit card companies need the support of cell phone handset manufacturers and carriers to successfully launch the programs. MasterCard is working with carrier Bell Canada and using handsets from an unnamed vendor. Visa is working with the Royal Bank of Canada (RBC) and is determining who the carrier and handset manufacturer will be.

MasterCard might have an advantage over Visa in that it already has significant penetration in this market with its PayPass technology. More than 28 million MasterCard PayPass cards are being used at more than 109,000 merchants worldwide. MasterCard is starting a trial with Bell Canada employees who will be paying with their phones and testing the benefits, which include faster checkouts and additional services. For example, financial services are provided that allow you to check your transaction history and bank balance and conduct online banking transactions. Another benefit is avoiding having to carry a wallet packed with plastic.

The big question in this race is whether consumers are interested in paying with their phones. A 2007 survey of 15–29 year-old Canadians showed that only 8.8 percent of those surveyed were interested in contactless payments via cell phone. Some people have voiced concern over security and privacy issues surrounding wireless payments. The technology might also decrease battery life in cell phones.

Providers are concerned about the legal risks of offering m-commerce services over near-field communication technologies. "Who's responsible for liability issues?" asked Anne Koski, head of payment innovations at the Royal Bank of Canada at a recent conference on the topic. If money is lost due to inefficiencies in the technology, who foots the bill—the handset manufacturer, the carrier, or the bank? Legal teams are devising the answer to this important question. Those in the industry know that convincing customers of the security of the new system is important in winning them over to the technology.

Data reliability, authentication, fraud, theft, and privacy protection are all issues that these companies are confronting as they begin planning their marketing campaigns. Some believe that creating a standard for mobile payments is the most important factor in launching mobile payments. Being able to advertise the stability and security features of an agreed upon standard would help in winning over consumers. However, a standard might mean that companies such as MasterCard have to overhaul their entire PayPass network.

Many believe that consumer education is also important. In the early days of Internet-based e-commerce, many people were afraid to purchase products and services on the Web due to concerns over privacy and security. Most of those people have overcome those fears in exchange for the convenience and opportunities that online shopping offers. Proponents of m-commerce and contactless payment systems are hoping that once educated, consumers will choose the convenience of paying by swiping your cell phone over any perceived risk.

Discussion Questions

1. What are the benefits and concerns associated with cell-phone payments?
2. Why does MasterCard have a head start in the cell-phone payment race taking place in Canada?

Critical Thinking Questions

1. What assurances would you need from the service provider prior to buying into a cell-phone wireless payment system?
2. Do you think that a cell phone payment system is inevitable in Canada and North America? If so, how long do you think it will take before it becomes the norm? If not, why?

SOURCES: Smith, Briony, "MasterCard Gets moving on mobile payments," *IT World Canada*, May 28, 2008, *www.itworldcanada.com/a/Enterprise-Business-Applications/36ee2d57–2fad-4a2a-8d41–03f6f6bb6a34.html*; Smith, Briony, "RBC, Visa prepare to jump mobile payment hurdles," *IT World Canada*, June 11, 2008, *www.itworldcanada.com/a/Enterprise-Business-Applications/d628b121–3309–47fd-a927-c8f9408c87e5.html*; Smith, Briony, "The biggest legal risks around mobile payments," *IT World Canada*, June 11, 2008, *www.itworldcanada.com/a/Enterprise-Business-Applications/e3425f71–90f6–416a-9121-edd7a4ccf154.html*.

Case Two
Delhi Government Embraces Enterprise Systems

Delhi is the second largest city in India, with a population of over 17 million. Located on the banks of the Yamuna River, Delhi was established in 1000 BC, making it one of the oldest cities in the world. Delhi is governed in part by the Municipal Corporation of Delhi (MCD), which has a staff of over 100,000 working in 107 offices across 12 geographic zones. Recently, the MCD has been working to move its staff online, migrating from a paper form-driven system to online automation. Those in charge of the project are learning that changing the work processes of an organization of this size is no small undertaking.

One of the MCD's most challenging tasks has been procuring contractors, services, and products for use on city projects. The traditional procedure for procurement involved getting the word out to let the community know that a project was planned and contractors were needed. Businesses that wanted to bid on the contract would then travel to the MCD, sometimes over hundreds of miles, to bid on the project. The process of bidding and negotiation might take months, until one contractor finally won the contract.

The time-honored bidding system, referred to as tendering, was far from convenient or fair. In many cases, eligible companies lost contracts because they missed the deadline by minutes due to travel delays. Corruption in the system also treated participants unfairly, with some companies using intimidation techniques to keep competitors from bidding on contracts. It was time to bring the MCD's tendering system into the digital age.

Arun Kumar, an executive engineer at the MCD, took on the responsibility of making that transition. Arun has completed the installation of an e-tendering system that automates much of the bidding process, and is working to go farther and establish a complete e-procurement system.

The new e-tendering system allows contractors to download and upload tender documents online, track the status of tenders, and receive e-mail alerts. The system was developed and deployed in stages. First, the MCD required companies to submit tenders and bids online. Secondly, the MCD set up an online and offline backup system to safeguard the tender information against equipment failure. The MCD also provided a telephone help desk available to contractors 24 hours a day.

The MCD leases data center space from the Center for Development of Advanced Computing (C-DAC). The Web-based e-tendering system was developed by information systems company Wipro, which has since installed the system in several other government agencies in India.

The new system has eliminated the need for companies to physically send representatives to the MCD headquarters. It has also provided bidders with privacy and the government with transparency. Bidders need no longer fear intimidation from competitors since no one knows who is bidding. Bids are placed in an anonymous fashion, freeing contracting decision makers from outside influences. Now contracts are awarded on a bidder's merit and bid, and not according to who knows whom.

The MCD went to great lengths to sell contractors and government staff on the new way of doing business. It contracted Wipro to train hundreds of users and thousands of contractors. At first, only 70 percent of the engineers used the system, with the remaining engineers not willing to touch a computer. After about six months, the advantages of the system won over the holdouts.

The new e-tendering system has been a huge success, with over 30,000 tenders placed over the system—the world's highest volumes in numbers by any government organization. The MCD increased the number of transactions it handles each week and reduced the time it takes to award a contract from 90 days to 30 days.

Still, Arun Kumar sees other areas that need improvement. He and his team are testing an e-procurement system that they hope will streamline approvals. The contract approval cycle currently takes two to three months because it's a manual task that involves a certifying authority auditing the process and paperwork. If Arun can reduce this time by two-thirds, that would really make a big difference.

Discussion Questions

1. What were some of the biggest challenges in implementing the new e-tendering system at the MCD?
2. What benefits does the new system provide for the MCD and its contractors?

Critical Thinking Questions

1. Compared to a business, what considerations might be different for a government agency designing an enterprise system?
2. Why do you think some engineers were hesitant to cooperate with the MCD in using the new system?

Sources: Talgeri, Kunal, "Delhi, India to set up 'comprehensive' e-procurement system," *itWorld Canada*, September 3, 2007, *www.itworldcanada.com/Pages/Docbase/ViewArticle.aspx?ID=idgml-a9370738-e371- 4cee-8dcb-d52207a095b1&Portal=2e6e7040-2373-432d-b393-91e487ee7d70&ParaStart=0&ParaEnd=15&direction=next&Next=Next*; MCD Online, *www.mcdonline.gov.in*, accessed June 30, 2008; Google Maps, *maps.google.com/maps?q=Delhi,+India&ie=UTF-8&oe=utf-8&rls=org.mozilla:en-US:official&client=firefox-a&um=1&sa=X&oi=geocode_result&resnum=1&ct=title*, accessed June 30, 2008; Delhi India City Guide Web site, *delhi.clickindia.com*, accessed June 30, 2008.

Questions for Web Case

See the Web site for this book to read about the Whitmann Price Consulting case for this chapter. Following are questions concerning this Web case.

Whitmann Price Consulting: Enterprise Systems

Discussion Questions

1. What would be the danger of Josh and Sandra developing the Advanced Mobile Communications and Information System without considering other systems within Whitmann Price?
2. What are the advantages of ERP systems that provide an integrated one-vendor approach over multiple systems from multiple vendors?

Critical Thinking Questions

1. What are the pros and cons of buying predesigned software from a vendor such as SAP instead of a company developing software itself to exactly meet its own needs?
2. How might a CRM system assist Whitmann Price consultants in the field?

NOTES

Sources for the opening vignette: Kahn, Michael, "Nike Says Just Do It Yourself," Reuters, May 30, 2005, *www.reuters.com*. Hallett, Vicky, "Satisfied Customizers," *U.S. News & World Report*, November 21, 2005, DIVERSIONS; Vol. 139, No. 19; Pg. D2, D4, or *www.lexis-nexis.com*. Nike Customization Web site, accessed May 4, 2006, *http://nikeid.nike.com/nikeid/*.

1   Chabrow, Eric, "E-Commerce Continues to Grow Very Nicely," *Web Design & Technology News*, May 1, 2003.

2   Messmer, Ellen, "Dot-Com Survivor Covisint Finds B2B Niche, *Network World*, November 1, 2007.

3   Broache, Anne, "Tax-Free Internet Shopping Days Could Be Numbered," *CNET News.com*, April 15, 2008.

4   "American Eagle Outfitters Launches E-Commerce Web site for Martin + OSA," *DataMonitor*, April 4, 2008.

5   "eBay Reports Fourth Quarter and Full Year 2007 Results," *http://files.shareholder.com/downloads/ebay,* accessed April 7, 2008.

6   Regan, Keith, "Sprint Cuts Ribbon on Mobile Shopping Service," *E-Commerce Times*, September 13, 2007.

7   "40% of Big U.S. Companies Boast Mobile Sites," *Internet Retailer*, January 18, 2008.

8   Offner, Jim, "Yahoo to Mobile Searchers: Talk to the Handset," *E-Commerce Times*, April 3, 2008.

9   Regan, Keith, "Amazon Aims to Light M-Commerce Fire with TextBuyIt," *E-Commerce Times*, April 2, 2008.

10   Siwicki, Bill, "Merchant Beware," *Internet Retailer*, January 2008.

11   Jones, Sandra M., "Sears' Quiet E-Commerce Revolution," *E-commerce Times*, April 13, 2008.

12   Regan, Keith, "Detroit Trading Exchange: Swinging the Door Wide Open for Auto Leads," *E-Commerce Times*, April 4, 2008.

13   Crum, Rachelle, "ATG's Cliff Conneighton: E-Tailers 'Can't Wing It Anymore'," *E-Commerce Times*, April 6, 2008.

14   Offner, Jim, "Report: Paying Bills Online Saves Gas, Trees, Water," *E-Commerce Times*, March 27, 2008.

15   "ePower Systems Executive Summary," Center for Business Planning Web site, *www.businessplans.org/ePower/ePower00.html*, accessed May 3, 2008.

16   "Zillow Goes After an Area That Real Estate Agents Can't Touch," *TechDirt*, July 11, 2007,

17   "Redfin," Crunchbase, *www.crunchbase.com/company/redfin*, accessed May 5, 2008.

18   "Blair Uses e-Sourcing to Contain Costs," eDyanQuote Case Study, *www.edynaquote.com*, accessed May 24, 2008.

19   "Walk 100 Yards North, Turn Right, Enter Store," *Internet Retailer*, December 31, 2007.

20   Kharif, Olga, "Building a Case for Mobile Advertising," *E-Commerce Times*, March 16, 2008.

21   "40% of Big U.S. Companies Boast Mobile Sites," *Internet Retailer*, January 18, 2008.

22   Rosencrance, Linda, "5 Companies to Test Mobile Coupons at Grocery Chain," *Computerworld*, April 18, 2008

23   UPS: Trade Tools and Terminology, UPS Web site, *www.ups.com/content/us/en/resources/advisor/tools_terms/index.html*, accessed June 2, 2008.

24   About Us page, PayPal Web site, *www.paypal.com*, accessed May 31, 2008.

25   "Barclaycard to Test Cobranded Transit Card on NFC Phones," *Card Technology*, November 2007.

26   Sowa, Tom, "Wading into the M-Commerce Waters," *E-Commerce Times*, April 14, 2008.

27   Cummings, Joanne, "TiVo's Disaster Recovery Plan," *NetworkWorld*, August 21, 2007.

28   Staff, "American Eagle Outfitters Selects Terradata's CRM Site," *CBR*, May 21, 2008.

29   Staff, "American of Martinsville Implements Microsoft Dynamics AX and CRM Solutions," *CBR*, May 23, 2008.

30   Staff , "Central Michigan University Improves Recruitment and Retention with Talisma," *CBR*, May 2, 2008.

31   Staff, "Kabel Selects SAP's CRM Application," *CBR*, May 20, 2008.

32   Staff, "Smart USA Selects IBS Automotive ERP System, *CBR*, May 15, 200.

33   Student Resource Glossary, http://www.associatedcontent.com/article/526265/accounting_financial_managerial.html?cat=3, accessed July 9, 2008.

34   "Top 5 SaaS ERP Software Applications," ERPSoftware 360 Web site at *www.erpsoftware360.com*, accessed June 11, 2008.

35   Fonseca, Brian, "SaaS Benefits Starting to Outweigh Risks for Some," *Computerworld*, May 5, 2008.

CHAPTER · 6 ·

Information and Decision Support Systems

PRINCIPLES	LEARNING OBJECTIVES
• Good decision-making and problem-solving skills are the key to developing effective information and decision support systems.	• Define the stages of decision making. • Discuss the importance of implementation and monitoring in problem solving.
• The management information system (MIS) must provide the right information to the right person in the right format at the right time.	• Explain the uses of MISs and describe their inputs and outputs. • Discuss information systems in the functional areas of business organizations.
• Decision support systems (DSSs) are used when problems are unstructured.	• List and discuss important characteristics of DSSs that give them the potential to be effective management support tools. • Identify and describe the basic components of a DSS.
• Specialized support systems, such as group support systems (GSSs) and executive support systems (ESSs), use the overall approach of a DSS in situations such as group and executive decision making.	• State the goals of a GSS and identify the characteristics that distinguish it from a DSS. • Identify the fundamental uses of an ESS and list the characteristics of such a system. • List and discuss the use of other special-purpose systems.

Information Systems in the Global Economy
General Mills, United States

Food Giant General Mills Relies on Management Information Systems to Maintain Product Specifications

General Mills markets many well-known food brands including Betty Crocker, Pillsbury, Green Giant, Yoplait, Häagen-Dazs, and Cheerios. One of the company's greatest challenges is consistently providing high quality in every product it produces, packages, and delivers around the world, while adhering to strict regulations imposed by various governments. Management information systems (MISs) allow General Mills to do just that.

General Mills stores the specifications for each of its products in a huge database. The data includes specifications for ingredients, formulas, processing, and packaging. In total, the company stores over 22,000 product specifications, which are released to all of its manufacturing sites on a regular schedule. Over the course of a year, the company makes over 10,000 modifications to product specifications.

The process of modifying specifications is much more complicated than it sounds. Many specifications are shared across multiple products. For example, if General Mills changes a design feature of its cereal box, it affects the specifications for dozens of products. If Germany changes its regulations regarding the use of partially hydrogenated oils, the change could affect the ingredients, formulas, and processing specifications of dozens of products.

Because of the interrelated nature of General Mills' product specifications and the size of its database, one change to specifications might require several hours of design work followed by a lengthy review and approval process. What the company needed was an MIS that could automate the process of updating specifications and undo changes as necessary to return to the previously approved specifications.

General Mills worked with IS professionals from an outside MIS development company to design "mass-change and undo" functionality for their product specifications system. The MIS professionals worked with General Mills employees in the United States and abroad to fully understand the nature of the challenge. After rigorous testing and user training, the new features were integrated into the existing system with impressive results. The first modification to a product affected 332 related specifications in the database. Prior to the system improvement, the change would have required over 5,000 keystrokes and more than one day's work to implement. With the new system, the change took six keystrokes and five minutes.

Not only does the new system save a significant amount of time, but it reduces errors. Reducing keystrokes from 5,000 to five obviously reduces data entry errors. The system also reduces logical errors; by requiring changes to be made in a specific order following instructions provided in the software, operators are much less likely to overlook important considerations when changing specifications.

The development of General Mills' new mass-change and undo system gives the company an advantage in the highly competitive and closely regulated food industry. When market conditions change due to new regulations or public opinion, the first company to adjust and bring the changes to market achieves a competitive advantage. General Mills now has a much greater chance of being the first to introduce products that customers want. Management information systems and decision support systems give businesses in all industries a path to success through more efficient and effective operations.

Why Learn About Information and Decision Support Systems?

You have seen throughout this book how information systems can make you more efficient and effective through database systems, the Internet, e-commerce, transaction processing systems, and many other technologies. The true potential of information systems, however, is in helping you and your coworkers make more informed decisions. This chapter shows you how to slash costs, increase profits, and uncover new opportunities for your company using management information and decision support systems. Transportation coordinators can use management information reports to find the least expensive way to ship products to market and to solve bottlenecks. A loan committee at a bank or credit union can use a group support system to help them determine who should receive loans. Store managers can use decision support systems to help them decide what and how much inventory to order to meet customer needs and increase profits. An entrepreneur who owns and operates a temporary storage company can use vacancy reports to help determine what price to charge for new storage units. Everyone wants to be a better problem solver and decision maker. This chapter shows you how information systems can help. It begins with an overview of decision making and problem solving.

As shown in the opening vignette, information and decision support are the lifeblood of today's organizations. Thanks to information and decision support systems, managers and employees can obtain useful information in real time. As discussed in Chapter 5, TPS and ERP systems capture a wealth of data. When this data is filtered and manipulated, it can provide powerful support for managers and employees. The ultimate goal of management information and decision support systems is to help managers and executives at all levels make better decisions and solve important problems. The result can be increased revenues, reduced costs, and the realization of corporate goals. Many of today's information and decision support systems are built into the organization's TPS or ERP systems. In other cases, they are developed separately. No matter what type of information and decision support system you use, a system's primary goal is to help you and others become better decision makers and problem solvers.

DECISION MAKING AND PROBLEM SOLVING

Every organization needs effective decision making. The U.S. Coast Guard, for example, uses a formal decision process called the Ports and Waterways Safety Assessment (PAWSA) model to determine what resources it needs to secure the nation's coastlines and waterways.[1] As a result of its formal decision-making process, the Coast Guard demonstrated it needed four additional vessel-traffic centers.

In most cases, strategic planning and the overall goals of the organization set the course for decision making, helping employees and business units achieve their objectives and goals. Often, information systems also assist with problem solving, helping people make better decisions and save lives. For example, an information system at Hackensack University Medical Center (*www.humed.com*) in New Jersey analyzes possible drug interactions. In one case, an AIDS patient taking drugs for depression avoided therapeutic medication that could have dangerously interacted with the depression medication. The hospital has invested millions of dollars into its information system.

Decision Making as a Component of Problem Solving

In business, one of the highest compliments you can receive is to be recognized by your colleagues and peers as a "real problem solver." Problem solving is a critical activity for any business organization. After identifying a problem, you begin solving it with decision making. A well-known model developed by Herbert Simon divides the **decision-making phase** of the problem-solving process into three stages: intelligence, design, and choice. This model was later incorporated by George Huber into an expanded model of the entire problem-solving process (see Figure 6.1).

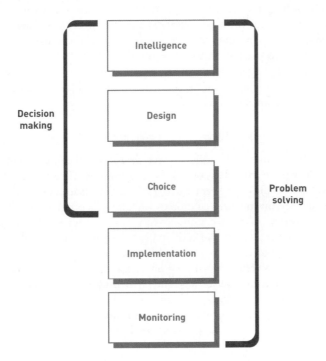

decision-making phase
The first part of problem solving, including three stages: intelligence, design, and choice.

Figure 6.1

How Decision Making Relates to Problem Solving

The three stages of decision making—intelligence, design, and choice—are augmented by implementation and monitoring to result in problem solving.

The first stage in the problem-solving process is the **intelligence stage**. During this stage, you identify and define potential problems or opportunities. You also investigate resource and environmental constraints. For example, if you were a Hawaiian farmer, during the intelligence stage you would explore the possibilities of shipping tropical fruit from your farm in Hawaii to stores in Michigan. The perishability of the fruit and the maximum price that consumers in Michigan are willing to pay for the fruit are problem constraints. In addition, you must consider federal and state regulations regarding the shipment of food products.

In the **design stage**, you develop alternative solutions to the problem and evaluate their feasibility. In the tropical fruit example, you would consider the alternative methods of shipment, including the transportation times and costs associated with each. During this stage, you might determine that shipment by freighter to California and then by truck to Michigan is not feasible because the fruit would spoil.

The last stage of the decision-making phase, the **choice stage**, requires selecting a course of action. In the tropical fruit example, you might select the method of shipping fruit by air from your Hawaiian farm to Michigan as the solution. The choice stage would then conclude with selection of an air carrier. As you will see later, various factors influence choice; the act of choosing is not as simple as it might first appear.

Problem solving includes and goes beyond decision making. It also includes the **implementation stage**, when the solution is put into effect. For example, if your decision is to ship tropical fruit to Michigan as air freight using a specific carrier, implementation involves informing your field staff of the new activity, getting the fruit to the airport, and actually shipping the product to Michigan. As another example, the Operations Research Group at British Airways used quantitative problem-solving techniques to help the airline achieve better departure statistics for its 750 flights per day to about 130 destinations.[2] On-time departure is a complex task, involving the coordination of cabin crews, airline cleaners,

intelligence stage
The first stage of decision making, in which potential problems or opportunities are identified and defined.

design stage
The second stage of decision making, in which alternative solutions to the problem are developed.

choice stage
The third stage of decision making, which requires selecting a course of action.

problem solving
A process that goes beyond decision making to include the implementation stage.

implementation stage
A stage of problem solving in which a solution is put into effect.

maintenance, catering, cargo, baggage, passengers, pilots, fuel, and air traffic clearance. British Airway's quantitative analysis helped them achieve a better departure record.

monitoring stage
The final stage of the problem-solving process, in which decision makers evaluate the implementation.

The final stage of the problem-solving process is the **monitoring stage**. In this stage, decision makers evaluate the implementation to determine whether the anticipated results were achieved and to modify the process in light of new information. Monitoring can involve feedback and adjustment. For example, after the first shipment of fruit from Hawaii to Michigan, you might learn that the flight of your chosen air freight firm routinely stops in Phoenix, Arizona, where the plane sits on the runway for a number of hours while loading additional cargo. If this unforeseen fluctuation in temperature and humidity adversely affects the fruit, you might have to readjust your solution to include a new carrier that does not make such a stop, or perhaps you would consider a change in fruit packaging.

Good decision makers monitor their decisions and make changes as necessary. After monitoring a decision to place its video programming on its Innertube, CBS decided to change course and place its sports, news, and entertainment content on a wide range of video Web sites to get wider coverage.[3] One person joked that the address for the Innertube site should have been CBS.com/nobodycomeshere.

Programmed Versus Nonprogrammed Decisions

programmed decision
A decision made using a rule, procedure, or quantitative method.

In the choice stage, various factors influence the decision maker's selection of a solution. One such factor is whether the decision can be programmed. **Programmed decisions** are made using a rule, procedure, or quantitative method. For example, to say that inventory should be ordered when inventory levels drop to 100 units is a programmed decision because it adheres to a rule. Programmed decisions are easy to computerize using traditional information systems. For example, you can easily program a computer to order more products when inventory levels for a certain item reach 100 units or less. Most of the processes automated through enterprise resource planning or transaction processing systems share this characteristic: The relationships between system elements are fixed by rules, procedures, or numerical relationships. Management information systems can also reach programmed decisions by providing reports on problems that are routine and in which the relationships are well defined. (In other words, they are structured problems.)

Ordering more products when inventory levels drop to specified levels is an example of a programmed decision.

(Source: © Andersen Ross/Getty Images.)

Nonprogrammed decisions deal with unusual or exceptional situations. In many cases, these decisions are difficult to quantify. Determining the appropriate training program for a new employee, deciding whether to develop a different product line, and weighing the benefits and drawbacks of installing a pollution control system are examples. Each of these decisions contains unique characteristics, and standard rules or procedures might not apply to them. Today, decision support systems help solve many nonprogrammed decisions, in which the problem is not routine and rules and relationships are not well defined (unstructured or ill-structured problems). These problems can include deciding the best location for a manufacturing plant or whether to rebuild a hospital that was severely damaged from a hurricane or tornado.

nonprogrammed decision
A decision that deals with unusual or exceptional situations.

Optimization, Satisficing, and Heuristic Approaches

In general, computerized decision support systems can either optimize or satisfice. An **optimization model** finds the best solution, usually the one that will best help the organization meet its goals. For example, an optimization model can find the appropriate number of products that an organization should produce to meet a profit goal, given certain conditions and assumptions. Optimization models use problem constraints. A limit on the number of available work hours in a manufacturing facility is an example of a problem constraint. Some spreadsheet programs, such as Excel, have optimizing features (see Figure 6.2). A business such as an appliance manufacturer can use an optimization program to reduce the time and cost of manufacturing appliances and increase profits by millions of dollars. The Scheduling Appointments at Trade Events (SATE) software package is an optimization program that schedules appointments between buyers and sellers at trade shows and meetings. Optimization software also allows decision makers to explore various alternatives.[4]

optimization model
A process to find the best solution, usually the one that will best help the organization meet its goals.

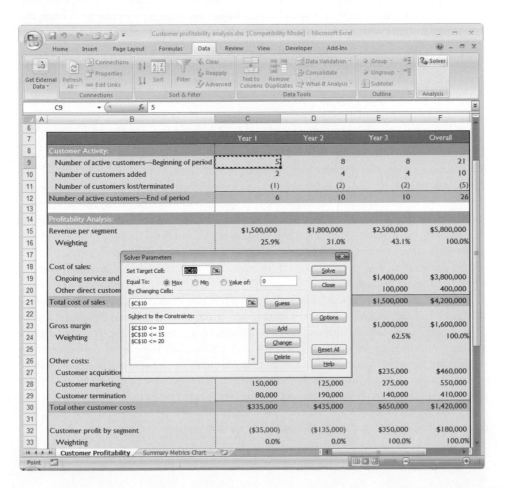

Figure 6.2

Optimization Software

Some spreadsheet programs, such as Microsoft Excel, have optimizing routines. This figure shows Solver, which can find an optimal solution given certain constraints.

Consider a few examples of how you can use optimization to achieve huge savings. Coca-Cola, for example, used optimization to schedule and route about 10,000 trucks used to deliver its soft drinks and products to save about $45 million annually.[5] Bombardier Flexjet (*www.flexjet.com*), a company that sells fractional ownership of jets, used an optimization program to better schedule its aircraft and crews, saving almost $30 million annually. Hutchinson Port Holdings (*www.hph.com.hk*), the world's largest container terminal, saved even more—over $50 million annually. The company processes 10,000 trucks and 15 ships every day, and used optimization to maximize the use of its fleet. Deere & Company, a manufacturer of commercial vehicles and equipment, increased shareholder value by over $100 million annually by using optimization to minimize inventory levels and enhance customer satisfaction.

satisficing model
A model that will find a good—but not necessarily the best—problem solution.

A **satisficing model** is one that finds a good—but not necessarily the best—problem solution. Satisficing is usually used because modeling the problem properly to get an optimal decision would be too difficult, complex, or costly. Satisficing normally does not look at all possible solutions but only at those likely to give good results. Consider a decision to select a location for a new manufacturing plant. To find the optimal (best) location, you must consider all cities in the United States or the world. A satisficing approach is to consider only five or ten cities that might satisfy the company's requirements. Limiting the options might not result in the best decision, but it will likely result in a good decision, without spending the time and effort to investigate all cities. Satisficing is a good alternative modeling method because it is sometimes too expensive to analyze every alternative to find the best solution.

heuristics
Commonly accepted guidelines or procedures that usually find a good solution.

Heuristics, often referred to as "rules of thumb"—commonly accepted guidelines or procedures that usually find a good solution—are often used in decision making. A heuristic that baseball team managers use is to place batters most likely to get on base at the top of the lineup, followed by the power hitters who can drive them in to score. An example of a heuristic used in business is to order four months' supply of inventory for a particular item when the inventory level drops to 20 units or less; although this heuristic might not minimize total inventory costs, it can serve as a good rule of thumb to avoid stockouts without maintaining excess inventory. Trend Micro (*www.trendmicro.com*), a provider of antivirus software, has developed an antispam product that is based on heuristics. The software examines e-mails to find those most likely to be spam. It doesn't examine all e-mails.

The Benefits of Information and Decision Support Systems

The information and decision support systems covered in this and the next chapter help individuals, groups, and organizations make better decisions, solve problems, and achieve their goals.[6] These systems include management information systems, decision support systems, group support systems, executive support systems, knowledge management systems, and a variety of special-purpose systems. As shown in Figure 6.3, the benefits are a measure of increased performance of these systems versus the cost to deliver them. The plus sign (+) by the arrow from *performance* to *benefits* indicates that increased performance has a positive impact on benefits. The minus sign (-) from *cost* to *benefits* indicates that increased cost has a negative impact on benefits.

The performance of these systems is typically a function of decision quality and problem complexity. Decision quality can result in increased effectiveness, increased efficiency, higher productivity, and many other measures first introduced in Chapter 1. Problem complexity depends on how hard the problem is to solve and implement. The cost of delivering these systems are the expenditures of the information technology components covered in Part II of this book, including hardware, software, databases, networks and the Internet, people, and procedures. But how do these systems actually deliver benefits to the individuals, groups, and organizations that use them? It depends on the type of information system. We begin our discussion with traditional management information systems.

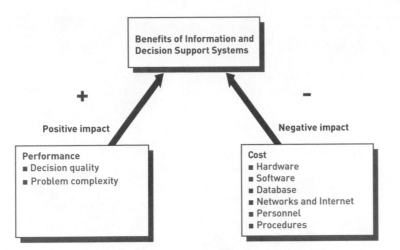

Figure 6.3

The Benefits of Information and Decision Support Systems

AN OVERVIEW OF MANAGEMENT INFORMATION SYSTEMS

A management information system (MIS) is an integrated collection of people, procedures, databases, and devices that provides managers and decision makers with information to help achieve organizational goals. MISs can often give companies and other organizations a competitive advantage by providing the right information to the right people in the right format and at the right time. For example, a shipping department could develop a spreadsheet to generate a report on possible delays that must be addressed to increase the number of on-time deliveries for the day. A music store might use a database system to develop a report that summarizes profits and losses for the month to make sure that the store is on track to make a 10 percent profit for the year.

Management Information Systems in Perspective

The primary purpose of an MIS is to help an organization achieve its goals by providing managers with insight into the regular operations of the organization so that they can control, organize, and plan more effectively. In short, an MIS provides managers with information, typically in reports, that supports effective decision making and provides feedback on daily operations. Figure 6.4 shows the role of MISs within the flow of an organization's information. Note that business transactions can enter the organization through traditional methods or via the Internet or an extranet connecting customers and suppliers to the firm's ERP or transaction processing systems. The use of MISs spans all levels of management. That is, they provide support to and are used by employees throughout the organization.

Inputs to a Management Information System

As shown in Figure 6.4, data that enters an MIS originates from both internal and external sources, including a company's supply chain, first discussed in Chapter 1. The most significant internal data sources for an MIS are the organization's various TPS and ERP systems and related databases. As discussed in Chapter 3, companies also use data warehouses and data marts to store valuable business information. Business intelligence, also discussed in Chapter 3, can be used to turn a database into useful information throughout the organization. Other internal data comes from specific functional areas throughout the firm.

External sources of data can include customers, suppliers, competitors, and stockholders, whose data is not already captured by the TPS, as well as other sources, such as the Internet. In addition, many companies have implemented extranets to link with selected suppliers and other business partners to exchange data and information.

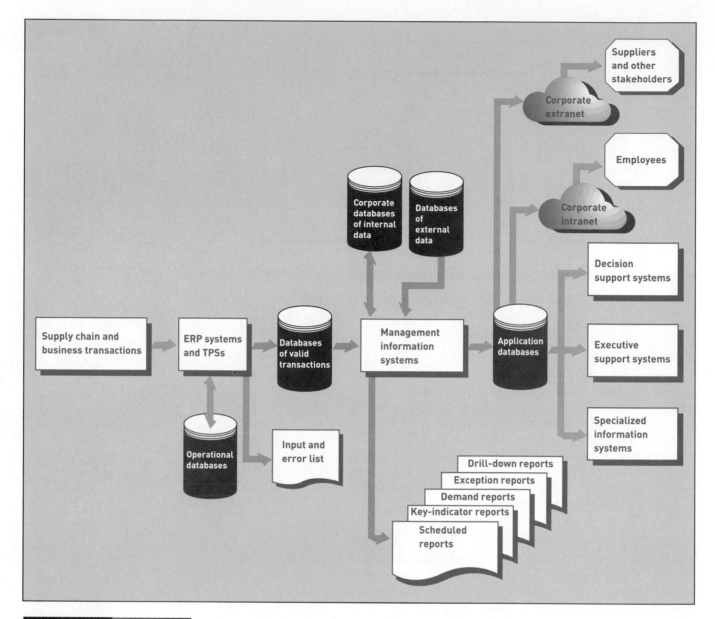

Figure 6.4

Sources of Managerial
Information

The MIS is just one of many sources
of managerial information. Decision
support systems, executive support
systems, and expert systems also
assist in decision making.

ETHICAL AND SOCIETAL ISSUES

Web 2.0 MIS Finds Compromise Between Service and Privacy

WhitePages.com maintains an MIS fed by a huge database that provides information on 180 million adults in the United States—80 percent of the U.S. adult population. Look up an old friend on WhitePages.com and you may find his new address, the name of his wife, his age, e-mail address, phone number, even a map to his house. Using associated services, you could find police records for your old friend and find out how much his house is worth. You can guess the obvious concern that many people have about WhitePages.com—privacy!

Information systems that post personal information on the Web without a person's consent run the risk of negative press and lawsuits. Consider the Beacon program provided by social networking giant Facebook. Beacon was designed to collect information about a Facebook user's activities on partner Web sites and use those members to endorse products. For example, you might see an announcement in your Facebook news feed informing you that your friend Shannon just rented the movie "Iron Man" from Blockbuster Online.

When Facebook users learned about how the Beacon system worked, many were enraged. Facebook quickly responded by giving users the ability to opt out of the Beacon system. The difference between Facebook and WhitePages.com is that Facebook is set up to serve registered users, and they have certain expectations of the Facebook service. WhitePages.com, however, was not designed to service members, but rather the entire Internet population. WhitePages.com does not obtain information about Internet users through an online profile that the user submits. It collects information from freely available public records on the Internet. This makes WhitePages.com less liable to privacy violations than Facebook. However, that level of liability might be changing.

WhitePages.com sees many similarities between today's popular social networks and its own directory service. It also sees the potential for substantial profits. In moving to a design that reflects a social network, WhitePages.com now includes a way for those listed in its directory to correct and add information about themselves. It is also setting up a service that allows members to fill out a profile and find others in the directory with similar interests. Users can send an anonymous message to others listed in the directory to find out if they are interested in striking up a friendship or renewing an old one. Now WhitePages.com finds itself in the same quandary as the big social networks: what information to share and what to keep private.

WhitePages.com decided to allow those listed in its directory to select the information they want to make public—if any. The company has decided to give its users control over their information. In doing so, WhitePages hopes to transform its directory service into an Internet-wide social network. While some users are likely to omit information that WhitePages previously made available, others might add to their information. By being conscious of privacy concerns, WhitePages hopes to improve its reputation and draw more visitors. The company has also released a software development platform that will allow developers to publish useful applications based on the WhitePages directory.

Businesses such as WhitePages.com need to control access to the private information in their information systems. Whether it's an information system available only to employees of the company, or one that's accessible on the public Web, the reputation of a business depends on the trust of its clientele. If a customer doesn't trust one company due to mismanagement of private information, they are likely to select a competitor that promises better security and privacy.

Discussion Questions

1. Why is WhitePages.com more concerned about customer privacy today than it has been in the past?
2. What are the differences between the services offered by WhitePages.com and Facebook.com?

Critical Thinking Questions

1. What benefits does WhitePages.com have compared to Facebook.com in regards to its customer base?
2. What types of applications might be developed using the WhitePages.com software development platform?

SOURCES: Vaughn-Nichols, Steven, "WhitePages.com grapples with privacy in a Web 2.0 world," *Computerworld*, May 16, 2008, *http://computerworld.com/action/article.do?command=viewArticleBasic&taxonomyName=security&articleId=9085718&taxonomyId=17&intsrc=kc_feat*; Cheng, Jacqui, "Facebook reevaluating Beacon after privacy outcry, possible FTC complaint," *Ars Technica*, November 29, 2007, *http://arstechnica.com/news.ars/post/20071129-facebook-reevaluating-beacon-after-privacy-outcry-possible-ftc-complaint.html*; WhitePages.com Web site, *www.whitepages.com*, accessed May 19, 2008.

Outputs of a Management Information System

The output of most management information systems is a collection of reports that are distributed to managers. These reports can be tailored for each user and delivered in a timely fashion. Providence Washington Insurance Company, for example, uses ReportNet from Cognos (*www.cognos.com*) to reduce the number of paper reports they produce and the associated costs.[7] The new reporting system creates an "executive dashboard" that shows current data, graphs, and tables to help managers make better real-time decisions.[8] Executives from Dunkin' Donuts use a dashboard to see the status of new stores.[9] The dashboard displays geographic areas and the new stores that are being developed. By clicking on a store, executives can see the details of how new stores are being constructed and if any stores are being delayed. The company hopes to grow to 15,000 franchises around the globe in the next several years. The city of Atlanta, Georgia also uses Cognos to measure the performance of its various departments and to keep track of its expenditures and budgets.[10] See Figure 6.5 for an example of an executive dashboard. In 2007, IBM announced that it would acquire Cognos.[11] Microsoft makes a reporting system called Business Scorecard Manager to give decision makers timely information about sales and customer information.[12] The software, which competes with Business Objects and Cognos, can integrate with other Microsoft software products, including Microsoft Office Excel. Hewlett-Packard's OpenView Dashboard is another MIS package that can quickly and efficiently render pictures, graphs, and tables that show how a business is functioning. In addition, some software packages and the Internet can be used to produce, gather, and distribute reports from different computer systems. Ace Hardware, for example, decided to use a more flexible report system called WebFocus from Information Builders (*www.informationbuilders.com*).[13] Referring to the old reporting system, one executive said, "People were getting tied in knots trying to develop reports in that tool. The tool was very rigid and had a lot of requirements as far as the way you did reporting." The new reporting system helped overcome some of these problems.

Figure 6.5

An Executive Dashboard

This MIS reporting system puts many kinds of real-time information at managers' fingertips to aid in decision making.

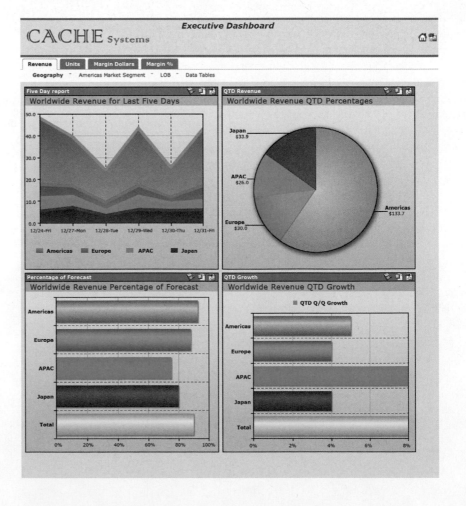

Management reports can come from various company databases, data warehouses, and other sources. These reports include scheduled reports, key-indicator reports, demand reports, exception reports, and drill-down reports (see Figure 6.6).

(a) Scheduled Report

Daily Sales Detail Report

Prepared: 08/10/08

Order #	Customer ID	Salesperson ID	Planned Ship Date	Quantity	Item #	Amount
P12453	C89321	CAR	08/12/08	144	P1234	$3,214
P12453	C89321	CAR	08/12/08	288	P3214	$5,660
P12454	C03214	GWA	08/13/08	12	P4902	$1,224
P12455	C52313	SAK	08/12/08	24	P4012	$2,448
P12456	C34123	JMW	08/13/08	144	P3214	$720
....	....	....	....	....	....	....

(b) Key-Indicator Report

Daily Sales Key-Indicator Report

	This Month	Last Month	Last Year
Total Orders Month to Date	$1,808	$1,694	$1,914
Forecasted Sales for the Month	$2,406	$2,224	$2,608

(c) Demand Report

Daily Sales by Salesperson Summary Report

Prepared: 08/10/08

Salesperson ID	Amount
CAR	$42,345
GWA	$38,950
SAK	$22,100
JWN	$12,350
....	....
....	....

(d) Exception Report

Daily Sales Exception—Orders Over $10,000

Prepared: 08/10/08

Order #	Customer ID	Salesperson ID	Planned Ship Date	Quantity	Item #	Amount
P12345	C89321	GWA	08/12/08	576	P1234	$12,856
P22153	C00453	CAR	08/12/08	288	P2314	$28,800
P23023	C32832	JMN	08/11/08	144	P2323	$14,400
....	....	....	....	....	....	....
....	....	....	....	....	....	....

(e) First-Level Drill-Down Report

Earnings by Quarter (Millions)

		Actual	Forecast	Variance
2nd Qtr.	2008	$12.6	$11.8	6.8%
1st Qtr.	2008	$10.8	$10.7	0.9%
4th Qtr.	2008	$14.3	$14.5	-1.4%
3rd Qtr.	2008	$12.8	$13.3	-3.8%

(f) Second-Level Drill-Down Report

Sales and Expenses (Millions)

Qtr: 2nd Qtr. 2008	Actual	Forecast	Variance
Gross Sales	$110.9	$108.3	2.4%
Expenses	$ 98.3	$ 96.5	1.9%
Profit	$ 12.6	$ 11.8	6.8%

(g) Third-Level Drill-Down Report

Sales by Division (Millions)

Qtr: 2nd Qtr. 2008	Actual	Forecast	Variance
Beauty Care	$ 34.5	$ 33.9	1.8%
Health Care	$ 30.0	$ 28.0	7.1%
Soap	$ 22.8	$ 23.0	-0.9%
Snacks	$ 12.1	$ 12.5	-3.2%
Electronics	$ 11.5	$ 10.9	5.5%
Total	$110.9	$108.3	2.4%

(h) Fourth-Level Drill-Down Report

Sales by Product Category (Millions)

Qtr: 2nd Qtr. 2008 Division: Health Care	Actual	Forecast	Variance
Toothpaste	$12.4	$10.5	18.1%
Mouthwash	$ 8.6	$ 8.8	-2.3%
Over-the-Counter Drugs	$ 5.8	$ 5.3	9.4%
Skin Care Products	$ 3.2	$ 3.4	-5.9%
Total	$30.0	$28.0	7.1%

Scheduled Reports

Scheduled reports are produced periodically, or on a schedule, such as daily, weekly, or monthly. For example, a production manager could use a weekly summary report that lists total payroll costs to monitor and control labor and job costs. A manufacturing report generated once per day to monitor the production of a new item is another example of a scheduled report. Other scheduled reports can help managers control customer credit, performance of sales representatives, inventory levels, and more.

A **key-indicator report** summarizes the previous day's critical activities and is typically available at the beginning of each workday. These reports can summarize inventory levels, production activity, sales volume, and the like. Key-indicator reports are used by managers and executives to take quick, corrective action on significant aspects of the business.

Demand Reports

Demand reports are developed to provide certain information upon request. In other words, these reports are produced on demand. Like other reports discussed in this section, they often come from an organization's database system. For example, an executive might want to know the production status of a particular item—a demand report can be generated to provide the requested information by querying the company's database. Suppliers and customers can also use demand reports. FedEx, for example, provides demand reports on its Web site to allow customers to track packages from their source to their final destination. Other examples of demand reports include reports requested by executives to show the hours worked by a

Figure 6.6

Reports Generated by an MIS

The types of reports are (a) scheduled, (b) key indicator, (c) demand, (d) exception, and (e–h) drill down.

(Source: George W. Reynolds, *Information Systems for Managers*, Third Edition. St. Paul, MN: West Publishing Co., 1995.)

scheduled report
A report produced periodically, or on a schedule, such as daily, weekly, or monthly.

key-indicator report
A summary of the previous day's critical activities; typically available at the beginning of each workday.

demand report
A report developed to give certain information at someone's request.

exception report
A report automatically produced when a situation is unusual or requires management action.

drill-down report
A report providing increasingly detailed data about a situation.

particular employee, total sales to date for a product, and so on. Many companies are putting some medical records on the Internet to make them available on demand.[14] Software and Internet companies like Microsoft are developing systems that allow people to search the Internet to get medical information including lab tests, drug records, and X-rays from different sources. Doctors use these systems to make a diagnosis and prescribe treatment from remote locations.

Exception Reports

Exception reports are reports that are automatically produced when a situation is unusual or requires management action. For example, a manager might set a parameter that generates a report of all inventory items with fewer than the equivalent of five days of sales on hand. This unusual situation requires prompt action to avoid running out of stock on the item. The exception report generated by this parameter would contain only items with fewer than five days of sales in inventory. Exception reports are used by businesses and nonprofit organizations. British Petroleum (BP), for example, uses a variety of data sources to get exception reports about potential damage from hurricanes to their facilities in the Gulf of Mexico.[15] The company collects data from geospatial map Internet sites and its own company resources. It then mashes up, or integrates, the different data sources into exception reports that show the location of its oil facilities that may have been damaged from a hurricane or severe storm. According to one company executive, "If there is any kind of damage after a storm, we want to know about it very quickly." Some companies, universities, and law enforcement agencies use cell phones and text messages to send exception reports to employees, students, or neighborhood residents. If a campus threat is detected or a crime committed, a text message can be sent to students and staff with the details. The University of Texas, for example, sent a text message to faculty, staff, and students about a potential ice storm, warning them that the university would be closed the next day and to stay home.[16] The next day, most of the university was deserted and no storm-related injuries were reported on campus. After police in the Netherlands sent a text message to neighborhood residents about a stolen boat, a resident called to say that she found a boat that met the description of the text message.[17] The police found the boat and arrested the criminal.

Drill-Down Reports

Drill-down reports provide increasingly detailed data about a situation. Using these drill-down reports, analysts can see data at a high level first (such as sales for the entire company), then at a more detailed level (such as the sales for one department of the company), and then a very detailed level (such as sales for one sales representative). Boehringer Ingelheim (*www.boehringer-ingelheim.com/corporate/home/home.asp*), a large German drug company with over $7 billion in revenues and thousands of employees in 60 countries, uses a variety of drill-down reports so it can respond rapidly to changing market conditions. Managers can drill down into more levels of detail to individual transactions if they want. Companies and organizations of all sizes and types use drill-down reports.[18] The military, for example, is using software from Business Objects to determine if a defective battery could explode and damage or destroy military vehicles. According to an Army spokesman, "We can drill down, go into every contract we ordered that battery on … and get them away from units so no one will get hurt."

Characteristics of a Management Information System

Scheduled, key-indicator, demand, exception, and drill-down reports have all helped managers and executives make better, more timely decisions. In general, MISs perform the following functions:

- Provide reports with fixed and standard formats.
- Produce hard-copy and soft-copy reports.
- Use internal data stored in the computer system.
- Allow users to develop their own custom reports.
- Require user requests for reports developed by systems personnel.

FUNCTIONAL ASPECTS OF THE MIS

Most organizations are structured along functional lines or areas. This functional structure is usually apparent from an organization chart, which typically shows a hierarchy in roles or positions. Some traditional functional areas include finance, manufacturing, marketing, human resources, and other specialized information systems. The MIS can also be divided along those functional lines to produce reports tailored to individual functions (see Figure 6.7).

Figure 6.7

An Organization's MIS

The MIS is an integrated collection of functional information systems, each supporting particular functional areas.

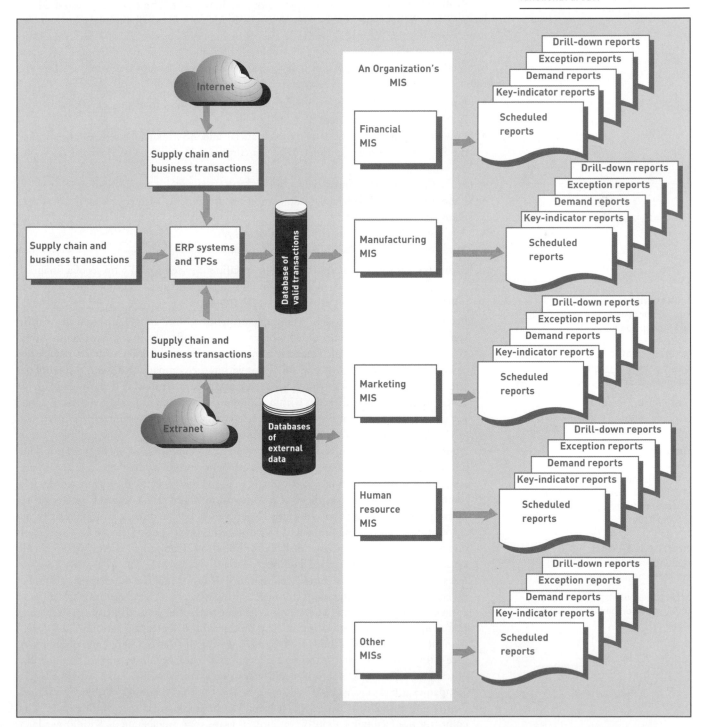

Financial Management Information Systems

A **financial MIS** provides financial information not only for executives but also for a broader set of people who need to make better decisions on a daily basis. Reuters, for example, has developed an automated reporting system that scans articles about companies for its stock traders to determine if the news is favorable or unfavorable.[19] The reports can result in buy orders if the news is positive or sell orders if the news is negative. Eventually, the system will be tied into machine trading that doesn't require trade orders generated by people. Web sites can also provide financial information. For example, Web sites like *www.kiva.com* can provide information for people seeking small loans, called microloans.[20] Many microloans are for $100 or less, and are made for a period of less than a year. The Internet is also used for larger loans. When Jeff Walsh wanted to get funds for his small business, he went to *www.prosper.com*.[21] According to Mr. Walsh, "I just bought a house in 2007 and was a little nervous about what the bank would say about my debt-to-income ratio." Financial MISs are often used to streamline reports of transactions. Most financial MISs perform the following functions:

- Integrate financial and operational information from multiple sources, including the Internet, into a single system.
- Provide easy access to data for both financial and nonfinancial users, often through the use of a corporate intranet to access corporate Web pages of financial data and information.
- Make financial data immediately available to shorten analysis turnaround time.
- Enable analysis of financial data along multiple dimensions—time, geography, product, plant, and customer.
- Analyze historical and current financial activity.
- Monitor and control the use of funds over time.

Figure 6.8 shows typical inputs, function-specific subsystems, and outputs of a financial MIS, including profit and loss, auditing, and uses and management of funds. The following are some of the financial MIS subsystems and outputs.

- **Profit/loss and cost systems.** Many departments within an organization are **profit centers**, which means that they focus on generating profits. An investment division of a large insurance or credit card company is an example of a profit center. Other departments can be **revenue centers**, which are divisions within the company that focus primarily on sales or revenues, such as a marketing or sales department. Still other departments can be **cost centers**, which are divisions within a company that do not directly generate revenue, such as manufacturing or research and development. In most cases, information systems are used to compute revenues, costs, and profits.

- **Auditing. Auditing** involves analyzing the financial condition of an organization and determining whether financial statements and reports produced by the financial MIS are accurate. **Internal auditing** is performed by individuals within the organization. For example, the finance department of a corporation might use a team of employees to perform an audit. **External auditing** is performed by an outside group, usually an accounting or consulting firm such as PricewaterhouseCoopers, Deloitte & Touche, or one of the other major, international accounting firms. Computer systems are used in all aspects of internal and external auditing.

- **Uses and management of funds.** Internal uses of funds include purchasing additional inventory, updating plants and equipment, hiring new employees, acquiring other companies, buying new computer systems, increasing marketing and advertising, purchasing raw materials or land, investing in new products, and increasing research and development. External uses of funds are typically investment related. Companies often invest excess funds in such external revenue generators as bank accounts, stocks, bonds, bills, notes, futures, options, and foreign currency using financial MISs. Some individuals and companies are exploring making loans over the Internet. Lending Club, for example, facilitates loans made between people using Facebook, the social-networking site.[22] The company has facilitated about $1 million in loans that average about $5,000 per loan with interest rates that vary from about 7 to 17 percent. The loan default rate has been less than 1 percent.

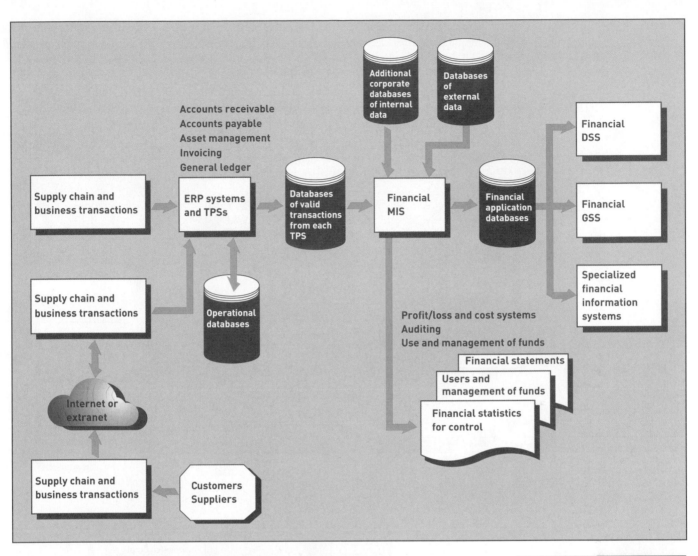

Overview of a Financial MIS

Financial institutions use information systems to shorten turnaround time for loan approvals.

(Source: © Royalty-Free/Corbis.)

Manufacturing Management Information Systems

More than any other functional area, advances in information systems have revolutionized manufacturing.[23] As a result, many manufacturing operations have been dramatically improved over the last decade. Also, with the emphasis on greater quality and productivity, having an effective manufacturing process is becoming even more critical. The use of computerized systems is emphasized at all levels of manufacturing—from the shop floor to the executive suite. Increasingly, companies are outsourcing the manufacturing process. With almost 300,000 employees, the Hon Hai company in China is one of the world's largest manufacturers of electronic products, including music players, cell phones, game consoles, and many other electronic products.[24] Some believe the company is China's largest exporter. Dell Computer has used both optimization and heuristic software to help it manufacture a larger variety of products.[25] Dell was able to double its product variety, while saving about $1 million annually in manufacturing costs. Figure 6.9 gives an overview of some of the manufacturing MIS inputs, subsystems, and outputs.

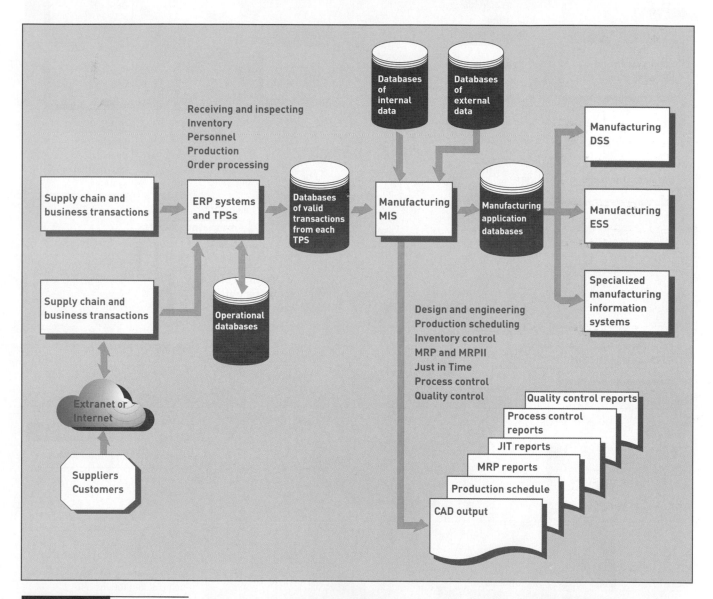

Figure 6.9

Overview of a Manufacturing MIS

The subsystems and outputs in a manufacturing MIS monitor and control the flow of materials, products, and services through the organization. As raw materials are converted to finished goods, the manufacturing MIS monitors the process at almost every stage. New technology could make this process easier. Using specialized computer chips and tiny radio transmitters, companies can monitor materials and products through the entire manufacturing process. Procter & Gamble, Wal-Mart, and Target have funded research into this manufacturing MIS. Car manufacturers, which convert raw steel, plastic, and other materials into a finished automobile, also monitor their manufacturing processes. Auto manufacturers add thousands of dollars of value to the raw materials they use in assembling a car. If the manufacturing MIS also lets them provide additional service, such as customized paint colors, it has added further value for customers. In doing so, the MIS helps provide the company the edge that can differentiate it from competitors. The success of an organization can depend on the manufacturing function. Some common information subsystems and outputs used in manufacturing are discussed next.

- **Design and engineering.** Manufacturing companies often use computer-aided design (CAD) with new or existing products. For example, Boeing (*www.boeing.com*) uses a CAD system to develop a complete digital blueprint of an aircraft before it begins the manufacturing process. As mock-ups are built and tested, the digital blueprint is constantly revised to reflect the most current design. Using such technology helps Boeing reduce manufacturing costs and the time to design a new aircraft.

Computer-aided design (CAD) is used in the development and design of complex products or structures.

(Source: © Kim Steele/Getty Images.)

- **Master production scheduling and inventory control.** Scheduling production and controlling inventory are critical for any manufacturing company.[26] The overall objective of master production scheduling is to provide detailed plans for both short-term and long-range scheduling of manufacturing facilities. Some companies hire outside companies to help them with inventory control. Delta Airlines, for example, has a long-term, $1 billion agreement with Chromalloy Gas Turbine to help in providing inventory parts and maintenance of its jet engines.[27] Apparel company Tween Brands, Inc. uses a number of software packages to help it control inventory and reduce costs.[28]

    Most techniques are used to minimize inventory costs. They determine when and how much inventory to order. One method of determining the amount of inventory to order is called the **economic order quantity** (**EOQ**). This quantity is calculated to minimize the total inventory costs. The "When to order?" question is based on inventory usage over time. Typically, the question is answered in terms of a **reorder point** (**ROP**), which is a critical inventory quantity level. When the inventory level for a particular item falls to the reorder point, or critical level, the system generates a report so that an order is immediately placed for the EOQ of the product. Another inventory technique used when demand for one item depends on the demand for another is called **material requirements planning** (**MRP**). The basic goal of MRP is to determine when finished products, such as automobiles or airplanes, are needed and then to work backward to determine deadlines and resources needed, such as engines and tires, to complete the final product on schedule. **Just-in-time** (**JIT**) **inventory** and manufacturing is an approach that maintains inventory at the lowest levels without sacrificing the availability of finished products. With this approach, inventory and materials are delivered just before they are used in a product. A JIT inventory system would arrange for a car windshield to be delivered to the assembly line only a few moments before it is secured to the automobile, rather than storing it in the manufacturing facility while the car's other components are being assembled. JIT, however, can result in some organizations running out of inventory when demand exceeds expectations. Even so, companies like Toyota continue to embrace JIT.[29] According to the president of the company, "We've been implementing this strategy for decades, and

economic order quantity (EOQ)
The quantity that should be reordered to minimize total inventory costs.

reorder point (ROP)
A critical inventory quantity that determines when to order more inventory.

material requirements planning (MRP)
A set of inventory-control techniques that help coordinate thousands of inventory items when the demand for one item is dependent on the demand for another.

just-in-time (JIT) inventory
A philosophy of inventory management in which inventory and materials are delivered just before they are used in manufacturing a product.

computer-assisted manufacturing (CAM)

A system that directly controls manufacturing equipment.

computer-integrated manufacturing (CIM)

Using computers to link the components of the production process into an effective system.

flexible manufacturing system (FMS)

An approach that allows manufacturing facilities to rapidly and efficiently change from making one product to making another.

Computer-assisted manufacturing systems control complex processes on the assembly line and provide users with instant access to information.

(Source: © PHOTOTAKE Inc. / Alamy.)

we will keep on with it." The car manufacturer had to momentarily close more than 10 plants when an earthquake prevented a supplier from producing $1.50 piston ring parts.

- **Process control.** Managers can use a number of technologies to control and streamline the manufacturing process. For example, computers can directly control manufacturing equipment, using systems called **computer-assisted manufacturing (CAM)**. CAM systems can control drilling machines, assembly lines, and more. **Computer-integrated manufacturing (CIM)** uses computers to link the components of the production process into an effective system. CIM's goal is to tie together all aspects of production, including order processing, product design, manufacturing, inspection and quality control, and shipping. A **flexible manufacturing system** (FMS) is an approach that allows manufacturing facilities to rapidly and efficiently change from making one product to another. In the middle of a production run, for example, the production process can be changed to make a different product or to switch manufacturing materials. By using an FMS, the time and cost to change manufacturing jobs can be substantially reduced, and companies can react quickly to market needs and competition.

- **Quality control and testing.** With increased pressure from consumers and a general concern for productivity and high quality, today's manufacturing organizations are placing more emphasis on **quality control**, a process that ensures that the finished product meets the customers' needs. Information systems are used to monitor quality and take corrective steps to eliminate possible quality problems.

quality control

A process that ensures that the finished product meets the customers' needs.

INFORMATION SYSTEMS @ WORK

Pharmaceutical Company Reduces Time-to-Market

AstraZeneca is one of the world's leading pharmaceutical companies, and truly a global corporation. With a presence in over 100 countries, AstraZeneca is based in London, England, with research and development sites in Sweden, the United States, and the United Kingdom. The company has more than 67,000 employees, which mostly work in Europe. AstraZeneca totaled $29.6 billion in sales in 2007.

The pharmaceutical industry is highly competitive, with many companies racing to be the first to market with drug remedies for common ailments and diseases. It typically takes 8–12 years to bring a new drug to market. Furthermore, a drug patent lasts 20–25 years, much of which is taken up in development time. The less time a company spends developing a drug, the more years it can reap in profits before generic alternatives are made available. Every day saved in development can mean millions of dollars in profits. The key to shortening the development time of drugs lies in efficient project management.

AstraZeneca developed a project management system to allow its research facilities around the world to share research and development (R&D) information. The system is called Matrix, and it gathers and analyzes research and development information stored in a large corporate data warehouse. Approximately 5,000 researchers working at the six AstraZeneca R&D sites have access to the Matrix system. The ability to access research information is transforming how research is conducted. Researchers can collaborate on projects from different sites, tracking each other's progress. Matrix has also eliminated wasteful duplication of effort caused by miscommunication.

AstraZeneca researchers create a high volume of project data every day. The Matrix system allows researchers and managers to track, understand, and manage that data. Top-level executives use an executive dashboard manager to view key performance indicators and keep their finger on the pulse of the business to make quick, confident decisions. The system helps the finance department get a clear picture of project costs. Managers can easily manage project scheduling, budgeting, and resource allocation. Matrix makes AstraZeneca more nimble, which is a significant achievement considering the size of the company. AstraZeneca can launch products quickly, often faster than the competition, which lets them establish leadership in a variety of specialty areas.

AstraZeneca business manager David Scanion believes that the new MIS has improved the company's project management, cost control, and resource usage improving the company's ability to compete with world class research done in record-breaking timescales.

Discussion Questions

1. What time factors affect the product life of new drugs?
2. What benefits does the Matrix system provide to researchers at AstraZeneca?

Critical Thinking Questions

1. How does the Matrix system improve project management at AstraZeneca?
2. What financial savings and benefits are provided to the company from the Matrix system?

SOURCES: Business Objects Staff, "AstraZeneca", Business Objects Customers in the Spotlight, 2008, *www.businessobjects.com/company/customers/spotlight/astrazeneca.asp*; AstraZeneca Web Site, accessed May 20, 2008, www.astrazeneca.com.

Marketing Management Information Systems

marketing MIS
An information system that supports managerial activities in product development, distribution, pricing decisions, and promotional effectiveness.

A **marketing MIS** supports managerial activities in product development, distribution, pricing decisions, promotional effectiveness, and sales forecasting. Marketing functions are increasingly being performed on the Internet.[30] When an English teacher in Japan put a number of fun family projects for children and their families on *www.Digg.com*, a popular Web site, it had an unexpected marketing consequence.[31] In only a few days, it dramatically increased sales of rivets used to complete the projects. According to Andy McGrew, owner of a company that sells blueprints for many home projects, "It would have taken me a year to sell that many rivets." Many companies are developing Internet marketplaces to advertise and sell products. The amount spent on online advertising is worth billions of dollars annually. Newer marketing companies, such as AdMob, Inc. (*www.admob.com*), are placing ads on cell phones and mobile devices with Internet access.[32] According to an executive of Ad-Mob, "Everybody's just trying to dip their toes in the water and figure out what's going to work." Software can measure how many customers see the advertising. Some companies use a software product called SmartLoyalty to analyze customer loyalty.

Some marketing departments are actively using the Internet to advertise their products and services and keep customers happy. Companies, for example, are starting to advertise their products and services on Facebook (*www.facebook.com*), a social-networking site.[33] YouTube, the video-sharing Internet site, sells video ads on its site to companies, including Ford, BMW, Time Warner, and others.[34] After about 10 seconds, the video ad disappears unless the user clicks it.[35] Corporate marketing departments also use social-networking sites, such as Second Life (*www.secondlife.com*), to advertise their products and perform marketing research.[36]

Customer relationship management (CRM) programs, available from some ERP vendors, help a company manage all aspects of customer encounters. CRM software can help a company collect customer data, contact customers, educate customers on new products, and sell products to customers through a Web site. An airline, for example, can use a CRM system to notify customers about flight changes. New Zealand's Jade Stadium uses CRM software from GlobalTech Solutions to give a single entry point to its marketing efforts and customer databases, instead of using about 20 spreadsheets. The CRM software will help Jade Stadium develop effective marketing campaigns, record and track client contacts, and maintain an accurate database of clients. Yet, not all CRM systems and marketing sites on the Internet are successful. Customization and ongoing maintenance of a CRM system can be expensive. Figure 6.10 shows the inputs, subsystems, and outputs of a typical marketing MIS.

Subsystems for the marketing MIS include marketing research, product development, promotion and advertising, and product pricing. These subsystems and their outputs help marketing managers and executives increase sales, reduce marketing expenses, and develop plans for future products and services to meet the changing needs of customers.

- **Marketing research.** The purpose of marketing research is to conduct a formal study of the market and customer preferences.[37] Computer systems are used to help conduct and analyze the results of surveys, questionnaires, pilot studies, and interviews. eCourier, for example, uses Crystal Reports from Business Objects to determine customer habits and preferences.[38] The company can perform marketing research using its Web site to help determine which customers are happy and still buying and which ones might switch to another company. According to an executive at eCourier, "We know of 10 cases where there were problems, for whatever reason, and we got in there early. As a result, we were able to save the client." In addition to knowing what you buy, market research can determine where you buy.[39] This can help in developing new products and services and tailoring ads and promotions. With the use of GPS positioning systems, marketing firms can promote products to you over cell phones and other mobile devices by knowing your location.

- **Product development.** Product development involves the conversion of raw materials into finished goods and services and focuses primarily on the physical attributes of the product. Many factors, including plant capacity, labor skills, engineering factors, and

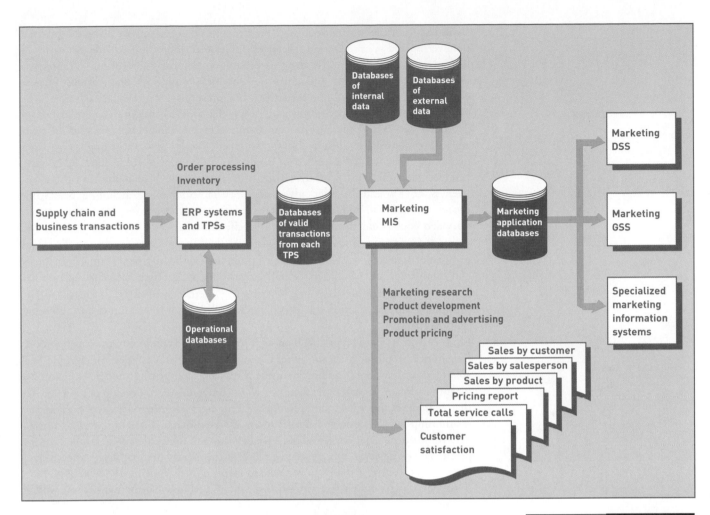

Figure 6.10

Overview of a Marketing MIS

Marketing research data yields valuable information for the development and marketing of new products.

(Source: © Michael Newman/Photo Edit.)

materials are important in product development decisions. In many cases, a computer program analyzes these various factors and selects the appropriate mix of labor, materials, plant and equipment, and engineering designs. Make-or-buy decisions can also be made with the assistance of computer programs. To get additional revenues, some TV programs and movies promote products and services during their programs.[40] Movies, for example, can show actors driving luxury cars and wearing expensive watches. Food companies can sponsor food and cooking TV programs that give their products more exposure. The approach is called "branded entertainment."

• **Promotion and advertising.** One of the most important functions of any marketing effort is promotion and advertising. Product success is a direct function of the types of advertising and sales promotion done. Increasingly, organizations are using the Internet to advertise and sell products and services. Johnson & Johnson used Internet cartoons instead of extensive TV advertising to promote a popular baby lotion.[41] Yahoo has launched Brand Universe and advertises specific products around various interest groups.[42] According to a Yahoo executive, "We can speak to a particular audience." The goal is to target ads to a specific group of people that are likely to purchase the advertised goods and services. Companies are also trying to measure the effectiveness of different advertising approaches, such as TV and Internet advertising.[43] According to a Toyota marketing executive, "We wanted to be able to have a tool to really start judging the Internet compared to TV." Toyota uses IAG Research to help it measure the effectiveness of TV and Internet advertising. Several companies, including ScanScout (*www.scanscout.com*) and YuMeNetworks (*www.yumenetworks.com*), are trying match up video content on the Internet with specific ads that cater to those watching the videos.[44] Companies, such as furniture maker Ikea, are increasingly hiring digital ad companies to make sure their products get promoted on the Internet.[45] Some individuals and companies are willing to tolerate advertising to get free software or Internet service.[46] Companies are also using blogs on the Internet to advertise products.[47] eBay, the popular Internet auction site, is working with Bid4Spots to auction radio advertising spots to clients.[48]

- **Product pricing.** Product pricing is another important and complex marketing function. Retail price, wholesale price, and price discounts must be set. Chrysler, for example, was able to save about $500 million by using a sophisticated pricing model that analyzed incentives, financing, and other factors.[49] Most companies try to develop pricing policies that will maximize total sales revenues. Computers are often used to analyze the relationship between prices and total revenues. Traditionally, executives used costs to determine prices. They simply added a profit margin to total costs to guarantee a decent profit. Today, however, more executives look at the market place to determine product pricing. In one case, a company was able to increase its revenue by about $200 million in part by using a more aggressive pricing policy based on what the market was willing to pay.[50]

- **Sales analysis.** Computerized sales analysis is important to identify products, sales personnel, and customers that contribute to profits and those that do not. Several reports can be generated to help marketing managers make good sales decisions (see Figure 6.11). The sales-by-product report lists all major products and their sales for a specified period of time. This report shows which products are doing well and which need improvement or should be discarded altogether. The sales-by-salesperson report lists total sales for each salesperson for each week or month. This report can also be subdivided by product to show which products are being sold by each salesperson. The sales-by-customer report is a tool that can be used to identify high- and low-volume customers.

Human Resource Management Information Systems

human resource MIS

An information system that is concerned with activities related to employees and potential employees of an organization, also called a personnel MIS.

A **human resource MIS (HRMIS)**, also called the *personnel MIS*, is concerned with activities related to previous, current, and potential employees of the organization. Because the personnel function relates to all other functional areas in the business, the human resource (HR) MIS plays a valuable role in ensuring organizational success. Some of the activities performed by this important MIS include workforce analysis and planning, hiring, training, job and task assignment, and many other personnel-related issues. An effective human resource MIS allows a company to keep personnel costs at a minimum, while serving the required business processes needed to achieve corporate goals. Although human resource information systems focus on cost reduction, many of today's HR systems concentrate on hiring and managing existing employees to get the total potential of the human talent in the organization. According to the High Performance Workforce Study conducted by Accenture, the most important HR initiatives include improving worker productivity, improving adaptability to new opportunities, and facilitating organizational change. Figure 6.12 shows some of the inputs, subsystems, and outputs of the human resource MIS.

(a) Sales by Product

Product	August	September	October	November	December	Total
Product 1	34	32	32	21	33	152
Product 2	156	162	177	163	122	780
Product 3	202	145	122	98	66	633
Product 4	345	365	352	341	288	1,691

(b) Sales by Salesperson

Salesperson	August	September	October	November	December	Total
Jones	24	42	42	11	43	162
Kline	166	155	156	122	133	732
Lane	166	155	104	99	106	630
Miller	245	225	305	291	301	1,367

(c) Sales by Customer

Customer	August	September	October	November	December	Total
Ang	234	334	432	411	301	1,712
Braswell	56	62	77	61	21	277
Celec	1,202	1,445	1,322	998	667	5,634
Jung	45	65	55	34	88	287

Reports Generated to Help Marketing Managers Make Good Decisions

(a) This sales-by-product report lists all major products and their sales for the period from August to December. (b) This sales-by-salesperson report lists total sales for each salesperson for the same time period. (c) This sales-by-customer report lists sales for each customer for the period. Like all MIS reports, totals are provided automatically by the system to show managers at a glance the information they need to make good decisions.

Human resource subsystems and outputs range from the determination of human resource needs and hiring through retirement and outplacement. Most medium and large organizations have computer systems to assist with human resource planning, hiring, training and skills inventorying, and wage and salary administration. Outputs of the human resource MIS include reports, such as human resource planning reports, job application review profiles, skills inventory reports, and salary surveys discussed next.

- **Human resource planning.** One of the first aspects of any human resource MIS is determining personnel and human needs. The overall purpose of this MIS subsystem is to put the right number and types of employees in the right jobs when they are needed, including internal employees that work exclusively for the organization and outside workers that are hired when they are needed. Some experts believe that workers should be managed like a supply chain, using supply chain management (SCM) and just-in-time techniques, first discussed in Chapter 1.[51] Effective human resource planning often requires computer programs, such as SPSS and SAS, to forecast the future number of employees needed and anticipating the future supply of people for these jobs. IBM used an HR pilot program, called Professional Marketplace, to plan for workforce needs, including the supplies and tools the workforce needs to work efficiently. Professional Marketplace helps IBM to catalog employees into a glossary of skills and abilities. Like many other companies, HR and workforce costs are IBM's biggest expense.
- **Personnel selection and recruiting.** If the human resource plan reveals that additional personnel are required, the next logical step is recruiting and selecting personnel. Companies seeking new employees often use computers to schedule recruiting efforts and trips and to test potential employees' skills. Many companies now use the Internet to screen for job applicants. Applicants use a template to load their résumé onto the

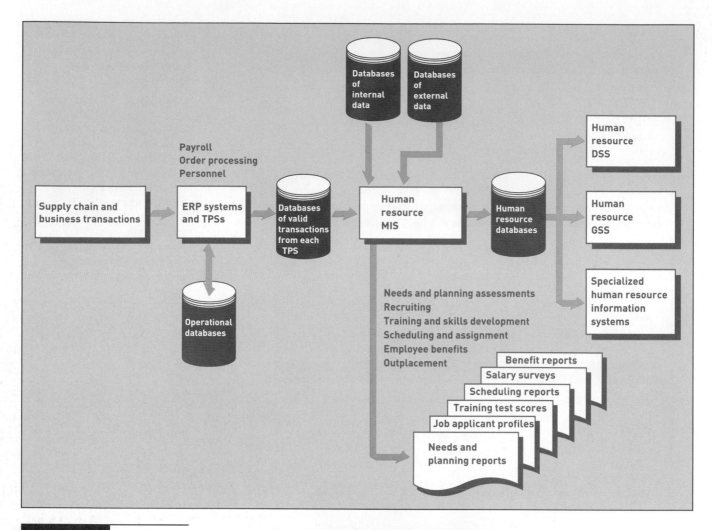

Figure 6.12

Overview of a Human Resource MIS

Human resource MIS subsystems help to determine personnel needs and match employees to jobs.

(Source: © Bambu Productions/ Getty Images.)

Internet site. HR managers can then access these résumés and identify applicants they are interested in interviewing.

- **Training and skills inventory.** Some jobs, such as programming, equipment repair, and tax preparation, require very specific training for new employees.[52] Other jobs may require general training about the organizational culture, orientation, dress standards, and expectations of the organization. When training is complete, employees often take computer-scored tests to evaluate their mastery of skills and new material.

- **Scheduling and job placement.** Employee schedules are developed for each employee, showing his job assignments over the next week or month. Job placements are often determined based on skills inventory reports, showing which employee might be best suited to a particular job. Sophisticated scheduling programs are often used in the airline industry, the military, and many other areas to get the right people assigned to the right jobs at the right time.

- **Wage and salary administration.** Another human resource MIS subsystem involves determining wages, salaries, and benefits, including medical payments, savings plans, and retirement accounts. Wage data, such as industry averages for positions, can be taken from the corporate database and manipulated by the human resource MIS to provide wage information and reports to higher levels of management.

- **Outplacement.** Employees leave a company for a variety of reasons. Outplacement services are offered by many companies to help employees make the transition. *Outplacement* can include job counseling and training, job and executive search, retirement and financial planning, and a variety of severance packages and options. Many employees use the Internet to plan their future retirement or to find new jobs, using job sites such as *www.monster.com*.

Other Management Information Systems

In addition to finance, manufacturing, marketing, and human resource MISs, some companies have other functional management information systems. For example, most successful companies have well-developed accounting functions and a supporting accounting MIS. Also, many companies use geographic information systems for presenting data in a useful form.

Accounting MISs

In some cases, accounting works closely with financial management. An **accounting MIS** performs a number of important activities, providing aggregate information on accounts payable, accounts receivable, payroll, and many other applications. The organization's enterprise resource planning and transaction processing system captures accounting data, which is also used by most other functional information systems.

Some smaller companies hire outside accounting firms to assist them with their accounting functions. These outside companies produce reports for the firm using raw accounting data. In addition, many excellent integrated accounting programs are available for personal computers in small companies. Depending on the needs of the small organization and its staff's computer experience, using these computerized accounting systems can be a very cost-effective approach to managing information.

accounting MIS
An information system that provides aggregate information on accounts payable, accounts receivable, payroll, and many other applications.

Geographic Information Systems

Increasingly, managers want to see data presented in graphical form. A **geographic information system (GIS)** is a computer system capable of assembling, storing, manipulating, and displaying geographically referenced information, that is, data identified according to its location. A GIS enables users to pair maps or map outlines with tabular data to describe aspects of a particular geographic region. For example, sales managers might want to plot total sales for each county in the states they serve. Using a GIS, they can specify that each county be shaded to indicate the relative amount of sales—no shading or light shading represents no or little sales, and deeper shading represents more sales. Staples Inc., the large office supply store chain, used a geographic information system to select new store locations. Finding the best location is critical. It can cost up to $1 million for a failed store because of a poor location. Staples uses a GIS tool from Tactician Corporation (*www.tactician.com*) along with software from SAS. Although many software products have seen declining revenues, the use of GIS software is increasing.

We saw earlier in this chapter that management information systems (MISs) provide useful summary reports to help solve structured and semistructured business problems. Decision support systems (DSSs) offer the potential to assist in solving both semistructured and unstructured problems. These systems are discussed next.

geographic information system (GIS)
A computer system capable of assembling, storing, manipulating, and displaying geographic information, that is, data identified according to its location.

AN OVERVIEW OF DECISION SUPPORT SYSTEMS

A DSS is an organized collection of people, procedures, software, databases, and devices used to help make decisions that solve problems. The focus of a DSS is on decision-making effectiveness when faced with unstructured or semistructured business problems. As with a TPS and an MIS, a DSS should be designed, developed, and used to help an organization achieve its goals and objectives. Decision support systems offer the potential to generate higher profits, lower costs, and better products and services. For example, healthcare organizations use DSSs to improve patient care and reduce costs.

Decision support systems, although skewed somewhat toward the top levels of management, are used at all levels. To some extent, today's managers at all levels are faced with less structured, nonroutine problems, but the quantity and magnitude of these decisions increase as a manager rises higher in an organization. Many organizations contain a tangled web of complex rules, procedures, and decisions. DSSs are used to bring more structure to these problems to aid the decision-making process. In addition, because of the inherent flexibility of decision support systems, managers at all levels are able to use DSSs to assist in some relatively routine, programmable decisions in lieu of more formalized management information systems. DSSs are also used in government, law enforcement, and nonprofit organizations. See Figure 6.13.

Figure 6.13

Decision support systems are also used by nonprofit organizations and in government, such as in police departments.

(Source: © Spencer C. Grant/Photo Edit.)

Capabilities of a Decision Support System

Developers of decision support systems strive to make them more flexible than management information systems and to give them the potential to assist decision makers in a variety of situations. Table 6.1 lists a few DSS applications. DSSs can assist with all or most problem-solving phases, decision frequencies, and different degrees of problem structure. DSS approaches can also help at all levels of the decision-making process. A single DSS might provide only a few of these capabilities, depending on its uses and scope.

Support for Problem-Solving Phases

The objective of most decision support systems is to assist decision makers with the phases of problem solving. As previously discussed, these phases include intelligence, design, choice, implementation, and monitoring. A specific DSS might support only one or a few phases. By supporting all types of decision-making approaches, a DSS gives the decision maker a great deal of flexibility in getting computer support for decision-making activities.

Table 6.1

Selected DSS Applications

Company or Application	Description
ING Direct	The financial services company uses a DSS to summarize the bank's financial performance. The bank needed a measurement and tracking mechanism to determine how successful it was and to make modifications to plans in real time.
Cinergy Corporation	The electric utility developed a DSS to reduce lead time and effort required to make decisions in purchasing coal.
U.S. Army	It developed a DSS to help recruit, train, and educate enlisted forces. The DSS uses a simulation that incorporates what-if features.
National Audubon Society	It developed a DSS called Energy Plan (EPLAN) to analyze the impact of U.S. energy policy on the environment.
Hewlett-Packard	The computer company developed a DSS called Quality Decision Management to help improve the quality of its products and services.
State of Virginia	The State of Virginia developed the Transportation Evacuation Decision Support System (TEDSS) to determine the best way to evacuate people in case of a nuclear disaster at its nuclear power plants.

Support for Different Decision Frequencies

Decisions can range on a continuum from one-of-a-kind to repetitive decisions. One-of-a-kind decisions are typically handled by an **ad hoc DSS**. An ad hoc DSS is concerned with situations or decisions that come up only a few times during the life of the organization; in small businesses, they might happen only once. For example, a company might need to decide whether to build a new manufacturing facility in another area of the country. Repetitive decisions are addressed by an institutional DSS. An **institutional DSS** handles situations or decisions that occur more than once, usually several times per year or more. An institutional DSS is used repeatedly and refined over the years. Examples of institutional DSSs include systems that support portfolio and investment decisions and production scheduling. These decisions might require decision support numerous times during the year. For example, DSSs are used to help solve computer-related problems that can occur multiple times throughout the day. With this approach, the DSS monitors computer systems second by second for problems and takes action to prevent problems, such as slowdowns and crashes, and to recover from them when they occur. One IBM engineer believes that this approach, called *autonomic computing*, is the key to the future of computing. Between these two extremes are decisions managers make several times, but not regularly or routinely.

Support for Different Problem Structures

As discussed previously, decisions can range from highly structured and programmed to unstructured and nonprogrammed. **Highly structured problems** are straightforward, requiring known facts and relationships. **Semistructured** or **unstructured problems**, on the other hand, are more complex. The relationships among the pieces of data are not always clear, the data might be in a variety of formats, and it is often difficult to manipulate or obtain. In addition, the decision maker might not know the information requirements of the decision in advance. For example, a DSS has been used to support sophisticated and unstructured investment analysis and make substantial profits for traders and investors. Some DSS trading software is programmed to place buy and sell orders automatically without a trader manually entering a trade, based on parameters set by the trader.

ad hoc DSS
A DSS concerned with situations or decisions that come up only a few times during the life of the organization.

institutional DSS
A DSS that handles situations or decisions that occur more than once, usually several times per year or more. An institutional DSS is used repeatedly and refined over the years.

highly structured problems
Problems that are straightforward and require known facts and relationships.

semistructured or unstructured problems
More complex problems in which the relationships among the pieces of data are not always clear, the data might be in a variety of formats, and the data is often difficult to manipulate or obtain.

Support for Various Decision-Making Levels

Decision support systems can provide help for managers at different levels within the organization. Operational managers can get assistance with daily and routine decision making. Tactical decision makers can use analysis tools to ensure proper planning and control. At the strategic level, DSSs can help managers by providing analysis for long-term decisions requiring both internal and external information (see Figure 6.14).

Figure 6.14

Decision-Making Level

Strategic managers are involved with long-term decisions, which are often made infrequently. Operational managers are involved with decisions that are made more frequently.

Table 6.2

Comparison of DSSs and MISs

A Comparison of DSS and MIS

A DSS differs from an MIS in numerous ways, including the type of problems solved, the support given to users, the decision emphasis and approach, and the type, speed, output, and development of the system used. Table 6.2 lists brief descriptions of these differences.

Factor	DSS	MIS
Problem Type	A DSS can handle unstructured problems that cannot be easily programmed.	An MIS is normally used only with structured problems.
Users	A DSS supports individuals, small groups, and the entire organization. In the short run, users typically have more control over a DSS.	An MIS supports primarily the organization. In the short run, users have less control over an MIS.
Support	A DSS supports all aspects and phases of decision making; it does not replace the decision maker—people still make the decisions.	Some MIS systems make automatic decisions and replace the decision maker.
Emphasis	A DSS emphasizes actual decisions and decision-making styles.	An MIS usually emphasizes information only.
Approach	A DSS is a direct support system that provides interactive reports on computer screens.	An MIS is typically an indirect support system that uses regularly produced reports.
System	The computer equipment that provides decision support is usually online (directly connected to the computer system) and related to real time (providing immediate results). Computer terminals and display screens are examples—these devices can provide immediate information and answers to questions.	An MIS, using printed reports that might be delivered to managers once per week, cannot provide immediate results.
Speed	Because a DSS is flexible and can be implemented by users, it usually takes less time to develop and is better able to respond to user requests.	An MIS's response time is usually longer.
Output	DSS reports are usually screen oriented, with the ability to generate reports on a printer.	An MIS typically is oriented toward printed reports and documents.
Development	DSS users are usually more directly involved in its development. User involvement usually means better systems that provide superior support. For all systems, user involvement is the most important factor for the development of a successful system.	An MIS is frequently several years old and often was developed for people who are no longer performing the work supported by the MIS.

COMPONENTS OF A DECISION SUPPORT SYSTEM

At the core of a DSS are a database and a model base. In addition, a typical DSS contains a user interface, also called **dialogue manager**, allows decision makers to easily access and manipulate the DSS and to use common business terms and phrases. Finally, access to the Internet, networks, and other computer-based systems permits the DSS to tie into other powerful systems, including the TPS or function-specific subsystems. Internet software agents, for example, can be used in creating powerful decision support systems. Figure 6.15 shows a conceptual model of a DSS. Specific DSSs might not have all the components shown in Figure 6.15.

dialogue manager
A user interface that allows decision makers to easily access and manipulate the DSS and to use common business terms and phrases.

Figure 6.15

Conceptual Model of a DSS

DSS components include a model base; database; external database access; access to the Internet and corporate intranet, networks, and other computer systems; and a user interface or dialogue manager.

The Database

The database management system allows managers and decision makers to perform *qualitative analysis* on the company's vast stores of data in databases, data warehouses, and data marts, discussed in Chapter 3. A *data-driven DSS* primarily performs qualitative analysis based on the company's databases. Data-driven DSSs tap into vast stores of information contained in the corporate database, retrieving information on inventory, sales, personnel, production, finance, accounting, and other areas. Tween Brands, Inc. specialty retail store uses the Oracle database to provide decision support to reduce inventory costs.[53] Jo-Ann Stores uses its database to support decision making. According to a vice president and CIO of the company, "The people in the business need to understand what to do with the data and how the data impacts other parts of the organization." Data mining and business intelligence, introduced in Chapter 5, are often used in a data-driven DSS.[54] A casino can use a data-driven DSS to search large databases to get detailed information on patrons. It can tell how much each patron spends per day on gambling, and more. Swiss telecommunications company Cablecom uses a data-driven DSS from SPSS to identify customers that might leave the company.[55] Data-driven DSSs can also be used in emergency medical situations to make split-second, life-or-death treatment decisions. Data-driven medical DSSs allow doctors to access the complete medical records of a patient. Some medical record systems also allow patients to enter their own health information into the database, such as medicines, allergies, and family health histories. WebMD, iHealthRecord, Walgreens, and PersonalHealthKey allow people to put their medical records online for rapid access.

Not everyone is satisfied with data-driven DSSs, however. One survey revealed that many middle-level managers spend about two hours each workday trying to find the data they need to perform their jobs.[56] The biggest problems are being deluged with too much data, other departments not sharing their data, and not knowing if the data they get is current and accurate. Some people also have privacy concerns. A few firms mine personal data on purchasing habits and then sell the information to online advertisers that want you to buy their products and services.[57] Some people believe this is an invasion of their privacy. Today, companies are spending over $500 million on online ads. This amount is expected to grow to about $4 billion over the next 10 years. Data mining firms help these companies target their advertising to people likely to buy their products and services

A database management system can also connect to external databases to give managers and decision makers even more information and decision support. External databases can include the Internet, libraries, government databases, and more. The combination of internal and external database access can give key decision makers a better understanding of the company and its environment. Schumacher Group, for example, uses software to mash up, or tie together, information from TV reports, maps, computerized phone books, and other sources to analyze the impact of hurricanes on how doctors are to be scheduled at different emergency rooms in Lafayette and other cities in Louisiana.[58] Other companies, including Audi and AccuWeather, are using similar software packages to integrate data from different sources into data-driven DSSs.

The Model Base

model base
Part of a DSS that provides decision makers access to a variety of models and assists them in decision making.

The **model base** allows managers and decision makers to perform *quantitative analysis* on both internal and external data. A *model-driven DSS* primarily performs mathematical or quantitative analysis. The model base gives decision makers access to a variety of models so that they can explore different scenarios and see their effects. Ultimately, it assists them in the decision-making process. Procter & Gamble, maker of Pringles potato chips, Pampers diapers, and hundreds of other consumer products, uses a model-driven DSS to streamline how raw materials and products flow from suppliers to customers. The model-driven DSS has saved the company hundreds of millions of dollars in supply chain-related costs. Model-driven DSSs are excellent at predicting customer behaviors. LoanPerformance (*www.loanperformance.com*), for example, uses models to help it forecast which customers might be late with payments or might default on their loans.[59] Other financial service and insurance firms, such as health insurer HighMark, use model-driven DSSs to predict fraud. Some stock trading and investment firms use sophisticated model-driven DSSs to make trading decisions and huge profits.[60] Some experts believe that a slight time advantage in computerized trading programs can result in millions of dollars of trading profits.

model management software
Software that coordinates the use of models in a DSS.

Model management software (MMS) can coordinate the use of models in a DSS, including financial, statistical analysis, graphical, and project-management models. Depending on the needs of the decision maker, one or more of these models can be used (see Table 6.3).

Table 6.3

Model Management Software

DSSs often use financial, statistical, graphical, and project-management models.

Model Type	Description	Software
Financial	Provides cash flow, internal rate of return, and other investment analysis	Spreadsheet, such as Microsoft Excel
Statistical	Provides summary statistics, trend projections, hypothesis testing, and more	Statistical programs, such as SPSS or SAS
Graphical	Assists decision makers in designing, developing, and using graphic displays of data and information	Graphics programs, such as Microsoft PowerPoint
Project Management	Handles and coordinates large projects; also used to identify critical activities and tasks that could delay or jeopardize an entire project if they are not completed in a timely and cost-effective fashion	Project management software, such as Microsoft Project

The User Interface or Dialogue Manager

The user interface or dialogue manager allows users to interact with the DSS to obtain information. It assists with all aspects of communications between the user and the hardware and software that constitute the DSS. In a practical sense, to most DSS users, the user interface is the DSS. Upper-level decision makers are often less interested in where the information came from or how it was gathered than that the information is both understandable and accessible.

GROUP SUPPORT SYSTEMS

The DSS approach has resulted in better decision making for all levels of individual users. However, many DSS approaches and techniques are not suitable for a group decision-making environment. Although not all workers and managers are involved in committee meetings and group decision-making sessions, some tactical and strategic-level managers can spend more than half their decision-making time in a group setting. Such managers need assistance with group decision making. A **group support system** (GSS), also called a *group decision support system* and a *computerized collaborative work system*, consists of most of the elements in a DSS, plus software to provide effective support in group decision-making settings (see Figure 6.16).

group support system (GSS)
Software application that consists of most elements in a DSS, plus software to provide effective support in group decision making; also called *group decision support system* or *computerized collaborative work system*.

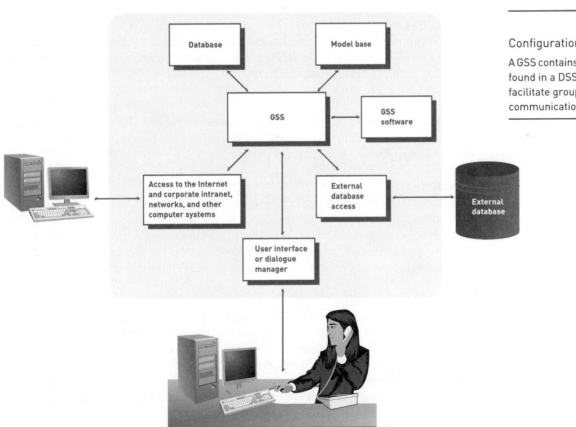

Figure 6.16

Configuration of a GSS

A GSS contains most of the elements found in a DSS, plus software to facilitate group member communications.

Group support systems are used in most industries, governments, and the military.[61] Architects are increasingly using GSSs to help them collaborate with other architects and builders to develop the best plans and to compete for contracts. Manufacturing companies use GSSs to link raw material suppliers to their own company systems. Engineers can use

Mathcad Enterprise, another GSS. The software allows engineers to create, share, and reuse calculations.

Social-networking Internet sites can be used to support group decision making.[62] Serena, a software company in California, uses the Facebook social-networking site to collaborate on projects and exchange documents.[63] The company believes this type of collaboration is so important that it has instituted "Facebook Fridays" to encourage its employees to use the social-networking site to collaborate and make group decisions.[64] Facebook Fridays also help the company work with clients and recruit new employees. Many other organizations have used Facebook or have developed their own social-networking sites to help their employees collaborate on important projects.[65] Popular snow skier Bode Miller helped found Ski Space (*www.skispace.com*), a social networking site for people interested in skiing and winter sports.[66] Some executives, however, believe that social-networking Internet sites waste time and corporate resources.[67]

Characteristics of a GSS That Enhance Decision Making

It is often said that two heads are better than one. When it comes to decision making, a GSS's unique characteristics have the potential to result in better decisions. Developers of these systems try to build on the advantages of individual support systems while adding new approaches unique to group decision making. For example, some GSSs can allow the exchange of information and expertise among people without direct face-to-face interaction, although some face-to-face meeting time is usually beneficial.[68] The following sections describe some characteristics that can improve and enhance decision making.

Special Design
The GSS approach acknowledges that special procedures, devices, and approaches are needed in group decision-making settings. These procedures must foster creative thinking, effective communications, and good group decision-making techniques.

Ease of Use
Like an individual DSS, a GSS must be easy to learn and use. Systems that are complex and hard to operate will seldom be used. Many groups have less tolerance than do individual decision makers for poorly developed systems.

Flexibility
Two or more decision makers working on the same problem might have different decision-making styles and preferences. Each manager makes decisions in a unique way, in part because of different experiences and cognitive styles. An effective GSS not only has to support the different approaches that managers use to make decisions, but also must find a means to integrate their different perspectives into a common view of the task at hand.

Decision-Making Support
A GSS can support different decision-making approaches, including the **delphi approach**, in which group decision makers are geographically dispersed throughout the country or the world. This approach encourages diversity among group members and fosters creativity and original thinking in decision making. Another approach, called **brainstorming**, in which members offer ideas "off the top of their heads," fosters creativity and free thinking. The **group consensus approach** forces members in the group to reach a unanimous decision. The Shuttle Project Engineering Office at the Kennedy Space Center has used the consensus-ranking organizational-support system (CROSS) to evaluate space projects in a group setting. The group consensus approach analyzes the benefits of various projects and their probabilities of success. CROSS is used to evaluate and prioritize advanced space projects. With the **nominal group technique**, each decision maker can participate; this technique encourages feedback from individual group members, and the final decision is made by voting, similar to a system for electing public officials.

delphi approach
A decision-making approach in which group decision makers are geographically dispersed; this approach encourages diversity among group members and fosters creativity and original thinking in decision making.

brainstorming
A decision-making approach that often consists of members offering ideas "off the top of their heads."

group consensus approach
A decision-making approach that forces members in the group to reach a unanimous decision.

nominal group technique
A decision-making approach that encourages feedback from individual group members, and the final decision is made by voting, similar to the way public officials are elected.

Anonymous Input

Many GSSs allow anonymous input, where the person giving the input is not known to other group members. For example, some organizations use a GSS to help rank the performance of managers. Anonymous input allows the group decision makers to concentrate on the merits of the input without considering who gave it. In other words, input given by a top-level manager is given the same consideration as input from employees or other members of the group. Some studies have shown that groups using anonymous input can make better decisions and have superior results compared with groups that do not use anonymous input. Anonymous input, however, can result in flaming, where an unknown team member posts insults or even obscenities on the GSS.

Reduction of Negative Group Behavior

One key characteristic of any GSS is the ability to suppress or eliminate group behavior that is counterproductive or harmful to effective decision making. In some group settings, dominant individuals can take over the discussion, which can prevent other members of the group from presenting creative alternatives. In other cases, one or two group members can sidetrack or subvert the group into areas that are nonproductive and do not help solve the problem at hand. Other times, members of a group might assume they have made the right decision without examining alternatives—a phenomenon called *groupthink*. If group sessions are poorly planned and executed, the result can be a tremendous waste of time. Today, many GSS designers are developing software and hardware systems to reduce these types of problems. Procedures for effectively planning and managing group meetings can be incorporated into the GSS approach. A trained meeting facilitator is often employed to help lead the group decision-making process and to avoid groupthink. See Figure 6.17.

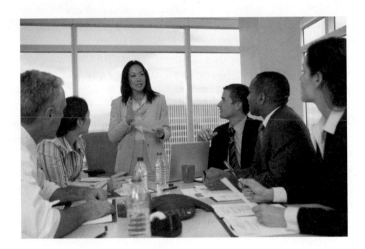

Figure 6.17

Using the GSS Approach

A trained meeting facilitator can help lead the group decision-making process and avoid groupthink.

(Source: © Bill Bachmann/Getty Images.)

Parallel and Unified Communication

With traditional group meetings, people must take turns addressing various issues. One person normally talks at a time. With a GSS, every group member can address issues or make comments at the same time by entering them into a PC or workstation. These comments and issues are displayed on every group member's PC or workstation immediately. *Parallel communication* can speed meeting times and result in better decisions. Increasingly, organizations are using unified communications to support group decision making. *Unified communications* ties together and integrates different communication systems, including traditional phones, cell phones, email, text messages, the Internet, and more.[69] With unified communications, members of a group decision-making team use a wide range of communications methods to help them collaborate and make better decisions. According to Microsoft founder Bill Gates, "Unified communications jumped out as one of those great opportunities to integrate something into the software you use for everything else you do. Even by Microsoft standards, it's a huge opportunity."

Automated Recordkeeping

Most GSSs can keep detailed records of a meeting automatically. Each comment that is entered into a group member's PC or workstation can be anonymously recorded. In some cases, literally hundreds of comments can be stored for future review and analysis. In addition, most GSS packages have automatic voting and ranking features. After group members vote, the GSS records each vote and makes the appropriate rankings.

GSS Software

GSS software, often called *groupware* or *workgroup software*, helps with joint work group scheduling, communication, and management. Software from Autodesk, for example, has GSS capabilities that allow groups to work together on the design. Designers, for example, can use Autodesk's Buzzsaw Professional Online Collaboration Service, which works with AutoCAD, a design and engineering software product from Autodesk. The U.S. Navy uses Virtual Office from Groove Networks to help it manage critical information in delivering humanitarian relief in disaster areas. The software is used for collaboration and communications in transmitting critical information between field offices. Virtual Office also has encryption capabilities to keep sensitive information safe and secure.

One popular package, Lotus Notes, can capture, store, manipulate, and distribute memos and communications that are developed during group projects. It can also incorporate knowledge management, discussed in Chapter 3, into the Lotus Notes Package. Some companies standardize on messaging and collaboration software, such as Lotus Notes. Lotus Connections is a newer feature of Lotus Notes that allows people to post documents and information on the Internet.[70] The new feature is similar to popular social-networking sites like Facebook and MySpace, but is designed for business use.[71] Microsoft has invested billions of dollars in GSS software to incorporate collaborative features into its Office suite and related products. Office Communicator, for example, is a Microsoft product developed to allow better and faster collaboration. Other companies are also heavily investing in GSS software. In addition to Lotus Notes, IBM has developed Workplace to allow workers to collaborate more efficiently in doing their jobs. Microsoft's NetMeeting product supports application sharing in multiparty calls. NetDocuments Enterprise can be used for Web collaboration. The groupware is intended for legal, accounting, and real-estate businesses. A Breakout Session feature allows two people to take a copy of a document to a shared folder for joint revision and work. The software also permits digital signatures and the ability to download and work on shared documents on handheld computers. Other GSS software packages include Collabnet, Collabra Share, OpenMind, and TeamWare. All of these tools can aid in group decision making. *Shared electronic calendars* can be used to coordinate meetings and schedules for decision-making teams.[72] Using electronic calendars, team leaders can block out time for all members of the decision-making team. Some employees, however, don't like the use of shared electronic calendars. A member of one team said, "It's an intrusion. It's just a theft of your time."

A number of additional collaborative tools are available on the Internet.[73] Sharepoint (*www.microsoft.com*), WebOffice (*www.weboffice.com*), and BaseCamp (*www.basecamphq.com*) are just a few examples.[74] Twitter (*www.twitter.com*) and Jaiku (*www.jaiku.com*) are Internet sites that some organizations use to help people and groups stay connected and coordinate work schedules.[75] Sermo (*www.sermo.com*) is a social-networking site used by doctors to collaborate with other doctors, share their medical experiences, and even help make diagnoses.[76] Many of these Internet packages embrace the use of Web 2.0. Some executives, however, worry about security and corporate compliance issues in adopting Web 2.0 technologies.[77]

In addition to stand-alone products, GSS software is increasingly being incorporated into existing software packages. Today, some transaction processing and enterprise resource planning packages include collaboration software. Some ERP producers (see Chapter 5), for example, have developed groupware to facilitate collaboration and to allow users to integrate applications from other vendors into the ERP system of programs. Today, groupware can interact with wireless devices. Research In Motion, the maker of BlackBerry software, offers mobile communications, access to group information, meeting schedules, and other services that can be directly tied to groupware software and servers. In addition to groupware, GSSs use a number of tools discussed previously, including the following:

- E-mail, instant messaging (IM), and text messaging (TM)
- Videoconferencing
- Group scheduling
- Project management
- Document sharing

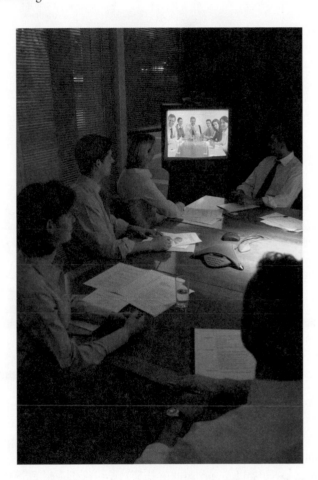

GSS software allows work teams to collaborate and reach better decisions—even if they work across town, in another region, or on the other side of the globe.

(Source: © Flying Colours Ltd./Getty Images.)

GSS Alternatives

Group support systems can take on a number of network configurations, depending on the needs of the group, the decision to be supported, and the geographic location of group members. GSS alternatives include a combination of decision rooms, local area networks, teleconferencing, and wide area networks.

- The **decision room** is ideal for situations in which decision makers are located in the same building or geographic area and the decision makers are occasional users of the GSS approach. In some cases, the decision room might have a few computers and a projector for presentations. In other cases, the decision room can be fully equipped with a network of computers and sophisticated GSS software. A typical decision room is shown in Figure 6.18.
- The *local area decision network* can be used when group members are located in the same building or geographic area and under conditions in which group decision making is frequent. In these cases, the technology and equipment for the GSS approach is placed directly into the offices of the group members.

decision room
A room that supports decision making, with the decision makers in the same building, combining face-to-face verbal interaction with technology to make the meeting more effective and efficient.

Figure 6.18

The GSS Decision Room

For group members who are in the same location, the decision room is an optimal GSS alternative. This approach can use both face-to-face and computer-mediated communication. By using networked computers and computer devices, such as project screens and printers, the meeting leader can pose questions to the group, instantly collect their feedback, and, with the help of the governing software loaded on the control station, process this feedback into meaningful information to aid in the decision-making process.

Meeting leader   **Control station**

Participant stations

- *Teleconferencing* is used when the decision frequency is low and the location of group members is distant. These distant and occasional group meetings can tie together multiple GSS decision-making rooms across the country or around the world.
- The *wide area decision network* is used when the decision frequency is high and the location of group members is distant. In this case, the decision makers require frequent or constant use of the GSS approach. This GSS alternative allows people to work in **virtual workgroups**, where teams of people located around the world can work on common problems.

virtual workgroups
Teams of people located around the world working on common problems.

EXECUTIVE SUPPORT SYSTEMS

executive support system (ESS)
Specialized DSS that includes all hardware, software, data, procedures, and people used to assist senior-level executives within the organization.

Because top-level executives often require specialized support when making strategic decisions, many companies have developed systems to assist executive decision making. This type of system, called an **executive support system (ESS)**, is a specialized DSS that includes all hardware, software, data, procedures, and people used to assist senior-level executives within the organization. In some cases, an ESS, also called an *executive information system (EIS)*, supports decision making of members of the board of directors, who are responsible to stockholders. These top-level decision-making strata are shown in Figure 6.19.

An ESS can also be used by individuals at middle levels in the organizational structure. Once targeted at the top-level executive decision makers, ESSs are now marketed to—and used by—employees at other levels in the organization. In the traditional view, ESSs give top executives a means of tracking critical success factors. Today, all levels of the organization share information from the same databases. However, for our discussion, assume ESSs remain in the upper-management levels, where they highlight important corporate issues, indicate new directions the company might take, and help executives monitor the company's progress.

Executive Support Systems in Perspective

An ESS is a special type of DSS, and, like a DSS, is designed to support higher-level decision making in the organization. The two systems are, however, different in important ways. DSSs provide a variety of modeling and analysis tools to enable users to thoroughly analyze problems—that is, they allow users to *answer* questions. ESSs present structured information about aspects of the organization that executives consider important. The characteristics of an ESS are summarized in the following list.

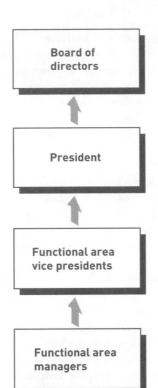

Figure 6.19

The Layers of Executive
Decision Making

- Are tailored to individual executives
- Are easy to use
- Have drill-down abilities
- Support the need for external data
- Can help with situations that have a high degree of uncertainty
- Have a future orientation
- Are linked with value-added business processes

Capabilities of Executive Support Systems

The responsibility given to top-level executives and decision makers brings unique problems and pressures to their jobs. This section discusses some of the characteristics of executive decision making that are supported through the ESS approach. ESSs take full advantage of data mining, the Internet, blogs, podcasts, executive dashboards, and many other technological innovations. As you will note, most of these decisions are related to an organization's overall profitability and direction. An effective ESS should have the capability to support executive decisions with components such as strategic planning and organizing, crisis management, and more.

Support for Defining an Overall Vision

One of the key roles of senior executives is to provide a broad vision for the entire organization. This vision includes the organization's major product lines and services, the types of businesses it supports today and in the future, and its overriding goals.

Support for Strategic Planning

ESSs also support strategic planning. **Strategic planning** involves determining long-term objectives by analyzing the strengths and weaknesses of the organization, predicting future trends, and projecting the development of new product lines. It also involves planning the acquisition of new equipment, analyzing merger possibilities, and making difficult decisions concerning downsizing and the sale of assets if required by unfavorable economic conditions.

strategic planning
Determining long-term objectives by analyzing the strengths and weaknesses of the organization, predicting future trends, and projecting the development of new product lines.

Support for Strategic Organizing and Staffing

Top-level executives are concerned with organizational structure. For example, decisions concerning the creation of new departments or downsizing the labor force are made by top-level managers. Overall direction for staffing decisions and effective communication with labor unions are also major decision areas for top-level executives. ESSs can be employed to help analyze the impact of staffing decisions, potential pay raises, changes in employee benefits, and new work rules.

Support for Strategic Control

Another type of executive decision relates to strategic control, which involves monitoring and managing the overall operation of the organization. Goal seeking can be done for each major area to determine what performance these areas need to achieve to reach corporate expectations. Effective ESS approaches can help top-level managers make the most of their existing resources and control all aspects of the organization.

Support for Crisis Management

Even with careful strategic planning, a crisis can occur. Major incidents, including natural disasters, fires, and terrorist activities, can totally shut down major parts of the organization. Handling these emergencies is another responsibility for top-level executives. In many cases, strategic emergency plans can be put into place with the help of an ESS. These contingency plans help organizations recover quickly if an emergency or crisis occurs.

Decision making is a vital part of managing businesses strategically. IS systems such as information and decision support, group support, and executive support systems help employees by tapping existing databases and providing them with current, accurate information. The increasing integration of all business information systems—from TPSs to MISs to DSSs—can help organizations monitor their competitive environment and make better-informed decisions. Organizations can also use specialized business information systems, discussed in Chapter 7, to achieve their goals.

SUMMARY

Principle

Good decision-making and problem-solving skills are the key to developing effective information and decision support systems.

Every organization needs effective decision making and problem solving to reach its objectives and goals. Problem solving begins with decision making. A well-known model developed by Herbert Simon divides the decision-making phase of the problem-solving process into three stages: intelligence, design, and choice. During the intelligence stage, potential problems or opportunities are identified and defined. Information is gathered that relates to the cause and scope of the problem. Constraints on the possible solution and the problem environment are investigated. In the design stage, alternative solutions to the problem are developed and explored. In addition, the feasibility and implications of these alternatives are evaluated. Finally, the choice stage involves selecting the best course of action. In this stage, the decision makers evaluate the implementation of the solution to determine whether the anticipated results were achieved and to modify the process in light of new information learned during the implementation stage.

Decision making is a component of problem solving. In addition to the intelligence, design, and choice steps of decision making, problem solving also includes implementation and monitoring. Implementation places the solution into effect. After a decision has been implemented, it is monitored and modified if necessary.

Decisions can be programmed or nonprogrammed. Programmed decisions are made using a rule, procedure, or quantitative method. Ordering more inventory when the level drops to 100 units or fewer is an example of a programmed decision. A nonprogrammed decision deals with unusual or exceptional situations. Determining the best training program for a new employee is an example of a nonprogrammed decision.

Decisions can use optimization, satisficing, or heuristic approaches. Optimization finds the best solution. Optimization problems often have an objective such as maximizing profits given production and material constraints. When a problem is too complex for optimization, satisficing is often used. Satisficing finds a good, but not necessarily the best, decision. Finally, a heuristic is a "rule of thumb" or commonly used guideline or procedure used to find a good decision.

Principle

The management information system (MIS) must provide the right information to the right person in the right format at the right time.

A management information system is an integrated collection of people, procedures, databases, and devices that provides managers and decision makers with information to help achieve organizational goals. An MIS can help an organization achieve its goals by providing managers with insight into the regular operations of the organization so that they can control, organize, and plan more effectively and efficiently. The primary difference between the reports generated by the TPS and those generated by the MIS is that MIS reports support managerial decision making at the higher levels of management.

Data that enters the MIS originates from both internal and external sources. The most significant internal sources of data for the MIS are the organization's various TPSs and ERP systems. Data warehouses and data marts also provide important input data for the MIS. External sources of data for the MIS include extranets, customers, suppliers, competitors, and stockholders.

The output of most MISs is a collection of reports that are distributed to managers. These reports include scheduled reports, key-indicator reports, demand reports, exception reports, and drill-down reports. Scheduled reports are produced periodically, or on a schedule, such as daily, weekly, or monthly. A key-indicator report is a special type of scheduled report. Demand reports are developed to provide certain information at a manager's request. Exception reports are automatically produced when a situation is unusual or requires management action. Drill-down reports provide increasingly detailed data about situations.

Management information systems have a number of common characteristics, including producing scheduled, demand, exception, and drill-down reports; producing reports with fixed and standard formats; producing hard-copy and soft-copy reports; using internal data stored in organizational computerized databases; and having reports developed and implemented by IS personnel or end users. Increasingly, MIS reports are being delivered over the Internet and through mobile devices, such as cell phones.

Most MISs are organized along the functional lines of an organization. Typical functional management information systems include financial, manufacturing, marketing, human resources, and other specialized systems. Each system is composed of inputs, processing subsystems, and outputs. The primary sources of input to functional MISs include the corporate strategic plan, data from the ERP system and TPS, information from supply chain and business transactions, and external sources including the Internet and extranets. The primary output of these functional MISs are summary reports that assist in managerial decision making.

A financial management information system provides financial information to all financial managers within an organization, including the chief financial officer (CFO). Subsystems are profit/loss and cost systems, auditing, and use and management of funds.

A manufacturing MIS accepts inputs from the strategic plan, the ERP system and TPS, and external sources, such as supply chain and business transactions. The systems involved support the business processes associated with the receiving and inspecting of raw material and supplies; inventory tracking of raw materials, work in process, and finished goods; labor and personnel management; management of assembly lines, equipment, and machinery; inspection and maintenance; and order processing. The subsystems involved are design and engineering, master production scheduling and inventory control, process control, and quality control and testing.

A marketing MIS supports managerial activities in the areas of product development, distribution, pricing decisions, promotional effectiveness, and sales forecasting. Subsystems include marketing research, product development, promotion and advertising, and product pricing.

A human resource MIS is concerned with activities related to employees of the organization. Subsystems include human resource planning, personnel selection and recruiting, training and skills inventories, scheduling and job placement, wage and salary administration, and outplacement.

An accounting MIS performs a number of important activities, providing aggregate information on accounts payable, accounts receivable, payroll, and many other applications. The organization's ERP system or TPS captures accounting data, which is also used by most other functional information systems. Geographic information systems provide regional data in graphical form.

Principle

Decision support systems (DSSs) are used when problems are unstructured.

A decision support system (DSS) is an organized collection of people, procedures, software, databases, and devices working to support managerial decision making. DSS characteristics include the ability to handle large amounts of data; obtain and process data from different sources; provide report and presentation flexibility; support drill-down analysis; perform complex statistical analysis; offer textual and graphical orientations; support optimization, satisficing,

and heuristic approaches; and perform what-if, simulation, and goal-seeking analysis.

DSSs provide support assistance through all phases of the problem-solving process. Different decision frequencies also require DSS support. An ad hoc DSS addresses unique, infrequent decision situations; an institutional DSS handles routine decisions. Highly structured problems, semistructured problems, and unstructured problems can be supported by a DSS. A DSS can also support different managerial levels, including strategic, tactical, and operational managers. A common database is often the link that ties together a company's TPS, MIS, and DSS.

The components of a DSS are the database, model base, user interface or dialogue manager, and a link to external databases, the Internet, the corporate intranet, extranets, networks, and other systems. The database can use data warehouses and data marts. A data-driven DSS primarily performs qualitative analysis based on the company's databases. Data-driven DSSs tap into vast stores of information contained in the corporate database, retrieving information on inventory, sales, personnel, production, finance, accounting, and other areas. Data mining is often used in a data-driven DSS. The model base contains the models used by the decision maker, such as financial, statistical, graphical, and project-management models. A model-driven DSS primarily performs mathematical or quantitative analysis. Model management software (MMS) is often used to coordinate the use of models in a DSS. The user interface provides a dialogue management facility to assist in communications between the system and the user. Access to other computer-based systems permits the DSS to tie into other powerful systems, including the TPS or function-specific subsystems.

Principle

Specialized support systems, such as group support systems (GSSs) and executive support systems (ESSs), use the overall approach of a DSS in situations such as group and executive decision making.

A group support system (GSS), also called a *computerized collaborative work system*, consists of most of the elements in a DSS, plus software to provide effective support in group decision-making settings. GSSs are typically easy to learn and use and can offer specific or general decision-making support. GSS software, also called *groupware*, is specially designed to help generate lists of decision alternatives and perform data analysis. These packages let people work on joint documents and files over a network. Newer Web 2.0 technologies are being used to a greater extent in delivering group decision-making support. Text messages and the Internet are also commonly used in a GSS.

The frequency of GSS use and the location of the decision makers will influence the GSS alternative chosen. The decision room alternative supports users in a single location who meet infrequently. Local area networks can be used when group members are located in the same geographic area and users meet regularly. Teleconferencing is used when decision frequency is low and the location of group members is distant. A wide area network is used when the decision frequency is high and the location of group members is distant.

Executive support systems (ESSs) are specialized decision support systems designed to meet the needs of senior management. They serve to indicate issues of importance to the organization, indicate new directions the company might take, and help executives monitor the company's progress.

ESSs are typically easy to use, offer a wide range of computer resources, and handle a variety of internal and external data. In addition, the ESS performs sophisticated data analysis, offers a high degree of specialization, and provides flexibility and comprehensive communications capabilities. An ESS also supports individual decision-making styles. Some of the major decision-making areas that can be supported through an ESS are providing an overall vision, strategic planning and organizing, strategic control, and crisis management.

CHAPTER 6: SELF-ASSESSMENT TEST

Good decision-making and problem-solving skills are the key to developing effective information and decision support systems.

1. Developing decision alternatives is done during what decision-making stage?
   a. initiation stage
   b. intelligence stage
   c. design stage
   d. choice stage

2. Problem solving is one of the stages of decision making. True or False?

3. The final stage of problem solving is _____.

4. A decision that products should be ordered when inventory levels drop to 500 units is an example of a(n) _____.
   a. synchronous decision
   b. asynchronous decision
   c. nonprogrammed decision
   d. programmed decision

5. A(n) _____ model will find the optimal solution, usually the one that will best help the organization meet its goals.

6. A satisficing model is one that will find a good problem solution, but not necessarily the best problem solution. True or False?

The management information system (MIS) must provide the right information to the right person in the right format at the right time.

7. What summarizes the previous day's critical activities and is typically available at the beginning of each workday?
   a. key-indicator report
   b. demand report
   c. exception report
   d. database report

8. MRP and JIT are a subsystem of the _____.

   a. marketing MIS
   b. financial MIS
   c. manufacturing MIS
   d. auditing MIS

9. Another name for the _____ MIS is the personnel MIS because it is concerned with activities related to employees and potential employees of the organization.

Decision support systems (DSSs) are used when problems are unstructured.

10. The focus of a decision support system is on decision-making effectiveness when faced with unstructured or semistructured business problems. True or False?

11. _____ is used to find the best solution.

12. What component of a decision support system allows decision makers to easily access and manipulate the DSS and to use common business terms and phrases?
   a. the knowledge base
   b. the model base
   c. the user interface or dialogue manager
   d. the expert system

Specialized support systems, such as group support systems (GSSs) and executive support systems (ESSs), use the overall approach of a DSS in situations such as group and executive decision making.

13. What decision-making technique allows voting group members to arrive at a final group decision?
   a. groupthink
   b. anonymous input
   c. nominal group technique
   d. delphi

14. A type of software that helps with joint work group scheduling, communication, and management is called _____.

15. The local area decision network is the ideal GSS alternative for situations in which decision makers are located in the

same building or geographic area and the decision makers are occasional users of the GSS approach. True or False?

16. A(n) _____ supports the actions of members of the board of directors, who are responsible to stockholders.

REVIEW QUESTIONS

1. What is a satisficing model? Describe a situation when it should be used.
2. What is the difference between intelligence and design in decision making?
3. What is the difference between a programmed decision and a nonprogrammed decision? Give several examples of each.
4. What are the basic kinds of reports produced by an MIS?
5. How can text messaging be used to develop MIS reports?
6. What are the functions performed by a financial MIS?
7. Describe the functions of a marketing MIS.
8. List and describe some other types of MISs.
9. What are the stages of problem solving?
10. What is the difference between decision making and problem solving?
11. What is a geographic information system?
12. Describe the difference between a structured and an unstructured problem and give an example of each.
13. Define *decision support system*. What are its characteristics?
14. Describe the difference between a data-driven and a model-driven DSS.
15. What is the difference between what-if analysis and goal-seeking analysis?
16. What are the components of a decision support system?
17. State the objective of a group support system (GSS) and identify three characteristics that distinguish it from a DSS.
18. How can social-networking sites be used in a GSS?
19. Identify three group decision-making approaches often supported by a GSS.
20. What is an executive support system? Identify three fundamental uses for such a system.

DISCUSSION QUESTIONS

1. Select an important problem you had to solve during the last two years. Describe how you used the decision-making and problem-solving steps discussed in this chapter to solve the problem.
2. Describe how a GSS can be used at your school or university.
3. How can management information systems be used to support the objectives of the business organization?
4. Describe a financial MIS for a *Fortune* 1000 manufacturer of food products. What are the primary inputs and outputs? What are the subsystems?
5. How can a strong financial MIS provide strategic benefits to a firm?
6. Why is auditing so important in a financial MIS? Give an example of an audit that failed to disclose the true nature of the financial position of a firm. What was the result?
7. Describe two industries where a marketing MIS is critical to sales and success.
8. You have been hired to develop a management information system and a decision support system for a manufacturing company. Describe what information you would include in printed reports and what information you would provide using a screen-based decision support system.
9. Pick a company and research its human resource management information system. Describe how the system works. What improvements could be made to the company's human resource MIS?
10. You have been hired to develop a DSS for a car company such as Ford or GM. Describe how you would use both data-driven and model-driven DSSs.
11. Describe how you would develop a social-networking Internet site to help a company collaborate on an important project or decision. What features would you include in the Internet site?
12. What functions do decision support systems support in business organizations? How does a DSS differ from a TPS and an MIS?
13. How is decision making in a group environment different from individual decision making, and why are information systems that assist in the group environment different? What are the advantages and disadvantages of making decisions as a group?

14. You have been hired to develop group support software. Describe the features you would include in your new GSS software.

15. Imagine that you are the vice president of manufacturing for a *Fortune* 1000 manufacturing company. Describe the features and capabilities of your ideal ESS.

PROBLEM-SOLVING EXERCISES

1. Use the Internet to research the use of Web 2.0 technologies in collaborative group decision making. Use a word processor to describe what you discovered. Develop a set of slides using a graphics program to deliver a presentation on the use of Web 2.0 in GSS.

2. Review the summarized consolidated statement of income for the manufacturing company whose data is shown here. Use graphics software to prepare a set of bar charts that shows the data for this year compared with the data for last year.

   a. This year, operating revenues increased by 3.5 percent, while operating expenses increased 2.5 percent.
   b. Other income and expenses decreased to $13,000.
   c. Interest and other charges increased to $265,000.

Operating Results (in millions)

Operating Revenues	$2,924,177
Operating Expenses (including taxes)	2,483,687
Operating Income	440,490
Other Income and Expenses	13,497
Income Before Interest and Other Charges	453,987
Interest and Other Charges	262,845
Net Income	191,142
Average Common Shares Outstanding	147,426
Earnings per Share	1.30

If you were a financial analyst tracking this company, what detailed data might you need to perform a more complete analysis? Write a brief memo summarizing your data needs.

3. As the head buyer for a major supermarket chain, you are constantly being asked by manufacturers and distributors to stock their new products. Over 50 new items are introduced each week. Many times, these products are launched with national advertising campaigns and special promotional allowances to retailers. To add new products, the amount of shelf space allocated to existing products must be reduced or items must be eliminated altogether. Develop a marketing MIS that you can use to estimate the change in profits from adding or deleting an item from inventory. Your analysis should include input such as estimated weekly sales in units, shelf space allocated to stock an item (measured in units), total cost per unit, and sales price per unit. Your analysis should calculate total annual profit by item and then sort the rows in descending order based on total annual profit.

TEAM ACTIVITIES

1. Use a social-networking site to help your team collaborate on a decision of your choice. Each team member should write a brief report on his or her experiences in using the social-networking site. In addition to the individual reports, your team should collaborate on a group report that describes the different perceptions of each team member and makes recommendations about the use of social-networking sites in group decision making.

2. Have your team make a group decision about how to solve the most frustrating aspect of college or university life. Appoint one or two members of the team to disrupt the meeting with negative group behavior. After the meeting, have your team describe how to prevent this negative group behavior. What GSS software features would you suggest to prevent the negative group behavior your team observed?

3. Imagine that you and your team have decided to develop an ESS software product to support senior executives in the music recording industry. What are some of the key decisions these executives must make? Make a list of the capabilities that such a system must provide to be useful. Identify at least six sources of external information that will be useful to its users.

WEB EXERCISES

1. Use a search engine, such as Yahoo! or Google, to explore two or more companies that produce and sell MIS or DSS software. Describe what you found and any problems you had in using search engines on the Internet to find information. You might be asked to develop a report or send an e-mail message to your instructor about what you found.
2. Use the Internet to explore two or more software packages described in this chapter. Summarize your findings in a report.
3. Software, such as Microsoft Excel, is often used to find an optimal solution to maximize profits or minimize costs. Search the Internet using Yahoo!, Google, or another search engine to find other software packages that offer optimization features. Write a report describing one or two of the optimization software packages. What are some of the features of the package?

CAREER EXERCISES

1. What decisions are critical for success in a career that interests you? What specific types of reports could help you make better decisions on the job? Give three specific examples.
2. Use two or more social-networking sites to explore careers that interest you. Describe what you found and the differences between the sites. What features would you like to see in a social-networking site to help you find a good job?

CASE STUDIES

Case One
Enterprise Rent-A-Car and Business Process Management

Enterprise Rent-A-Car is based in St. Louis, Missouri, and runs approximately 6,900 branch offices around the world. Enterprise routes all of its information system requests through its information systems group housed in the home office. Until recently, Enterprise used an outdated process for managing corporate requests. The Enterprise Requests Online system assisted the Requests department in handling product and service requests. The system automated everything from setting up a new laptop to opening a new Enterprise Rent-A-Car office. To make a request, Enterprise employees would navigate to a corporate Web site where a system request form was located. Fifteen categories of requests were provided for employees, who entered the details of the problem into a text box. The Web form generated an e-mail to the Requests department that managed its job queue from its e-mail inbox.

While such a system was considered state of the art ten years ago, it has inherent problems that more modern systems have addressed. For one thing, e-mail is a difficult communications medium to manage. Individual messages must be opened to examine details of the request. Messages are also easy to lose. Those seeking help cannot see how or whether their request is being handled. Maintaining a history of work requests over many years is next to impossible with such a system.

Enterprise decided to revamp and improve its work request process to streamline the process and improve service to its branch offices. The information system that Enterprise wanted is called a business process management (BPM) system—an automated method of streamlining business processes. Enterprise turned to APPIAN, a company that specializes in BPM systems. The two worked together to produce a powerful BPM system for managing incoming information system requests.

Enterprise's new Requests Online system provides users with detailed options to narrow a request to one of 200 request types. The system recognizes the user and lists only options for that particular branch office. For example, the system detects software used by that branch office and provides options only for that software. This focuses the options to those that interest the user, saving time. The software also fills in the user data, such as name, phone, and location, saving even more time. Once submitted, users of the system can view the progress of their request in a job queue page.

On the back end, Requests staff uses an executive dashboard application to keep track of their work. The system tracks all jobs in the queue and produces useful reports, which provide graphical information to the dashboard that indicates how smoothly operations are running. Using these visual cues, employees can tell if they are keeping up with the work, and managers can decide how many workers are required to meet the load. Using other reports, managers can determine which months of the year are busiest, and which weeks, days, and hours require the most or least amount of staffing.

The new system is expected to save Enterprise between 15 and 20 percent in costs and time in administrative and data entry activities. Already the savings are being felt. Enterprise has redeployed its staff from maintaining the old legacy system to jobs that are more "strategically valuable."

Discussion Questions

1. What is the purpose of business process management systems? What benefits do they provide?
2. What problems did the original Enterprise Requests Online system have?

Critical Thinking Questions

1. What attributes do you think make up a system that is optimized for the greatest convenience to users?
2. What factors would lead a business to decide that it is time to improve its business processes with a new MIS?

SOURCES: Ruffolo, Rafael, "Enterprise Rent-A-Car drives service request app," *ITWorldCanada.com*, March 27, 2008, *www.itworldcanada.com//Pages/Docbase/ViewArticle.aspx?ID=idgml-924c5e06-673e-4b68-86de-a83dbfcb0f37*; Appian Corporate Web site, *www.appian.com*, accessed May 20, 2008; Enterprise Rent-A-Car Web site, *www.enterprise.com*, accessed May 20, 2008.

Case Two

Keiper Watches Production Like a Hawk

Keiper GmbH & Co. KG is a leading manufacturer of the metal components of car seats. The company runs 11 production sites scattered around the world and employs 6,000 workers. Keiper struggles with the common challenges facing all international manufacturing businesses. One of its largest problems is synchronizing production across manufacturing plants separated by many miles and time zones.

A few years ago, Keiper made the wise move to connect its production sites over the Internet using production management software. The software allowed system specialists and production managers at Keiper headquarters in Kaiserslautern, Germany, to monitor production systems at all 11 sites. This ability made all the difference in the world to Keiper's production quality. Managers can troubleshoot problems as they arise, upgrade system software, and make database entries at the same time across all locations. In short, the system made it seem as though the 11 production facilities were actually one big manufacturing plant.

More recently, Keiper upgraded the system to allow it to react to emergencies more quickly. One of the biggest problems in manufacturing occurs when flaws are introduced into the process. Flaws might be the result of a defective part received from a supplier and used on the assembly line. Sometimes these defects are not noticed until many products have been manufactured and shipped. Keiper wanted a traceability system so it could track the car seat components and assembled car seats through the production line to their destinations.

Keiper partnered with an information system company that specialized in traceability systems for the automotive industry. The company designed a system that collected and connected information from production facilities around the world. The information included specific information about the parts used in the manufacturing process. The system tracks each step in the production process at all 11 manufacturing plants in real time. If an employee on the line notices that a certain type of screw is defective, an alarm is sounded and an investigation immediately launched. The defective screw is traced back to the batch from which it came. Each screw used from that batch is traced to a specific seat in the production line, shipment center, or automotive plant where the cars are assembled. Recalls of seats containing the bad screw can occur within hours of the discovery.

Using MISs to control production lines around the world provides Keiper management with more control over its business. The ability to quickly catch defects in its products minimizes the extent of the damage they cause. Keiper customers appreciate the corporation's ability to minimize problems before they grow to an unmanageable scope.

Discussion Questions

1. What is one of the biggest production challenges facing global manufacturing corporations?
2. How did Keiper management gain more control over its 11 manufacturing facilities?

Critical Thinking Questions

1. What type of management information system does Keiper use, and what functional unit of the company is it designed for?
2. What information components do you think were used in Keiper's new traceability system?

SOURCES: Staff SAP, "Keiper GmbH & Co. KG," SAP Customer Success Story, *http://download.sap.com/solutions/manufacturing/customersuccess/download.epd?cont ext=BAC27560689AE-F8040ED89D61B269843913D4B9555561819698A5E0F5A5BF40 C229A297AA3C9C20AA4384FDE88B9A968C706CA6FF70055CD*, accessed May 20, 2008.

Questions for Web Case

See the Web site for this book to read about the Whitmann Price Consulting case for this chapter. The following are questions concerning this Web case.

Whitmann Price Consulting: MIS and DSS Considerations

Discussion Questions

1. What different types of needs can MISs and DSSs fulfill in Whitmann Price's new system?

2. Why did Whitmann Price decide to design its own Calendar and Contacts MIS rather than using standard BlackBerry software?

Critical Thinking Questions

1. How does the source of input differ between the MISs being designed for all consulting areas and the DSSs uniquely designed for each consulting area?
2. How might a GSS be useful for consultants at Whitmann Price?

NOTES

Sources for the opening vignette: SAP Staff, "General Mills," SAP Customer Success Story, *http://download.sap.com/solutions/business-suite/plm/customersuccess/download.epd?context=DF359538DBDBF-B3739D0C646A8 BF36C42C2999E7D28C06CB095A5648B35773-CA7A03AFD504A5C7CA79A98EB720 AFFA28877CC33DEAF2BD52*, accessed May 18, 2008; General Mills Web site, *www.generalmills.com/corporate/company/index.aspx*, accessed May 18, 2008.

1   Merrick, Jason, and Harrald, John, "Making Decisions About Safety in US Ports and Waterways," *Interfaces,* June 2007, p. 240.
2   Ross, Alex and Swain, Alison, "Fighting Flight Delays," *OR/MS Today,* April 2007, p. 22.
3   Barnes, Brooks, "Can CBS Put the Net into the Network?" *The Wall Street Journal,* May 14, 2007, p. B1.
4   *www.cmis.csiro.au/OR/MEDIA/sate.htm*, accessed November 16, 2007.
5   Kant, Goos, "Coca-Cola Enterprises Optimizes Vehicle Routes for Efficient Product Delivery," *Interfaces,* January 2008, p. 40.
6   Clark, T., et al., "The Dynamic Structure of Management Support Systems," *MIS Quarterly,* September 2007, p. 579.
7   *www.cognos.com*, accessed November 14, 2007.
8   Biddick, Michael, "Hunting the Elusive CIO Dashboard," *InformationWeek,* March 3, 2008, p. 47.
9   Hayes Weier, Mary, "A Steaming Cup of Insight," *InformationWeek,* May 7, 2007, p. 47.
10 Havenstein, Heather, "Atlanta to Roll Out Cognos Analysis Software to City Agency Workers," *Computerworld,* January 15, 2007, p. 15.
11 *www.cognos.com/news/releases/2007/1112.html?mc=-web_hp*, accessed November 15, 2007.
12 *www.microsoft.com/dynamics/product/business_scorecard_manager.mspx*, accessed November 14, 2007.
13 Havenstein, Heather, "Ace Hardware Shifts BI Tools After Early Struggles," *Computerworld,* April 23, 2007, p. 11.
14 Greene, Jay, "Microsoft Wants Your Health Records," *Business Week,* October 15, 2007, p. 44.
15 Havenstein, Heather, "Military, Oil Firm Use BI to Avert Disaster," *Computerworld,* October 1, 2007, p. 8.
16 Yuan, Li and Prada, Paulo, "Texting When There's Trouble," *The Wall Street Journal,* April 18, 2007, p. B1.
17 Yuan, Li, "Murder, She Texted," *The Wall Street Journal,* July 2, 2007, p. B1.
18 Havenstein, Heather, "Military, Oil Firm Use BI to Avert Disaster," *Computerworld,* October 1, 2007, p. 8.
19 Patrick, Aaron O., "Reuters to Launch Trade Software," *The Wall Street Journal,* April 30, 2007, p. C2.
20 Ferraro, Nicole, "Lending and Philanthropy 2.0," *InformationWeek,* February 4, 2008, p. 40.
21 Kim, Jane, "Where Either a Borrower Or a Lender Can Be," *The Wall Street Journal,* March 12, 2008, p. D1.
22 Brown, Erika, "Widget Work," *Forbes,* October 29, 2007, p. 66.
23 Nagali, Jerry, et al., "Procurement Risk Management at Hewlett-Packard Company," *Interfaces,* January 2008, p. 51.
24 Dean, Jason, "The Forbidden City of Terry Gou," *The Wall Street Journal,* August 11, 2007, p. A1.
25 Loveland, Jennifer, et al., "Dell Uses a New Production Scheduling Algorithm," *Interfaces,* May 2007, p. 209.
26 Khawam, John, et al., "Warranty Inventory Optimization for Hitachi Global Storage," *Interfaces,* September 2007, p. 455.
27 Prada, Paulo, "Delta Maintains Its Maintenance," *The Wall Street Journal,* November 7, 2007, p. A16.
28 Songini, Marc, "BI Helping Retailers Control Inventory," *Computerworld,* January 22, 2007, p. 14.
29 Chozick, Amy, "Toyota Sticks by Just In Time Strategy After Quake," *The Wall Street Journal,* July 24, 2007, p. A2.
30 Vranica, Suzanne, "Internet Reshapes Role of Media Buyers," *The Wall Street Journal,* February 28, 2007, p. B3.
31 Warren, Jamin and Jurgensen, John, "A New Kind of Web Site," *The Wall Street Journal,* February 10, 2007, p. P1.
32 Sharma, Amol, "Companies Vie for Ad Dollars on Mobile Web," *The Wall Street Journal,* January 17, 2007, p. D1.
33 Vara, Vauhini, "Facebook Gets Personal with Ad Targeting Plan," *The Wall Street Journal,* August 23, 2007, p. B1.
34 Steel, Emily, "YouTube to Start Selling Ads in Video," *The Wall Street Journal,* August 22, 2007, p. B3.
35 Jesdanun, Anick, "YouTube's Overlay Ads Aim to Retain Viewership," *Rocky Mountain News,* August 8, 2007, p. 4.
36 Fass, Allison, "Sex, Pranks, and Reality," *Forbes,* July 2, 2007, p. 48.
37 McMillan, Robert, "Every Click You Make …" *Computerworld,* January 7, 2008, p. 8.
38 Hayes Weier, Mary, "Dear Customer: Please Don't Leave," *Information Week,* June 18, 2007, p. 49.
39 Copeland, Michael, "Location, Location, Location," *Fortune,* November 12, 2007, p. 147.
40 Vranica, Suzanne, "Hellmann's Targets Yahoo for Its Spread," *The Wall Street Journal,* June 27, 2007, p. B4.
41 Steel, Emily, "J&J's Web Ads Depart from TV Formula," *The Wall Street Journal,* February 12, 2008, p. B3.

42 Gonsalves, Antone, "Yahoo's Brand Grab Bag," *Information Week,* February 5, 2007, p. 25.

43 Steel, Emily, "Web vs. TV," *The Wall Street Journal,* March 19, 2008, p. B3.

44 Delaney, Kevin, "Start-ups Seek to Cash in on Web-Video Advertising," *The Wall Street Journal,* March 2, 2007, p. B1.

45 Patrick, Aaron, "Internet Ad Shops Are Crossing the Digital Divide," *The Wall Street Journal,* February 9, 2007, p. B3.

46 Vara, Vauhini, "Companies Tolerate Ads to Get Free Software," *The Wall Street Journal,* March 27, 2007, p. B1.

47 Banjo, Shelly, "Attention, Bloggers," *The Wall Street Journal,* March 17, 2008, p. R5.

48 Vara, Vauhini, "Ebay to Broker Radio Ad Time," *The Wall Street Journal,* June 6, 2007, p. A13.

49 Silva-Risso, Jorge, et al., "Chrysler and J.D. Powers: Pioneering Scientific Price Customization in the Automotive Industry," *Interfaces,* January 2008, p. 26.

50 Aeppel, Timothy, "Seeking Perfect Prices, CEO Tears Up the Rules," *The Wall Street Journal,* March 27, 2007, p. A1.

51 Melymuka, Kathleen, "Just-In-Time Talent," *Computerworld,* March 24, 2008, p. 34.

52 Needleman, Sarah, "Demand Rises for Talent-Management Software," *The Wall Street Journal,* January 15, 2008, p. B8.

53 Songini, Marc, "BI Helping Retailers Control Inventory," *Computerworld,* January 22, 2007, p. 14.

54 Howson, Cindi, "The Road to Pervasive BI," *Information Week,* February 25, 2008, p. 39.

55 Hayes Weier, Mary, "Dear Customer: Please Don't Leave," *Information Week,* June 18, 2007, p. 49.

56 Kolbasuk McGee, Marianne, "The Useless Hunt for Data," *Information Week,* January 8, 2007, p. 19.

57 Delaney, Devin and Steel, Emily, "Firm Mines Offline Data to Target Online Ads," *The Wall Street Journal,* October 17, 2007, p. B1.

58 Worthen, Ben, "'Mashups' Sew Data Together," *The Wall Street Journal,* July 31, 2007, p. B4.

59 LoanPerformance Web site, *www.loanperformance.com,* accessed November 15, 2007.

60 Martin, Richard, "Business at Light Speed," *Information Week,* April 23, 2007, p. 42.

61 Retchless, Todd, et al., "A Group Decision-Making Application of the Analytic Hierarchy Process," *Interfaces,* April 2007, p. 163.

62 Shepherd, Lauren, "MySpace Becoming a Business Launch Pad," *Rocky Mountain News,* June 18, 2007, p. 4.

63 Green, Heather, "Now, Social Networking Fridays," *Business Week,* November 12, 2007, p. 24.

64 Gomes, Lee, "More Firms Create Own Social Networks," *The Wall Street Journal,* February 19, 2008, p. B3.

65 Green, Heather, "The Water Cooler Is Now on the Web," *Business Week,* October 1, 2007, p. 78.

66 Lehman, Paula, "Social Networks That Break a Sweat," *Business Week,* February 4, 2008, p. 66.

67 Hoover, Nicholas, "Social Experiment," *Information Week,* September 24, 2007, p. 40.

68 Cowen, Tyler, "In Favor of Face Time," *Forbes,* October 1, 2007, p. 30.

69 Schlender, Brent, "And What Does Mr. Gates Think?" *Fortune,* November 1, 2007, p. 102.

70 Bulkeley, William, "IBM Plans to Join Social-Networking Field," *The Wall Street Journal,* January 22, 2007, p. B5.

71 Hoover, Nicholas, "Lotus' Leap," *Information Week,* January 29, 2007, p. 29.

72 Sandberg, Jared, "Shared Calendars Mean Never Getting to Fib, I'm Booked," *The Wall Street Journal,* June 19, 2007, p. B1.

73 Babcock, Charles, "Microsoft Gets BI Focused," *Information Week,* September 24, 2007, p. 28.

74 Spanbauer, Scott, "Comparing Collaborative Web Services," *PC World,* January 2007, p. 61.

75 Dejean, David, "Pick Your Presence Tool," *Information Week,* May 21, 2007, p. 51.

76 Vascellaro, Jessica, "Social Networking Goes Professional," *The Wall Street Journal,* August 28, 2007, p. D1.

77 Havenstein, Heather, "Advocates Overcoming IT Resistance to Web 2.0," *Computerworld,* October 1, 2007, p. 14.

CHAPTER · 7 ·

Knowledge Management and Specialized Information Systems

PRINCIPLES	LEARNING OBJECTIVES
• Knowledge management allows organizations to share knowledge and experience among their managers and employees.	• Discuss the differences among data, information, and knowledge. • Describe the role of the chief knowledge officer (CKO). • List some of the tools and techniques used in knowledge management.
• Artificial intelligence systems form a broad and diverse set of systems that can replicate human decision making for certain types of well-defined problems.	• Define the term *artificial intelligence* and state the objective of developing artificial intelligence systems. • List the characteristics of intelligent behavior and compare the performance of natural and artificial intelligence systems for each of these characteristics. • Identify the major components of the artificial intelligence field and provide one example of each type of system.
• Expert systems can enable a novice to perform at the level of an expert but must be developed and maintained very carefully.	• List the characteristics and basic components of expert systems. • Identify at least three factors to consider in evaluating the development of an expert system. • Outline and briefly explain the steps for developing an expert system. • Identify the benefits associated with the use of expert systems.
• Virtual reality systems can reshape the interface between people and information technology by offering new ways to communicate information, visualize processes, and express ideas creatively.	• Define the term *virtual reality* and provide three examples of virtual reality applications.
• Specialized systems can help organizations and individuals achieve their goals.	• Discuss examples of specialized systems for organizational and individual use.

Information Systems in the Global Economy
Ericsson, Sweden

Telecom Giant Uses Knowledge Management and Expert Systems

A corporation's success depends on the knowledge it maintains and uses through its employees, executives, board of directors, and information systems. Building that body of corporate knowledge is a primary goal of most businesses—a goal that requires a concerted effort. If left untended, corporate knowledge pools in certain people, who gain it through time and experience, and is no longer accessible when those people leave the organization. Information systems, such as knowledge management systems and expert systems, can store knowledge gained by those within a corporation over time. For expert systems, artificial intelligence (AI) techniques can automate expert reasoning and activities. Consider how an expert system allows telecom giant Ericsson to monitor telecom networks in a manner that would overwhelm a human expert.

Established in 1876 and headquartered in Stockholm, Sweden, Ericsson is a world leader in telecommunications services, networking, multimedia solutions, and core technologies for mobile handsets. In fact, ten of the world's largest mobile phone operators use Ericsson technologies and 40 percent of all mobile traffic travels through Ericsson systems. Ericsson's 50 percent share of Sony Ericsson Mobile Communications has further expanded its power in the market.

Ericsson's busy telecom networks, as with all complex networks, encounter many obstacles over the course of the day. Lines may go down, some percentage of the thousands of routers, junction boxes, transmitters, and other telecom hardware devices used may fail, and systems software may experience bugs. Each time a problem, or fault, occurs, network administrators are alerted through dashboards that test and monitor the network.

Ericsson network administrators work in a stressful environment, where streams of alarms indicating network faults need to be evaluated and acted upon, sometimes immediately. Therein lays the challenge. It is difficult, if not impossible, for network administrators to determine which alarms are important and need immediate attention and which can wait. Given time, the administrators could read, interpret, and analyze each alarm to determine its level of importance, but with hundreds of thousands of alarms flowing in each day, there isn't even enough time to read one before ten more have arrived. This type of problem is well suited for an expert system.

Ericsson worked with an expert system company to assist with network fault management. The expert system company used real-time rule technology to automate the process of monitoring network alarms and determining which alarms needed immediate attention and which could wait. The human experts at Ericsson assisted the company in designing the software by sharing the secrets to the complicated process of alarm interpretation. With more than 50 types of equipment sending more than 500,000 alarms each day, the expert knowledge that was collected needed to be complete and executed quickly.

Today the expert system for managing network faults is deployed in 500 Ericsson systems in over 100 countries. The stream of alarms faced by human experts has been slowed to a tolerable amount of only the most important problems that require immediate intervention. The performance and quality of Ericsson networks has improved, its administrators can focus on priorities, and important corporate knowledge has been digitized to benefit the company over generations of employees.

As you read this chapter, consider the following:

- What steps can a business take to retain corporate knowledge within the business?
- How does computer intelligence compare to human intelligence?
- How can people and businesses make the best use of artificial intelligence and other specialized systems?

Why Learn About Knowledge Management and Specialized Information Systems?

Knowledge management and specialized information systems are used in almost every industry. If you are a manager, you might use a knowledge management system to support decisive action to help you correct a problem. If you are a production manager at an automotive company, you might oversee robots that attach windshields to cars or paint body panels. As a young stock trader, you might use a special system called a *neural network* to uncover patterns and make millions of dollars trading stocks and stock options. As a marketing manager for a PC manufacturer, you might use virtual reality on a Web site to show customers your latest laptop and desktop computers. If you are in the military, you might use computer simulation as a training tool to prepare you for combat. In a petroleum company, you might use an expert system to determine where to drill for oil and gas. You will see many additional examples of using these specialized information systems throughout this chapter. Learning about these systems will help you discover new ways to use information systems in your day-to-day work.

Like other aspects of an information system, the overall goal of knowledge management and the specialized systems discussed in this chapter is to help people and organizations achieve their goals. In some cases, knowledge management and these specialized systems can help an organization achieve a long-term, strategic advantage. In this chapter, we explore knowledge management, artificial intelligence, and many other specialized information systems, including expert systems, robotics, vision systems, natural language processing, learning systems, neural networks, genetic algorithms, intelligent agents, and virtual reality.

KNOWLEDGE MANAGEMENT SYSTEMS

Chapter 1 defines and discusses data, information, and knowledge. Recall that *data* consists of raw facts, such as an employee number, number of hours worked in a week, inventory part numbers, or sales orders. A list of the quantity available for all items in inventory is an example of data. When these facts are organized or arranged in a meaningful manner, they become information. *Information* is a collection of facts organized so that they have additional value beyond the value of the facts themselves. An exception report of inventory items that might be out of stock in a week because of high demand is an example of information. *Knowledge* is the awareness and understanding of a set of information and the ways that information can be made useful to support a specific task or reach a decision. Knowing the procedures for ordering more inventory to avoid running out is an example of knowledge. In a sense, information tells you what has to be done (low inventory levels for some items), while knowledge tells you how to do it (make two important phone calls to the right people to get the needed inventory shipped overnight). See Figure 7.1.

Figure 7.1

The Differences Among Data, Information, and Knowledge

Data	There are 20 PCs in stock at the retail store.
Information	The store will run out of inventory in a week unless more is ordered today.
Knowledge	Call 800-555-2222 to order more inventory.

According to Carol Csanda, director of knowledge management at State Farm Insurance, "We feel at State Farm that everybody's role is in some way about managing and transferring knowledge."[1] A *knowledge management system (KMS)* is an organized collection of people, procedures, software, databases, and devices used to create, store, share, and use the organization's knowledge and experience.[2] KMSs cover a wide range of systems from software that contains some KMS components to dedicated systems designed specifically to capture, store, and use knowledge.[3] According to a software engineer for the Tata Group, a large Indian company, "We have learned that every time staff leave our companies, their knowledge has gone with them and our knowledge is gradually reduced. Hence, we put knowledge management in the group's strategy to ensure our staff at all levels are able to maintain knowledge at all time."[4] Using KMS, the steel, software, coffee, chemicals, watches, and power company was able to win the Most Admired Knowledge Enterprise (MAKE) award in Asia. Other organizations also face the loss of knowledge when workers leave or retire. According to a spokesperson for the Aerospace and Defense (A&D) organization, "Knowledge management represents a major issue for A&D organizations, as they stand to lose decades of accumulated institutional knowledge and expertise each time an employee retires. This is particularly true for program managers and engineers, whose retirement rate nearly doubled between 2004 and 2005."[5]

Overview of Knowledge Management Systems

Like the other systems discussed throughout the book, including information and decision support systems, knowledge management systems attempt to help organizations achieve their goals.[6] For businesses, this usually means increasing profits or reducing costs. For nonprofit organizations, it can mean providing better customer service or providing special needs to people and groups. Many types of firms use KMSs to increase profits or reduce costs. Pratt & Whitney, for example, uses knowledge management systems to help it deliver information and knowledge about its jet engine parts to airlines, including Delta and United.[7] The World Bank uses knowledge management to obtain, share, and use knowledge to help fight worldwide poverty and disease.[8] According to a survey of CEOs, firms that use KMSs are more likely to innovate and perform better.

Pratt & Whitney uses knowledge management systems to help it deliver information and knowledge about its jet engine parts to airlines, including Delta and United.

(Source: Courtesy of Pratt & Whitney, a United Technologies Company.)

A KMS can involve different types of knowledge.[9] *Explicit knowledge* is objective and can be measured and documented in reports, papers, and rules. For example, knowing the best road to take to minimize drive time from home to the office when a major highway is closed due to an accident is explicit knowledge. It can be documented in a report or a rule, as in "If I-70 is closed, take Highway 6 to town and the office." *Tacit knowledge*, on the other hand, is hard to measure and document and typically is not objective or formalized. Knowing the best way to negotiate with a foreign government about nuclear disarmament or a volatile hostage situation often requires a lifetime of experience and a high level of skill. These are examples of tacit knowledge. It is difficult to write a detailed report or a set of rules that would always work in every hostage situation. Many organizations actively attempt to convert tacit knowledge to explicit knowledge to make the knowledge easier to measure, document, and share with others.

Data and Knowledge Management Workers and Communities of Practice

The personnel involved in a KMS include data workers and knowledge workers. Secretaries, administrative assistants, bookkeepers, and similar data-entry personnel are often called *data workers*. As mentioned in Chapter 1, *knowledge workers* are people who create, use, and disseminate knowledge. They are usually professionals in science, engineering, or business, and work in offices and belong to professional organizations. Other examples of knowledge workers include writers, researchers, educators, and corporate designers.[10] See Figure 7.2.

Figure 7.2

Knowledge Workers

Knowledge workers are people who create, use, and disseminate knowledge, including professionals in science, engineering, business, and other areas.

(Source: Eliza Snow/iStockphoto.)

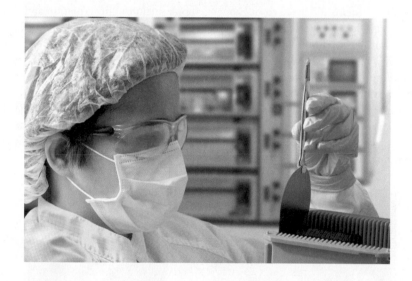

chief knowledge officer (CKO)

A top-level executive who helps the organization use a KMS to create, store, and use knowledge to achieve organizational goals.

The **chief knowledge officer (CKO)** is a top-level executive who helps the organization work with a KMS to create, store, and use knowledge to achieve organizational goals.[11] The CKO is responsible for the organization's KMS, and typically works with other executives and vice presidents, including the chief executive officer (CEO), chief financial officer (CFO), and others. According to Jay Kostrzewa, assistant vice president of knowledge management at CNA, "My role as leader of the knowledge management area is to make certain the company has the right tools, the right information, and the right processes in place to share information."[12] CNA is a Chicago financial services company.

Some organizations and professions use *communities of practice (COP)* to create, store, and share knowledge. A COP is a group of people dedicated to a common discipline or practice, such as open-source software, auditing, medicine, or engineering. A group of oceanographers investigating climate change or a team of medical researchers looking for new ways to treat lung cancer are examples of COPs. COPs excel at obtaining, storing, sharing, and using knowledge. A COP from the International Conference on Knowledge Management in Nuclear Facilities investigates the use of knowledge management systems in the development and control of nuclear facilities.[13] According to Yuri Sokolov, IAEA deputy director general and head of the Department of Nuclear Energy, "All applications of nuclear technology are based on nuclear knowledge, so managing, preserving, and building on the knowledge we have accumulated is both wise [in the] near term and an important intergenerational responsibility."

Obtaining, Storing, Sharing, and Using Knowledge

Obtaining, storing, sharing, and using knowledge is the key to any KMS.[14] MWH Global, located in Colorado, uses a KMS to create, disseminate, and use knowledge specializing in environmental engineering, construction, and management activities worldwide.[15] The company has about 7,000 employees and 170 offices around the world. A KMS can help an organization increase profits or achieve its goals, but obtaining, storing, sharing, and using knowledge can be difficult.[16] In one survey, almost 60 percent of the respondents indicated that they couldn't find the information and knowledge that they need to do their jobs every day. Using a KMS often leads to additional knowledge creation, storage, sharing, and usage. According to Richard Cantor, knowledge management team manager for Chubb Commercial Insurance, "At Chubb, we're focusing on using our intranet as the vehicle that delivers shared knowledge. Many of our knowledge management efforts are packaged within that veil."[17] Business professors often conduct research in marketing strategies, management practices, corporate and individual investments and finance, effective accounting and auditing practices, and much more. Drug companies and medical researchers invest billions of dollars in creating knowledge on cures for diseases. Although knowledge workers can act alone, they often work in teams to create or obtain knowledge. See Figure 7.3.

Figure 7.3

Knowledge Management System

Obtaining, storing, sharing, and using knowledge is the key to any KMS.

After knowledge is created, it is often stored in a *knowledge repository* that includes documents, reports, files, and databases. The knowledge repository can be located both inside the organization and outside. Some types of software can store and share knowledge contained in documents and reports. Adobe Acrobat PDF files, for example, allow you to store corporate reports, tax returns, and other documents and send them to others over the Internet. This publisher and the authors of this book used PDF files to store, share, and collaborate on each chapter. Traditional databases, data warehouses, and data marts discussed in Chapter 3 often store the organization's knowledge. Specialized knowledge bases in expert systems, discussed later in this chapter, can also be used.

Because knowledge workers often work in groups or teams, they can use collaborative work software and group support systems discussed in Chapter 6 to share knowledge, such as groupware, meeting software, and collaboration tools.[18] Intranets and password-protected Internet sites also provide ways to share knowledge. The social services department of the Surrey County Council in the United Kingdom, for example, used an intranet to help it create and manipulate knowledge. Because knowledge can be critical in maintaining a competitive advantage, businesses should be careful in how they share knowledge. Although they want important decision makers inside and outside the organization to have complete and easy access to knowledge, they also need to protect knowledge from competitors, hackers, and others who shouldn't obtain the organization's knowledge. As a result, many businesses use patents, copyrights, trade secrets, Internet firewalls, and other measures to keep prying eyes from seeing important knowledge that is often expensive and hard to create.

Using a knowledge management system begins with locating the organization's knowledge. This is often done using a *knowledge map* or directory that points the knowledge worker to the needed knowledge. Drug companies have sophisticated knowledge maps that include database and file systems to allow scientists and drug researchers to locate previous medical studies. The Army Defense Ammunition Center has signed an $8 million contract with SI International to provide it with knowledge management tools to help evaluate its training courses.[19] Medical researchers, university professors, and even textbook authors use Lexis-Nexis to locate important knowledge. Corporations often use the Internet or corporate Web portals to help their knowledge workers find knowledge stored in documents and reports.

Technology to Support Knowledge Management

KMSs use a number of tools discussed throughout the book. In Chapter 1, for example, we explored the importance of *organizational learning* and *organizational change*. An effective KMS is based on learning new knowledge and changing procedures and approaches as a result. A manufacturing company, for example, might learn new ways to program robots on the factory floor to improve accuracy and reduce defective parts. The new knowledge will likely cause the manufacturing company to change how it programs and uses its robots. In Chapter 3, we investigated the use of *data mining* and *business intelligence*. These powerful tools can be important in capturing and using knowledge. Enterprise resource planning tools, such as SAP, include knowledge management features. In Chapter 6, we showed how *groupware* could improve group decision-making and collaboration. Groupware can also be used to help capture, store, and use knowledge. Of course, hardware, software, databases, telecommunications, and the Internet, discussed in Part 2, are important technologies used to support most knowledge management systems.

Hundreds of organizations provide specific KM products and services. See Figure 7.4. In addition, researchers at colleges and universities have developed tools and technologies to support knowledge management. The University of South Carolina, for example, has joined with Collexis to develop and deliver new knowledge management software, based on Collexis's Knowledge Discovery Platform.[20] It has been estimated that American companies will pay about $70 billion on knowledge management technology in 2007.[21] This amount is expected to increase by more than 15 percent in 2008. Companies such as IBM have many knowledge management tools in a variety of products, including Lotus Notes and Domino. Lotus Notes is a collection of software products that help people work together to create, share, and store important knowledge and business documents. Its knowledge management features include domain search, content mapping, and Lotus Sametime. Domain search allows people to perform sophisticated searches for knowledge in Domino databases using a single, simple query. Content mapping organizes knowledge by categories, like a table of contents for a book. Lotus Sametime helps people communicate, collaborate, and share ideas in real time. Lotus Domino Document Manager, formerly called Lotus Domino, helps people and organizations store, organize, and retrieve documents. The software can be used to write, review, archive, and publish documents throughout the organization.

Figure 7.4

Knowledge Management Technology

Lotus Notes helps people communicate, collaborate, and streamline their work.

(Source: Courtesy of IBM Corporation)

Microsoft offers a number of knowledge management tools, including Digital Dashboard, which is based on the Microsoft Office suite. Digital Dashboard integrates information from different sources, including personal, group, enterprise, and external information and documents. Other tools from Microsoft include Web Store Technology, which uses wireless technology to deliver knowledge to any location at any time; Access Workflow Designer, which helps database developers create effective systems to process transactions and keep work flowing through the organization; and related products. In addition to these tools, several artificial intelligence and special-purpose technologies and tools, discussed next, can be used in a KMS.

AN OVERVIEW OF ARTIFICIAL INTELLIGENCE

At a Dartmouth College conference in 1956, John McCarthy proposed the use of the term **artificial intelligence (AI)** to describe computers with the ability to mimic or duplicate the functions of the human brain. For example, advances in AI have led to systems that work like the human brain to recognize complex patterns.

Many AI pioneers attended this first conference; a few predicted that computers would be as "smart" as people by the 1960s. The prediction has not yet been realized, but the benefits of artificial intelligence in business and research can be seen today, and research continues.

artificial intelligence (AI)
The ability of computers to mimic or duplicate the functions of the human brain.

Science fiction movies give us a glimpse of the future, but many practical applications of artificial intelligence exist today, among them medical diagnostics and development of computer systems.

(Source: WALL-E, 2008. © Walt Disney Studios Motion Pictures/ Courtesy Everett Collection.)

Artificial Intelligence in Perspective

Artificial intelligence systems include the people, procedures, hardware, software, data, and knowledge needed to develop computer systems and machines that demonstrate characteristics of intelligence.[22] Artificial intelligence can be used by most industries and applications. According to University of California-Santa Cruz professor Michael Mateas, "As graphics improvements top out, artificial intelligence will [drive] game innovation." Researchers, scientists, and experts on how human beings think are often involved in developing these systems.

artificial intelligence systems
People, procedures, hardware, software, data, and knowledge needed to develop computer systems and machines that demonstrate the characteristics of intelligence.

The Nature of Intelligence

From the early AI pioneering stage, the research emphasis has been on developing machines with intelligent behavior.[23] In a book called *The Singularity Is Near* and articles by and about him, Ray Kurzweil predicts computers will have humanlike intelligence in 20 years.[24] The author also foresees that by 2045 human and machine intelligence might merge. According to Kurzweil, "The Singularity Institute for Artificial Intelligence (SIAI) is playing a critical role in advancing humanity's understanding of the profound promise and peril of strong AI."[25] Machine intelligence, however, is hard to achieve.

The *Turing Test* attempts to determine whether the responses from a computer with intelligent behavior are indistinguishable from responses from a human being. No computer has passed the Turing Test, developed by Alan Turing, a British mathematician. The Loebner Prize offers money and a gold medal for anyone developing a computer that can pass the

intelligent behavior
The ability to learn from experiences and apply knowledge acquired from experience, handle complex situations, solve problems when important information is missing, determine what is important, react quickly and correctly to a new situation, understand visual images, process and manipulate symbols, be creative and imaginative, and use heuristics.

Turing Test (see *www.loebner.net*). Some of the specific characteristics of **intelligent behavior** include the ability to do the following:

- **Learn from experience and apply the knowledge acquired from experience.** Learning from past situations and events is a key component of intelligent behavior and is a natural ability of humans, who learn by trial and error. This ability, however, must be carefully programmed into a computer system. Today, researchers are developing systems that can learn from experience. For instance, computerized AI chess software can learn to improve while playing human competitors. In one match, Garry Kasparov competed against a personal computer with AI software developed in Israel, called Deep Junior. This match was a 3–3 tie, but Kasparov picked up something the machine would have no interest in—$700,000. The 20 questions (20q) Web site, *www.20q.net*, is another example of a system that learns.[26] The Web site is an artificial intelligence game that learns as people play.

20Q is an online game where users play the popular game, *Twenty Questions*, against an artificial intelligence foe.

(Source: *www.20q.net*.)

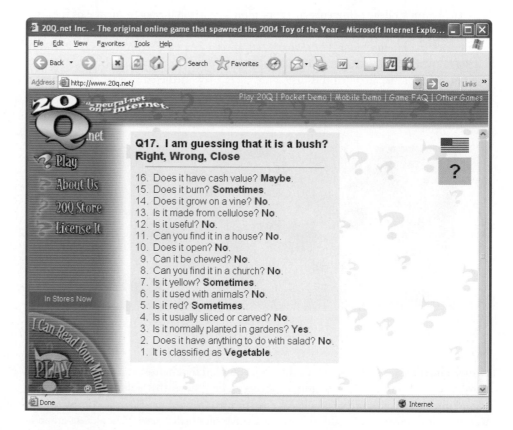

- **Handle complex situations.** People are often involved in complex situations. World leaders face difficult political decisions regarding terrorism, conflict, global economic conditions, hunger, and poverty. In a business setting, top-level managers and executives must handle a complex market, challenging competitors, intricate government regulations, and a demanding workforce. Even human experts make mistakes in dealing with these situations. Developing computer systems that can handle perplexing situations requires careful planning and elaborate computer programming.
- **Solve problems when important information is missing.** The essence of decision making is dealing with uncertainty. Often, decisions must be made with little information or inaccurate information because obtaining complete information is too costly or impossible. Today, AI systems can make important calculations, comparisons, and decisions even when information is missing.
- **Determine what is important.** Knowing what is truly important is the mark of a good decision maker. Developing programs and approaches to allow computer systems and machines to identify important information is not a simple task.

- **React quickly and correctly to a new situation.** A small child, for example, can look over a ledge or a drop-off and know not to venture too close. The child reacts quickly and correctly to a new situation. Computers, on the other hand, do not have this ability without complex programming.
- **Understand visual images.** Interpreting visual images can be extremely difficult, even for sophisticated computers. Moving through a room of chairs, tables, and other objects can be trivial for people but extremely complex for machines, robots, and computers. Such machines require an extension of understanding visual images, called a **perceptive system**. Having a perceptive system allows a machine to approximate the way a person sees, hears, and feels objects. Military robots, for example, use cameras and perceptive systems to conduct reconnaissance missions to detect enemy weapons and soldiers. Detecting and destroying them can save lives.

perceptive system
A system that approximates the way a person sees, hears, and feels objects.

- **Process and manipulate symbols.** People see, manipulate, and process symbols every day. Visual images provide a constant stream of information to our brains. By contrast, computers have difficulty handling symbolic processing and reasoning. Although computers excel at numerical calculations, they aren't as good at dealing with symbols and three-dimensional objects. Recent developments in machine-vision hardware and software, however, allow some computers to process and manipulate symbols on a limited basis.
- **Be creative and imaginative.** Throughout history, some people have turned difficult situations into advantages by being creative and imaginative. For instance, when shipped defective mints with holes in the middle, an enterprising entrepreneur decided to market these new mints as LifeSavers instead of returning them to the manufacturer. Ice cream cones were invented at the St. Louis World's Fair when an imaginative store owner decided to wrap ice cream with a waffle from his grill for portability. Developing new and exciting products and services from an existing (perhaps negative) situation is a human characteristic. Few computers can be imaginative or creative in this way, although software has been developed to enable a computer to write short stories.
- **Use heuristics.** For some decisions, people use heuristics (rules of thumb arising from experience) or even guesses. In searching for a job, you might rank the companies you are considering according to profits per employee. Today, some computer systems, given the right programs, obtain good solutions that use approximations instead of trying to search for an optimal solution, which would be technically difficult or too time consuming.

This list of traits only partially defines intelligence. Unlike the terminology used in virtually every other field of IS research, in which the objectives can be clearly defined, the term *intelligence* is a formidable stumbling block. One of the problems in AI is arriving at a working definition of real intelligence against which to compare the performance of an AI system.

The Difference Between Natural and Artificial Intelligence

Since the term *artificial intelligence* was defined in the 1950s, experts have disagreed about the difference between natural and artificial intelligence. Can computers be programmed to have common sense? Profound differences separate natural from artificial intelligence, but they are declining in number (see Table 7.1). One of the driving forces behind AI research is an attempt to understand how people actually reason and think. Creating machines that can reason is possible only when we truly understand our own processes for doing so.

The Major Branches of Artificial Intelligence

AI is a broad field that includes several specialty areas, such as expert systems, robotics, vision systems, natural language processing, learning systems, and neural networks (see Figure 7.5). Many of these areas are related; advances in one can occur simultaneously with or result in advances in others.

Table 7.1

A Comparison of Natural and Artificial Intelligence

Ability to	Natural Intelligence (Human)		Artificial Intelligence (Machine)	
	Low	**High**	**Low**	**High**
Use sensors (eyes, ears, touch, smell)		√	√	
Be creative and imaginative		√	√	
Learn from experience		√	√	
Adapt to new situations		√	√	
Afford the cost of acquiring intelligence		√	√	
Acquire a large amount of external information		√		√
Use a variety of information sources		√		√
Make complex calculations	√			√
Transfer information	√			√
Make a series of calculations rapidly and accurately	√			√

Figure 7.5

A Conceptual Model of Artificial Intelligence

Expert Systems

expert system
Hardware and software that stores knowledge and makes inferences, similar to a human expert.

An **expert system** consists of hardware and software that stores knowledge and makes inferences, similar to those of a human expert.[27] Because of their many business applications, expert systems are discussed in more detail in the next several sections of the chapter.

Robotics

robotics
Mechanical or computer devices that perform tasks requiring a high degree of precision or that are tedious or hazardous for humans.

Robotics involves developing mechanical or computer devices that can paint cars, make precision welds, and perform other tasks that require a high degree of precision or are tedious or hazardous for human beings.[28] The word "robot" comes from a play by Karel Capek in the 1920s, when he used the word "robota" to describe factory machines that do drudgery work and revolt.[29] The use of robots has expanded and is likely to increase in the future.

According to Takeo Kande, "Someday, robots will do more than vacuum your floors. They'll train you and advise you—and maybe even help out with the cooking."[30] Some robots are mechanical devices that don't use the AI features discussed in this chapter. Others are sophisticated systems that use one or more AI features or characteristics, such as vision systems, learning systems, or neural networks discussed later in the chapter. For many businesses, robots are used to do the three Ds—dull, dirty, and dangerous jobs. Manufacturers use robots to assemble and paint products.[31] The NASA shuttle crash of 2003 has led some people to recommend using robots instead of people to explore space and perform scientific research. Some robots, such as the ER series by Intelitek (*www.intelitek.com*), can be used for training or entertainment. Contemporary robotics combine both high-precision machine capabilities and sophisticated controlling software. The controlling software in robots is what is most important in terms of AI.

The field of robotics has many applications, and research into these unique devices continues. The following are a few examples:

- The Robot Learning Laboratory, part of the computer science department and the Robotics Institute at Carnegie Mellon University (*www.ri.cmu.edu*), conducts research into the development and use of robotics.[32]
- IRobot (*www.irobot.com*) is a company that builds a number of robots, including the Roomba Floorvac for cleaning floors and the PackBot, an unmanned vehicle used to assist and protect soldiers.
- Robots are used in a variety of ways in medicine. The Porter Adventist Hospital (*www.porterhospital.org*) in Denver, Colorado, uses a $1.2 million Da Vinci Surgical System to perform surgery on prostate cancer patients.[33] The robot has multiple arms that hold surgical tools. According to one doctor at Porter, "The biggest advantage is it improves recovery time. Instead of having an eight-inch incision, the patient has a 'band-aid' incision. It's much quicker." The Heart Lander is a very small robot that is inserted below the rib cage and used to perform delicate heart surgery.[34] Cameron Riviere at the Carnegie Mellon Robotics Institute (*www.ri.cmu.edu*) developed the robot along with help from John Hopkins University.
- DARPA (The Defense Advanced Research Project Agency) sponsors the DARPA Grand Challenge (*www.darpagrandchallenge.com*), a 132-mile race over rugged terrain for computer-controlled cars. The agency also sponsors other races and challenges.[35]
- The Hybrid Assisted Limb (HAL) lab is developing a robotic suit to help paraplegics and stroke victims move and perform basic functions.[36] The suit helps with lifting heavy objects, walking long distances, or performing other basic movements that can't be done otherwise. HAL was also the name of an artificial-intelligence computer in the classic movie *2001: A Space Odyssey.* The letters in HAL are one letter up from the letters in IBM.
- In the military, robots are moving beyond movie plots to become real weapons.[37] The Air Force is developing a smart robotic jet fighter. Often called *unmanned combat air vehicles (UCAVs)*, these robotic war machines, such as the X-45A, will be able to identify and destroy targets without human pilots. UCAVs send pictures and information to a central command center and can be directed to strike military targets. These machines extend the current Predator and Global Hawk technologies the military used in Afghanistan after the September 11 terrorist attacks and Iraq. Big Dog, made by Boston Dynamics (*www.bostondynamics.com*), is a robot that can carry up to 200 pounds of military gear in field conditions.

Although most of today's robots are limited in their capabilities, future robots will find wider applications in banks, restaurants, homes, doctors' offices, and hazardous working environments such as nuclear stations. The Repliee Q1 and Q2 robots from Japan are ultra-humanlike robots or androids that can blink, gesture, speak, and even appear to breathe (*www.ed.ams.eng.osaka-u.ac.jp/development/Android_ReplieeQ2_e.html*). See Figure 7.6. Microrobotics, also called *micro-electro-mechanical systems (MEMS)*, are also being developed (*www.memsnet.org/mems/what-is.html*). MEMS can be used in a person's blood to monitor the body, and for other purposes in air bags, cell phones, refrigerators, and more.

Big Dog, manufactured by Boston Dynamics, is a robot that can carry up to 200 pounds of military gear in field conditions.

(Source: Courtesy of Boston Dynamics.)

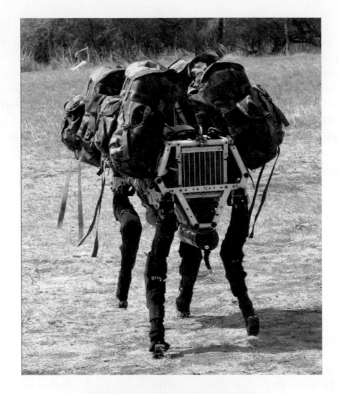

Figure 7.6

The Repliee Q2 Robot from Japan

(Source: AP Photo/Katsumi Kasahara.)

Vision Systems

vision systems
The hardware and software that permit computers to capture, store, and manipulate visual images.

Another area of AI involves vision systems. **Vision systems** include hardware and software that permit computers to capture, store, and manipulate visual images. The U.S. Justice Department uses vision systems to perform fingerprint analysis with almost the same level of precision as human experts. The speed with which the system can search a huge database of fingerprints has brought quick resolution to many long-standing mysteries. Vision systems are also effective at identifying people based on facial features. In another application, a California wine bottle manufacturer uses a computerized vision system to inspect wine bottles for flaws. The company produces about 2 million wine bottles per day, and the vision system saves the bottle producer both time and money. According to Takeo Kanade, Professor of Computer Science and Robotics at Carnegie Mellon University, "The trend toward computer vision is clear, and it will accelerate. In ten years, I wouldn't be surprised to see computers recognizing certain levels of emotions, expressions, gestures, and behaviors, all through vision."[38]

Natural Language Processing and Voice Recognition

As discussed in Chapter 2, **natural language processing** allows a computer to understand and react to statements and commands made in a "natural" language, such as English. Google, for example, has a service called Google Voice Local Search that allows you to dial a toll-free number and search for local businesses using voice commands and statements.[39] Restoration Hardware (*www.restorationhardware.com*) has developed a Web site that uses natural language processing to allow its customers to quickly find what they want. The natural language processing system corrects spelling mistakes, converts abbreviations into words and commands, and allows people to ask questions in English.

In some cases, voice recognition is used with natural language processing. *Voice recognition* involves converting sound waves into words.[40] After converting sounds into words, natural language processing systems react to the words or commands by performing a variety of tasks. Brokerage services are a perfect fit for voice-recognition and natural language processing technology to replace the existing "press 1 to buy or sell a stock" touchpad telephone menu system. People buying and selling stock use a vocabulary too varied for easy access through menus and touchpads, but still small enough for software to process in real time. Several brokerages—including Charles Schwab & Company, Fidelity Investments, DLJdirect, and TD Waterhouse Group—offer these services. These systems use voice recognition and natural language processing to let customers access retirement accounts, check balances, and find stock quotes. Eventually, the technology will allow people to make transactions using voice commands over the phone and to use search engines to have their questions answered through the brokerage firm's call center. Using voice recognition to convert recordings into text is also possible.[41] Some companies claim that voice-recognition and natural language processing software is so good that customers forget they are talking to a computer and start discussing the weather or sports scores.

Learning Systems

Another part of AI deals with **learning systems**, a combination of software and hardware that allows a computer to change how it functions or reacts to situations based on feedback it receives.[42] For example, some computerized games have learning abilities. If the computer does not win a game, it remembers not to make the same moves under the same conditions again. DARPA is investing about $10 million into a learning system called Bootstrapped Learning that will help military computers learn from human instructors.[43] If successful, the project could help the development and control of unmanned aircraft. The Center for Automated Learning and Discovery at Carnegie Mellon University (*www.cald.cs.cmu.edu*) is experimenting with two learning software packages that help each other learn. The hope is that two learning software packages that cooperate are better than separate learning packages.

Learning systems software requires feedback on the results of actions or decisions. At a minimum, the feedback needs to indicate whether the results are desirable (winning a game) or undesirable (losing a game). The feedback is then used to alter what the system will do in the future.

natural language processing
Processing that allows the computer to understand and react to statements and commands made in a "natural" language, such as English.

Dragon Systems' Naturally Speaking 9 Essentials uses continuous voice recognition, or natural speech, allowing the user to speak to the computer at a normal pace without pausing between words. The spoken words are transcribed immediately onto the computer screen.

(Source: Courtesy of Nuance Communications, Inc.)

learning systems
A combination of software and hardware that allows the computer to change how it functions or reacts to situations based on feedback it receives.

Neural Networks

neural network
A computer system that can simulate the functioning of a human brain.

An increasingly important aspect of AI involves neural networks, also called neural nets. A **neural network** is a computer system that can act like or simulate the functioning of a human brain.[44] The systems use massively parallel processors in an architecture that is based on the human brain's own mesh-like structure. In addition, neural network software simulates a neural network using standard computers. Neural networks can process many pieces of data at the same time and learn to recognize patterns. A chemical company, for example, can use neural network software to analyze a large amount of data to control chemical reactors. Neural network analysis has also helped some medical clinics diagnose cardiovascular disease.[45] Some oil and gas exploration companies use a program called the Rate of Penetration based on neural networks to monitor and control drilling operations.[46] The neural network program helps engineers slow or speed drilling operations to help increase drilling accuracy and reduce costs. Some of the specific abilities of neural networks include the following:

- Retrieving information even if some of the neural nodes fail
- Quickly modifying stored data as a result of new information
- Discovering relationships and trends in large databases
- Solving complex problems for which all the information is not present

A particular skill of neural nets is analyzing detailed trends.[47] Large amusement parks and banks use neural networks to determine staffing needs based on customer traffic—a task that requires precise analysis, down to the half-hour. Increasingly, businesses are firing up neural nets to help them navigate ever-thicker forests of data and make sense of a myriad of customer traits and buying habits. Computer Associates has developed Neugents (*www.neugents.com*), neural intelligence agents that "learn" patterns and behaviors and predict what will happen next. For example, Neugents can track the habits of insurance customers and predict which ones will not renew an automobile policy, for example. They can then suggest to an insurance agent what changes to make in the policy to persuade the consumer to renew it. The technology also can track individual users at e-commerce sites and their online preferences so that they don't have to enter the same information each time they log on—their purchasing history and other data is recalled each time they access a Web site.

AI Trilogy, available from the Ward Systems Group (*www.wardsystems.com*), is a neural network software program that can run on a standard PC. The software can make predictions with NeuroShell Predictor and classify information with NeuroShell Classifier. See Figure 7.7. The software package also contains GeneHunter, which uses a special type of algorithm called a genetic algorithm to get the best result from the neural network system. (Genetic algorithms are discussed later in this chapter.) Some pattern-recognition software uses neural networks to analyze hundreds of millions of bank, brokerage, and insurance accounts involving a trillion dollars to uncover money laundering and other suspicious money transfers.

Other Artificial Intelligence Applications

genetic algorithm
An approach to solving large, complex problems in which a number of related operations or models change and evolve until the best one emerges.

A few other artificial intelligence applications exist in addition to those just discussed. A **genetic algorithm**, also called a genetic program, is an approach to solving large, complex problems in which many repeated operations or models change and evolve until the best one emerges. The approach is based on the theory of evolution that requires (1) variation and (2) natural selection. The first step is to change or vary competing solutions to the problem. This can be done by changing the parts of a program or by combining different program segments into a new program, mimicking the evolution of species, in which the genetic makeup of a plant or animal mutates or changes over time. The second step is to select only the best models or algorithms, which continue to evolve. Programs or program segments that are not as good as others are discarded, similar to natural selection or "survival of the fittest," in which only the best species survive and continue to evolve. This process of variation and natural selection continues until the genetic algorithm yields the best possible solution to the original problem. For example, some investment firms use genetic algorithms to help select the best stocks or bonds. Genetic algorithms can help companies control inventory levels and get the best usage of warehouse space.[48] Genetic algorithms are also being used to monitor patient health.[49]

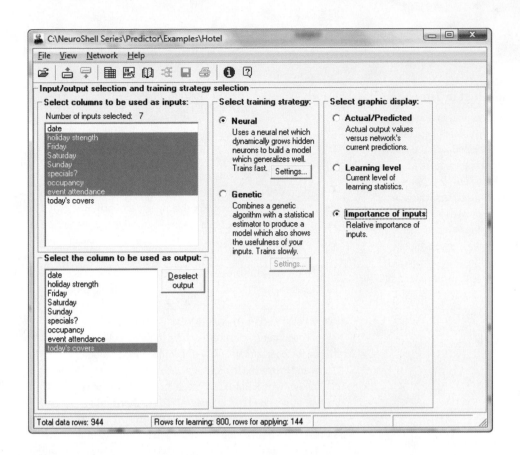

Figure 7.7

Neural Network Software

NeuroShell Predictor uses recognized forecasting methods to look for future trends in data.

(Source: Courtesy of Ward Systems Group, Inc.)

An **intelligent agent** (also called an *intelligent robot* or *bot*) consists of programs and a knowledge base used to perform a specific task for a person, a process, or another program.[50] Like a sports agent who searches for the best endorsement deals for a top athlete, an intelligent agent often searches to find the best price, schedule, or solution to a problem. The programs used by an intelligent agent can search large amounts of data as the knowledge base refines the search or accommodates user preferences. Often used to search the vast resources of the Internet, intelligent agents can help people find information on an important topic or the best price for a new digital camera. Intelligent agents can also be used to make travel arrangements, monitor incoming e-mail for viruses or junk mail, and coordinate meetings and schedules of busy executives. The U.S. Army uses intelligent agents to help in its recruiting efforts.[51] Called Sgt. Star, the intelligent agent personalizes responses to visitors and potential recruits to its Web site, *www.goarmy.com*.

intelligent agent
Programs and a knowledge base used to perform a specific task for a person, a process, or another program; also called *intelligent robot* or *bot*.

Some prosthetic limbs use AI to improve "virtual touch," which improves sensation and mobility. The Power Knee, for example, receives information from a sensor on the shoe of the sound leg to accurately mimic movement.

(Source: AP Photo/Dima Gavrysh.)

Providing Knowledge to Physicians Just in Time

Few professions are more complex and continuously changing and expanding than medicine. The quality of healthcare provided to a community depends on physicians being equipped with the latest medical knowledge on wide-ranging ailments and treatments. Information systems that feed this knowledge to physicians are the foundation on which life-and-death decisions are based. The responsibility of acquiring and managing the knowledge of the most up-to-date discoveries by the greatest minds in medicine is a daunting task, one that Partners Healthcare takes very seriously.

Partners HealthCare is an integrated healthcare system founded by Brigham and Women's Hospital and Massachusetts General Hospital. The system includes primary care and specialty physicians, community hospitals, academic medical centers, and other health-related entities. For years, Partners has invested heavily in medical knowledge management systems that provide physicians with information about the latest drugs and treatments for illnesses and diseases. In recent years, the amount of knowledge needed to make healthcare decisions has become so immense and changes so frequently that it has become unmanageable through traditional systems that rely on committee meetings and e-mail. In the near future, as physicians begin to practice personalized gene-based medicine, the amount of information to manage will explode in size.

Partners HealthCare's main objective is to maintain the quality of knowledge and information in medical systems. Partners involves hundreds of physicians in the process of storing and checking information in its medical knowledge management system. Gathering physicians together to build a knowledge base is difficult enough. Establishing a way to keep the content updated is even more challenging. Rather than focusing on the knowledge, Partners Healthcare began focusing on improving the efficiency of acquiring and maintaining the knowledge.

Starting with one person, Partners built a knowledge management (KM) team that has grown to more than 50 people in the last five years. Those involved include analysts, project managers, knowledge engineers, and software developers. The goal was to develop policies and processes for maintaining clinical knowledge content. The team focused on building a collaborative system that allowed domain knowledge experts to communicate without attending meetings or conference calls.

The team created a central repository for knowledge based on a product from EMC called Documentum. Documentum is a content management platform accessed through eRoom collaboration software. Together these products provide physicians with a robust, Web-based content management infrastructure that is flexible and scalable.

Prior to the Documentum system, physicians organized medical documents in folders on file systems. The location of the file and its name provided all the hierarchical and organizational information for storing and retrieving the file. Files were often updated and over time the organization of the system degraded. Files were also lost and mismanaged. With the Documentum system, knowledge is stored in a database. All interaction with the data is tracked and archived. For example, if a pharmacist reads about new findings related to dosages of ibuprofen for geriatric patients, she can share that article with colleagues, making it available through the database management system. Colleagues can then comment on the article and work to a consensus decision on what dosage is best for patients. The article, the discussion, and the vote are all catalogued in the database and can be referenced in a few years if someone wants to re-evaluate the dosage.

The Documentum and eRoom system has substantially reduced the cost of maintaining the knowledge management system and increased the speed at which Partners Healthcare can acquire knowledge. Physicians have more confidence in the information provided by the system. Rather than attending monthly meetings, clinicians are spending time poring over the information provided by the knowledge management system. To maintain the quality of data, participants log on to the system at the end of the day to comment on or approve new guidelines. Allowing physicians to work at their convenience saves everyone time and makes an unmanageable amount of information manageable.

Discussion Questions

1. What was the main challenge facing Partners Healthcare for managing clinical knowledge and information?
2. What functionality does Documentum and eRoom provide that was missing in Partners Healthcare's previous system?

Critical Thinking Questions

1. How does the quality of a medical knowledge management system affect a community?
2. How might the Partners Healthcare knowledge management system be expanded to benefit medical organizations nationwide or even worldwide in developing countries?

SOURCES: *Computerworld* Staff, "Managing clinical evidence at the speed of change," *Computerworld*—Honors Program, 2008, *www.cwhonors.org/viewCaseStudy2008.asp?NominationID=365*; Partners Healthcare Web site, *www.partners.org* accessed July 4, 2008; EMC Web site, *www.emc.com*, accessed July 4, 2008.

AN OVERVIEW OF EXPERT SYSTEMS

As mentioned earlier, an expert system behaves similarly to a human expert in a particular field. Computerized expert systems have been developed to diagnose problems, predict future events, and solve energy problems. Like human experts, computerized expert systems use heuristics, or rules of thumb, to arrive at conclusions or make suggestions. The research conducted in AI during the past two decades is resulting in expert systems that explore new business possibilities, increase overall profitability, reduce costs, and provide superior service to customers and clients. Blagg & Johnson uses the Lantek expert system to cut and fabricate metal into finished products for the automotive, construction, and mining industries.[52] The expert system helps reduce raw material waste and increase profits. The U.S. Army uses the Knowledge and Information Fusion Exchange (KnIFE) expert system to help soldiers in the field make better military decisions based on successful decisions made in previous military engagements.[53]

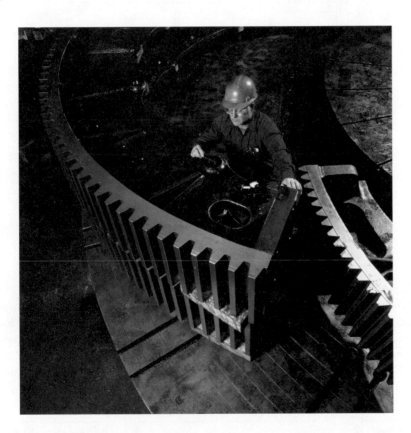

Expert systems are used in metal fabrication plants to aid in decision making.

(Source: © H. Mark Weidman Photography/Alamy.)

When to Use Expert Systems

Sophisticated expert systems can be difficult, expensive, and time consuming to develop. This is especially true for large expert systems implemented on mainframes. The following is a list of factors that normally make expert systems worth the expenditure of time and money. People and organizations should develop an expert system if it can do any of the following:

- Provide a high potential payoff or significantly reduce downside risk
- Capture and preserve irreplaceable human expertise
- Solve a problem that is not easily solved using traditional programming techniques
- Develop a system more consistent than human experts
- Provide expertise needed at a number of locations at the same time or in a hostile environment that is dangerous to human health
- Provide expertise that is expensive or rare

- Develop a solution faster than human experts can
- Provide expertise needed for training and development to share the wisdom and experience of human experts with many people

Components of Expert Systems

An expert system consists of a collection of integrated and related components, including a knowledge base, an inference engine, an explanation facility, a knowledge base acquisition facility, and a user interface. A diagram of a typical expert system is shown in Figure 7.8. In this figure, the user interacts with the interface, which interacts with the inference engine. The inference engine interacts with the other expert system components. These components must work together to provide expertise. This figure shows the inference engine coordinating the flow of knowledge to other components of the expert system. Note that an expert system can have different knowledge flows, depending on what the system is doing and the specific expert system involved.

Figure 7.8

Components of an Expert System

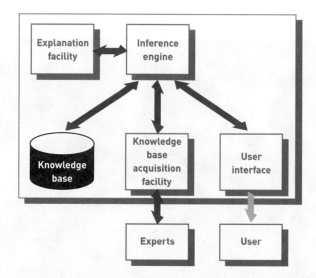

The Knowledge Base

knowledge base
A component of an expert system that stores all relevant information, data, rules, cases, and relationships used by the expert system.

The **knowledge base** stores all relevant information, data, rules, cases, and relationships that the expert system uses. As shown in Figure 7.9, a knowledge base is a natural extension of a database (presented in Chapter 3) and an information and decision support system (presented in Chapter 6). A knowledge base must be developed for each unique application. For example, a medical expert system contains facts about diseases and symptoms. The following are some tools and techniques that can be used to create a knowledge base.

Figure 7.9

The Relationships Among Data, Information, and Knowledge

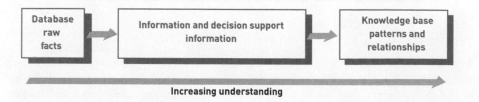

- **Assembling human experts.** One challenge in developing a knowledge base is to assemble the knowledge of multiple human experts. Typically, the objective in building a knowledge base is to integrate the knowledge of people with similar expertise (for example, many doctors might contribute to a medical diagnostics knowledge base).
- **Using fuzzy logic.** Another challenge for designers and developers of expert systems is capturing knowledge and relationships that are not precise or exact. Instead of the black-and-white, yes/no, or true/false conditions of typical computer decisions, fuzzy logic

allows shades of gray, or what is known as "fuzzy sets." Fuzzy logic rules help computers evaluate the imperfect or imprecise conditions they encounter and make educated guesses based on the probability of correctness of the decision.

- **Using rules.** A **rule** is a conditional statement that links conditions to actions or outcomes. In many instances, these rules are stored as **IF-THEN statements**, such as "If a certain set of network conditions exists, then a certain network problem diagnosis is appropriate." In an expert system for a weather forecasting operation, for example, the rules could state that if certain temperature patterns exist with a given barometric pressure and certain previous weather patterns over the last 24 hours, then a specific forecast will be made, including temperatures, cloud coverage, and wind-chill factor. Figure 7.10 shows how to use expert system rules in determining whether a person should receive a mortgage loan from a bank. These rules can be placed in almost any standard program language discussed in Chapter 2 using "IF-THEN" statements or into special expert systems shells and products, discussed later in the chapter. In general, as the number of rules that an expert system knows increases, the precision of the expert system also increases.

rule
A conditional statement that links conditions to actions or outcomes.

IF-THEN statements
Rules that suggest certain conclusions.

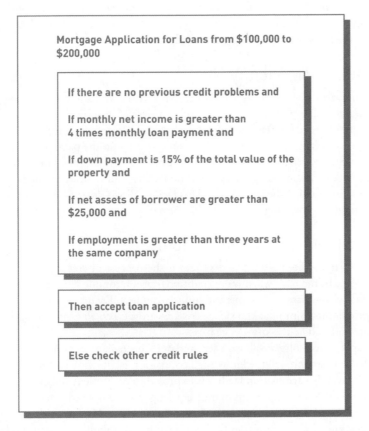

Mortgage Application for Loans from $100,000 to $200,000

If there are no previous credit problems and

If monthly net income is greater than 4 times monthly loan payment and

If down payment is 15% of the total value of the property and

If net assets of borrower are greater than $25,000 and

If employment is greater than three years at the same company

Then accept loan application

Else check other credit rules

Figure 7.10

Rules for a Credit Application

- **Using cases.** An expert system can use cases in developing a solution to a current problem or situation. This process involves (1) finding cases stored in the knowledge base that are similar to the problem or situation at hand and (2) modifying the solutions to the cases to fit or accommodate the current problem or situation. For example, a company might use an expert system to determine the best location for a new service facility in the state of New Mexico. The expert system might identify two previous cases involving the location of a service facility where labor and transportation costs were also important—one in the state of Colorado and the other in the state of Nevada. The expert system can modify the solution to these two cases to determine the best location for a new facility in New Mexico.

The Inference Engine

inference engine
Part of the expert system that seeks information and relationships from the knowledge base and provides answers, predictions, and suggestions similar to the way a human expert would.

backward chaining
The process of starting with conclusions and working backward to the supporting facts.

forward chaining
The process of starting with the facts and working forward to the conclusions.

The overall purpose of an **inference engine** is to seek information and relationships from the knowledge base and to provide answers, predictions, and suggestions similar to the way a human expert would. In other words, the inference engine is the component that delivers the expert advice. To provide answers and give advice, expert systems can use backward and forward chaining. **Backward chaining** is the process of starting with conclusions and working backward to the supporting facts. If the facts do not support the conclusion, another conclusion is selected and tested. This process is continued until the correct conclusion is identified. **Forward chaining** starts with the facts and works forward to the conclusions. Consider the expert system that forecasts future sales for a product. Forward chaining starts with a fact such as "The demand for the product last month was 20,000 units." With the forward-chaining approach, the expert system searches for rules that contain a reference to product demand. For example, "IF product demand is over 15,000 units, THEN check the demand for competing products." As a result of this process, the expert system might use information on the demand for competitive products. Next, after searching additional rules, the expert system might use information on personal income or national inflation rates. This process continues until the expert system can reach a conclusion using the data supplied by the user and the rules that apply in the knowledge base.

The Explanation Facility

explanation facility
Component of an expert system that allows a user or decision maker to understand how the expert system arrived at certain conclusions or results.

An important part of an expert system is the **explanation facility**, which allows a user or decision maker to understand how the expert system arrived at certain conclusions or results. A medical expert system, for example, might reach the conclusion that a patient has a defective heart valve given certain symptoms and the results of tests on the patient. The explanation facility allows a doctor to find out the logic or rationale of the diagnosis made by the expert system. The expert system, using the explanation facility, can indicate all the facts and rules that were used in reaching the conclusion. This facility allows doctors to determine whether the expert system is processing the data and information correctly and logically.

The Knowledge Acquisition Facility

A difficult task in developing an expert system is the process of creating and updating the knowledge base. In the past, when more traditional programming languages were used, developing a knowledge base was tedious and time consuming. Each fact, relationship, and rule had to be programmed into the knowledge base. In most cases, an experienced programmer had to create and update the knowledge base.

knowledge acquisition facility
Part of the expert system that provides convenient and efficient means of capturing and storing all the components of the knowledge base.

Today, specialized software allows users and decision makers to create and modify their own knowledge bases through the knowledge acquisition facility (see Figure 7.11). The purpose of the **knowledge acquisition facility** is to provide a convenient and efficient means for capturing and storing all components of the knowledge base. Knowledge acquisition software can present users and decision makers with easy-to-use menus. After entering the appropriate attributes, the knowledge acquisition facility correctly stores information and relationships in the knowledge base, making the knowledge base easier and less expensive to set up and maintain. Knowledge acquisition can be a manual process or a mixture of manual and automated procedures. Regardless of how the knowledge is acquired, the knowledge base must be validated and updated frequently to make sure that it is still accurate.

Figure 7.11

Knowledge Acquisition Facility

The knowledge acquisition facility acts as an interface between experts and the knowledge base.

The User Interface

Specialized user interface software is employed for designing, creating, updating, and using expert systems. The main purpose of the user interface is to make an expert system easier for users and decision makers to develop and use. At one time, skilled computer personnel created and operated most expert systems; today, the user interface permits decision makers to develop and use their own expert systems. Because expert systems place more emphasis on directing user activities than do other types of systems, text-oriented user interfaces (using menus, forms, and scripts) might be more common in expert systems than the graphical interfaces often used with DSSs. A typical user interface between a user and the expert system for the credit application shown in Figure 7.10 follows.

EXPERT SYSTEM: Are there any previous credit problems?

USER: No.

EXPERT SYSTEM: Is the applicant's net income greater than four times the monthly loan payment?

USER: Yes.

EXPERT SYSTEM: Is the down payment greater than 15% of the property value?

USER: Yes.

EXPERT SYSTEM: Are the net assets of the borrower greater than $25,000?

USER: Yes.

EXPERT SYSTEM: Has the borrower worked more than three years at the same company?

USER: Yes.

EXPERT SYSTEM: **Accept the loan application.**

Participants in Developing and Using Expert Systems

Typically, several people are involved in developing and using an expert system (see Figure 7.12).

Domain expert   Knowledge engineer   Knowledge user

Figure 7.12

Participants in Expert Systems Development and Use

The Domain Expert

Because of the time and effort involved in the task, an expert system is developed to address only a specific area of knowledge. This area of knowledge is called the **domain**. The **domain expert** is the person or group with the expertise or knowledge the expert system is trying to capture. In most cases, the domain expert is a group of human experts. The domain expert (individual or group) usually can do the following:

- Recognize the real problem
- Develop a general framework for problem solving

domain
The area of knowledge addressed by the expert system.

domain expert
The person or group who has the expertise or knowledge the expert system is trying to capture.

- Formulate theories about the situation
- Develop and use general rules to solve a problem
- Know when to break the rules or general principles
- Solve problems quickly and efficiently
- Learn from experience
- Know what is and is not important in solving a problem
- Explain the situation and solutions of problems to others

knowledge engineer
A person who has training or experience in the design, development, implementation, and maintenance of an expert system.

knowledge user
The person or group who uses and benefits from the expert system.

The Knowledge Engineer and Knowledge Users

A **knowledge engineer** is a person who has training or experience in the design, development, implementation, and maintenance of an expert system, including training or experience with expert system shells. The **knowledge user** is the person or group who uses and benefits from the expert system. Knowledge users do not need any previous training in computers or expert systems.

Expert Systems Development Tools and Techniques

Theoretically, expert systems can be developed from any programming language. Since the introduction of computer systems, programming languages have become easier to use, more powerful, and increasingly able to handle specialized requirements. In the early days of expert systems development, traditional high-level languages, including Pascal, FORTRAN, and COBOL, were used (see Figure 7.13). LISP was one of the first special languages developed and used for artificial intelligence applications. PROLOG was also developed for AI applications. Since the 1990s, however, other expert system products (such as shells) have become available that remove the burden of programming, allowing nonprogrammers to develop and benefit from the use of expert systems.

Figure 7.13

Expert Systems Development

Software for expert systems development has evolved greatly since 1980, from traditional programming languages to expert system shells.

Expert System Shells and Products

An *expert system shell* is a collection of software packages and tools used to design, develop, implement, and maintain expert systems. Expert system shells are available for both personal computers and mainframe systems. Some shells are inexpensive, costing less than $500. In addition, off-the-shelf expert system shells are complete and ready to run. The user enters the appropriate data or parameters, and the expert system provides output to the problem or situation. Table 7.2 lists a few expert system products.

Table 7.2

Popular Expert System Products

Name of Product	Application and Capabilities
G2	Assists in oil and gas operations. Transco, a British company, uses it to help in the transport of gas to more than 20 million commercial and domestic customers.
HazMat Loader	Analyzes hazardous materials in truck shipments (*http://hazmat.dot.gov*).
Imprint Business Systems	This company has an expert system that helps printing and packaging companies manage their businesses (*www.imprint-mis.co.uk*).
Lantek Expert System	Helps metal fabricators reduce waste and increase profits (*www.lantek.es*).
RAMPART	Developed by Sandia National Laboratories, the U.S. General Services Administration (GSA) uses it to analyze risk to the approximately 8,000 federal buildings it manages (*www.sandia.gov*).

Applications of Expert Systems and Artificial Intelligence

Expert systems and artificial intelligence have wide applications in business and government. A few additional applications of expert systems that are being used today or have been used in the past are summarized next:

- **Credit granting and loan analysis.** KPMG Peat Marwick uses an expert system called Loan Probe to review its reserves to determine whether sufficient funds have been set aside to cover the risk of some uncollectible loans.
- **Catching cheats and terrorists.** Some gambling casinos use expert system software to catch gambling cheats.
- **Plant layout and manufacturing.** FLEXPERT was an expert system that uses fuzzy logic to perform plant layout. The software helped companies determine the best placement for equipment and manufacturing facilities.
- **Hospitals and medical facilities.** Hospitals, pharmacies, and other healthcare providers can use Alineo by MEDecision to determine possible high-risk or high-cost patients. MYCIN is an expert system developed at Stanford University to analyze blood infections. UpToDate is another expert system used to diagnose patients. To help doctors in the diagnosis of thoracic pain, MatheMEDics has developed THORASK, a straightforward, easy-to-use program, requiring only the input of carefully obtained clinical information. The program helps the less experienced to distinguish the three principal categories of chest pain from each other.
- **Employee performance evaluation.** An expert system developed by Austin-Hayne, called Employee Appraiser, provides managers with expert advice for use in employee performance reviews and career development.
- **Repair and maintenance.** ACE is an expert system used by AT&T to analyze the maintenance of telephone networks. IET-Intelligent Electronics uses an expert system to diagnose maintenance problems related to aerospace equipment. General Electric Aircraft Engine Group uses an expert system to enhance maintenance performance levels at all sites and improve diagnostic accuracy.
- **Shipping.** CARGEX cargo expert system is used by Lufthansa, a German airline, to help determine the best shipping routes.
- **Marketing.** CoverStory is an expert system that extracts marketing information from a database and automatically writes marketing reports.

Computer manufacturers often use expert systems to monitor manufacturing processes and improve yields.

(Source: © Colorblind/Getty Images.)

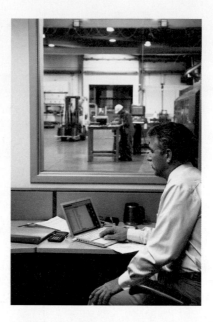

VIRTUAL REALITY

The term *virtual reality* was initially coined by Jaron Lanier, founder of VPL Research, in 1989. Originally, the term referred to *immersive virtual reality* in which the user becomes fully immersed in an artificial, three-dimensional world that is completely generated by a computer. Immersive virtual reality can represent any three-dimensional setting, real or abstract, such as a building, an archaeological excavation site, the human anatomy, a sculpture, or a crime scene reconstruction. Through immersion, the user can gain a deeper understanding of the virtual world's behavior and functionality. The Media Grid at Boston College has a number of initiatives in the use of immersive virtual reality in education.[54]

virtual reality system
A system that enables one or more users to move and react in a computer-simulated environment.

A **virtual reality system** enables one or more users to move and react in a computer-simulated environment. Virtual reality simulations require special interface devices that transmit the sights, sounds, and sensations of the simulated world to the user. These devices can also record and send the speech and movements of the participants to the simulation program, enabling users to sense and manipulate virtual objects much as they would real objects. This natural style of interaction gives the participants the feeling that they are immersed in the simulated world. For example, an auto manufacturer can use virtual reality to help it simulate and design factories.

INFORMATION SYSTEMS @ WORK

Realtors Rely on Virtual Reality

Virtual reality allows you to experience places, to some degree, without the inconvenience of travel. Although a trip to a virtual location is not as rich of an experience as actually being there, virtual reality sometimes provides valuable information all the same. For example, consider the information provided by virtual earth software from Microsoft and Google.

In its initial release, virtual earth software stitched together high-resolution satellite imagery available from commercial providers to let users scroll and pan around the Earth from a satellite view. The effect was breathtaking, though flat. Virtual earth developers began building photorealistic, geospecific, 3-D landscapes that allowed users to zoom in on satellite images and fly horizontally through virtual landscapes that replicated the landscape of the Earth and its cities. The effort to virtualize the world is ongoing, with thousands of people providing assistance in adding cities and buildings to the virtual landscape.

It didn't take long for virtual earth applications to move from novelty to serious business tool. Professionals in the real-estate industry were quick to acknowledge the value of visiting neighborhoods virtually. Seain Conover points out that while a real-estate agent's photo may show a quaint house in the country, virtual earth would let you see that it's actually in the shadow of a five-story apartment building.

Conover works for Terasoft Corporation, a Canadian company based in British Columbia, that specializes in Multiple Listing System (MLS) systems. The MLS allows realtors to list houses for other Realtors and house-hunters to find. Through the MLS, realtors can communicate their needs and recommend properties to others in the business. Until now, the MLS has provided home photos and specifications, but virtual earth is changing all of that.

Terasoft and other software companies around the world are working with virtual earth providers such as Microsoft and Google to build commercial applications using virtual earth as a foundation. Using an application developed by Terasoft on Microsoft Virtual Earth, a realtor can take a client to a prospective property virtually, zoom down to the rooftop, and then turn up 45 degrees to view the building, property, and neighborhood from all sides. What used to take days of driving around is now condensed into an hour. Clients and realtors can quickly narrow the market to a few houses that match the client's interests.

Terasoft has built an overlay for Microsoft Virtual Earth that specifies items of interest such as school districts, demographics, and crime rate. Using color-coding, home shoppers can find neighborhoods they desire. Marking a geographic area on the map with the mouse quickly displays the available houses in that area in the shopper's price range. Rather than scrolling through thousands of listings in the MLS, realtors can search the virtual landscape for specific needs, see them as pinpoints on a map, zoom in, and view the house in 3-D while analyzing information about the house and its neighborhood.

Terasoft chose Microsoft for this project over Google due to Microsoft's long history of developing software for the real-estate industry. Another contributing factor is that Microsoft's mapping and virtual earth software has clearly been defined as an enterprise platform.

The use of virtual reality in real estate doesn't end at a house's front door. Now through 360-degree photography, customers can inspect a home's interior as well. A virtual walkthrough combined with a floor plan reveals the layout and condition of the interior. As virtual reality technologies mature, finding the home of your dreams from your computer screen or even a VR headset is quickly becoming possible. You only need to visit your dream house to make sure that the virtual-reality experience is a true one.

Real-estate professionals all over the world are turning to virtual earth software to revolutionize their business. Professionals in the public sector, hospitality and travel, retail, financial services, manufacturing, utilities, oil and gas, and media and entertainment are also applying virtual earth technology to their industries. The ability to view remote locations through virtual reality is proving to help boost productivity and build customer satisfaction.

Discussion Questions

1. What conveniences does a virtual reality system such as the one developed by Terasoft provide for realtors and home shoppers?
2. Describe the work required to develop and maintain a virtual model of the Earth and its towns and cities.

Critical Thinking Questions

1. If home shoppers can access tools such as the one provided by Terasoft, why are realtors necessary? How might realtors change their job description to maintain their value to customers?
2. Many industries use Virtual Earth. Provide a few examples of how you think they might apply its technology. How might they customize Virtual Earth to their needs?

SOURCES: Lau, Kathleen, "ISV aims Virtual Earth at realtors," *IT World Canada*, December 12, 2007, *www.itworldcanada.com/a/Enterprise-Business-Applications/cf429ae4-e618-4281-90b5-0cc5596cd234.html*; Terasoft Web site, *www.terasoft.com/company*, accessed July 4, 2008; Microsoft Virtual Earth Web site, *www.microsoft.com/virtualearth/industry/realestate.aspx*, accessed July 4, 2008.

Interface Devices

To see in a virtual world, often the user wears a head-mounted display (HMD) with screens directed at each eye. The HMD also contains a position tracker to monitor the location of the user's head and the direction in which the user is looking. Using this information, a computer generates images of the virtual world—a slightly different view for each eye—to match the direction that the user is looking, and displays these images on the HMD. Many companies sell or rent virtual-reality interface devices, including Virtual Realities (*www.vrealities.com*), Amusitronix (*www.amusitronix.com*), I-O Display Systems (*www.i-glassesstore.com*), and others.

The PowerWall is a virtual reality system that displays large models in accurate dimensions.

(Source: Courtesy of Fakespace Systems, Inc.)

The Electronic Visualization Laboratory at the University of Illinois at Chicago introduced a room constructed of large screens on three walls and the floor on which the graphics are projected. The CAVE, as this room is called, provides the illusion of immersion by projecting stereo images on the walls and floor of a room-sized cube (*http://cave.ncsa.uiuc.edu*). Several persons wearing lightweight stereo glasses can enter and walk freely inside the CAVE. A head-tracking system continuously adjusts the stereo projection to the current position of the leading viewer.

Military personnel train in an immersive CAVE system.

(Source: Courtesy of Fakespace Systems, Inc.)

Users hear sounds in the virtual world through earphones. The information reported by the position tracker is also used to update audio signals. When a sound source in virtual space is not directly in front of or behind the user, the computer transmits sounds to arrive at one ear a little earlier or later than at the other and to be a little louder or softer and slightly different in pitch.

The *haptic* interface, which relays the sense of touch and other physical sensations in the virtual world, is the least developed and perhaps the most challenging to create.[55] Currently, with the use of a glove and position tracker, the computer locates the user's hand and measures finger movements. The user can reach into the virtual world and handle objects; however, it is difficult to generate the sensations of a person tapping a hard surface, picking up an object, or running a finger across a textured surface. Touch sensations also have to be synchronized with the sights and sounds users experience.

Forms of Virtual Reality

Aside from immersive virtual reality, which we just discussed, virtual reality can also refer to applications that are not fully immersive, such as mouse-controlled navigation through a three-dimensional environment on a graphics monitor, stereo viewing from the monitor via stereo glasses, stereo projection systems, and others.

Some virtual reality applications allow views of real environments with superimposed virtual objects. Motion trackers monitor the movements of dancers or athletes for subsequent studies in immersive virtual reality. Telepresence systems (such as telemedicine and telerobotics) immerse a viewer in a real world that is captured by video cameras at a distant location and allow for the remote manipulation of real objects via robot arms and manipulators. Many believe that virtual reality will reshape the interface between people and information technology by offering new ways to communicate information, visualize processes, and express ideas creatively.

Computer-generated image technology and simulation are used by companies to determine plant capacity, manage bottlenecks, and optimize production rates.

(Source: © Lester Lefkowitz/Getty Images.)

Virtual Reality Applications

You can find thousands of applications of virtual reality, with more being developed as the cost of hardware and software declines and people's imaginations are opened to the potential of virtual reality. The following are a few virtual reality applications in medicine, education and training, business and commerce, and entertainment. See Figure 7.14.

Figure 7.14

Virtual Reality Applications

Virtual reality has been used to increase real estate sales in several powerful ways. RealSpace Vision Communication, for example, helps real estate developers showcase their properties with virtual reality tours.

(Source: Courtesy of RealSpace Vision Communication Inc.)

Medicine

Barbara Rothbaum, the director of the Trauma and Recovery Program at Emory University School of Medicine and cofounder of Virtually Better, uses an immersive virtual reality system to help in the treatment of anxiety disorders.[56] "For most of our applications, we use a head-mounted display that's kind of like a helmet with a television screen in front of each eye and has position trackers and sensors," says Rothbaum. One VR program, called SnowWorld, helps treat burn patients.[57] Using VR, the patients can navigate through icy terrain and frigid waterfalls. VR helps because it gets a patient's mind off the pain.

Some virtual reality systems help train people to overcome their fear of public speaking.

(Source: Copyright © Virtually Better, Inc.)

Virtual reality technology can also link stroke patients to their physical therapists. Patients put on special gloves and other virtual reality devices at home that are linked to the physical therapist's office. The physical therapist can then see whether the patient is performing the correct exercises without having to travel to the patient's home or hospital room. In this way, using virtual reality can cut travel time and costs.

Education and Training

Virtual environments are used in education to bring exciting new resources into the classroom. Students can stroll among digital bookshelves, learn anatomy on a simulated cadaver, or participate in historical events—all virtually. Virtual reality, for example, has been used to help those with disabilities.[58] The Archaeology Technologies Laboratory at North Dakota State University has developed a three-dimensional virtual reality system that displays an eighteenth-century American Indian village.[59] Third-grade students at John Cotton Tayloe School in Washington, North Carolina, can take a virtual trip down the Nile for a cross-disciplinary lesson on ancient Egypt. This interactive virtual reality computer lesson integrates social studies, geography, music, art, science, math, and language arts.

Virtual technology has also been applied by the military. To help with aircraft maintenance, a virtual reality system has been developed to simulate an aircraft and give a user a sense of touch, while computer graphics provide a sense of sight and sound. The user sees, touches, and manipulates the various parts of the virtual aircraft during training. The Virtual Aircraft Maintenance System simulates real-world maintenance tasks that are routinely performed on the AV-8B vertical takeoff and landing aircraft used by the U.S. Marines. Also, the Pentagon is using a virtual reality training lab to prepare for a military crisis. The virtual reality system simulates various war scenarios.

Business and Commerce

Virtual reality has been used in all areas of business. Kimberly-Clark Corporation, for example, has developed a virtual reality system to view store aisles carrying its products.[60] This allows executives to see how Kimberly-Clark products look in stores aside competing products. The virtual reality view of store aisles should help executives monitor customer behavior and determine the best packaging and placement of their products on store aisles. Boeing uses virtual reality to help it design and manufacture airplane parts and new planes, including the 787 Dreamliner. Boeing uses 3D PLM from Dassault Systems.[61] Clothing and fashion companies, such as Neiman Marcus and Saks Fifth Avenue, are using virtual reality on the Internet to display and promote new products and fashions.[62] In another Web application, virtual reality was used to design a $90 million addition to the Denver Art Museum. The software can also show the picture, length, and diameter of the 50,000 bolts that are being used. Palomar Pomerado Heath used Second Life to create a virtual hospital when it started construction of a real $700 million hospital in California in 2007.[63] The purpose of the Second Life virtual hospital was to show clients and staff the layout and capabilities of the new hospital. Second Life has also been used in business and recruiting.

Boeing uses virtual reality to collaborate with customers during aircraft design.

(Source: AP Photo/Ted S. Warren.)

Entertainment

Computer-generated image technology, or CGI, has been around since the 1970s. Many movies use this technology to bring realism to the silver screen, including *Finding Nemo*, *Spider-Man II*, and *Star Wars Episode II—Attack of the Clones*. A team of artists rendered the roiling seas and crashing waves of *Perfect Storm* almost entirely on computers using weather reports, scientific formulas, and their imagination. Other films include *Dinosaur* with its realistic talking reptiles, *Titan A.E.*'s beautiful 3-D space-scapes, and the casts of computer-generated crowds and battles in *Gladiator* and *The Patriot*. CGI can also be used for sports simulation to enhance the viewers' knowledge and enjoyment of a game. SimCity (*http://simcity.ea.com/*), a virtual reality game, allows people to experiment with decisions related to urban planning. Natural and man-made disasters test decisions on designing buildings and the surrounding area. Other games can display a 3-D view of the world and allow people to interact with simulated people or avatars in the game. Second Life (*www.secondlife.com*) allows people to play games, interact with avatars, and build structures, such as homes.[64]

OTHER SPECIALIZED SYSTEMS

In addition to artificial intelligence, expert systems, and virtual reality, other interesting specialized systems have appeared. Segway, for example, is an electric scooter that uses sophisticated software, sensors, and gyro motors to transport people through warehouses, offices, downtown sidewalks, and other spaces (*www.segway.com*). Originally designed to transport people around a factory or around town, more recent versions are being tested by the military for gathering intelligence and transporting wounded soldiers to safety. The military and DARPA are developing energy-efficient, mechanical computers that have the ability to operate in environments that are too harsh for traditional chip-based computers.[65] A Japanese company is experimenting with using specially designed floor mats that contain wires and other electronic components to generate electricity when people step on them.[66] The 3VR Security (*www.3vr.com*) system makes a video face-recognition system to identify people from pictures or images.[67] According to a security officer for the Bank of Hawaii, "It seemed too good to be true, but since being installed at bank branches in December, it has reduced our surveillance time, and it's especially useful in tracking multiple transactions by an ID thief."

A number of special-purpose systems are now available in vehicles. Ford Motor Company and Microsoft have developed a voice-activated system called *Sync* that can play music, make phones calls, and more.[68] The Advanced Warning System by Mobileye warns drivers to keep a safe distance from other vehicles and drivers.[69] Automotive software allows cars and trucks to connect to the Internet. The software can track a driver's speed and location, allow gas stations to remotely charge for fuel and related services, and more.

Many new computing devices, such as Microsoft's Surface, are also becoming available.[70] The Surface is a touch-screen computer that uses a glass-top display. It looks like a coffee table or dining room table with a built-in computer. Microsoft's Smart Personal Objects Technology (SPOT) allows small devices to transmit data and messages over the air. SPOT is being used in wrist watches to transmit data and messages over FM radio broadcast bands. The new technology, however, requires a subscription to the Microsoft MSN Direct information service. Some manufacturing is also being done with inkjet printers to allow them to "print" 3-D parts. For example, the printer sprays layers of polymers onto circuit boards to form transistors and other electronic components. Some new computers can even be worn on your body. Smith Drug, for example, uses a wearable computer by Vocollect, Inc. (*www.vocollect.com*) to help its employees monitor inventory levels.[71] The waist-worn computer that includes a headset with a microphone and speaker dramatically increases productivity and helps eliminate errors. According to a corporate executive, "Previously, they had a clipboard with 25 items per sheet. Now, they don't have to look at the paper. Their hands are free, and all they have to do is listen and think."

Increasingly, companies are using special-purpose tracking devices, chips, and bar codes.[72] As mentioned previously, *Radio Frequency Identification (RFID)* tags that contain small chips with information about products or packages can be quickly scanned to perform inventory control or trace a package as it moves from a supplier to a company to its customers.[73] Many companies have used RFID tags to reduce costs, improve customer service, and achieve a competitive advantage.[74] When attached to clothing and worn close to a mirror, some RFID tags will display sizes, styles, colors, suggested accessories, and images of models wearing the clothing on the mirror or a display screen.[75] RFID tags are even used to help track lost airline luggage.[76] The state of Colorado uses RFID to track elk herds. Farmers are looking into using these tags to track cattle to help identify and control mad cow disease. An Italian cheese consortium uses RFID tags in the crust of cheese wheels. The RFID tags contain information about when and where the cheese was made to ensure freshness and avoid spoilage. Two German students have developed a smart beer mat, which uses sensor chips to help determine the weight or amount of beer in a glass or beer mug. When the chips sense that the beer mug is nearly empty, the sensor chip sends an alert to a computer monitor telling the bartender that a customer needs more beer. The endorsement of an electronic product code standard will likely make RFID even more popular.[77]

Special-purpose bar codes are also being introduced in a variety of settings. For example, to manage office space efficiently, a company gives each employee and office a bar code. Instead of having permanent offices, the employees are assigned offices and supplies as needed, and the bar codes help to make sure that an employee's work, mail, and other materials are routed to the right place. Companies can save millions of dollars by reducing office space and supplies. Another technology is being used to create "smart containers" for ships, railroads, and trucks. NaviTag (*http://navitag.com/*) and other companies are developing communications systems that allow containers to broadcast the contents, location, and condition of shipments to shipping and cargo managers. A railroad company can use standard radio messages to generate shipment and tracking data for customers and managers.

Navitag, an electronic security device, is attached to a cargo container door and monitors whether the door opens and whether light, radiation, or carbon monoxide enters the container.

(Source: AP Photo/Mike Derer.)

One special application of computer technology is derived from a branch of mathematics called game theory. **Game theory** involves the use of information systems to develop competitive strategies for people, organizations, or even countries. Two competing businesses in the same market can use game theory to determine the best strategy to achieve their goals. The military could also use game theory to determine the best military strategy to win a conflict against another country, and individual investors could use game theory to determine the best strategies when competing against other investors in a government auction of bonds. Groundbreaking work on game theory was pioneered by John Nash, the mathematician whose life was profiled in the book and film *A Beautiful Mind*. Game theory has also been used to develop approaches to deal with terrorism. The Los Angeles airport is experimenting with the use of game theory to help security guards do a better job patrolling sensitive areas.[78]

game theory
The use of information systems to develop competitive strategies for people, organizations, or even countries.

Informatics, another specialized system, combines traditional disciplines, such as science and medicine, with information systems and technology. *Bioinformatics*, for example, combines biology and computer science. Also called *computational biology*, bioinformatics has been used to help map the human genome and conduct research on biological organisms. Using sophisticated databases and artificial intelligence, bioinformatics helps unlock the secrets of the human genome, which could eventually prevent diseases and save lives. Stanford University has a course on bioinformatics and offers a bioinformatics certification. Medical informatics combines traditional medical research with computer science. Journals, such as *Healthcare Informatics*, report current research on applying computer systems and technology to reduce medical errors and improve healthcare. The University of Edinburgh even has a School of Informatics (*www.ed.ac.uk/about/structure/informatics.html*). The school has courses on the structure, behavior, and interactions of natural and artificial computational systems. The program combines artificial intelligence, computer science, engineering, and science.

informatics
A specialized system that combines traditional disciplines, such as science and medicine, with computer systems and technology.

SUMMARY

Principle

Knowledge management allows organizations to share knowledge and experience among their managers and employees.

Knowledge is an awareness and understanding of a set of information and the ways that information can be made useful to support a specific task or reach a decision. A knowledge management system (KMS) is an organized collection of people, procedures, software, databases, and devices used to create, store, share, and use the organization's knowledge and experience. Explicit knowledge is objective and can be measured and documented in reports, papers, and rules. Tacit knowledge is hard to measure and document and is typically not objective or formalized.

Knowledge workers are people who create, use, and disseminate knowledge. They are usually professionals in science, engineering, business, and other areas. The chief knowledge officer (CKO) is a top-level executive who helps the organization use a KMS to create, store, and use knowledge to achieve organizational goals. Some organizations and professions use communities of practice (COP) to create, store, and share knowledge. A COP is a group of people or community dedicated to a common discipline or practice, such as open-source software, auditing, medicine, engineering, and other areas.

Obtaining, storing, sharing, and using knowledge is the key to any KMS. The use of a KMS often leads to additional knowledge creation, storage, sharing, and usage. Many tools and techniques can be used to create, store, and use knowledge. These tools and techniques are available from IBM, Microsoft, and other companies and organizations.

Principle

Artificial intelligence systems form a broad and diverse set of systems that can replicate human decision making for certain types of well-defined problems.

The term *artificial intelligence* is used to describe computers with the ability to mimic or duplicate the functions of the human brain. The objective of building AI systems is not to replace human decision making completely but to replicate it for certain types of well-defined problems.

Intelligent behavior encompasses several characteristics, including the abilities to learn from experience and apply this knowledge to new experiences; handle complex situations and solve problems for which pieces of information might be missing; determine relevant information in a given situation, think in a logical and rational manner, and give a quick and correct response; and understand visual images and process symbols. Computers are better than people at transferring information, making a series of calculations rapidly and accurately, and making complex calculations, but human beings are better than computers at all other attributes of intelligence.

Artificial intelligence is a broad field that includes several key components, such as expert systems, robotics, vision systems, natural language processing, learning systems, and neural networks. An expert system consists of the hardware and software used to produce systems that behave as a human expert would in a specialized field or area (e.g., credit analysis). Robotics uses mechanical or computer devices to perform tasks that require a high degree of precision or are tedious or hazardous for humans (e.g., stacking cartons on a pallet). Vision systems include hardware and software that permit computers to capture, store, and manipulate images and pictures (e.g., face-recognition software). Natural language processing allows the computer to understand and react to statements and commands made in a "natural" language, such as English. Learning systems use a combination of software and hardware to allow a computer to change how it functions or reacts to situations based on feedback it receives (e.g., a computerized chess game). A neural network is a computer system that can simulate the functioning of a human brain (e.g., disease diagnostics system). A genetic algorithm is an approach to solving large, complex problems in which a number of related operations or models change and evolve until the best one emerges. The approach is based on the theory of evolution, which requires variation and natural selection. Intelligent agents consist of programs and a knowledge base used to perform a specific task for a person, a process, or another program.

Principle

Expert systems can enable a novice to perform at the level of an expert but must be developed and maintained very carefully.

An expert system consists of a collection of integrated and related components, including a knowledge base, an inference engine, an explanation facility, a knowledge acquisition facility, and a user interface. The knowledge base is an extension of a database, discussed in Chapter 3, and an information and decision support system, discussed in Chapter 6. It contains all the relevant data, rules, and relationships used in the expert system. The rules are often composed of if-then statements, which are used for drawing conclusions. Fuzzy logic allows expert systems to incorporate facts and relationships into expert system knowledge bases that might be imprecise or unknown.

The inference engine processes the rules, data, and relationships stored in the knowledge base to provide answers, predictions, and suggestions the way a human expert would. Two common methods for processing include backward and forward chaining. Backward chaining starts with a conclusion, then searches for facts to support it; forward chaining starts with a fact, then searches for a conclusion to support it.

The explanation facility of an expert system allows the user to understand what rules were used in arriving at a decision. The knowledge acquisition facility helps the user add or update knowledge in the knowledge base. The user interface makes it easier to develop and use the expert system.

The people involved in the development of an expert system include the domain expert, the knowledge engineer, and the knowledge users. The domain expert is the person or group who has the expertise or knowledge being captured for the system. The knowledge engineer is the developer whose job is to extract the expertise from the domain expert. The knowledge user is the person who benefits from the use of the developed system.

The steps involved in the development of an expert system include determining requirements, identifying experts, constructing expert system components, implementing results, and maintaining and reviewing the system.

Expert systems can be implemented in several ways. Previously, traditional high-level languages, including Pascal, FORTRAN, and COBOL, were used. LISP and PROLOG are two languages specifically developed for creating expert systems from scratch. A faster and less-expensive way to acquire an expert system is to purchase an expert system shell or existing package. The shell program is a collection of software packages and tools used to design, develop, implement, and maintain expert systems.

The benefits of using an expert system go beyond the typical reasons for using a computerized processing solution. Expert systems display "intelligent" behavior, manipulate symbolic information and draw conclusions, provide portable knowledge, and can deal with uncertainty. Expert systems can be used to solve problems in many fields or disciplines and can assist in all stages of the problem-solving process. Past successes have shown that expert systems are good at strategic goal setting, planning, design, decision making, quality control and monitoring, and diagnosis.

Applications of expert systems and artificial intelligence include credit granting and loan analysis, catching cheats and terrorists, budgeting, games, information management and retrieval, AI and expert systems embedded in products, plant layout, hospitals and medical facilities, help desks and assistance, employee performance evaluation, virus detection, repair and maintenance, shipping, and warehouse optimization.

Principle

Virtual reality systems can reshape the interface between people and information technology by offering new ways to communicate information, visualize processes, and express ideas creatively.

A virtual reality system enables one or more users to move and react in a computer-simulated environment. Virtual reality simulations require special interface devices that transmit the sights, sounds, and sensations of the simulated world to the user. These devices can also record and send the speech and movements of the participants to the simulation program. Thus, users can sense and manipulate virtual objects much as they would real objects. This natural style of interaction gives the participants the feeling that they are immersed in the simulated world.

Virtual reality can also refer to applications that are not fully immersive, such as mouse-controlled navigation through a three-dimensional environment on a graphics monitor, stereo viewing from the monitor via stereo glasses, stereo projection systems, and others. Some virtual reality applications allow views of real environments with superimposed virtual objects. Virtual reality applications are found in medicine, education and training, real estate and tourism, and entertainment.

Principle

Specialized systems can help organizations and individuals achieve their goals.

A number of specialized systems have recently appeared to assist organizations and individuals in new and exciting ways. Segway, for example, is an electric scooter that uses sophisticated software, sensors, and gyro motors to transport people through warehouses, offices, downtown sidewalks, and other spaces. Originally designed to transport people around a factory or around town, more recent versions are being tested by the military for gathering intelligence and transporting wounded soldiers to safety. Radio Frequency Identification (RFID) tags are used in a variety of settings. Game theory involves the use of information systems to develop competitive strategies for people, organizations, and even countries. Informatics combines traditional disciplines, such as science and medicine, with computer science. Bioinformatics and medical informatics are examples. A number of special-purpose telecommunications systems can be placed in products for varied uses.

CHAPTER 7: SELF-ASSESSMENT TEST

Knowledge management allows organizations to share knowledge and experience among their managers and employees.

1. _____ are people who create, use, and disseminate knowledge and are typically professionals in business, science, engineering, or another area.

2. What type of knowledge is objective and can be measured and documented in reports, papers, and rules?
   a. tacit
   b. descriptive
   c. prescriptive
   d. explicit

3. A community of practice (COP) is a group of people or community dedicated to a common discipline or practice, such as open-source software, auditing, medicine, engineering, and other areas. True or False?

Artificial intelligence systems form a broad and diverse set of systems that can replicate human decision making for certain types of well-defined problems.

4. The Turing Test attempts to determine whether the responses from a computer with intelligent behavior are indistinguishable from responses from a human. True or False?

5. _____ are rules of thumb arising from experience or even guesses.

6. What is *not* an important attribute for artificial intelligence?
   a. the ability to understand visual images
   b. the ability to learn from experience
   c. the ability to be creative
   d. the ability to make complex calculations

7. _____ involves mechanical or computer devices that can paint cars, make welds, and perform other tasks that require a high degree of precision or are tedious or hazardous for human beings.

8. What branch of artificial intelligence involves a computer understanding and reacting to statements in English or another language?
   a. expert systems
   b. neural networks
   c. natural language processing
   d. vision systems

9. A(n) _____ is a combination of software and hardware that allows the computer to change how it functions or reacts to situations based on feedback it receives.

Expert systems can enable a novice to perform at the level of an expert but must be developed and maintained very carefully.

10. What is a disadvantage of an expert system?
    a. the inability to solve complex problems
    b. the inability to deal with uncertainty
    c. limitations to relatively narrow problems
    d. the inability to draw conclusions from complex relationships

11. A(n) _____ is a collection of software packages and tools used to develop expert systems that can be implemented on most popular PC platforms to reduce development time and costs.

12. A heuristic consists of a collection of software and tools used to develop an expert system to reduce development time and costs. True or False?

13. What stores all relevant information, data, rules, cases, and relationships used by the expert system?
    a. the knowledge base
    b. the data interface
    c. the database
    d. the acquisition facility

14. A disadvantage of an expert system is the inability to provide expertise needed at a number of locations at the same time or in a hostile environment that is dangerous to human health. True or False?

15. What allows a user or decision maker to understand how the expert system arrived at a certain conclusion or result?
    a. domain expert
    b. inference engine
    c. knowledge base
    d. explanation facility

16. An important part of an expert system is the _____, which allows a user or decision maker to understand how the expert system arrived at certain conclusions or results.

17. In an expert system, the domain expert is the individual or group who has the expertise or knowledge one is trying to capture in the expert system. True or False?

Virtual reality systems can reshape the interface between people and information technology by offering new ways to communicate information, visualize processes, and express ideas creatively.

18. A(n) _____ enables one or more users to move and react in a computer-simulated environment.

19. What type of virtual reality is used to make human beings feel as though they are in a three-dimensional setting, such as a building, an archaeological excavation site, the human anatomy, a sculpture, or a crime scene reconstruction?
    a. chaining
    b. relative
    c. immersive
    d. visual

Specialized systems can help organizations and individuals achieve their goals.

20. _____ involves the use of information systems to develop competitive strategies for people, organizations, or even countries.

CHAPTER 7: SELF-ASSESSMENT TEST ANSWERS

(1) knowledge workers (2) d (3) True (4) True (5) Heuristics (6) d (7) Robotics (8) c (9) learning system (10) c (11) expert system shell (12) False (13) a (14) False (15) d (16) explanation facility (17) True (18) virtual reality system (19) c (20) game theory

REVIEW QUESTIONS

1. What is a knowledge management system?
2. What is a community of practice?
3. What is the difference between knowledge and information?
4. What is a vision system? Discuss two applications of such a system.
5. What is natural language processing? What are the three levels of voice recognition?
6. Describe three examples of the use of robotics. How can a microrobot be used?
7. What is a learning system? Give a practical example of such a system.
8. What is a neural network? Describe two applications of neural networks.
9. Under what conditions is the development of an expert system likely to be worth the effort?
10. Identify the basic components of an expert system and describe the role of each.
11. What is fuzzy logic?
12. What is virtual reality? Give several examples of its use.
13. Expert systems can be built based on rules or cases. What is the difference between the two?
14. Describe the roles of the domain expert, the knowledge engineer, and the knowledge user in expert systems.
15. What is informatics? Give a few examples.
16. Describe three applications of expert systems or artificial intelligence.
17. Identify three special interface devices developed for use with virtual reality systems.
18. Identify and briefly describe three specific virtual reality applications.
19. What is informatics? How is it used?
20. Give three examples of other specialized systems.

DISCUSSION QUESTIONS

1. What are the requirements for a computer to exhibit human-level intelligence? How long will it be before we have the technology to design such computers? Do you think we should push to try to accelerate such a development? Why or why not?
2. You work for an insurance company as an entry-level manager. The company contains both explicit and tacit knowledge. Describe the types of explicit and tacit knowledge that might exist in your insurance company. How you would capture each type of knowledge?
3. Describe a knowledge management system for a college or university.
4. What are some of the tasks at which robots excel? Which human tasks are difficult for them to master? What fields of AI are required to develop a truly perceptive robot?
5. Describe how natural language processing could be used in a university setting.
6. Discuss how learning systems can be used in a military war simulation to train future officers and field commanders.
7. You have been hired to develop an expert system for a university career placement center. Develop five rules a student could use in selecting a career.
8. What is the relationship between a database and a knowledge base?

9. Imagine that you are developing the rules for an expert system to select the strongest candidates for a medical school. What rules or heuristics would you include?
10. Describe how informatics can be used in a business setting.
11. Which interface is the least developed and most challenging to create in a virtual reality system? Why do you think this is so?

12. What application of virtual reality has the most potential to generate increased profits in the future?
13. Describe a situation where game theory would be appropriate and could be used.

PROBLEM-SOLVING EXERCISES

1. You are a senior vice president of a company that manufactures kitchen appliances. You are considering using robots to replace up to ten of your skilled workers on the factory floor. Using a spreadsheet, analyze the costs of acquiring several robots to paint and assemble some of your products versus the cost savings in labor. How many years would it take to pay for the robots from the savings in fewer employees? Assume that the skilled workers make $20 per hour, including benefits.

2. Assume that you have just won a lottery worth $100,000. You have decided to invest half the amount in the stock market. Develop a simple expert system to pick ten stocks to consider. Using your word processing program, create seven or more rules that could be used in such an expert system. Create five cases and use the rules you developed to determine the best stocks to pick.
3. Using a graphics program, develop a diagram that shows a KMS for a college or university.

TEAM ACTIVITIES

1. Do research with your team to identify KMSs in three different businesses or nonprofit organizations. Describe the types of tacit and explicit knowledge that would be needed by each organization or business.
2. Form a team and debate other teams from your class on the following topic: "Are expert systems superior to human

beings when it comes to making objective decisions?" Develop several points supporting either side of the debate.
3. Have your team members explore the use of a special-purpose system in an industry of your choice. Describe the advantages and disadvantages of this special-purpose system.

WEB EXERCISES

1. Use the Internet to find information about the use of robotics. Describe three examples of how this technology is used.
2. This chapter discussed several examples of expert systems. Search the Internet for two examples of the use of expert

systems. Which one has the greatest potential to increase profits for the firm? Explain your choice.
3. Use the Internet to get information about the application of game theory in business or the military. Write a report about what you found.

CAREER EXERCISES

1. Describe how a COP can be used to help advance your career.

2. Describe the future of artificial intelligence in a career area of your choice.

CASE STUDIES

Case One
Bird & Bird Have Knowledge in Hand

Bird & Bird (B&B) is an international commercial law firm that focuses on industries including aviation and aerospace, financial services, communications, e-commerce, IT, life sciences, media, and sport. The firm has offices in Beijing, Brussels, Dusseldorf, Frankfurt, The Hague, Helsinki, Hong Kong, London, Lyon, Madrid, Milan, Paris, Rome, and Stockholm.

B&B lawyers wanted to be able to access case histories and other legal reference materials online from any of the firm's 14 offices. The firm had a system called Solutions Lab designed to perform this task, but the system had failed to keep up with improvements in search technologies. B&B lawyers wanted a more powerful system.

The knowledge management (KM) team at B&B collaborated with the information systems specialists to incorporate cutting-edge technologies into a new knowledge management system for the firm. They began by conducting focus groups to learn about the needs of the lawyers. All 14 offices were involved and the results provided innovative ideas for serving the law firm's needs.

Next, the team evaluated off-the-shelf KM products to see if any would suit their needs—unfortunately, none did. The KM team decided to custom design the KM system in-house and commission an external company to build it. They selected UC Logic, a document and knowledge management systems company.

B&B lawyers wanted more powerful searching capabilities for finding topic-related content within documents stored in the KM system. The KM team found a search technology called conceptSearching that provides more flexibility than traditional keyword search. ConceptSearching allows the user to enter natural sentences that might include several key terms or topics, and then applies artificial intelligence for impressive search results.

After a few months of testing and refining the conceptSearching technology to the favor of the firm, the team designed the user interface for the new KM system. The resulting system was designed by the KM team, implemented by UCLogic, and incorporated conceptSearching. The KM team saved time and effort by using the internal and external repositories designed for use in the previous document management system. The new system was implemented gradually to ensure a smooth transition.

Using the new "know-how" KM system, lawyers can now search in two areas: the firm's own internal document repositories, holding the relevant experience of those working in the firm, and external sources to which the firm subscribes such as LexisNexis. Using powerful AI technologies applied through conceptSearching, the system yields tabulated results ranked by relevance. A second list of related topics is presented in a sidebar. Articles are ranked by "know-who"—the amount of hours invested in the work. The system allows the users to adjust search relevancy rankings to further refine the quality of results.

Lawyers in the firm receive a one-page guide on using the system along with one-on-one training sessions as needed. Besides basic guides to the steps, the training has allowed the development team to meet with the lawyers and solicit new suggestions for the system. It has also increased overall usage by encouraging some lawyers who do not use online systems to try it at least once. Due to the increased contact between developers and users, the number of documents being submitted to the system is reaching record highs. Shortly after its successful roll out, the KM team is already hard at work on improvements to add in the next round of development.

Discussion Questions

1. How do knowledge management systems assist B&B lawyers in their research for cases?
2. What additional power does B&B's new "know-how" system provide for its lawyers?

Critical Thinking Questions

1. Why do you think the lawyers at B&B found traditional keyword searching insufficient for their needs?
2. What further benefits might the KM team at B&B design into its next system?

SOURCES: McQuay, Martha, "Know-how: Ushering in the next generation," *PLC Magazine*, July 24, 2007, *http://plc.practicallaw.com/0-374-0976* and *www.uclogic.com/Articles/TwoBirds.pdf*; UCLogic Web site, http://www.uclogic.com, accessed July 5, 2008; Bird & Bird Web site, *www.twobirds.com*, accessed July 5, 2008; ConceptSearching Web site, *www.conceptsearching.com/web*, accessed July 5, 2008.

Case Two
Where Virtual Worlds and AI Collide

The use of information systems saves workers and businesses countless hours of tedious labor. Still, users of information systems often become frustrated with the system's inability to grasp simple common sense knowledge. "Why can't you understand what I need?" is a typical response to a computer incapable of working outside the framework for which it was designed.

The solution is to create computer systems that use AI and guess what the user needs provided with little input or prompting. However, true artificial intelligence has challenged computer scientists for decades. Most researchers believe that to create a thinking machine, the machine must have a physical presence with which to experience its environment. Researchers have used robotics to try this approach, but it is costly, especially since robotics technology is still in its infancy.

Many AI researchers are turning to virtual worlds to provide their systems with an environment from which to experience life. Virtual worlds such as Second Life provide a virtual landscape for people to explore through the use of avatars, characters within the environment that the user controls. Second Life allows users to build houses and businesses and even sell products to other avatars. It strives to mirror the physical, social, and economic sense of the real world. Because Second Life is not bound by the laws of physics, users perform actions that aren't possible in real life, such as fly.

Second Life presents the perfect environment for AI systems to experience the world, and then learn from those experiences. Novamente LLC is one AI company that has created AI-driven avatars in a virtual world. These avatars appear as animals that are eager to learn. For example, a Novamente dog avatar can be taught to play soccer. Through the use of praise and correction, the AI system will learn how to play the game, including its rules and strategies. Novamente also has a Parrot avatar that is learning language skills by talking with people.

While in Second Life, you might run into Edd, an AI avatar created by researchers at Rensselaer Polytechnic Institute. Edd can converse and reason, although he has the intelligence of a four year old. Even that much intelligence takes an immense amount of complex calculus to accomplish. Still, Edd can communicate and influence a real user's actions.

Many AI avatars are moving to Second Life and other virtual worlds. AI researchers find it to be an ideal environment for training AI systems and allowing them to interact with real people through virtual avatars. Michael Mateas, a computer science professor at the University of California, Santa Cruz, says, "It's a fantastic sweet spot—not too simple, not too complicated, high cultural value." But how will these AI systems service people and businesses?

Selmer Bringsjord, head of Rensselaer's Cognitive Science Department and leader of the research project, sees the research applying to practical needs in other virtual environments such as entertainment and gaming, as well as immersive training and education. "The apps, frankly, are endless," Bringsjord said. "Imagine being able to step into a simulation environment in which you interact with synthetic characters as sophisticated as those seen in Star Trek's holodeck."

Consider other uses of AI avatars in virtual worlds. AI systems might be used to work on the behalf of businesses. For example, a salesperson could create a thousand avatars of himself and send the team out to sell products. Other AI systems could be used to collect information and survey the population for marketing or other uses. At the same time, rules must be developed to govern the use of virtual reality in business. For example, an avatar should identify itself as a virtual being so that people do not assume it is human.

Discussion Questions

1. Why are AI features and avatars proliferating in Second Life and other virtual worlds?
2. What types of research are being conducted using AI avatars?

Critical Thinking Questions

1. Why might AI avatars eventually make Second Life users uncomfortable? What might be done to calm the fears?
2. What types of business applications might be provided by AI avatars in Second Life?

SOURCES: Tay, Liz, "Child-like intelligence created in Second Life," *ITNews*, March 14, 2008, *www.itnews.com.au/News/72057,childlike-intelligence-created-in-second-life.aspx*; Hill, Michael, "'Second Life' is frontier for AI research," MSNBC, May 18, 2008, *www.msnbc.msn.com/id/24668099*; Havenstein, Heather, "Virtual worlds making artificial intelligence apps 'smarter'," *Computerworld*, September 13, 2007, *www.computerworld.com/action/article.do?command=viewArticleBasic&taxonomyId=11&articleId=9036438&intsrc=hm_topic*.

Questions for Web Case

See the Web site for this book to read about the Whitmann Price Consulting case for this chapter. The following are questions concerning this Web case.

Whitmann Price Consulting: Knowledge Management and Specialized Information Systems

Discussion Questions

1. List three forms of AI that are being considered for the AMCI system and how they will be used.
2. List the advantages and disadvantages of implementing the AI systems in the AMCI system.

Critical Thinking Questions

1. What types of considerations might Josh and Sandra take into account when deciding which AI system to include?
2. How might Whitmann Price consultants react when they learn about the Presence system that will track their location? Why?

NOTES

Sources for the opening vignette: Gensym Staff, "Ericsson", Gensym Success Stories, 2008, *www.gensym.com/?p=success_stories&id=13*; Gensym Web site, *www.gensym.com*, accessed July 2, 2008; Ericsson Web site, *www.wricsson.com* accessed July 2, 2008.

1 Vorro, Alex, "Knowledge Management: Searching For Knowledge," *Insurance Networking News,* October 1, 2007, p. 1.

2 Chua, Alton, "The Curse of Success," *The Wall Street Journal,* April 28, 2007, p. R8.

3 King, William, "Knowledge Management and Organizational Learning," *Omega,* April 2007, p. 167.

4 Asawanipont, Nitida, "Timely Tips from Tata," *The Nation,* January 22, 2007, p. 1B.

5 Short, Tim and Schwendinger, Jim, "How To Win the War For Talent," *Aviation Week,* February 5, 2007, p. 54.

6 Seneviratne, Vienna, "IAEA Team Finds Canada Plants Using Comprehensive KM Practices," *Inside NRC,* May 28, 2007, p. 16.

7 McCormick, John, "5 Big Companies That Got Knowledge Management Right," *CIO Magazine,* October 5, 2007.

8 McCormick, John, "5 Big Companies That Got Knowledge Management Right," *CIO Magazine,* October 5, 2007.

9 Nguyen, Le, et al., "Acquiring Tacit and Explicit Marketing Knowledge from Foreign Partners in IJVs," *Journal of Business Research,* November 2007, p. 1152.

10 McGregor, Jena and Hamm, Steve, "Managing the Global Workforce," *Business Week,* January 28, 2008, p. 34.

11 Staff, "Remedy Interactive Adds CFO and CKO to Management Team," *Market Wire,* March 12, 2007.

12 Vorro, Alex, "Knowledge Management: Searching for Knowledge," *Insurance Networking News,* October 1, 2007, p. 1.

13 Seneviratne, Gamini, "Knowledge Management Now Seen as a Priority," *Nuclear News,* September 2007, p. 59.

14 King, William, "Motivating Knowledge Sharing Through a Knowledge Management System," *Omega,* February 2007, p. 131.

15 Davenport, T., et al., "Knowledge Management Can Make a Difference," *The Wall Street Journal,* March 10, 2008, p. R11.

16 Blackman, Andrew, "Dated and Confused," *The Wall Street Journal,* May 14, 2007, p. R5.

17 Vorro, Alex, "Knowledge Management: Searching for Knowledge," *Insurance Networking News,* October 1, 2007, p. 1.

18 Martin, Richard, "Collaboration Cisco Style," *Information Week,* January 28, 2008, p. 30.

19 Beizer, Doug, "SI Gets Army Knowledge Management Deal," *Washington Technology,* September 26, 2007.

20 Staff, "University of South Carolina, Collexis Announce Collaboration on Knowledge Management Applications," *Biotech Business Week,* February 19, 2007, p. 168.

21 McCormick, John, "5 Big Companies That Got Knowledge Management Right," *CIO Magazine,* October 5, 2007.

22 Snider, Mike, "AI is OK in New Games," *USA Today,* September 25, 2007, p. 3D.

23 Miller, Stephen, "MIT's Professor's Work Led Him to Preach the Evils of Computers," *The Wall Street Journal,* March 15, 2008, p. A6.

24 O'Keefe, Brian, "The Smartest, the Nuttiest Futurist on Earth," *Fortune,* May 14, 2007, p. 60.

25 Staff, "Ray Kurzweil Joins the Singularity Institute's Board of Directors," *Business Wire,* May 20, 2007.

26 *www.20q.net,* accessed November 26, 2007.

27 Staff, "VIASPACE Leveraging SHINE Technology," *PR Newswire US,* February 14, 2007.

28 Anders, George, "The Winding Road to the Robotic Future," *The Wall Street Journal,* March 16, 2007, p. W6.

29 Abate, Tom, "Future Moving From I, Robot to My Robot," *Rocky Mountain News,* February 26, 2007, p. 8.

30 Anthes, Gary "I, Coach," *Computerworld,* May 21, 2007, p. 38.

31 *www.irobot.com,* accessed November 23, 2007.

32 *www.cs.cmu.edu/~rll,* accessed July 18, 2006.

33 *www.porterhospital.org,* accessed November 23, 2007.

34 Staff, "Worming Its Way into Our Hearts," *Business Week,* May 7, 2007, p. 79.

35 *www.darpagrandchallenge.com,* accessed November 23, 2007.

36 Kelly, Tim, "Robot Race," *Forbes,* September 3, 2007, p. 39.

37 Staff, "The Military Machine," *Rocky Mountain News,* July 14, 2007, p. 27.

38 Anthes, Gary "I, Coach," *Computerworld,* May 21, 2007, p. 38.

39 Clark, Don, "Google Targets Voice Searches," *The Wall Street Journal,* April 12, 2007, p. B3.

40 Gomes, Lee, "Voice Recognition Is Starting to Work," *The Wall Street Journal,* January 10, 2007, p. B1.

41 Ali, Sarmad, "New Services Turn Recordings into Text," *The Wall Street Journal,* May 24, 2007, p. B1.

42 Staff, "Designed to Meet the Special Needs of Cooperative Learning," *School Planning and Management,* September 1, 2007, p. 170.

43 Johnson, Colin, "DARPA Contract Tries Improved Take on AI," *Electronic Engineering Times,* September 10, 2007, p. 18.

44 Srinivasan, D., "Freeway Incident Detection Using Hybrid Fuzzy Neural Network," *IET Intelligent Transportation Systems,* December 2007, p. 249.

45 Ninad, Patil and Smith, Timothy, "Neural Network Analysis Speeds Disease Risk Predictions," *Scientific Computing,* July 1, 2007, p. 36.

46 Mendes, J., et al., "Applying a Genetic Neuro-Model Reference Adaptive Controller in Drilling Optimization," *World Oil,* October 2007.

47 Preminger, Arie, "Forecasting Exchange Rates," *International Journal of Forecasting,* January 2007, p. 71.

48 Ming-Jong, Yao, "A Genetic Algorithm for Determining Optimal Replenishment Cycles," *Omega,* August 2008, p. 619.

49 Kuang, K., "An Application of a Plastic Optical Fiber Sensor and Genetic Algorithm for Structural Health Monitoring," *Fiber Optic Sensors and Systems,* June 1, 2007, p. 5.

50 Staff, "Intelligent Agent Technology; Proceedings," *SciTech Book News,* March 1, 2007.

51 Staff, "Meet Sgt Star, The U.S. Army's Software Recruiting Agent," *COMMWEB,* January 3, 2007.

52 Staff, "Less Waste, More Profits," *Metalworking Production,* July 31, 2007, p. 72.

53 Rendleman, John, "Joint Command, Microsoft Team on R&D," *Government Computer News,* May 11, 2007.

54 *www.mediagrid.org,* accessed November 24, 2007.

55 *http://osl-www.colorado.edu/Research/haptic/hapticInterface.shtml,* accessed November 24, 2007.

56 *www.emory.edu/EMORY_MAGAZINE/winter96/rothbaum.html,* accessed November 24, 2007.

57 *www.temple.edu/ispr/examples/ex03_07_23.html,* accessed November 24, 2007.

58 Walsh, Aaron, "Advocate of Virtual Reality," *Computerworld,* July 9, 2007, p. 52.

59 Staff, "History Is Virtually Alive," *Rocky Mountain News,* May 30, 2007, p. D7.

60 Byron, Ellen, "A Virtual View of Store Aisle," *The Wall Street Journal,* October 3, 2007, B1.

61 *www.3ds.com/home,* accessed November 24, 2007.

62 Peng Tina, "Why Trunk Shows Are Going Virtual," *The Wall Street Journal,* August 11, 2007, p. P1.

63 Gaudin, Sharon, "Real-World Hospital Makes Virtual Debut in Second Life," *Computerworld,* March 3, 2008, p. 12.

64 Anthes, Gary, "Second Life: Is There Any There?" *Computerworld,* December 3, 2007, p. 30.

65 Anthes, Gary, "Different Engines," *Computerworld,* March 10, 2008, p. 34.

66 Bergstein, Brian, "New Wrinkle," *Rocky Mountain News,* February 18, 2008, .3.

67 Bulkeley, William, "Recognizing the Face of Fraud," *The Wall Street Journal,* March 22, 2007, p. B3.

68 White, Joseph, "Digital Driving a Reality," *The Wall Street Journal,* August 28, 2007, p. D5.

69 Ramirez, Elva, "A Warning System As Your Co-Pilot," *The Wall Street Journal,* June 21, 2007, p. D3.

70 Guth, Robert and Clark, Don, "The Shape of Computers to Come," *The Wall Street Journal,* May 30, 2007, p. B1.

71 Wood, Lamont, "Wearable Computers," *Computerworld,* December 17, 2007, p. 32.

72 Weier-Hayes, Mary, "Wal-Mart Gets Tough on RFID," *Information Week,* January 21, 2008, p. 26.

73 Weier-Hayes, Mary, "RFID Tags Are on the Menu," *Information Week,* February 5, 2007, p. 49.

74 Mitchell, Robert, "No Contact," *Computerworld,* June 11, 2007, p. 20.

75 Woyke, Elizabeth, "Mirror, Mirror, Talk to Me," *Business Week,* July 9, 2007, p. 12.

76 McCartney, Scott, "A New Way to Prevent Lost Luggage," *The Wall Street Journal,* February 22, 2007, p. D1.

77 Weier-Hayes, Mary, "Will This Rev Up RFID," *Information Week,* April 23, 2007, p. 30.

78 Staff, "Airport Uses Game Theory to Boost Security," *New Scientist,* October 13, 2007, p. 27.

Systems Development and Social Issues

CHAPTER
· 8 ·

Systems Development

PRINCIPLES	LEARNING OBJECTIVES
■ Effective systems development requires a team effort of stakeholders, users, managers, systems development specialists, and various support personnel, and it starts with careful planning.	■ Identify the key participants in the systems development process and discuss their roles. ■ Define the term *information systems planning* and discuss the importance of planning a project.
■ Systems development often uses different approaches and tools such as traditional development, prototyping, rapid application development, end-user development, computer-aided software engineering, and object-oriented development to select, implement, and monitor projects.	■ Discuss the key features, advantages, and disadvantages of the traditional, prototyping, rapid application development, and end-user systems development life cycles. ■ Discuss the use of computer-aided software engineering (CASE) tools and the object-oriented approach to systems development.
■ Systems development starts with investigation and analysis of existing systems.	■ State the purpose of systems investigation. ■ Discuss the importance of performance and cost objectives. ■ State the purpose of systems analysis and discuss some of the tools and techniques used in this phase of systems development.
■ Designing new systems or modifying existing ones should always be aimed at helping an organization achieve its goals.	■ State the purpose of systems design and discuss the differences between logical and physical systems design. ■ Discuss the issues involved in environmental design. ■ Define the term RFP and discuss how this document is used to drive the acquisition of hardware and software.
■ The primary emphasis of systems implementation is to make sure that the right information is delivered to the right person in the right format at the right time.	■ State the purpose of systems implementation and discuss the various activities associated with this phase of systems development.
■ Maintenance and review add to the useful life of a system but can consume large amounts of resources, so they benefit from the same rigorous methods and project management techniques applied to systems development.	■ State the importance of systems and software maintenance and discuss the activities involved. ■ Describe the systems review process.

Information Systems in the Global Economy
GRUMA, Mexico

Systems Development South of the Border

GRUMA is a global business based in Monterrey, Nuevo Leòn, Mexico. GRUMA is the world leader in corn flour and tortilla production. It runs operations in the United States, Mexico, Central America, Venezuela, Europe, and China. About 19,000 employees work for GRUMA, and the company earns $3.2 billion in revenue.

GRUMA has only recently expanded its operation to Europe and China. One factor that allowed the expansion is a complete redesign of GRUMA's core information systems. The systems development process contributes to the success of the new systems.

GRUMA began operations in Mexico in 1949 based on progressive ideas. The company set goals to revolutionize the corn flour and tortilla industry through modern industrialization that was ecologically sound and efficient. Before long, the company grew to be the largest in the country and planned to expand beyond its borders. After establishing branches in the United States and South America, GRUMA faced challenges to further growth.

The company's information systems were designed to handle only one country's currency, taxes, and regulations. Different systems were designed for different countries—GRUMA in Mexico used one system, another system in the U.S., and others in South American countries. To expand further, GRUMA needed a flexible centralized system that adapted to the economic requirements of many countries. Information systems are designed and implemented to meet the primary goals and strategic plans of a business. During the systems investigation phase, GRUMA discovered that its current system could not support its goal of becoming a global business. The company decided to conduct research into developing a new system that could help to achieve its goals.

Through a process called systems analysis, GRUMA studied their existing systems to discover what changes they needed. GRUMA's information systems team interviewed employees who interacted with the system and others who were otherwise affected by the system. These people are called stakeholders. Through the interviews, the information systems team learned how the current system was used, what operations were effective, and what operations needed an overhaul. Realizing the system needed significant changes, GRUMA decided to contract with an information systems firm to design the new system. It developed a request for proposals (RFP) to find a company to provide assistance for a reasonable price.

After negotiations with several companies, GRUMA selected the information systems company SAP to develop an ERP system for the company. SAP dedicated systems analysts to work with GRUMA's information systems team to design a new system on which to base GRUMA's operations. At this point, the systems development process progressed from investigation to analysis, and then to design. Together the two companies designed a system that accommodated "country-specific variances for taxes, product requirements, and currencies, languages, and cultural differences," according to SAP.

The team designed a template that GRUMA could distribute to international companies it acquired to standardize operations in all corporate facilities. The template would minimize work to be done setting up new facilities. The system also supported a variety of character sets so it could work with international languages including Chinese.

During the systems implementation stage, members of the project team at GRUMA tested several prototypes of the system before rolling out the new system for use. The team decided to use a phase-in approach to implement the new system gradually. If the system or users had any problems, business would not be interrupted. The new system was deployed at GRUMA's headquarters in Mexico for testing. After success in Mexico, the system was deployed to GRUMA operations worldwide.

GRUMA management is thrilled with the new system, saying it allows them to better control growth and more quickly merge new acquisitions into in-house resources. The company can manage global operations in real time, reacting to market changes and new customer requirements as they arise.

Now in its operations and maintenance phase, GRUMA's new system is a huge success. GRUMA's information systems team is reviewing the system to measure its success and identify areas that can be improved. The team is looking to fine-tune features to better coordinate with systems managed by its business partners. It also plans to expand customer relationship management (CRM) tools. The work by GRUMA's systems team and SAP systems analysts has so impressed the executives at GRUMA that they now view the team as a value-creation unit with many more projects to investigate.

As you read this chapter, consider the following:

- What situations can arise in a business to trigger new systems development initiatives?
- What are the best methods for a business to use in approaching new systems development projects?

Why Learn About Systems Development?

Throughout this book, you have seen many examples of the use of information systems in a variety of careers. A manager at a hotel chain can use an information system to look up client preferences. An entrepreneur can use systems development to build a new information systems and a new business. An accountant at a manufacturing company can use an information system to analyze the costs of a new plant. A sales representative for a music store can use an information system to determine which CDs to order and which to discount because they are not selling. Information systems have been designed and implemented for almost every career and industry. But where do you start to acquire these systems or have them developed? How can you work with IS personnel, such as systems analysts and computer programmers, to get what you need to succeed on the job? This chapter gives you the answer. You will see how you can initiate the systems development process and analyze your needs with the help of IS personnel. In this chapter, you will learn how your project can be planned, aligned with corporate goals, rapidly developed, and much more. We start with an overview of the systems development process.

When an organization needs to accomplish a new task or change a work process, how does it do so? It develops a new system or modifies an existing one. Systems development is the activity of creating or modifying systems. It refers to all aspects of the process—from identifying problems to solve or opportunities to exploit to implementing and refining the chosen solution.

AN OVERVIEW OF SYSTEMS DEVELOPMENT

This chapter provides you with a deeper appreciation of the systems development process for individuals and organizations. Individuals can use systems development to their advantage.[1] When Marc Mallow couldn't find off-the-shelf software to schedule workers, he took a few years to develop his own program. The software he created became the core of a New York-based company he founded. Corporations and nonprofit organizations use systems development to achieve their goals. First Health of the Carolinas upgraded its old imaging system to slash costs and provide better health care for patients. The nonprofit health organization reduced costs by more than 30 percent and offered doctors better radiological images to improve patient care.[2] Regardless of your career, you will likely be a participant in the systems development process.

Participants in Systems Development

Effective systems development requires a team effort. The team usually consists of stakeholders, users, managers, systems development specialists, and various support personnel. This team, called the *development team*, is responsible for determining the objectives of the information system and delivering a system that meets these objectives. Many development teams use a project manager to head the systems development effort combined with the project management approach to help coordinate the systems development process. A *project* is a planned collection of activities that achieves a goal, such as constructing a new manufacturing plant or developing a new decision support system. All projects have a defined starting point and ending point, normally expressed as dates such as August 4 and November 12. Most have a budget, such as $150,000. A *project manager* is responsible for coordinating all people and resources needed to complete a project on time. The project manager can make the difference between project success and failure. According to Tyrone Howard, founder of BizNova Consulting, "A project management system is just a tool. It is like this: A carpenter can buy a hammer, but the hammer won't build a house.... In IT, it's the people who do the building, not the technology."[3] In systems development, the project manager can be an IS person inside the organization or an external consultant hired to complete the project. Project managers need technical, business, and people skills. In addition to completing the project on time and within the specified budget, the project manager is usually responsible for controlling project quality, training personnel, facilitating communications, managing risks, and acquiring any necessary equipment, including office supplies and sophisticated computer systems. Project escalation, where the size and scope of a new systems development effort greatly expands over time, is a major problem for project managers.[4] Project escalation often causes projects to go over budget and behind schedule.

In the context of systems development, **stakeholders** are people who, either themselves or through the area of the organization they represent, ultimately benefit from the systems development project. **Users** are people who will interact with the system regularly. They can be employees, managers, or suppliers. For large-scale systems development projects, where the investment in and value of a system can be high, it is common for senior-level managers, including the functional vice presidents (of finance, marketing, and so on), to be part of the development team.

stakeholders
People who, either themselves or through the organization they represent, ultimately benefit from the systems development project.

users
People who will interact with the system regularly.

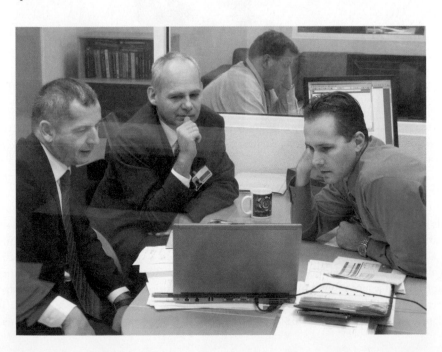

Because stakeholders ultimately benefit from the systems development project, they often work with others in developing a computer application.

(Source: © Reza Estakhrian/Getty Images.)

systems analyst
A professional who specializes in analyzing and designing business systems.

programmer
A specialist responsible for modifying or developing programs to satisfy user requirements.

Depending on the nature of the systems project, the development team might include systems analysts and programmers, among others. A **systems analyst** is a professional who specializes in analyzing and designing business systems. Systems analysts play various roles while interacting with the stakeholders and users, management, vendors and suppliers, external companies, programmers, and other IS support personnel (see Figure 8.1). Like an architect developing blueprints for a new building, a systems analyst develops detailed plans for the new or modified system. The **programmer** is responsible for modifying or developing programs to satisfy user requirements. Like a contractor constructing a new building or renovating an existing one, the programmer takes the plans from the systems analyst and builds or modifies the necessary software. The demand for systems analysts and computer programmers is expected to increase.[5] In Canada, the unemployment rate for IS professionals is about one-third the national average. According to the chairman of the Computer Science Department at the University of Toronto, "The numbers are quite stark. It's clear the demand in the workforce is there."

Figure 8.1

Role of the Systems Analyst

The systems analyst plays an important role in the development team and is often the only person who sees the system in its totality. The one-way arrows in this figure do not mean that there is no direct communication between other team members. These arrows just indicate the pivotal role of the systems analyst—a person who is often called on to be a facilitator, moderator, negotiator, and interpreter for development activities.

Managers

External companies

Software programmers

System stakeholders

SYSTEMS ANALYST

Vendors and suppliers

Technical specialists

Users

Information Systems Planning and Aligning Corporate and IS Goals

Information systems planning and aligning corporate and IS goals are important aspects of any systems development project.[6] Achieving a competitive advantage is often the overall objective of systems development.

Information Systems Planning

information systems planning
Translating strategic and organizational goals into systems development initiatives.

The term **information systems planning** refers to translating strategic and organizational goals into systems development initiatives (see Figure 8.2).[7] Proper IS planning ensures that specific systems development objectives support organizational goals. Long range planning can also be important and result in getting the most from a systems development effort. It can also align IS goals with corporate goals and culture, discussed next.[8] Hess Corporation, a large energy company with over 1,000 retail gasoline stations, uses long-range planning to determine what computer equipment they need and the IS personnel needed to run it.[9] According to Hess's CIO, "It became pretty clear that we needed to lay out a long-term strategy that would allow us to figure out how IT could support our business strategy over the next five years."

Figure 8.2

Information Systems Planning

Information systems planning transforms organizational goals outlined in the strategic plan into specific systems development activities.

Aligning Corporate and IS Goals

Aligning organizational goals and IS goals is critical for any successful systems development effort. Because information systems support other business activities, IS staff and people in other departments need to understand each other's responsibilities and tasks. Determining whether organizational and IS goals are aligned can be difficult, so researchers have increasingly tackled the problem. Most corporations, for example, have profits and return on investment (ROI), first introduced in Chapter 1, as primary goals. Procter & Gamble (P&G) uses ROI to measure the success of its projects and systems development efforts.[10] P&G produces Tide, Pringles, Pampers, and many other consumer products. The huge consumer-products company has a $76 billion annual supply chain. ROI calculations help companies like P&G prioritize systems development projects and align them with corporate goals. Providing outstanding service is another important corporate goal.[11] Coca-Cola Enterprises, which is Coca Cola's largest bottler and distributor, decided to use online services from Microsoft and SharePoint to speed its systems development process.[12] According to the company CIO, "This is not a head-count reduction for us. Services are complementary to our IT strategy."

Specific systems development initiatives can spring from the IS plan, but the IS plan must also provide a broad framework for future success. The IS plan should guide development of the IS infrastructure over time. Another benefit of IS planning is that it ensures better use of IS resources—including funds, personnel, and time for scheduling specific projects. The steps of IS planning are shown in Figure 8.3.

Figure 8.3

The Steps of IS Planning

Some projects are identified through overall IS objectives, whereas additional projects, called *unplanned projects*, are identified from other sources. All identified projects are then evaluated in terms of their organizational priority.

Investigating Conversion at Art.com

Art.com was an early Web pioneer, launching in 1995 with the purpose of selling all kinds of visual art online. Since then, Art.com has assisted over four million customers in decorating their walls by providing a virtual gallery of approximately 400,000 images. The company operates in both the United States and Europe and employs more than 500 worldwide.

Art.com has over 12 million visitors to its Web sites per month, most of whom visit the site without making a purchase. Art.com wanted to increase the percentage of visitors that make purchases—known as the conversion rate—by improving its Web site. With 12 million visitors, even a small improvement could mean a major increase in profits. The challenge was that the Web site had been continuously revised during the company's many years in business, and Art.com's management did not know what changes would improve the visitor's experience. They certainly didn't want to risk changes that might inadvertently turn visitors away.

With the goal clearly articulated to "increase the conversion rate by offering the best customer experience," Art.com systems analysts began to investigate what portions of the current system worked well, and what portions could be improved.

The systems investigation proved to be no small task. Art.com draws thousands of images from product lines offered by many online properties. The company had been using a traditional Web analytics information system that recorded information such as number of visitors and which products were most popular; however, the system did not evaluate information on site obstacles that might be discouraging sales. Art.com needed a system that could provide more telling information such as key performance indicators (KPI) that suggested what customers did not like about its site.

The systems team found an off-the-shelf solution that performed more detailed Web analysis. The "online customer experience management solution" allowed systems investigators to view key performance indicators and then review the qualitative details of individual customer sessions on the site. Viewing the basic analytics allowed the investigators to quickly find trends in customer activity. Drilling down into those trends allowed the investigators to "play back" a customer's activities on the site to determine where the customer experienced problems or decided to leave the site. Rather than having to guess what was happening on the site, investigators could track the action in real-time.

Using the new online customer experience management solution allowed Art.com to make several improvements to its Web site that contributed to a significant increase in conversion rate and prevented possible disasters.

One example of disaster recovery took place when Art.com sent sale coupons to many of its customers. Unfortunately, the coupon numbers were not entered into the back-end system, so when customers with coupons checked out, they received an "invalid coupon code" error message. Most abandoned their purchase at that point. Art.com's new Web analytics tool alerted management to the problem within hours. The coupon codes were added to the back-end system, and because user data was collected by the system, Art.com contacted those who were frustrated by the error and entice them back. The quick correction of the problem probably saved Art.com $25,000 of revenue per day.

In another example, the checkout process at Art.com's French site was displaying error messages to customers using outdated browsers. Art.com's new system caught the problem when the alarm was raised and management corrected the problem within days. Again, customers who experienced difficulties were contacted and enticed to return.

In a third example, Art.com's new system showed investigators that up to 20,000 visitors referred by Web search engines were greeted with a page that informed them that "sorry, this product is no longer available." Web developers at Art.com changed the message to be less negative and more inviting by providing alternative products that might interest those visitors.

Systems analysts depend on tools to provide them with information on which portions of systems are working and which are not. Analyzing a Web site such as Art.com is like analyzing pedestrian traffic in a major city—it's impossible without appropriate tools. Using powerful Web analytics and an online customer experience management system, systems analysts can continuously review and investigate the effect that the system is having on Web site visitors, launching systems development projects as needed.

Discussion Questions

1. What was Art.com's biggest challenge in improving their customer's online experience?
2. How does the new online system allow Art.com to launch systems development projects that can improve sales?

Critical Thinking Questions

1. What are some useful functions of a good Web analytics and online customer experience management system?
2. During which stages of the systems development life cycle can Web analytics be useful, and why?

SOURCES: Tealeaf staff, "Art.com: Purveyor of the World's Largest Selection of Wall Décor," Computerworld/TeaLeaf, 2007, *http://zones.computerworld.com/tealeaf_customer_exp/registration.php?item=13&from= cw&src=cwlp*; Art.com Web site, *www.art.com*, accessed July 12, 2008; Tealeaf Web site, *www.tealeaf.com*, accessed July 12, 2008.

SYSTEMS DEVELOPMENT LIFE CYCLES

The systems development process is also called a *systems development life cycle (SDLC)* because the activities associated with it are ongoing. As each system is built, the project has timelines and deadlines, until at last the system is installed and accepted. The life of the system continues as it is maintained and reviewed. If the system needs significant improvement beyond the scope of maintenance, if it needs to be replaced because of a new generation of technology, or if the IS needs of the organization change significantly, a new project will be initiated and the cycle will start over.

The Traditional Systems Development Life Cycle

Traditional systems development efforts can range from a small project, such as purchasing an inexpensive computer program, to a major undertaking. The steps of traditional systems development might vary from one company to the next, but most approaches have five common phases: investigation, analysis, design, implementation, and maintenance and review (see Figure 8.4).

Figure 8.4

The Traditional Systems Development Life Cycle

Sometimes, information learned in a particular phase requires cycling back to a previous phase.

systems investigation
The systems development phase during which problems and opportunities are identified and considered in light of the goals of the business.

systems analysis
The systems development phase that determines what the information system must do to solve the problem by studying existing systems and work processes to identify strengths, weaknesses, and opportunities for improvement.

systems design
The systems development phase that defines how the information system will do what it must do to obtain the problem solution.

In the **systems investigation** phase, potential problems and opportunities are identified and considered in light of the goals of the business. Systems investigation attempts to answer the questions "What is the problem, and is it worth solving?" The primary result of this phase is a defined development project for which business problems or opportunity statements have been created, to which some organizational resources have been committed, and for which systems analysis is recommended. **Systems analysis** attempts to answer the question "What must the information system do to solve the problem?" This phase involves studying existing systems and work processes to identify strengths, weaknesses, and opportunities for improvement. The major outcome of systems analysis is a list of requirements and priorities. **Systems design** seeks to answer the question "How will the information system do what it must do to obtain the problem solution?" The primary result of this phase is a technical design that either describes the new system or describes how existing systems will be modified.

The systems design details system outputs, inputs, and user interfaces; specifies hardware, software, database, telecommunications, personnel, and procedure components; and shows how these components are related. **Systems implementation** involves creating or acquiring the various system components detailed in the systems design, assembling them, and placing the new or modified system into operation.[13] An important task during this phase is to train the users. Systems implementation results in an installed, operational information system that meets the business needs for which it was developed. It can also involve phasing out or removing old systems, which can be difficult for existing users, especially when the systems are free. In 2005, Walt Disney developed the *Virtual Magic Kingdom (VMK)* game to celebrate the fiftieth anniversary of Disneyland.[14] The VMK game used Disney avatars and offered virtual rewards to game players. When Disney decided to remove or terminate the game, some players were outraged and protested outside Disney offices in California.

The purpose of **systems maintenance and review** is to ensure that the system operates and to modify the system so that it continues to meet changing business needs. As shown in Figure 8.4, a system under development moves from one phase of the traditional SDLC to the next.

Prototyping

Prototyping takes an iterative approach to the systems development process. During each iteration, requirements and alternative solutions to the problem are identified and analyzed, new solutions are designed, and a portion of the system is implemented.[15] Users are then encouraged to try the prototype and provide feedback (see Figure 8.5). Prototyping begins with creating a preliminary model of a major subsystem or a scaled-down version of the entire system. For example, a prototype might show sample report formats and input screens. After they are developed and refined, the prototypical reports and input screens are used as models for the actual system, which can be developed using an end-user programming language such as Visual Basic. The first preliminary model is refined to form the second- and third-generation models, and so on until the complete system is developed.

Rapid Application Development, Agile Development, Joint Application Development, and Other Systems Development Approaches

Rapid application development (RAD) employs tools, techniques, and methodologies designed to speed application development. Vendors, such as Computer Associates International, IBM, and Oracle, market products targeting the RAD market. Rational Software, a division of IBM, has a RAD tool, called Rational Rapid Developer, to make developing large Java programs and applications easier and faster. Locus Systems, a program developer, used a RAD tool called OptimalJ to generate more than 60 percent of the computer code for three applications it developed. Royal Bank of Canada used OptimalJ to develop some customer-based applications. According to David Hewick, group manager of application architecture for the bank, "It was an opportunity to improve the development life cycle, reduce costs, and bring consistency." Advantage Gen, formerly known as COOL:Gen, is a RAD tool from Computer Associates International. It can be used to rapidly generate computer code from business models and specifications.[16]

Other approaches to rapid development, such as *agile development* or *extreme programming* (*XP*), allow the systems to change as they are being developed. Agile development requires frequent face-to-face meetings with the systems developers and users as they modify, refine, and test how the system meets users' needs and what its capabilities are. Microsoft, for example, has adopted a more agile development process in its server development division.[17] According to a Microsoft senior vice president, "We just realized that we're building products for customers, not just for technology's sake. So the sooner we could engage with our customers, the better we could make it from an architecture, feature, quality, and scalability perspective." BT Group, a large British telecommunications company, uses agile systems development to substantially reduce development time and increase customer satisfaction.[18] According to BT's managing director of service design, "BT's shift from

Figure 8.5

Prototyping

Prototyping is an iterative approach to systems development.

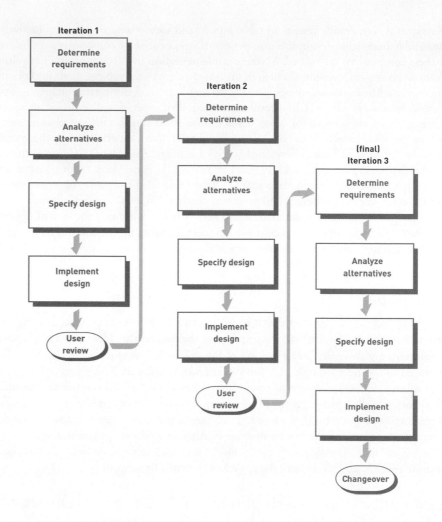

joint application development (JAD)

A process for data collection and requirements analysis in which users, stakeholders, and IS professionals work together to analyze existing systems, propose possible solutions, and define the requirements of a new or modified system.

traditional waterfall development techniques to an agile approach has led to significant productivity and business benefits, but it didn't happen overnight, nor was it easy for a company as massive and widespread as BT." Extreme programming (XP) uses pairs of programmers who work together to design, test, and code parts of the systems they develop.[19] The iterative nature of XP helps companies develop robust systems with fewer errors. Sabre Airline Solutions, a $2 billion computer company serving the airline travel industry, used XP to eliminate programming errors and shorten program development times.

RAD makes extensive use of the **joint application development (JAD)** process for data collection and requirements analysis. Originally developed by IBM Canada in the 1970s, JAD involves group meetings in which users, stakeholders, and IS professionals work together to analyze existing systems, propose possible solutions, and define the requirements of a new or modified system. Today, JAD often uses *group support systems (GSS)* software to foster positive group interactions, while suppressing negative group behavior. Boeing, for example, used RAD and JAD to help develop software for its airplanes.[20] Group support systems were introduced in Chapter 6.

In addition to the systems development approaches discussed previously, a number other systems development approaches are available, including adaptive software development, lean software development, Rational Unified Process (RUP), Feature-Driven Development (FDD), and dynamic systems development method. Often created by computer vendors and authors of systems development books, these approaches all attempt to deliver better systems. The Ohio Causality Corporation, for example, uses RUP from IBM and Rational Software. RUP uses an iterative approach to software development that concentrates on software quality as it is changed and updated over time.[21] Many other companies have also used RUP to their advantage.[22]

The End-User Systems Development

The term **end-user systems development** describes any systems development project in which business managers and users assume the primary effort. User-developed systems range from the very small (such as a software routine to merge form letters) to those of significant organizational value (such as customer contact databases for the Web). With end-user systems development, managers and other users can get the systems they want without having to wait for IS professionals to develop and deliver them.[23] End-user systems development, however, does have some disadvantages. Some end users don't have the training to effectively develop and test a system. Multimillion-dollar mistakes, for example, can be made using faulty spreadsheets that were never tested. Some end-user systems are also poorly documented. When these systems are updated, problems can be introduced that make the systems error-prone. In addition, some end users spend time and corporate resources developing systems that were already available.

end-user systems development
A systems development project in which business managers and users assume the primary effort.

Many end users today are demonstrating their systems development capability by designing and implementing their own PC-based systems.

(Source: © Daniel Allan/Getty Images.)

Outsourcing and On-Demand Computing

Many companies hire an outside consulting firm or computer company that specializes in systems development to take over some or all of its development and operations activities.[24] Some companies, such as General Electric, have their own outsourcing subunits or have spun off their outsourcing subunits as separate companies. As mentioned in Chapter 1, *outsourcing* and *on-demand computing* are often used.[25]

Increasingly, small and medium-sized firms are using outsourcing to cut costs and acquire needed technical expertise that would be difficult to afford with in-house personnel. Millennium Partners Sports Club Management, for example, used Center Beam to outsource many of its IS functions, including its help desk operations. The Boston-based company plans to spend about $30,000 a month on outsourcing services, which it estimates to be less than it would have to pay in salaries for additional employees.[26] According to a company vice president, "If we hadn't outsourced, I couldn't focus 100 percent on things that can drive the company forward." The market for outsourcing services for small and medium-sized firms is expected to increase by 15 percent annually through 2010 and beyond.

Reducing costs, obtaining state-of-the-art technology, eliminating staffing and personnel problems, and increasing technological flexibility are reasons that companies have used the outsourcing and on-demand computing approaches.[27] Reducing costs is a primary reason for outsourcing. One American computer company, for example, estimated that a programmer with three to five years of experience in China would cost about $13 per hour, while a programmer with similar experience in the United States would cost about $56 per hour. U.S. companies also provide outsourcing services. Aelera Corporation spent about six months looking for the best outsourcing deal and determined that a company in Savannah, Georgia, was the best. McKesson Corporation saved about $10 million by outsourcing jobs from San Francisco to Dubuque, Iowa. Mattel outsourced to rural Jonesboro, Arkansas. Increasingly, companies are looking to American outsourcing companies to reduce costs and increase services. Individuals, including students, are also outsourcing tasks they have to perform.

A number of companies and nonprofit organizations offer outsourcing and on-demand computing services—from general systems development to specialized services.[28] IBM's Global Services, for example, is one of the largest full-service outsourcing and consulting services.[29] IBM has consultants located in offices around the world. In India, IBM has increased its employees from less than 10,000 people to more than 30,000.[30] The company looks for skilled and talented workers. According to Amitabh Ray, a vice president of consulting and application services at IBM in India, "We're not a body shop. We need the right kind of people." Electronic Data Systems (EDS) is another large company that specializes in consulting and outsourcing.[31] EDS has approximately 140,000 employees in almost 60 countries and more than 9,000 clients worldwide. Accenture is another company that specializes in consulting and outsourcing.[32] The company has more than 75,000 employees in 47 countries. Amazon, the large online retailer of books and other products, will offer on-demand computing to individuals and other companies of all sizes, allowing them to use Amazon's computer expertise and database capacity.[33] Individuals and companies will only pay for the computer services they use. See Figure 8.6.

Figure 8.6

With more than 75,000 employees in 47 countries, Accenture specializes in consulting and outsourcing.

(Source: © Namas Bhojani/ Bloomberg News/Landov.)

Use of Computer-Aided Software Engineering (CASE) Tools

computer-aided software engineering (CASE)
Tools that automate many of the tasks required in a systems development effort and encourage adherence to the SDLC.

Computer-aided software engineering (CASE) tools automate many of the tasks required in a systems development effort and encourage adherence to the SDLC, thus instilling a high degree of rigor and standardization to the entire systems development process. Prover Technology has developed a CASE tool that searches for programming bugs. The CASE tool searches for all possible design scenarios to make sure that the program is error free. Other CASE tools include Visible Systems (*www.visible.com*), Popkin Software (*www.popkin.com*), Rational Rose (part of IBM), and Visio, a charting and graphics program from Microsoft. Companies that produce CASE tools include Accenture, Microsoft, and Oracle. Oracle Designer and Developer CASE tools, for example, can help systems analysts automate and simplify the development process for database systems. See Table 8.1 for a list of CASE tools and their providers. The advantages and disadvantages of CASE tools are listed in Table 8.2. CASE tools that focus on activities associated with the early stages of systems development are often called *upper-CASE* tools. These packages provide automated tools to assist with systems investigation, analysis, and design activities. Other CASE packages, called *lower-CASE* tools, focus on the later implementation stage of systems development, and can automatically generate structured program code.

CASE Tool	Vendor
Oracle Designer	Oracle Corporation *www.oracle.com*
Visible Analyst	Visible Systems Corporation *www.visible.com*
Rational Rose	Rational Software *www.ibm.com*
Embarcadero Describe	Embarcadero Describe *www.embarcadero.com*

Table 8.1

Typical CASE Tools

Advantages	Disadvantages
Produce systems with a longer effective operational life	Increase the initial costs of building and maintaining systems
Produce systems that more closely meet user needs and requirements	Require more extensive and accurate definition of user needs and requirements
Produce systems with excellent documentation	Can be difficult to customize
Produce systems that need less systems support	Require more training of maintenance staff
Produce more flexible systems	Can be difficult to use with existing systems

Table 8.2

Advantages and Disadvantages of CASE Tools

Object-Oriented Systems Development

The success of a systems development effort can depend on the specific programming tools and approaches used. As mentioned in Chapter 2, object-oriented (OO) programming languages allow the interaction of programming objects—that is, an object consists of both data and the actions that can be performed on the data. So, an object could be data about an employee and all the operations (such as payroll, benefits, and tax calculations) that might be performed on the data.

Developing programs and applications using OO programming languages involves constructing modules and parts that can be reused in other programming projects. DTE Energy, a $7 billion Detroit-based energy company, has set up a library of software components that can be reused by its programmers. Systems developers from the company reuse and contribute to software components in the library. DTE's developers meet frequently to discuss ideas, problems, and opportunities of using the library of reusable software components.

Object-oriented systems development (OOSD) combines the logic of the systems development life cycle with the power of object-oriented modeling and programming. OOSD follows a defined systems development life cycle, much like the SDLC. The life cycle phases are usually completed with many iterations. Object-oriented systems development typically involves the following tasks:

- **Identifying potential problems and opportunities within the organization that would be appropriate for the OO approach.** This process is similar to traditional systems investigation. Ideally, these problems or opportunities should lend themselves to the development of programs that can be built by modifying existing programming modules.
- **Defining what kind of system users require.** This analysis means defining all the objects that are part of the user's work environment (object-oriented analysis). The OO team must study the business and build a model of the objects that are part of the business (such as a customer, an order, or a payment). Many of the CASE tools discussed in the previous section can be used, starting with this step of OOSD.
- **Designing the system.** This process defines all the objects in the system and the ways they interact (object-oriented design). Design involves developing logical and physical models of the new system by adding details to the object model started in analysis.

object-oriented systems development (OOSD)
An approach to systems development that combines the logic of the systems development life cycle with the power of object-oriented modeling and programming.

- **Programming or modifying modules.** This implementation step takes the object model begun during analysis and completed during design and turns it into a set of interacting objects in a system. Object-oriented programming languages are designed to allow the programmer to create classes of objects in the computer system that correspond to the objects in the actual business process. Objects such as customer, order, and payment are redefined as computer system objects—a customer screen, an order entry menu, or a dollar sign icon. Programmers then write new modules or modify existing ones to produce the desired programs.
- **Evaluation by users.** The initial implementation is evaluated by users and improved. Additional scenarios and objects are added, and the cycle repeats. Finally, a complete, tested, and approved system is available for use.
- **Periodic review and modification.** The completed and operational system is reviewed at regular intervals and modified as necessary.

SYSTEMS INVESTIGATION

As discussed earlier in the chapter, systems investigation is the first phase in the traditional SDLC of a new or modified business information system. The purpose is to identify potential problems and opportunities and consider them in light of the goals of the company. In general, systems investigation attempts to uncover answers to the following questions:

- What primary problems might a new or enhanced system solve?
- What opportunities might a new or enhanced system provide?
- What new hardware, software, databases, telecommunications, personnel, or procedures will improve an existing system or are required in a new system?
- What are the potential costs (variable and fixed)?
- What are the associated risks?

Initiating Systems Investigation

systems request form
A document filled out by someone who wants the IS department to initiate systems investigation.

Because systems development requests can require considerable time and effort to implement, many organizations have adopted a formal procedure for initiating systems development, beginning with systems investigation. The **systems request form** is a document that is filled out by someone who wants the IS department to initiate systems investigation. This form typically includes the following information:

- Problems in or opportunities for the system
- Objectives of systems investigation
- Overview of the proposed system
- Expected costs and benefits of the proposed system

The information in the systems request form helps to rationalize and prioritize the activities of the IS department. Based on the overall IS plan, the organization's needs and goals, and the estimated value and priority of the proposed projects, managers make decisions regarding the initiation of each systems investigation for such projects.

Feasibility Analysis

feasibility analysis
Assessment of the technical, economic, legal, operational, and schedule feasibility of a project.

technical feasibility
Assessment of whether the hardware, software, and other system components can be acquired or developed to solve the problem.

A key step of the systems investigation phase is **feasibility analysis**, which assesses technical, economic, legal, operational, and schedule feasibility (see Figure 8.7). **Technical feasibility** is concerned with whether the hardware, software, and other system components can be acquired or developed to solve the problem.

Technical

Economic

Legal

Operational

Schedule

Figure 8.7

Technical, Economic, Legal, Operational, and Schedule Feasibility

Economic feasibility determines whether the project makes financial sense and whether predicted benefits offset the cost and time needed to obtain them. A securities company, for example, investigated the economic feasibility of sending research reports electronically instead of through the mail. Economic analysis revealed that the new approach could save the company up to $500,000 per year. Economic feasibility can involve cash flow analysis such as that done in net present value or internal rate of return (IRR) calculations.

Legal feasibility determines whether laws or regulations can prevent or limit a systems development project. For example, a Web site that allowed users to share music without paying musicians or music producers was sued. Legal feasibility should have identified this vulnerability during the Web site development project. Legal feasibility involves an analysis of existing and future laws to determine the likelihood of legal action against the systems development project and the possible consequences.

Operational feasibility is a measure of whether the project can be put into action or operation. It can include logistical and motivational (acceptance of change) considerations. Motivational considerations are important because new systems affect people and data flows and can have unintended consequences. As a result, power and politics might come into play, and some people might resist the new system. On the other hand, recall that a new system can help avoid major problems. For example, because of deadly hospital errors, a healthcare consortium looked into the operational feasibility of developing a new computerized physician order-entry system to require that all prescriptions and every order a doctor gives to staff be entered into the computer. The computer then checks for drug allergies and interactions between drugs. If operationally feasible, the new system could save lives and help avoid lawsuits.

Schedule feasibility determines whether the project can be completed in a reasonable amount of time—a process that involves balancing the time and resource requirements of the project with other projects.

economic feasibility
The determination of whether the project makes financial sense and whether predicted benefits offset the cost and time needed to obtain them.

legal feasibility
The determination of whether laws or regulations may prevent or limit a systems development project.

operational feasibility
The measure of whether the project can be put into action or operation.

schedule feasibility
The determination of whether the project can be completed in a reasonable amount of time.

Object-Oriented Systems Investigation

The object-oriented approach can be used during all phases of systems development, from investigation to maintenance and review. Consider a kayak rental business in Maui, Hawaii, where the owner wants to computerize its operations, including renting kayaks to customers and adding new kayaks into the rental program (see Figure 8.8). As you can see, the kayak rental clerk rents kayaks to customers and adds new kayaks to the current inventory of kayaks available for rent. The stick figure is an example of an *actor*, and the ovals each represent an event, called a *use case*. In our example, the actor (the kayak rental clerk) interacts with two use cases (rent kayaks to customers and add new kayaks to inventory). The use case diagram is part of the Unified Modeling Language (UML) that is used in object-oriented systems development.

Figure 8.8

Use Case Diagram for a Kayak Rental Application

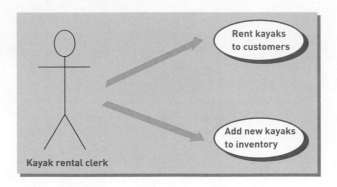

The Systems Investigation Report

systems investigation report
A summary of the results of the systems investigation and the process of feasibility analysis and recommendation of a course of action.

The primary outcome of systems investigation is a **systems investigation report**, also called a *feasibility study*. This report summarizes the results of systems investigation and the process of feasibility analysis and recommends a course of action: Continue on into systems analysis, modify the project in some manner, or drop it. A typical table of contents for the systems investigation report is shown in Figure 8.9.

Figure 8.9

A Typical Table of Contents for a Systems Investigation Report

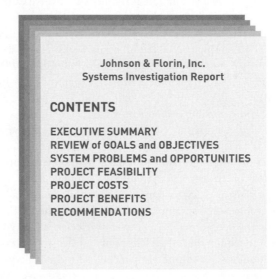

steering committee
An advisory group consisting of senior management and users from the IS department and other functional areas.

The systems investigation report is reviewed by senior management, often organized as an advisory committee, or **steering committee**, consisting of senior management and users from the IS department and other functional areas. These people help IS personnel with their decisions about the use of information systems in the business and give authorization to pursue further systems development activities. After review, the steering committee might agree with the recommendation of the systems development team or suggest a change in project focus to concentrate more directly on meeting a specific company objective. Another alternative is that everyone might decide that the project is not feasible and cancel the project.

SYSTEMS ANALYSIS

After a project has been approved for further study, the next step is to answer the question "What must the information system do to solve the problem?" The process needs to go beyond mere computerization of existing systems. The entire system, and the business process with which it is associated, should be evaluated. Often, a firm can make great gains if it restructures both business activities and the related information system simultaneously. The overall emphasis of analysis is gathering data on the existing system, determining the requirements for the new system, considering alternatives within these constraints, and investigating the feasibility of the solutions. The primary outcome of systems analysis is a prioritized list of systems requirements. During its systems analysis phase, Mobius Management Systems (*www.mobius.com*), a company that manages databases and data resources for other companies, determined that the physical size of its data centers was an important systems requirement. Its current data centers were simply too large.[34] According to one IS administrator, "We were taking over what formerly were people's offices and making them data centers." The company analyzed the impact of replacing more than 100 of its hardware servers for software virtualization that allowed multiple applications to run on a single server, saving a tremendous amount of space.

Data Collection

The purpose of data collection is to seek additional information about the problems or needs identified in the systems investigation report. During this process, the strengths and weaknesses of the existing system are emphasized.

Identifying Sources of Data

Data collection begins by identifying and locating the various sources of data, including both internal and external sources (see Figure 8.10).

Figure 8.10

Internal and External Sources of Data for Systems Analysis

Internal Sources

Users, stakeholders, and managers

Organization charts

Forms and documents

Procedure manuals and policies

Financial reports

IS manuals

Other measures of business process

External Sources

Customers

Suppliers

Stockholders

Government agencies

Competitors

Outside groups

Journals, etc.

Consultants

Collecting Data

After data sources have been identified, data collection begins. Figure 8.11 shows the steps involved. Data collection might require a number of tools and techniques, such as interviews, direct observation, and questionnaires.

Interviews can either be structured or unstructured. In a **structured interview**, the questions are written in advance. In an **unstructured interview**, the questions are not written in advance; the interviewer relies on experience in asking the best questions to uncover the inherent problems of the existing system. An advantage of the unstructured interview is that it allows the interviewer to ask follow-up or clarifying questions immediately. With **direct observation**, one or more members of the analysis team directly observe the existing system in action. When many data sources are spread over a wide geographic area, **questionnaires** might be the best method. Like interviews, questionnaires can be either structured or unstructured.

structured interview
An interview where the questions are written in advance.

unstructured interview
An interview where the questions are not written in advance.

direct observation
Watching the existing system in action by one or more members of the analysis team.

questionnaires
A method of gathering data where the data sources are spread over a wide geographic area.

Direct observation is a method of data collection. One or more members of the analysis team directly observe the existing system in action.

(Source: © Kriss Russell / iStockphoto.)

Data Analysis

The data collected in its raw form is usually not adequate to determine the effectiveness of the existing system or the requirements for the new system. The next step is to manipulate the collected data so that the development team members who are participating in systems analysis can use the data. This manipulation is called **data analysis**. Data and activity modeling and using data-flow diagrams and entity-relationship diagrams are useful during data analysis to show data flows and the relationships among various objects, associations, and activities. Other common tools and techniques for data analysis include application flowcharts, grid charts, CASE tools, and the object-oriented approach.

Data Modeling

Data modeling, first introduced in Chapter 3, is a commonly accepted approach to modeling organizational objects and associations that employ both text and graphics. How data modeling is employed, however, is governed by the specific systems development methodology.

Data modeling is most often accomplished through the use of entity-relationship (ER) diagrams. Recall from Chapter 3 that an entity is a generalized representation of an object type—such as a class of people (employee), events (sales), things (desks), or places (city)—and that entities possess certain attributes. Objects can be related to other objects in many ways. An entity-relationship diagram, such as the one shown in Figure 8.12a, describes a number of objects and the ways they are associated. An ER diagram (or any other modeling tool) cannot by itself fully describe a business problem or solution because it lacks descriptions of the related activities. It is, however, a good place to start because it describes object types and attributes about which data might need to be collected for processing.

data analysis
The manipulation of collected data so that the development team members who are participating in systems analysis can use the data.

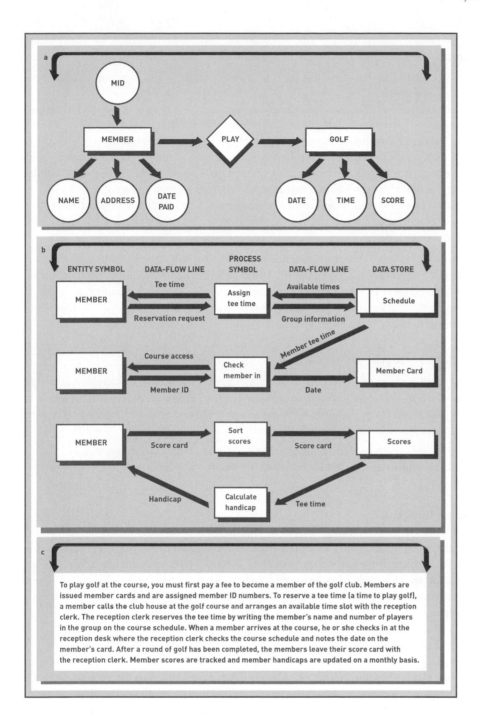

Figure 8.12

Data and Activity Modeling

(a) An entity-relationship diagram. (b) A data-flow diagram. (c) A semantic description of the business process.

(Source: G. Lawrence Sanders, *Data Modeling*, Boyd & Fraser Publishing, Danvers, MA: 1995.)

Activity Modeling

To fully describe a business problem or solution, the related objects, associations, and activities must be described. Activities in this sense are events or items that are necessary to fulfill the business relationship or that can be associated with the business relationship in a meaningful way.

Activity modeling is often accomplished through the use of data-flow diagrams. A **data-flow diagram (DFD)** models objects, associations, and activities by describing how data can flow between and around various objects. DFDs work on the premise that every activity involves some communication, transference, or flow that can be described as a data element. DFDs describe the activities that fulfill a business relationship or accomplish a business task, not how these activities are to be performed. That is, DFDs show the logical sequence of associations and activities, not the physical processes. A system modeled with a DFD could operate manually or could be computer based; if computer based, the system could operate with a variety of technologies.

data-flow diagram (DFD)
A model of objects, associations, and activities that describes how data can flow between and around various objects.

data-flow line
Arrows that show the direction of data element movement.

process symbol
Representation of a function that is performed.

entity symbol
Representation of either a source or destination of a data element.

data store
Representation of a storage location for data.

DFDs are easy to develop and easily understood by nontechnical people. Data-flow diagrams use four primary symbols, as illustrated in Figure 8.12b.

- **Data flow.** The **data-flow line** includes arrows that show the direction of data element movement.
- **Process symbol.** The **process symbol** reveals a function that is performed. Computing gross pay, entering a sales order, delivering merchandise, and printing a report are examples of functions that can be represented with a process symbol.
- **Entity symbol.** The **entity symbol** shows either the source or destination of the data element. An entity can be, for example, a customer who initiates a sales order, an employee who receives a paycheck, or a manager who receives a financial report.
- **Data store.** A **data store** reveals a storage location for data. A data store is any computerized or manual data storage location, including magnetic tape, disks, a filing cabinet, or a desk.

Comparing entity-relationship diagrams with data-flow diagrams provides insight into the concept of top-down design. Figure 8.12a and b show an entity-relationship diagram and a data-flow diagram for the same business relationship—namely, a member of a golf club playing golf. Figure 8.12c provides a brief description of the business relationship for clarification.

Requirements Analysis

requirements analysis
The determination of user, stakeholder, and organizational needs.

The overall purpose of **requirements analysis** is to determine user, stakeholder, and organizational needs.[35] For an accounts payable application, the stakeholders could include suppliers and members of the purchasing department. Questions that should be asked during requirements analysis include the following:

- Are these stakeholders satisfied with the current accounts payable application?
- What improvements could be made to satisfy suppliers and help the purchasing department?

Asking Directly

asking directly
An approach to gather data that asks users, stakeholders, and other managers about what they want and expect from the new or modified system.

One the most basic techniques used in requirements analysis is asking directly. **Asking directly** is an approach that asks users, stakeholders, and other managers about what they want and expect from the new or modified system. This approach works best for stable systems in which stakeholders and users clearly understand the system's functions. The role of the systems analyst during the analysis phase is to critically and creatively evaluate needs and define them clearly so that the systems can best meet them.

Critical Success Factors

Another approach uses critical success factors (CSFs). As discussed earlier, managers and decision makers are asked to list only the factors that are critical to the success of their area of the organization. A CSF for a production manager might be adequate raw materials from suppliers; a CSF for a sales representative could be a list of customers currently buying a certain type of product. Starting from these CSFs, the system inputs, outputs, performance, and other specific requirements can be determined.

The IS Plan

As we have seen, the IS plan translates strategic and organizational goals into systems development initiatives. The IS planning process often generates strategic planning documents that can be used to define system requirements. Working from these documents ensures that requirements analysis will address the goals set by top-level managers and decision makers (see Figure 8.13). There are unique benefits to applying the IS plan to define systems requirements. Because the IS plan takes a long-range approach to using information technology within the organization, the requirements for a system analyzed in terms of the IS plan are more likely to be compatible with future systems development initiatives.

Figure 8.13

Converting Organizational
Goals into Systems
Requirements

Requirements Analysis Tools

A number of tools can be used to document requirements analysis, including CASE tools. As requirements are developed and agreed on, entity-relationship diagrams, data-flow diagrams, screen and report layout forms, and other types of documentation are stored in the CASE repository. These requirements might also be used later as a reference during the rest of systems development or for a different systems development project.

Object-Oriented Systems Analysis

The object-oriented approach can also be used during systems analysis. Like traditional analysis, problems or potential opportunities are identified during object-oriented analysis. Identifying key participants and collecting data is still performed. But instead of analyzing the existing system using data-flow diagrams and flowcharts, an object-oriented approach is used.

In the section "Object-Oriented Systems Investigation," we introduced a kayak rental example. A more detailed analysis of that business reveals that there are two classes of kayaks: single kayaks for one person and tandem kayaks that can accommodate two people. With the OO approach, a class is used to describe different types of objects, such as single and tandem kayaks. The classes of kayaks can be shown in a generalization/specialization hierarchy diagram (see Figure 8.14). KayakItem is an object that will store the kayak identification number (ID) and the date the kayak was purchased (datePurchased).

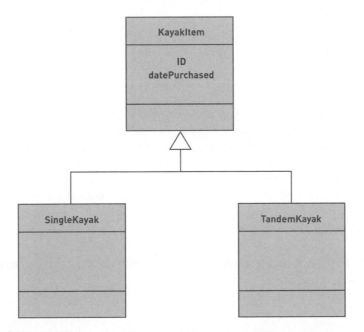

Figure 8.14

Generalization/Specialization
Hierarchy Diagram for Single
and Tandem Kayak Classes

Of course, there could be subclasses of customers, life vests, paddles, and other items in the system. For example, price discounts for kayak rentals could be given to seniors (people over 65 years) and students. Thus, the Customer class could be divided into regular, senior, and student customer subclasses.

The Systems Analysis Report

Systems analysis concludes with a formal systems analysis report. It should cover the following elements:

- The strengths and weaknesses of the existing system from a stakeholder's perspective
- The user/stakeholder requirements for the new system (also called the *functional requirements*)
- The organizational requirements for the new system
- A description of what the new information system should do to solve the problem

Suppose analysis reveals that a marketing manager thinks a weakness of the existing system is its inability to provide accurate reports on product availability. These requirements and a preliminary list of the corporate objectives for the new system will be in the systems analysis report. Particular attention is placed on areas of the existing system that could be improved to meet user requirements. The table of contents for a typical report is shown in Figure 8.15.

Figure 8.15

A Typical Table of Contents for a Report on an Existing System

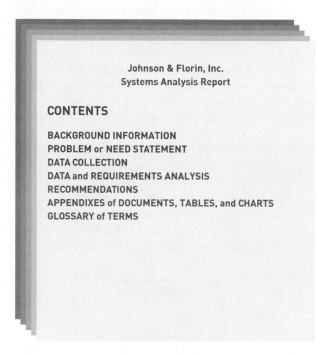

Johnson & Florin, Inc.
Systems Analysis Report

CONTENTS

BACKGROUND INFORMATION
PROBLEM or NEED STATEMENT
DATA COLLECTION
DATA and REQUIREMENTS ANALYSIS
RECOMMENDATIONS
APPENDIXES of DOCUMENTS, TABLES, and CHARTS
GLOSSARY of TERMS

The systems analysis report gives managers a good understanding of the problems and strengths of the existing system. If the existing system is operating better than expected or the necessary changes are too expensive relative to the benefits of a new or modified system, the systems development process can be stopped at this stage. If the report shows that changes to another part of the system might be the best solution, the development process might start over, beginning again with systems investigation. Or, if the systems analysis report shows that it will be beneficial to develop one or more new systems or to make changes to existing ones, systems design, which is discussed next.

SYSTEMS DESIGN

systems design
The stage of systems development that answers the question "How will the information system solve a problem?"

The purpose of **systems design** is to answer the question "How will the information system solve a problem?" The primary result of the systems design phase is a technical design that details system outputs, inputs, and user interfaces; specifies hardware, software, databases, telecommunications, personnel, and procedures; and shows how these components are related. The new or modified system should take advantage of the latest developments in technology.[36] Many companies, for example, are looking into cloud computing, where applications are run on the Internet instead of being developed and run within the company or organization. Cloud computing is allowing individuals, such as racecar driving instructor Tom Dyer, to work while traveling or in a car.[37] According to Dyer, "Anywhere I go, I can hook up to the Net for whatever reason. It makes life a lot easier." Increasingly, companies and individuals are developing or purchasing systems that allow them to take advantage of the Internet.[38] Microsoft's Live Mesh, for example, allows systems developers to seamlessly coordinate data among different devices and provide data backup on the Internet.

Systems design is typically accomplished using the tools and techniques discussed earlier in this chapter. Depending on the specific application, these methods can be used to support and document all aspects of systems design. Two key aspects of systems design are logical and physical design.

Logical and Physical Design

Design has two dimensions: logical and physical. The **logical design** refers to what the system will do. Logical design describes the functional requirements of a system.[39] That is, it conceptualizes what the system will do to solve the problems identified through earlier analysis. Without this step, the technical details of the system (such as which hardware devices should be acquired) often obscure the best solution. Logical design involves planning the purpose of each system element, independent of hardware and software considerations. The logical design specifications that are determined and documented include output, input, process, file and database, telecommunications, procedures, controls and security, and personnel and job requirements.

Security is always an important logical design issue for corporations and governments.[40] Rules published in September 2005, for example, require that federal agencies incorporate security procedures in the design of new or modified systems. In addition, the Federal Information Security Management Act, enacted in 2002, requires federal agencies to make sure that security protection measures are incorporated into systems provided by outside vendors and contractors. The Federal Rules of Civil Procedure requires that companies make emails, text messages, and other electronic communications available in some court hearings.[41] Failure to meet these electronic disclosure requirements in a timely fashion can result in executives and managers facing fines and jail time. This requirement has spurred new systems development projects that can search and find electronic communications to meet federal requirements.[42]

The **physical design** refers to how the tasks are accomplished, including how the components work together and what each component does. Physical design specifies the characteristics of the system components necessary to put the logical design into action. In this phase, the characteristics of the hardware, software, database, telecommunications, personnel, and procedure and control specifications must be described in detail. These physical design components were discussed in Part 2 on technology.

> **logical design**
> A description of the functional requirements of a system.

> **physical design**
> The specification of the characteristics of the system components necessary to put the logical design into action.

Object-Oriented Design

Logical and physical design can be accomplished using either the traditional approach or the object-oriented approach to systems development. Both approaches use a variety of design models to document the new system's features and the development team's understandings and agreements. Many organizations today are turning to OO development because of its increased flexibility. This section outlines a few OO design considerations and diagrams.[43]

Using the OO approach, you can design key objects and classes of objects in the new or updated system. This process includes considering the problem domain, the operating environment, and the user interface. The problem domain involves the classes of objects related to solving a problem or realizing an opportunity. In our Maui, Hawaii, kayak rental shop example and referring back to the generalization/specialization hierarchy showing classes we presented there, KayakItem in Figure 8.14 is an example of a problem domain object that will store information on kayaks in the rental program. The operating environment for the rental shop's system includes objects that interact with printers, system software, and other software and hardware devices. The user interface for the system includes objects that users interact with, such as buttons and scroll bars in a Windows program.

During the design phase, you also need to consider the sequence of events that must happen for the system to function correctly. For example, you might want to design the sequence of events for adding a new kayak to the rental program. A sequence of events is often called a *scenario*, and it can be diagrammed in a sequence diagram (see Figure 8.16).

Figure 8.16

A Sequence Diagram to Add a New KayakItem Scenario

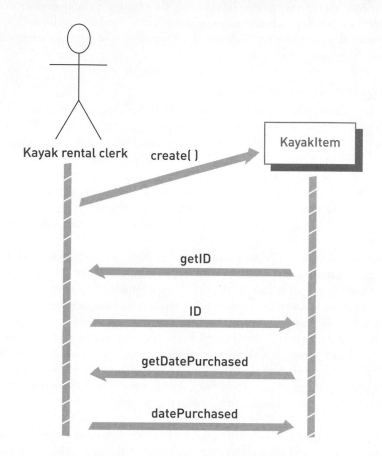

You read a sequence diagram starting at the top and moving down.

1. The Create arrow at the top is a message from the kayak rental clerk to the KayakItem object to create information on a new kayak to be placed into the rental program.
2. The KayakItem object knows that it needs the ID for the kayak and sends a message to the clerk requesting the information. See the getID arrow.
3. The clerk then types the ID into the computer. This is shown with the ID arrow. The data is stored in the KayakItem object.
4. Next, KayakItem requests the purchase date. This is shown in the getDatePurchased arrow.
5. Finally, the clerk types the purchase date into the computer. The data is also transferred to KayakItem object. This is shown in the datePurchased arrow at the bottom of Figure 8.16.

This scenario is only one example of a sequence of events. Other scenarios might include entering information about life jackets, paddles, suntan lotion, and other accessories. The same types of use case and generalization/specialization hierarchy diagrams discussed earlier in the chapter can be created for each event, and additional sequence diagrams will also be needed.

Going Green Saves Millions of Dollars for Nationwide

Nationwide is one of the largest insurance and financial services in the world. The company offers a full range of insurance products and financial services for home, car, and family. It has garnered over $161 billion in statutory assets.

With 36,000 employees managing 16 million policies, Nationwide requires a large data center to store and manipulate policy data. Actually, Nationwide has 20 data centers and a $250 million budget for its information system infrastructure. Nationwide's primary data center in Columbus, Ohio, supports roughly 400 million transactions per month for activities such as calculating policy quotes, making policy additions, changes and deletions, and processing claims.

Scott Miggo, vice president of Technology Solutions at Nationwide, manages the company's data centers. Scott continuously monitors demand on the data center's servers, power, and cooling. Scott has tracked a consistent five percent growth in data center processing from year to year. At this rate, he estimates that the data center demand will outstrip the power capacity of the company's primary data center by 2013.

Scott had a number of options for developing systems to meet future demand. He might expand by building a new data center to add to the processing power of the current center. Many companies would choose to begin construction on a new data center, retiring the old equipment in favor of using the latest energy-efficient technologies. If Nationwide began construction immediately, they could have the new data center online by 2013. However, a new data center would cost Nationwide hundreds of millions of dollars. Scott tried to find a solution that might forestall the inevitable.

Scott and his team found several solutions that would buy them two or more years beyond 2013 without having to invest in major construction. First, they began using virtualization with VMware. VMware allowed one large mainframe server to act as 20 virtual servers. By implementing virtual servers, Scott reduced the numbers of servers in the data center from 5,000 to 3,500. The VM servers were running at 65 percent usage, up from 10 percent. In essence, virtualization allowed Nationwide to get more work out of each server—freeing up space and lowering power and air conditioning demands.

Secondly, Scott and his team began replacing the oldest, energy-intensive servers with green servers. The new energy-efficient servers saved the data center $40,000 a year in electricity and cooling.

In another cost-saving effort, Scott and his team replaced tape silos with more modern, denser tapes and faster tape robots. The result was more efficient data storage and retrieval in a smaller amount of space. Although the savings from this upgrade were negligible compared to virtualization and server upgrades, every little bit helped. Scott says, "You've got to look at it holistically. We are looking at going to a massive array of idle disks that shut down and are brought up only when you need the data on a particular disk."

The total upgrade of the main data center set Nationwide back $30 million. This is a small amount compared to the hundreds of millions they would have spent constructing a new data center. Space is no longer an issue at the data center. However, Nationwide eventually needs to build an additional data center. They don't need the server space any more, but the infrastructure of the building will no longer support the power and cooling needs of the growing amount of servers—even the greenest, most energy-efficient models. Nationwide plans to continue virtualizing servers and upgrading to more energy-efficient models in its 20 data centers based on the model that Scott Miggo created.

Discussion Questions

1. What issues did Nationwide face with their data centers? What considerations determined the data center's processing capacity?
2. What three techniques did Scott Miggo implement to save Nationwide hundreds of millions of dollars?

Critical Thinking Questions

1. Why was it good for Scott Miggo to anticipate the needs of data processing ten years in advance? What luxuries did it afford him, and how did it pay off for Nationwide?
2. What other green technologies might be used to further extend the usefulness of Nationwide's primary data center and reduce the cost of operations?

SOURCES: Bartholomew, Doug, "Refurbishing Old Data Center Provides Big Savings," *CIO Insight*, September 11, 2007, *www.cioinsight.com/c/a/Case-Studies/Refurbishing-Old-Data-Center-Provides-Big-Savings*; Nationwide Web site, *www.nationwide.com*, accessed July 19, 2008.

ENVIRONMENTAL DESIGN CONSIDERATIONS

environmental design

Also called *green design*, it involves systems development efforts that slash power consumption, require less physical space, and result in systems that can be disposed in a way that doesn't negatively affect the environment.

Developing new systems and modifying existing ones in an environmentally sensitive way is becoming increasingly important for many IS departments.[44] **Environmental design**, also called *green design,* involves systems development efforts that slash power consumption, require less physical space, and result in systems that can be disposed in a way that doesn't negatively affect the environment. Today, companies are using innovative ways to design efficient systems and operations, including using virtual servers to save energy and space, pushing cold air under data centers to cool equipment, using software to efficiently control cooling fans, building facilities with more insulation, and even collecting rain water from roofs to cool equipment.[45] VistaPrint (*www.vistaprint.com*), a graphics design and printing company, switched from traditional servers to virtual servers and saved about $500,000 in electricity costs over a three-year period, representing a 75 percent reduction in energy usage.[46] A *Computerworld* survey revealed that over 80 percent of IS managers considered energy efficiency when selecting new computer equipment.[47] The Environmental Protection Agency (EPA) estimates that a 10 percent cut in data center electricity usage would be enough to power about a million U.S. homes every year.[48] According to a McKinsey & Co. study, the amount of greenhouse gases generated from data centers will increase by 400 percent by 2020 and become more than the greenhouse gases emitted by U.S. airlines.[49]

Many companies are developing products and services to help save energy. EMC, for example, has developed new disk drives that use substantially less energy.[50] Environmental design also involves developing software and systems that help organizations reduce power consumption for other aspects of their operations. Carbonetworks and Optimum Energy, for example, have developed software products to help companies reduce energy costs by helping them determine when and how to use electricity.[51] UPS developed its own software to reduce the miles its 90,000 trucks and other vehicles drive by routing them more efficiently. The new software helped UPS cut 30 million miles per year, slash fuel costs, and reduce carbon emissions by over 30,000 metric tons. Hewlett-Packard and Dell Computer have developed procedures and machines to dispose of old computers and computer equipment in environmentally friendly ways.[52] Old computers and computer equipment are fed into machines that shred them into small pieces and sort them into materials that can be reused. The process is often called *green death.*[53] One study estimates that more than 130,000 PCs in the United States are thrown out every day. The U.S. government is also involved in environmental design. It has a plan to require federal agencies to purchase energy-efficient computer systems and equipment.[54] The plan would require federal agencies to use the *Electronic Product Environmental Assessment Tool (EPEAT)* to analyze the energy usage of new systems. The U.S. Department of Energy rates products with the *Energy Star* designation to help people select products that save energy.[55]

Companies such as Hewlett-Packard and Dell Computer dispose of old computers and computer equipment in environmentally friendly ways.

(Source: Robin Beck/AFP/Getty Images.)

Generating Systems Design Alternatives

When people or organizations require a system to perform additional functions that an existing system cannot support, they often turn to outside vendors to design and supply their new systems. Whether an individual is purchasing a personal computer or a company is acquiring an expensive mainframe computer, the system can be obtained from a single vendor or multiple vendors. If the new system is complex, the original development team might want to involve other personnel in generating alternative designs. In addition, if new hardware and software are to be acquired from an outside vendor, a formal request for proposal (RFP) can be made.

Request for Proposal

The **request for proposal (RFP)** is an important document for many organizations involved with large, complex systems development efforts. Smaller, less-complex systems often do not require an RFP. A company that is purchasing an inexpensive piece of software that will run on existing hardware, for example, might not need to go through a formal RFP process.

In some cases, separate RFPs are developed for different needs. For example, a company might develop separate RFPs for hardware, software, and database systems. The RFP also communicates these needs to one or more vendors, and it provides a way to evaluate whether the vendor has delivered what was expected. In some cases, the RFP is part of the vendor contract. The Table of Contents for a typical RFP is shown in Figure 8.17.

request for proposal (RFP)
A document that specifies in detail required resources such as hardware and software.

Johnson & Florin, Inc.
Systems Investigation Report

Contents

COVER PAGE (with company name and contact person)
BRIEF DESCRIPTION of the COMPANY
OVERVIEW of the EXISTING COMPUTER SYSTEM
SUMMARY of COMPUTER-RELATED NEEDS and/or PROBLEMS
OBJECTIVES of the PROJECT
DESCRIPTION of WHAT IS NEEDED
HARDWARE REQUIREMENTS
PERSONNEL REQUIREMENTS
COMMUNICATIONS REQUIREMENTS
PROCEDURES to BE DEVELOPED
TRAINING REQUIREMENTS
MAINTENANCE REQUIREMENTS
EVALUATION PROCEDURES (how vendors will be judged)
PROPOSAL FORMAT (how vendors should respond)
IMPORTANT DATES (when tasks are to be completed)
SUMMARY

Figure 8.17

A Typical Table of Contents for a Request for Proposal

Evaluating and Selecting a Systems Design

The final step in systems design is to evaluate the various alternatives and select the one that will offer the best solution for organizational goals. Normally, evaluation and selection involves both a preliminary and a final evaluation before a design is selected. A *preliminary evaluation* begins after all proposals have been submitted. The purpose of this evaluation is to dismiss unwanted proposals. Several vendors can usually be eliminated by investigating their proposals and comparing them with the original criteria. The *final evaluation* begins with a detailed investigation of the proposals offered by the remaining vendors. The vendors should be asked to make a final presentation and to fully demonstrate the system. The demonstration should be as close to actual operating conditions as possible.

The Design Report

design report
The primary result of systems design, reflecting the decisions made and preparing the way for systems implementation.

System specifications are the final results of systems design. They include a technical description that details system outputs, inputs, and user interfaces, as well as all hardware, software, databases, telecommunications, personnel, and procedure components and the way these components are related. The specifications are contained in a **design report**, which is the primary result of systems design. The design report reflects the decisions made for systems design and prepares the way for systems implementation. The contents of the design report are summarized in Figure 8.18.

Figure 8.18

A Typical Table of Contents for a Systems Design Report

Johnson & Florin, Inc.
Systems Design Report

Contents

PREFACE
EXECUTIVE SUMMARY of SYSTEMS DESIGN
REVIEW of SYSTEMS ANALYSIS
MAJOR DESIGN RECOMMENDATIONS
      Hardware design
      Software design
      Personnel design
      Communications design
      Database design
      Procedures design
      Training design
      Maintenance design
SUMMARY of DESIGN DECISIONS
APPENDICES
GLOSSARY of TERMS
INDEX

SYSTEMS IMPLEMENTATION

systems implementation
A stage of systems development that includes hardware acquisition, software acquisition or development, user preparation, hiring and training of personnel, site and data preparation, installation, testing, start-up, and user acceptance. We start our discussion of systems implementation with hardware acquisition.

After the information system has been designed, a number of tasks must be completed before the system is installed and ready to operate. This process, called **systems implementation**, includes hardware acquisition, programming and software acquisition or development, user preparation, hiring and training of personnel, site and data preparation, installation, testing, start-up, and user acceptance. The typical sequence of systems implementation activities is shown in Figure 8.19.

Acquiring Hardware from an IS Vendor

To obtain the components for an information system, organizations can purchase, lease, or rent computer hardware and other resources from an IS vendor. An *IS vendor* is a company that offers hardware, software, telecommunications systems, databases, IS personnel, or other computer-related resources. Types of IS vendors include general computer manufacturers (such as IBM and Hewlett-Packard), small computer manufacturers (such as Dell and Toshiba), peripheral equipment manufacturers (such as Hewlett-Packard and Cannon), computer dealers and distributors (such as Radio Shack and Best Buy), and chip makers such as Intel and AMD.[56] Hardware vendors can provide very small or very large systems. The U.S. Census Bureau, for example, will acquire over 1,000 small handheld devices to help it collect census data.[57] On the other hand, the Defense Advanced Research Projects Agency (DARPA) purchased super computers and systems from IBM and Cray, valued at $500 million.[58] DARPA hopes its efforts to build a supercomputer will have commercial applications.

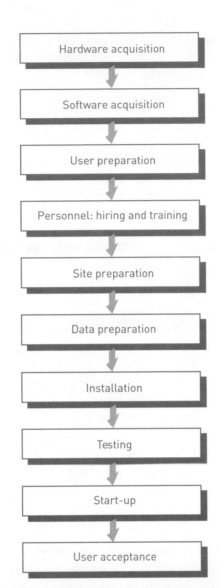

Figure 8.19

Typical Steps in Systems Implementation

Hardware acquisition

Software acquisition

User preparation

Personnel: hiring and training

Site preparation

Data preparation

Installation

Testing

Start-up

User acceptance

Computer dealers, such as Best Buy, manufacture build-to-order computer systems and sell computers and supplies from other vendors.

(Source: Justin Sullivan/Getty Images.)

As mentioned earlier, organizations often consider virtual machines, such as servers, in acquiring hardware.[59] Tellabs, for example, acquired virtualized Dell PowerEdge Servers for its operations.[60] According to a Tellabs representative, "In the past, we had been loading many of our servers with just one application, and their utilization rates were commonly 25 percent or less. We knew virtualization technology had evolved to where it could really help us consolidate." In addition to buying, leasing, or renting computer hardware, companies can pay only for the computing services that it uses. Called "pay-as-you-go," "on-demand," or "utility" computing, this approach requires an organization to pay only for the computer power it uses, as it would pay for a utility such as electricity.

Acquiring Software

make-or-buy decision
The decision regarding whether to obtain the necessary software from internal or external sources.

As with hardware, application software can be acquired in several ways. As previously mentioned, it can be purchased from external developers or developed in-house. This decision is often called the **make-or-buy decision**. Allstate Insurance, for example decided to develop a new software program, called Next Gen, to speed claims processing and reinforce its "You're in good hands" slogan.[61] The company is expected to spend over $100 million on the new software.

Businesses are using virtualization software such as VMware to safeguard private data.

(Source: Courtesy of VMware, Inc.)

As mentioned in Chapter 2, *Software as a Service (SaaS)* allows businesses to subscribe to Web-delivered application software by paying a monthly service charge or a per-use fee. Instead of acquiring software externally from a traditional software vendor, SaaS allows individuals and organizations to access needed software applications over the Internet. The Humane Society of the United States, for example, used SaaS to obtain and process credit-card contributions from donors.[62] The Humane Society used a SaaS product called QualysGuard by Qualys (*www.qualys.com*) to meet its needs. According to the CIO, "SaaS opened our eyes to a new way of doing things. With QualysGuard, we didn't need to install any software or infrastructure." Companies such as Google are using the cloud computing approach to deliver word processing, spreadsheet programs, and other software over the Internet.[63] According to the research director for AMR Research, "It's the beginning of the first major challenge to Microsoft as the default enterprise interface in the last ten years."

Acquiring Database and Telecommunications Systems

Because databases are a blend of hardware and software, many of the approaches discussed earlier for acquiring hardware and software also apply to database systems including open-source databases.[64] MasterCard International needed to acquire additional storage capacity.

Existing storage capacity was about to run out as the company expanded its business. Instead of adding incremental storage capacity, the company decided to use a large-scale storage area network (SAN). The results were immediate and apparent. *Virtual databases* and *database as a service (DaaS)* are also popular ways to acquire database capabilities.[65] Sirius XM Radio, Bank of America, and Southwest Airlines, for example, use the DaaS approach to manage many of their database operations from the Internet.[66] In another case, a brokerage company was able to reduce storage capacity by 50 percent by using database virtualization.[67]

With the increased use of e-commerce, the Internet, intranets, and extranets, telecommunications is one of the fastest-growing applications for today's organizations. Medical Missions for Children (MMC), for example, uses medical diagnosis through teleconferencing to treat children and improve their lives.[68] Like database systems, telecommunications systems require a blend of hardware and software.

User Preparation

User preparation is the process of readying managers, decision makers, employees, other users, and stakeholders for the new systems. This activity is an important but often ignored area of systems implementation. When a new operating system or application software package is implemented, user training is essential.[69] "It will take a lot more planning and definitely some thought toward training," said a TRW Automotive Holdings executive about installing Office 2007. In some cases, companies decide not to install the latest software because the amount of time and money needed to train employees is too much.[70] In one survey, over 70 percent of the respondents indicated that they were in no hurry to install a new operating system.[71] Additional user training was an important factor in delaying the installation of the new operating system for many companies. Because user training is so important, some companies provide training for their clients, including in-house training, training software, training videos, training on the Internet, and other training approaches.[72]

user preparation
The process of readying managers, decision makers, employees, other users, and stakeholders for new systems.

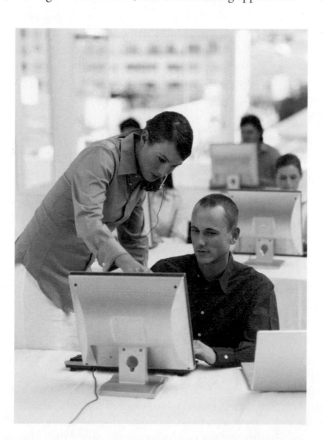

Providing users with proper training can help ensure that the information system is used correctly, efficiently, and effectively.

(Source: Stockbyte/Getty Images.)

IS Personnel: Hiring and Training

Depending on the size of the new system, an organization might have to hire and, in some cases, train new IS personnel.[73] An IS manager, systems analysts, computer programmers, data-entry operators, and similar personnel might be needed for the new or modified system.

Site Preparation

site preparation
Preparation of the location of a new system.

The location of the new system needs to be prepared, a process called **site preparation**. For a small system, site preparation can be as simple as rearranging the furniture in an office to make room for a computer. With a larger system, this process is not so easy because it can require special wiring and air conditioning. One or two rooms might have to be completely renovated, and additional furniture might have to be purchased. A special floor might have to be built, under which the cables connecting the various computer components are placed, and a new security system might be needed to protect the equipment. For larger systems, additional power circuits might also be required. Today, developing IS sites that are energy efficient is important for most systems development implementations.[74]

Data Preparation

data preparation, or data conversion
Ensuring all files and databases are ready to be used with new computer software and systems.

Data preparation, or **data conversion**, involves making sure that all files and databases are ready to be used with new computer software and systems. If an organization is installing a new payroll program, the old employee-payroll data might have to be converted into a format that can be used by the new computer software or system. After the data has been prepared or converted, the computerized database system or other software will then be used to maintain and update the computer files.

Installation

installation
The process of physically placing the computer equipment on the site and making it operational.

Installation is the process of physically placing the computer equipment on the site and making it operational. Although normally the manufacturer is responsible for installing computer equipment, someone from the organization (usually the IS manager) should oversee the process, making sure that all equipment specified in the contract is installed at the proper location. After the system is installed, the manufacturer performs several tests to ensure that the equipment is operating as it should.

Testing

Good testing procedures are essential to make sure that the new or modified information system operates as intended. Inadequate testing can result in mistakes and problems. A $13 million systems development effort to build a vehicle title and registration system had to be shut down because inaccurate data led to vehicles being pulled over or stopped by mistake.[75] According to one state official, "We couldn't have people out there having their cars impounded because of inaccurate information in the … database." Several forms of testing should be used, including testing each program (*unit testing*), testing the entire system of programs (*system testing*), testing the application with a large amount of data (*volume testing*), and testing all related systems together (*integration testing*), as well as conducting any tests required by the user (*acceptance testing*).

Start-Up

start-up (also called *cutover*)
The process of making the final tested information system fully operational.

Start-up, also called *cutover*, begins with the final tested information system. When start-up is finished, the system is fully operational. Start-up can be critical to the success of the organization. If not done properly, the results can be disastrous. One of the authors is aware of a small manufacturing company that decided to stop an accounting service used to send out bills on the same day they were going to start their own program to send out bills to customers. The manufacturing company wanted to save money by using their own billing program developed by an employee. The new program didn't work, the accounting service wouldn't help because they were upset about being terminated, and the manufacturing company wasn't able to send out any bills to customers for more than three months. The manufacturing company almost went bankrupt.

Various start-up approaches are available (see Figure 8.20). **Direct conversion** (also called *plunge* or *direct cutover*) involves stopping the old system and starting the new system on a given date. Direct conversion is usually the least desirable approach because of the potential for problems and errors when the old system is shut off and the new system is turned on at the same instant.

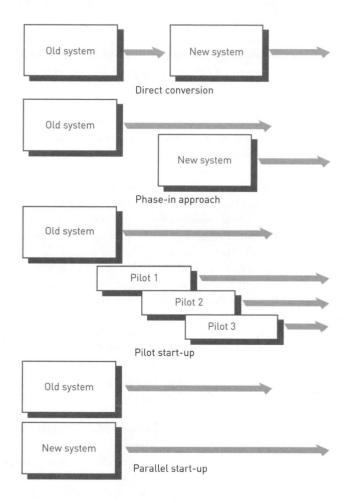

The **phase-in approach** is a popular technique preferred by many organizations. In this approach, sometimes called a *piecemeal approach*, components of the new system are slowly phased in while components of the old one are slowly phased out. When everyone is confident that the new system is performing as expected, the old system is completely phased out. This gradual replacement is repeated for each application until the new system is running every application. In some cases, the phase-in approach can take months or years.

Pilot start-up involves running the new system for one group of users rather than all users. For example, a manufacturing company with many retail outlets throughout the country could use the pilot start-up approach and install a new inventory control system at one of the retail outlets. When this pilot retail outlet runs without problems, the new inventory control system can be implemented at other retail outlets. Google's personal health records application was tested as a pilot project in a Cleveland health clinic before it was made available to more health care facilities.[76] This pilot start-up approach let Google install the application on a smaller scale before making the medical records application available to people around the world.

Parallel start-up involves running both the old and new systems for a period of time. The output of the new system is compared closely with the output of the old system, and any differences are reconciled. When users are comfortable that the new system is working correctly, the old system is eliminated.

User Acceptance

user acceptance document
A formal agreement signed by the user that states that a phase of the installation or the complete system is approved.

Most mainframe computer manufacturers use a formal **user acceptance document**—a formal agreement the user signs stating that a phase of the installation or the complete system is approved. This is a legal document that usually removes or reduces the IS vendor's liability for problems that occur after the user acceptance document has been signed. Because this document is so important, many companies get legal assistance before they sign the acceptance document. Stakeholders can also be involved in acceptance testing to make sure that the benefits to them are indeed realized.

SYSTEMS OPERATION AND MAINTENANCE

systems operation
Use of a new or modified system.

Systems operation involves all aspects of using the new or modified system in all kinds of operating conditions. Getting the most out of a new or modified system during its operation is the most important aspect of systems operations for many organizations. Throughout this book, we have seen many examples of information systems operating in a variety of settings and industries. Thus, we will not cover the operation of an information system in detail in this section. To provide adequate support, many companies use a formal help desk. A *help desk* consists of people with technical expertise, computer systems, manuals, and other resources needed to solve problems and give accurate answers to questions. With today's advances in telecommunications, help desks can be located around the world. If you are having trouble with your PC and call a toll-free number for assistance, you might reach a help desk in India, China, or another country. For most organizations, operations costs over the life of a system are much greater than the development costs.

systems maintenance
A stage of systems development that involves checking, changing, and enhancing the system to make it more useful in achieving user and organizational goals.

Systems maintenance involves checking, changing, and enhancing the system to make it more useful in achieving user and organizational goals. Systems maintenance is important for individuals, groups, and organizations. Individuals, for example, can use the Internet, computer vendors, and independent maintenance companies, including YourTechOnline.com (*www.yourtechonline.com*), Geek Squad (*www.geeksquad.com*), PC Pinpoint (*www.pcpinpoint.com*), and others. Organizations often have personnel dedicated to maintenance.

This maintenance process can be especially difficult for older software. A *legacy system* is an old system that might have been patched or modified repeatedly over time. An old payroll program in COBOL developed decades ago and frequently changed is an example of a legacy system. Legacy systems can be very expensive to maintain. At some point, it becomes less expensive to switch to new programs and applications than to repair and maintain the legacy system. Maintenance costs for older legacy systems can be 50 percent of total operating costs in some cases. Aspen Skiing Company, for example, decided to replace one of its legacy systems for a newer one.[77] According to the CIO, "We've been using a legacy application developed in-house. It's a very effective application from the standpoint of customer service. But it's extremely difficult to sustain it. So we had to go out and actually buy a new system to do what we do."

Relativity Technologies' Modernization Workbench is a PC-based software solution that enables companies to consolidate legacy or redundant systems into a single, more maintainable, and modern application.

(Source: Courtesy of Relativity Technologies.)

SYSTEMS REVIEW

Systems review, the final step of systems development, is the process of analyzing systems to make sure that they are operating as intended. This process often compares the performance and benefits of the system as it was designed with the actual performance and benefits of the system in operation. In some cases, a formal audit of the application can be performed, using internal and external auditors.[78] Systems review can be performed during systems development, resulting in halting the new systems while they are being built because of problems.[79] A payroll application being developed for a national health service, for example, was almost $170 million over budget. As a result, work on the application that serves about 37,000 workers was halted so the entire project could be reviewed in detail.

systems review
The final step of systems development, involving the analysis of systems to make sure that they are operating as intended.

System Performance Measurement

Systems review often involves monitoring the system, called **system performance measurement**. The number of errors encountered, the amount of memory required, the amount of processing or CPU time needed, and other problems should be closely observed. If a particular system is not performing as expected, it should be modified, or a new system should be developed or acquired.

System performance products have been developed to measure all components of the information system, including hardware, software, database, telecommunications, and network systems. IBM Tivoli OMEGAMON can monitor system performance in real time.[80] Precise Software Solutions has system performance products that provide around-the-clock performance monitoring for ERP systems, Oracle database applications, and other programs.[81] HP also offers a software tool called Business Technology Optimization (BTO) software to help companies analyze the performance of their computer systems, diagnose potential problems, and take corrective action if needed.[82] When properly used, system performance products can quickly and efficiently locate actual or potential problems.

Measuring a system is, in effect, the final task of systems development. The results of this process can bring the development team back to the beginning of the development life cycle, where the process begins again.

system performance measurement
Monitoring the system—the number of errors encountered, the amount of memory required, the amount of processing or CPU time needed, and other problems.

system performance products
Software that measures all components of the computer-based information system, including hardware, software, database, telecommunications, and network systems.

SUMMARY

Principle

Effective systems development requires a team effort of stakeholders, users, managers, systems development specialists, and various support personnel, and it starts with careful planning.

The systems development team consists of stakeholders, users, managers, systems development specialists, and various support personnel. The development team is responsible for determining the objectives of the information system and delivering to the organization a system that meets its objectives.

A systems analyst is a professional who specializes in analyzing and designing business systems. The programmer is responsible for modifying or developing programs to satisfy user requirements. Other support personnel on the development team include technical specialists, either IS department employees or outside consultants. Depending on the magnitude of the systems development project and the number of IS systems development specialists on the team, the team may also include one or more IS managers.

Information systems planning refers to the translation of strategic and organizational goals into systems development initiatives. Benefits of IS planning include a long-range view of information technology use and better use of IS resources. Planning requires developing overall IS objectives; identifying IS projects; setting priorities and selecting projects; analyzing resource requirements; setting schedules, milestones, and deadlines; and developing the IS planning document.

Principle

Systems development often uses different approaches and tools such as traditional development, prototyping, rapid application development, end-user development, computer-aided software engineering, and object-oriented development to select, implement, and monitor projects.

The five phases of the traditional SDLC are investigation, analysis, design, implementation, and maintenance and review. Systems investigation involves identifying potential problems and opportunities and considering them in light of organizational goals. Systems analysis seeks a general understanding of the solution required to solve the problem; the existing system is studied in detail and weaknesses are identified. Systems design involves creating new or modified system requirements. Systems implementation encompasses programming, testing, training, conversion, and operation of the system. Systems operation involves running the system once it is implemented. Systems maintenance and review entails monitoring the system and performing enhancements or repairs.

Prototyping is an iterative development approach that involves defining the problem, building the initial version, having users utilize and evaluate the initial version, providing feedback, and incorporating suggestions into the second version. Rapid application development (RAD) uses tools and techniques designed to speed application development. Its use reduces paper-based documentation, automates program source code generation, and facilitates user participation in development activities. An agile, or extreme programming, approach allows systems to change as they are being developed. RAD makes extensive use of the joint application development (JAD) process to gather data and perform requirements analysis. JAD involves group meetings in which users, stakeholders, and IS professionals work together to analyze existing systems, propose possible solutions, and define the requirements for a new or modified system.

The end-user SDLC is used to support projects where the primary effort is undertaken by a combination of business managers and users. End-user SDLC is becoming increasingly important as more users develop systems for their personal computers.

The use of automated tools enables detailed development, tracking, and control of the project schedule. Effective use of these tools enables a project manager to deliver a high-quality system and to make intelligent trade-offs among cost, schedule, and quality. CASE tools can automate many of the systems development tasks, thus reducing the time and effort required to complete them while ensuring good documentation. With the object-oriented systems development (OOSD) approach, a project can be broken down into a group of objects that interact. Instead of requiring thousands or millions of lines of detailed computer instructions or code, the systems development project might require a few dozen or maybe a hundred objects.

Principle

Systems development starts with investigation and analysis of existing systems.

In most organizations, a systems request form initiates the investigation process. This form typically includes the problems in or opportunities for the system, objectives of systems investigation, overview of the proposed system, and expected costs and benefits of the proposed system. The systems investigation is designed to assess the feasibility of implementing solutions for business problems. An investigation team follows up on the request and performs a feasibility analysis that addresses technical, economic, legal, operational, and schedule feasibility. Object-oriented systems investigation is being used to a greater extent today. As a final step in the investigation process, a systems investigation report should be prepared to document relevant findings.

Systems analysis is the examination of existing systems, which begins once approval for further study is received from management. Additional study of a selected system allows those involved to further understand the system's weaknesses and potential improvement areas. An analysis team is assembled to collect and analyze data on the existing system.

Data collection methods include observation, interviews, and questionnaires. Data analysis manipulates the collected data to provide information. Data modeling is used to model organizational objects and associations using text and graphical diagrams. It is most often accomplished through the use of entity-relationship (ER) diagrams. Activity modeling is often accomplished through the use of data-flow diagrams (DFDs), which model objects, associations, and activities by describing how data can flow between and around various objects. DFDs use symbols for data flows, processing, entities, and data stores. The overall purpose of requirements analysis is to determine user and organizational needs. Object-oriented systems analysis also involves diagramming techniques, such as a generalization/specialization hierarchy diagram.

Principle
Designing new systems or modifying existing ones should always be aimed at helping an organization achieve its goals.

The purpose of systems design is to prepare the detailed design needs for a new system or modifications to an existing system. Logical systems design refers to the way the various components of an information system will work together. Physical systems design refers to the specification of the actual physical components.

If new hardware or software will be purchased from a vendor, a formal request for proposal (RFP) is needed. The RFP outlines the company's needs; in response, the vendor provides a written reply. Organizations have three alternatives for acquiring computer systems: purchase, lease, or rent. RFPs from various vendors are reviewed and narrowed down to the few most likely candidates. Near the end of the design stage, an organization prohibits further changes in the design of the system. The design specifications are then said to be frozen. After the vendor is chosen, contract negotiations can begin. One of the most important steps in systems design is to develop a good contract if new computer facilities are being acquired. The final step is to develop a design report that details the outputs, inputs, and user interfaces. It also specifies hardware, software, databases, telecommunications, personnel, and procedure components and the way these components are related.

Environmental design, also called green design, involves systems development efforts that slash power consumption, take less physical space, and result in systems that can be disposed in a way that doesn't negatively affect the environ-

ment. TA number of companies are developing products and services to help save energy. Environmental design also deals with how companies are developing systems to dispose of old equipment. The U.S. government is also involved in environmental design. It has a plan to require federal agencies to purchase energy-efficient computer systems and equipment. The plan would require federal agencies to use the Electronic Product Environmental Assessment Tool (EPEAT) to analyze the energy usage of new systems. The U.S. Department of Energy rates products with the Energy Star designation to help people select products that save energy and are friendly to the environment.

Principle
The primary emphasis of systems implementation is to make sure that the right information is delivered to the right person in the right format at the right time.

The purpose of systems implementation is to install a system and make everything, including users, ready for its operation. Systems implementation includes hardware acquisition, software acquisition or development, user preparation, hiring and training of IS personnel, site and data preparation, installation, testing, start-up, and user acceptance. Hardware acquisition requires purchasing, leasing, or renting computer resources from a vendor. Increasingly, companies are using service providers to acquire software, Internet access, and other IS resources.

Software can be purchased from external vendors or developed in-house—a decision termed the *make-or-buy decision*. Implementation must also address database and telecommunications systems, user preparation, and IS personnel requirements. User preparation involves readying managers, employees, and other users for the new system. New IS personnel may need to be hired, and users must be well trained in the system's functions. The physical site of the system must be prepared, and any existing data to be used in the new system must be converted to the new format. Hardware is installed during the implementation step. Testing includes program (unit) testing, systems testing, volume testing, integration testing, and acceptance testing.

Start-up begins with the final tested information system. When start-up is finished, the system is fully operational. There are a number of different start-up approaches. Direct conversion (also called *plunge* or *direct cutover*) involves stopping the old system and starting the new system on a given date. With the phase-in approach, sometimes called a *piecemeal approach*, components of the new system are slowly phased in while components of the old one are slowly phased out. When everyone is confident that the new system is performing as expected, the old system is completely phased out. Pilot start-up involves running the new system for one group of users rather than all users. Parallel start-up involves running both the old and new systems for a period of time. The output of the new system is compared closely with the output of the old system, and any differences are reconciled. When users are comfortable that the new system is working correctly, the old system is eliminated. The final step of implementation is user acceptance.

Principle

Maintenance and review add to the useful life of a system but can consume large amounts of resources, so they benefit from the same rigorous methods and project management techniques applied to systems development.

Systems operation is the use of a new or modified system. Systems maintenance involves checking, changing, and enhancing the system to make it more useful in obtaining user and organizational goals. Maintenance is critical for the continued smooth operation of the system. Some major reasons for maintenance are changes in business processes; new requests from stakeholders, users, and managers; bugs or errors in the program; technical and hardware problems; corporate mergers and acquisitions; government regulations; change in the operating system or hardware; and unexpected events, such as terrorist attacks.

Systems review is the process of analyzing systems to make sure that they are operating as intended. It involves monitoring systems to be sure they are operating as designed. The two types of review procedures are event-driven and time-driven. An event-driven review is triggered by a problem or opportunity. A time-driven review is started after a specified amount of time.

CHAPTER 8: SELF-ASSESSMENT TEST

Effective systems development requires a team effort of stakeholders, users, managers, systems development specialists, and various support personnel, and it starts with careful planning.

1. _____ is the activity of creating or modifying existing business systems. It refers to all aspects of the process—from identifying problems to be solved or opportunities to be exploited to the implementation and refinement of the chosen solution.

2. Which of the following individuals ultimately benefit from a systems development project?
   a. Computer programmers
   b. Systems analysts
   c. Stakeholders
   d. Senior-level managers

3. Like a contractor constructing a new building or renovating an existing one, the programmer takes the plans from the systems analyst and builds or modifies the necessary software. True or False?

Systems development often uses different approaches and tools such as traditional development, prototyping, rapid application development, end-user development, computer-aided software engineering, and object-oriented development to select, implement, and monitor projects.

4. What employs tools, techniques, and methodologies designed to speed application development?
   a. rapid application development
   b. joint optimization
   c. prototyping
   d. extended application development

5. _____ takes an iterative approach to the systems development process. During each iteration, requirements and alternative solutions to the problem are identified and analyzed, new solutions are designed, and a portion of the system is implemented.

Systems development starts with investigation and analysis of existing systems.

6. Feasibility analysis is typically done during which systems development stage?
   a. Investigation
   b. Analysis
   c. Design
   d. Implementation

7. Rapid application development (RAD) employs tools, techniques, and methodologies designed to speed application development. True or False?

Designing new systems or modifying existing ones should always be aimed at helping an organization achieve its goals.

8. Scenarios and sequence diagrams are used with _____.
   a. object-oriented design
   b. point evaluation
   c. incremental design
   d. nominal evaluation

9. _____ involves systems development efforts that slash power consumption and require less physical space.

10. Near the end of the design stage, an organization prohibits further changes in the design of the system. This is called _____.

The primary emphasis of systems implementation is to make sure that the right information is delivered to the right person in the right format at the right time.

11. Software can be purchased from external developers or developed in house. This decision is often called the_____ decision.

12. The phase-in approach to conversion involves running both the old system and the new system for a three months or longer. True or False?

Maintenance and review add to the useful life of a system but can consume large amounts of resources, so they benefit from the same rigorous methods and project management techniques applied to systems development.

13. A systems review that is caused by a problem with an existing system is called
    a. Object review
    b. Structured review
    c. Event-driven review
    d. Critical factors review

14. Monitoring a system after it has been implemented to make it more useful in achieving user and organizational goals is called _____.

CHAPTER 8: SELF-ASSESSMENT TEST ANSWERS

(1) Systems development, (2) c (3) True (4) a (5) Prototyping (6) a (7) True (8) a (9) environmental design (10) freezing design specifications (11) make-or-buy (12) False (13) c (14) systems maintenance

REVIEW QUESTIONS

1. What is an information system stakeholder?
2. What is the goal of information systems planning? What steps are involved in IS planning?
3. What are the steps of the traditional systems development life cycle?
4. What is the difference between systems investigation and systems analysis? Why is it important to identify and remove errors early in the systems development life cycle?
5. What is the difference between a programmer and a systems analyst?
6. List the different types of feasibility.
7. What is the purpose of systems analysis?
8. How does the JAD technique support the RAD systems development life cycle?
9. What is prototyping?
10. What are the steps of object-oriented systems development?
11. What is an RFP? What is typically included in one? How is it used?
12. What is systems operation?
13. What activities go on during the user preparation phase of systems implementation?
14. Give three examples of a computer system vendor.
15. What are the financial options of acquiring hardware?
16. How can SaaS be used in software acquisition?
17. What are some of the reasons for program maintenance?
18. Describe how you back up the files you use at school.

DISCUSSION QUESTIONS

1. Why is it important for business managers to have a basic understanding of the systems development process?
2. Briefly describe the role of a system user in the systems investigation and systems analysis stages of a project.
3. For what types of systems development projects might prototyping be especially useful? What are the characteristics of a system developed with a prototyping technique?
4. Imagine that your firm has never developed an information systems plan. What sort of issues between the business functions and IS organization might exist?
5. You have been hired by your university to find an outsourcing company to perform the university's payroll function. What are your recommendations? Describe the advantages and disadvantages of the outsourcing approach for this application.
6. Briefly describe when you would use the object-oriented approach to systems development instead of the traditional systems development life cycle.
7. How important are communications skills to IS personnel? Consider this statement: "IS personnel need a combination of skills—one-third technical skills, one-third business skills, and one-third communications skills." Do you think this is true? How would this affect the training of IS personnel?
8. You have been hired to perform systems investigation for a French restaurant owner in a large metropolitan area.

She is thinking of opening a new restaurant with a state-of-the-art computer system that would allow customers to place orders on the Internet or at Kiosks at restaurant tables. Describe how you would determine the technical, economic, legal, operational, and schedule feasibility for the restaurant and its new computer system.

9. Identify some of the advantages and disadvantages of purchasing a database package versus the DaaS approach.

10. Assume that you are the owner of a company that is about to start marketing and selling bicycles over the Internet. Describe your top three objectives in developing a new Web site for this systems development project.

11. Assume that you want to start a new video-rental business for students at your college or university. Go through logical design for a new information system to help you keep track of the videos in your inventory.

12. Identify some of the advantages and disadvantages of purchasing versus leasing hardware.

13. Identify some of the advantages and disadvantages of purchasing versus developing software.

14. Identify the various forms of testing used. Why are there so many different types of tests?

15. What is the goal of conducting a systems review? What factors need to be considered during systems review?

16. How would you go about evaluating a software vendor?

17. Assume that you have a personal computer that is several years old. Describe the steps you would use to perform a systems review to determine whether you should acquire a new PC.

18. Describe how you would select the best admissions software for your college or university. What features would be most important for school administrators? What features would be most important for students?

PROBLEM-SOLVING EXERCISES

1. You are developing a new information system for The Fitness Center, a company that has five fitness centers in your metropolitan area, with about 650 members and 30 employees in each location. This system will be used by both members and fitness consultants to track participation in various fitness activities, such as free weights, volleyball, swimming, stair climbers, and aerobic and yoga classes. One of the performance objectives of the system is that it must help members plan a fitness program to meet their particular needs. The primary purpose of this system, as envisioned by the director of marketing, is to assist The Fitness Center in obtaining a competitive advantage over other fitness clubs. Use a graphics program to develop a flowchart or a grid chart to show the major components of your information system and how the components are tied together.

2. You have been hired to develop a new computer system for a video rental business using the object-oriented approach. Using a graphics program, develop a use case diagram for the business.

3. You have been hired to develop a payroll program for a medium sized company. At a minimum, the application should have an hours-worked table that contains how many hours each employee worked and an employee table that contains information about each employee, including hourly pay rate. Design and develop these tables that could be used in a database for the payroll program.

TEAM ACTIVITIES

1. Your team should interview people involved in systems development in a local business or at your college or university. Describe the process used. Identify the users, analysts, and stakeholders for a systems development project that has been completed or is currently under development.

2. Your team has been hired to determine the requirements of a new medium-cost coffee bar to compete with higher-priced coffee shops like Starbucks. The new coffee bar will offer computer kiosks for customers to surf the Internet or order coffee and other products from the coffee bar. Using RAD and JAD techniques with your team, develop requirements for this new coffee bar and its computer system.

3. Assume you work for a medium-sized company that trades treasury bonds in New York. Your firm has 500 employees in a downtown location. The firm is considering the purchase of a local area network (hardware and software) that is tied into a global trading network with other firms. Develop a brief request for proposal (RFP) using your word processing program for this new LAN system.

WEB EXERCISES

1. Use the Internet to find two different systems development projects that failed to meet cost or performance objectives. Summarize the problems and what should have been done. You might be asked to develop a report or send an e-mail message to your instructor about what you found.

2. Cloud computing, where applications like word processing and spreadsheet analysis are delivered over the Internet, is becoming more popular. You have been hired to analyze the potential of a cloud computing application that performs payroll and invoicing over the Internet from a large Internet company. Describe the systems development steps and procedures you would use to analyze the feasibility of this approach.

CAREER EXERCISES

1. Pick a career that you are considering. What type of information system would help you on the job? Perform technical, economic, legal, operational, and schedule feasibility for an information system you would like developed for you.

2. For a career you are considering, describe how you would interact with various IS personnel to acquire computer applications or a new information system that would be the most useful to you in your new career. Describe the IS personnel that would be needed to build the application and the best way to interact with them to get what you want.

CASE STUDIES

Case One

Information and Security Systems at the All England Lawn Tennis and Croquet Club

For 351 days a year, the All England Lawn Tennis and Croquet Club is a quiet private tennis club set in a sleepy suburb of London. For 14 days each year, half a million tennis enthusiasts arrive at the club from around the world to witness the Wimbledon Championships. Needless to say, much preparation is required as the club accommodate world tennis stars, their fans, and the press and television crews. Many information systems of all types are required to support the global sporting event. Information systems specialists at the club work to support the event, maintain the Club's culture, brand, and values, and support the Club's primary mission: "to blend tradition with innovation to substantially improve the quality of the Wimbledon experience for all the key stakeholders."

A primary goal for the annual two-week event is the security and safety of all in attendance. To accomplish this goal, the Club has invested in a new electronic security and surveillance system. With so many people to watch on a vast property, Club management knew that they would need state-of-the-art automated surveillance software. The Club management wanted a system that could integrate images from video cameras, intruder alarms, trip wires with identifying information such as license plates to provide real-time reports of suspicious activities. With these specifications in mind, the Club found a solution in a Digital and Video Security (DVS) solution designed by IBM. The system provides "real-time intelligence to automatically monitor trends and analyze events captured by security devices."

The system was tested at the 2007 Wimbledon Championships with success. At the 2008 event, the system was extended to hundreds of cameras. Because the system is scalable, it can grow each year as attendance grows at the event without any degradation in performance.

Besides the half-million attendees, millions of fans can now view the event and related information online thanks to another new information systems development project. Knowing that interest in the event was continuously growing, Wimbledon event coordinators looked for ways to provide remote coverage to more tennis fans. A system was developed called SlamTracker that provides live online scoring for matches in progress. Additional SlamTracker tools allow fans to track player progression and other important player and match statistics.

Those watching the event in person and on TV have noticed other high-tech improvements. Large LED screens have been installed on the main courts, providing the audience with statistics such as the speed of the serve. Other systems are provided to the athletes to display important statistics and trends captured during the last match so that the players can see what improvements are required and create strategies for the next match.

Systems analysts work year round to improve information systems for the next Wimbledon Championships. Security systems, media systems, Web systems, and a host of other types of systems are examined for strengths and weaknesses searching for ways to improve the experience. In 1990, Wimbledon decided to hire one company to manage all of its systems: IBM. While outsourcing its systems to IBM may be costly, other savings make it well worth the investment. With all of its systems managed by one company, the All England Lawn Tennis and Croquet Club can integrate its systems more easily, saving money by removing redundancy that often exists with multivendor systems. Furthermore, All England Lawn Tennis and Croquet Club is not in the business of information systems. For 351 days a year, the club would rather focus on its own membership rather than on the two-week event next summer.

Discussion Questions

1. How do information systems being implemented at the All England Lawn Tennis and Croquet Club support its primary goals?
2. How do information systems add to the enjoyment of viewing the Wimbledon Championships through multiple channels—in person, on TV, and online?

Critical Thinking Questions

1. What other sporting events might benefit from the technologies used at Wimbledon?
2. How might systems failure cause catastrophe for Wimbledon Championships organizers?

SOURCES: IBM Staff, "For two weeks a year, Wimbledon stops being a private members' club and starts welcoming the world," IBM Case Studies, June 19, 2008, *www-01.ibm.com/software/success/cssdb.nsf/cs/JGIL-7FRDYD? OpenDocument&Site=gicss67mdia&cty=en_us*; Wimbledon Web site, *http://aeltc.wimbledon.org/en_GB/about/guide/club.html*, accessed July 13, 2

Case Two
Rogers Pulls an All-Nighter

Rogers Communications is one of Canada's largest telecom companies. It offers home phone services, wireless phone services, Internet service, and cable TV service. Rogers provides small shops in 93 malls across Canada to service its customers. The stores sell a wide variety of phones and other telecom devices and services.

Until recently Rogers outsourced the information technology services for its mall stores to a third-party company. Francois Chevallier, vice-president of retail systems for Rogers Retail, thought that Rogers could improve its systems and business practices if it took control of its own systems. "When you have different stores with different systems and management structures, the experience cannot be consistent. Our goal was to achieve that consistency and raise the bar," says Chevallier.

Chevallier proposed a massive upgrade of Rogers' retail systems that would provide consistency in business practices and customer experience, and connect all data in a unified system accessible from headquarters. The project faced two big challenges. First, to avoid any interruption to service, systems in all 93 stores should be upgraded simultaneously when stores were closed, which is tricky when dealing with stores across four time zones. Second, the upgrade would take place in the middle of winter when weather was unpredictable, with periodic snow and wind storms and temperatures diving as low as -27 degrees Celsius (-17 F).

This challenging project would require more human resources than Chevallier had. He pulled in an outside company, Connections Canada Inc. (CCI), to assist with the project. The team decided that the upgrade would need to take place at all locations simultaneously over a six-hour period while the stores were closed, which meant working through the night. In order for the upgrade to go flawlessly, the team would need to invest in practice, training, and preparation. Unfortunately they had only four months to prepare.

They decided to create a virtual store, or staging facility, at CCI that mimicked the real Rogers' mall stores. The components of the new system were set up in the virtual store including all software and hardware: routers, computers, cash-registers, and PIN readers. The intention was to create and configure a system for each store using the staging facility, and then ship the preconfigured components of the system to each store—a kind of "store-in-a-box."

Experts from different fields including information systems, human resources, operations, finance, supply chain, real estate, marketing, inventory management, and internal communications assisted in configuring the system to meet all organizational needs. The finance expert set up the banking environment for each store. The supply chain expert modified the supply-chain elements of the new system. In other words, each expert worked on his or her area of specialty. The team used a SharePoint intranet to allow all experts involved to communicate online. They spent weeks developing, testing, and adjusting the new system in the virtual store. The development continued until all experts were satisfied.

Once the system was designed and running smoothly in the store prototype, four actual stores were selected for testing. One at a time, the system was installed in each store. With each installation, lessons were learned and problems became fewer. The installation at the fourth store was carried out flawlessly.

The new system was ready for installation in the remaining stores. One technician was hired and trained for each of the store installations. Many back-up technicians were trained as well in case the primary technician failed to show up. The store-in-a-box was shipped to each of the stores in shipping containers that could withstand the coldest Canadian winter temperatures. Communications were set up that would allow each technician to give a step-by-step report to the control center at headquarters. At headquarters, ten project managers would be tracking the progress of their districts and reporting to the primary project managers. If trouble arose, it could be addressed within minutes.

The installation went off without a hitch. The project was completed on schedule in four months at a cost of one million dollars. In the end, 500 Rogers employees were trained on the new system, which included an "intranet for resources and policies, a new supply chain model, integrated point-of-sale and merchandise management systems, and a foundation for good customer service." The team credits the success of this ambitious systems development project to its detailed preparation, especially the staging facility, and tight communications and cooperation throughout the process.

Discussion Questions

1. What challenges did Rogers face in the installation of its new retail system?
2. What implementation techniques did Rogers employ to assure a smooth transition to the new system?

Critical Thinking Questions

1. Why did Rogers feel it necessary to upgrade all stores at the same time? What are the benefits and risks of that decision?
2. What role did communications technologies play in the success of this system upgrade?

SOURCES: Lau, Kathleen, "Rogers' IT overhaul: All in a day's work," itWorldCanada, May 25, 2007, *www.itworldcanada.com/Pages/Docbase/ViewArticle.aspx?id=idgml-7bce5860-7681-4e37-86f7-7b6fa8244639&sub=472646*; Rogers Communications Web site, *www.rogers.com*, accessed July 20, 2008.

Questions for Web Case

See the Web site for this book to read about the Whitmann Price Consulting case for this chapter. Following are questions concerning this Web case.

WHITMANN PRICE CONSULTING: SYSTEMS INVESTIGATION AND ANALYSIS CONSIDERATIONS

Discussion Questions

1. How will the proposed AMCI system help to meet corporate goals and provide Whitmann Price with a competitive advantage?
2. Who are the stakeholders in this systems development project? Who are the primary systems analysts?

Critical Thinking Questions

1. Why did the systems investigation for the proposed AMSI system proceed so quickly and smoothly? What type of proposal might require a more time-consuming formal investigation?
2. What reasons do you think Josh and Sandra might have for interviewing division managers and not each individual consultant in their system review? What are the pros and cons of both approaches?

NOTES

Sources for the opening vignette: SAP staff, "GRUMA, Standardizing Business Processes to Support Growth," SAP Business Transformation Study, 2008, *http://download.sap.com/solutions/business-suite/erp/customersuccess/download.epd?context=41D7EC5F1B536250F4ECA519A1F6F2E3CA2CF88579D5D63A6E8D998182804C9A0A1304D4584991E5C21230651CD5EBA4D13732442AF0C1AB*; GRUMA Web site, *www.gruma.com*, accessed July 12, 2008.

1 Pratt, Mary, "Rolling the Dice in Your Career," *Computerworld*, February 25, 2008, p. 34.
2 Hoffman, Thomas, "Beyond Film," *Computerworld*, May 5, 2008, p. 32.
3 Demaitre, Eugene, "Project Manager Extraordinaire," *Computerworld*, July 9, 2007, p. 72.
4 Mahring, Magnus, et al., "Information Technology Project Escalation," *Decision Sciences*, May 2008, p. 239.
5 Chilton, David, "Demand for IT Workers Won't Meet Supply," *The Toronto Sun*, April 23, 2008, p. J2.
6 Smaltz, Detlev, "Planning, Early Support Key to E-Health Success," *Computerworld*, May 21, 2007, p. 10.
7 Oh, W., et al., "On the Assessment of Strategic Value of Information Technologies," *MIS Quarterly*, June, 2007, p. 239.
8 Iivari, J., et al., "The Relationship Between Organizational Culture and the Deployment of Systems Development Methodologies," *MIS Quarterly*, March 2007, p. 35.
9 Hoffman, Thomas, "Building an IT Project Pipeline," *Computerworld*, April 7, 2008, p. 34.
10 Anthes, Gary, "What's Your Project Worth?" *Computerworld*, March 10, 2008, p. 29.
11 Jia, Ronni, "IT Service Climate," *Journal of the Association of Information Systems*, May 2008, p. 294.
12 Weier, Mary Hayes, "SaaS at Scale," *Information Week*, May 19, 2008, p. 19.
13 Perkins, Bart, "Post-project Review," *Computerworld*, January 15, 2007, p. 30.
14 Sanders, Peter, "Fans Resist End of Virtual Disneyland," *The Wall Street Journal*, May 20, 2008, p. B1.
15 Konke, Robin, "Debugging Embedded C," *Embedded Systems Design*, January 1, 2008, p. 34.
16 Babcock, Charles, "Speedy Web Development," *Information Week*, March 17, 2008, p. 22.
17 Lai, Eric, "Microsoft Tries to Steer a More Agile Development Course," *Computerworld*, March 3, 2008.
18 Rangaswaimi, J. P., "The Roots of Agile Development," *Computerworld*, April 28, 2008, p. 31.

19 Bradbury, Danny, "Business and Technology," *Computer Weekly*, April 1, 2008.

20 Biesecker, Calvin, "Boeing Officials Describe Project 28 Shortcomings," *Defense Daily*, February 28, 2008.

21 The Rational Unified Process, *www-306.ibm.com/software/awdtools/rup/support...*, accessed June 2, 2008.

22 Rational Case Studies, *www-01.ibm.com/software/success/cssdb.nsf/softwareL2VW?OpenView&Start=1&Count=30&RestrictToCategory=rational_RationalUnifiedProcess*, accessed June 2, 2008.

23 Au, N., et al., "Extending the Understanding of End User Information Systems Satisfaction Formulation," *MIS Quarterly*, March 2008, p. 43.

24 Xiao, T., et al., "Strategic Outsourcing Decisions for Manufacturers That Produce Partially Substitutable Products," *Decision Sciences*, February 2007, p. 81.

25 Mithas, Sunil and Whitaker, Jonathan, "Is the World Flat or Spiky?" *Information Systems Research*, September 2007, p. 237.

26 Tam, Pui-Wing, "Outsourcing Finds New Niche," *The Wall Street Journal*, April 17, 2007, p. B5.

27 Hamm, Steve, "How Accenture One-Upped Bangalore," *Business Week*, April 23, 2007, p. 98.

28 Hamm, Steve, "Outsourcing Heads to the Outskirts, *Business Week*, January 22, 2007, p. 56.

29 IBM Web site, *www.ibm.com*, accessed June 2, 2008.

30 Weier, Mary Hayes, "As Hiring Soars in India, Good Managers Are Hard to Find," *Information Week*, February 5, 2007, p. 33.

31 EDS Web site, *www.eds.com*, accessed June 2, 2008.

32 Accenture Web site, *www.accenture.com*, accessed June 2, 2008.

33 Menchin, Scott, "Amazon Takes on IBM, Oracle, and HP," *Business Week*, April 21, 2008, p. 25.

34 Lawton, Christopher and Clark, Don, "Virtualization is Pumping Up Servers," *The Wall Street Journal*, March 6, 2007, p. B4.

35 Mathiassen, Lars, et al., "A Contingency Model of Requirements Development," *Journal of the Association of Information Systems*, November, 2007, p. 569.

36 Preston, Rob, "Will Cloud Computing Rain on IT's Parade?" *Information Week*, February 18, 2008, p. 52,

37 Cheng, Roger, "Those on the Go Get to Go Online," *The Wall Street Journal*, September 4, 2007, p. B3.

38 Hoover, Nicholas, "Microsoft Opens Live Mesh to More Users," *Information Week*, April 28, 2008, p. 22.

39 Gomes, Lee, "Good Site, Bad Site: Evolving Web Design," *The Wall Street Journal*, June 12, 2007. p. B3.

40 Shipley, Greg, "Risk=$," *Information Week*, March 31, 2008, p. 31.

41 Fonseca, Brian, "E-Discovery Rules Still Causing IT Headaches," *Computerworld*, January 7, 2008, p. 14.

42 Srivastava, Samar, "Search Software Gets Boost," *The Wall Street Journal*, May 16, 2007, p. B6.

43 Lei, Tao Ai, "Using Software to Save Time and Money," *The Straits Times,"* May 6, 2008.

44 Hagel, John, "The Grill," *Computerworld*, January 7, 2008, p. 18.

45 Brandel, Mary, "Green IT Companies," *Computerworld*, February 18, 2008, p. 29.

46 Collett, Stacy, "Slimmed-Down Servers," *Computerworld*, March 3, 2008, p. 28.

47 Mitchell, Robert, "Power Pinch," *Computerworld*, April 30, 2007, p. 19.

48 Thibodeau, Patrick, "EPA Helps Put Data Centers on an Energy Diet," *Computerworld*, March 24, 2008, p. 12.

49 Weier, Mary Hayes, "So Your Company's Going Green," *Information Week*, May 5, 2008, p. 24.

50 Wittmann, Art, "Think Thin, Green, and Virtual," *Information Week*, May 26, 2008, p. 19.

51 Carlton, Jim, "To Cut Fuel Bills, Try High-Tech Help," *The Wall Street Journal*, March 11, 2008, p. B3.

52 Chea, Terence, "Ramping Up Against E-Waste," *Rocky Mountain News*, March 4, 2007, p. 8K.

53 Boehret, Katherine, "Where Computers Go When They Die," *The Wall Street Journal*, April 11, 2007, p. D1.

54 Thibodeau, Patrick, "Green PC Push By Feds May Seed Wider Adoption," *Computerworld*, January 14, 2008, p. 14.

55 *www.energystar.gov/index.cfm?fuseaction=find_a_product*, accessed June 15, 2008.

56 Nam, Susan, "The Journey Machine," *Forbes*, November 26, 2007, p. 152.

57 Hamblen, Matt, "Census to Start Small on Handheld Rollout," *Computerworld*, January 8, 2007, p. 7.

58 Thibodeau, Patrick, "DARPA Pushes to Bring Supercomputer to the Masses," *Computerworld*, January 8, 2007, p. 35.

59 Thibodeau, Patrick, "Marines Look for a Few Less Servers, Via Virtualization," *Network World*, November 6, 2007.

60 Dell Staff, "Pushing the Virtualization Envelope," *Information Week*, May 28, 2007, p. NWC 8.

61 Weier, Mary Hayes, "Good Hands Aren't Enough," *Information Week*, April 23, 2007, p. 49.

62 Magda, Beverly, "SaaS to the Rescue," *Information Week*, May 19, 2008, p. 31.

63 Delaney, Kevin, "Google Further Tests Microsoft's Domain," *The Wall Street Journal*, February 22, 2007, p. A4.

64 Singer, Michael, "Sun Sticks To Its Open Agenda," *Information Week*, May 5, 2008, p. 20.

65 IT Redux Web site, *http://itredux.com/office-20/database/?family=Database*, accessed March 23, 2008.

66 Lai, Eric, "Cloud database vendors: What, us worry about Microsoft?" *Computerworld*, March 12, 2008, *www.computerworld.com/action/article.do?command=viewArticleBasic&articleId=9067979&pageNumber=1*.

67 Fonseca, Brian, "Virtualization Cutting Storage Costs for Some Large Firms," *Computerworld*, April 23, 2007, *www.computerworld.com/action/article.do?command=viewArticleBasic&taxonomyId=154&articleId=290230&intsrc=hm_topic*.

68 Pratt, Mary, "Goodwill Mission," *Computerworld*, April 28, 2008, p. 24.

69 Guth, Robert, "Office 2007 Poses Hurdle," *The Wall Street Journal*, January 23, 2007, p. B2.

70 Mossberg, Walter, "Bold Redesign Improves Office 2007," *The Wall Street Journal*, January 4, 2007, p. B1.

71 Dornan, Andy, "No Hurry with Vista," *Information Week*, December 3, 2007, p. AB10.

72 Stackpole, Beth, "IT's Top Five Training Mistakes," *Computerworld*, April 28, 2008, p. 20.

73 Hoover, Nicholas, "A Year Later, IT Still Grapples with Vista Deployment Timing," *Information Week*, December 17, 2007, p. 32.

74 Carlton, Jim, "Computer Rooms Feel the Heat," *The Wall Street Journal*, April 10, 2007, p. B4.

75 Weiss, Todd, "Colorado DMV Puts Brakes on $13 M Registration System," *Computerworld*, April 9, 2007, p. 8.

76 McGee, Marianne Kolbasuk, "Cleveland, Then the World," *Information Week*, May 26, 2008, p. 18.

77 Major, Paul, "The Grill," *Computerworld*, May 12, 2008, p. 18.

78 Thurman, Mathias, "An Audit Can Be an Opportunity," *Computerworld*, January 14, 2008, p. 36.

79 Perhins, Bart, "Pulling the Plug on a Project," *Computerworld*, March 10, 2008, p. 42.

80 *www-306.ibm.com/software/tivoli/products/omegamon-xe-cics*, accessed June 20, 2008.

81 *www.precise.com*, accessed June 20, 2008.

82 *https://h10078.www1.hp.com/cda/hpms/display/main/hpms_home.jsp?zn=bto&cp=1_4011_100__*, accessed June 20, 2008.

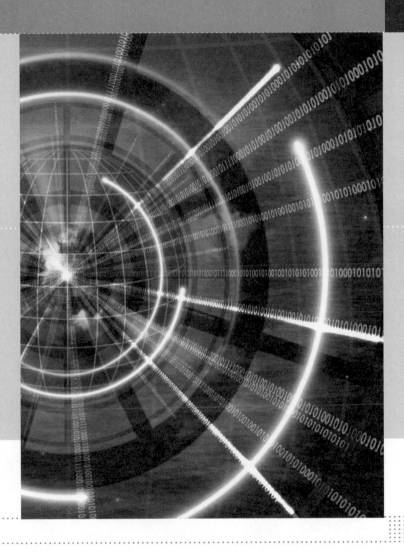

Information Systems in Business and Society

CHAPTER
· 9 ·

The Personal and Social Impact of Computers

PRINCIPLES	LEARNING OBJECTIVES

- Policies and procedures must be established to avoid computer waste and mistakes.

- Describe some examples of waste and mistakes in an IS environment, their causes, and possible solutions.
- Identify policies and procedures useful in eliminating waste and mistakes.
- Discuss the principles and limits of an individual's right to privacy.

- Computer crime is a serious and rapidly growing area of concern requiring management attention.

- Explain the types and effects of computer crime.
- Identify specific measures to prevent computer crime.

- Jobs, equipment, and working conditions must be designed to avoid negative health effects.

- List the important effects of computers on the work environment.
- Identify specific actions that must be taken to ensure the health and safety of employees.

- Practitioners in many professions subscribe to a code of ethics that states the principles and core values that are essential to their work.

- Outline criteria for the ethical use of information systems.

Information Systems in the Global Economy ▶▶
eBay, United States

Battling Hackers and Fraudsters Day in and Day out

Online giant eBay provides the world's largest online marketplace. Roughly 147 million people buy and sell all kinds of merchandise and services on eBay. It is estimated that more than $1,900 worth of goods is sold on eBay every second. Hundreds of thousands of people in the United States depend on eBay for their living. Some of those people earn their living illegally and unethically.

With so many transactions taking place on eBay, the company manages as much money as a global bank. In fact, eBay owns the largest online banking service, PayPal, which is used to facilitate transactions between online buyers and sellers. Unfortunately, managing the largest online bank and marketplace makes eBay a huge target for hackers and fraudsters. The level of information security implemented at eBay far exceeds the level of security used in a brick-and-mortar bank.

Fraud is a major challenge for eBay. Criminals gain illegal access to customer accounts and use the accounts and their good reputation to sell knock-offs—primarily imitations of high-quality items. Of all Internet-related crimes, auction fraud appears to be the biggest problem. The 2007 Internet Crime Report published by the FBI explains: "Internet auction fraud was by far the most reported offense, comprising 44.9 percent of referred complaints. Nondelivered merchandise and/or payment accounted for 19.0 percent of complaints. Check fraud made up 4.9 percent of complaints. Credit/debit card fraud, computer fraud, confidence fraud, and financial institutions fraud round out the top seven categories of complaints referred to law enforcement during the year."

In 2008, at least two eBay crimes made headlines. The first involves a Romanian hacker called Vladuz who hacked into eBay systems and masqueraded as an official eBay representative. The damage Vladuz caused is estimated at 1 million dollars. He was eventually apprehended and awaits trial. The second big story involves Jeremiah Mondello, a 23-year-old from Oregon. Mondello stole more than 40 eBay and PayPal accounts and used them to sell over a million dollars worth of counterfeit software, making more than $400,000 for himself. Mondello is now spending two years in prison, and must pay $225,000 in fines and devote 450 hours to community service once released from prison.

Even though many crimes are taking place on eBay as you read this sentence, eBay says that fraud remains a tiny fraction of the million or so transactions that take place each day. eBay keeps fraud under control by investing heavily in information security tools and practices.

eBay uses many types of security tools to address many kinds of threats. On its Web site, eBay states that PayPal uses "the world's most advanced proprietary fraud prevention systems to create a safe payment solution." The company also invests in an automated security system to keep hackers out of the network. The system uses more than a dozen scanning applications to monitor vulnerabilities on eBay's global network and on all partner networks that connect to eBay's extranet.

The security software that patrols eBay's systems provides continuous reports to security engineers. The software also creates reports for system administrators and executives that provide an overview of network conditions and illustrate the impact of information security investment. Additionally, the security software measures eBay's compliance with government regulations involving information security.

The battle to protect valuable and private information online is one in which all levels of management in businesses and governments are or should be fully engaged. Attacks are increasing from people around the world looking to benefit financially or politically. Businesses such as eBay enforce rigid policies and procedures to make sure that their networks, employees, and partners are operating in the most secure manner possible.

As you read this chapter, consider the following:

- What are the primary concerns of corporations regarding security, privacy, and ethics?
- What strategies can assist a company with issues of security and privacy, and at what cost?

Why Learn About the Personal and Social Impact of the Internet?

A wide range of nontechnical issues associated with the use of information systems and the Internet provide both opportunities and threats to modern organizations. The issues span the full spectrum—from preventing computer waste and mistakes, to avoiding violations of privacy, to complying with laws on collecting data about customers, to monitoring employees. If you become a member of a human resources, information systems, or legal department within an organization, you will likely be charged with leading the organization in dealing with these and other issues covered in this chapter. Also, as a user of information systems and the Internet, it is in your own self-interest to become well versed on these issues. You need to know about the topics in this chapter to help avoid or recover from crime, fraud, privacy invasion, and other potential problems. This chapter begins with a discussion of preventing computer waste and mistakes.

Earlier chapters detailed the significant benefits of computer-based information systems in business, including increased profits, superior goods and services, and higher quality of work life. Computers have become such valuable tools that today's businesspeople have difficulty imagining work without them. Yet the information age has also brought the following potential problems for workers, companies, and society in general:

- Computer waste and mistakes
- Computer crime
- Privacy
- Work environment
- Ethical issues

This chapter discusses some of the social and ethical issues as a reminder of these important considerations underlying the design, building, and use of computer-based information systems. No business organization, and, hence, no information system, operates in a vacuum. All IS professionals, business managers, and users have a responsibility to see that the potential consequences of IS use are fully considered.

Managers and users at all levels play a major role in helping organizations achieve the positive benefits of IS. These people must also take the lead in helping to minimize or eliminate the negative consequences of poorly designed and improperly utilized information systems. For managers and users to have such an influence, they must be properly educated. Many of the issues presented in this chapter, for example, should cause you to think back to some of the systems design and systems control issues discussed previously. They should also help you look forward to how these issues and your choices might affect your future use of information systems.

COMPUTER WASTE AND MISTAKES

Computer-related waste and mistakes are major causes of computer problems, contributing as they do to unnecessarily high costs and lost profits. Computer waste involves the inappropriate use of computer technology and resources. Computer-related mistakes refer to errors, failures, and other computer problems that make computer output incorrect or not useful, caused mostly by human error. This section explores the damage that can be done as a result of computer waste and mistakes.

Computer Waste

The U.S. government is the largest single user of information systems in the world. It should come as no surprise, then, that it is also perhaps the largest abuser. The government is not unique in this regard—the same type of waste and misuse found in the public sector also exists in the private sector. Some companies discard old software and computer systems when they still have value. Others waste corporate resources to build and maintain complex systems that are never used to their fullest extent.

A less-dramatic, yet still relevant, example of waste is the amount of company time and money employees can waste playing computer games, sending unimportant e-mail, or accessing the Internet. Junk e-mail, also called *spam*, and junk faxes also cause waste. People receive hundreds of e-mail messages and faxes advertising products and services not wanted or requested. Not only does this waste time, but it also wastes paper and computer resources. Worse yet, spam messages often carry attached files with embedded viruses that can cause networks and computers to crash or allow hackers to gain unauthorized access to systems and data.

A spam filter is software that attempts to block unwanted e-mail. One approach to filtering spam involves building lists of acceptable and unacceptable e-mail addresses. The lists can be created manually or automatically based on how the users keep or discard their e-mail. Another approach is automatic rejection of e-mail based on the content of the message or the appearance of keywords in the message. Rejected e-mail automatically goes to the spam or junk e-mail folder of your e-mail service. CA Anti Spam, SpamEater Pro, ChoiceMail One, and Spam Buster are among the most highly rated anti-spam software and cost from around $20 to $50.[1] Many e-mail programs have built-in spam filters.

A word of caution: some spam filters might require first-time e-mailers to be verified before their e-mails are accepted. This can be disastrous for people in sales or customer service who are frequently receiving e-mails from people they do not know. In one case, a spam filter blocked e-mail to close a real estate deal valued at around $175,000. The deal never closed and the real estate was sold to someone else because of the blocked e-mail.

Image-based spam is a new tactic spammers use to circumvent spam-filtering software that rejects e-mail based on the content of messages and the use of keywords. The message is presented in a graphic form that can be read by people but not computers. The images in this form of spam can be quite offensive.

When waste is identified, it typically points to one common cause: the improper management of information systems and resources.

Computer-Related Mistakes

Despite many people's distrust of them, computers rarely make mistakes. Yet even the most sophisticated hardware cannot produce meaningful output if users do not follow proper procedures. Mistakes can be caused by unclear expectations and a lack of feedback. A programmer might also develop a program that contains errors. In other cases, a data-entry clerk might enter the wrong data. Unless errors are caught early and prevented, the speed of computers can intensify mistakes. As information technology becomes faster, more complex, and more powerful, organizations and computer users face increased risks of experiencing the results of computer-related mistakes. Consider these examples from recent news.

- Shares of Moody's Corporation rating agency fell more than 20 percent following a *Financial Times* report alleging a computer coding error boosted the investment ratings of a particular class of debt instrument four levels to Aaa (the highest possible rating) and wasn't immediately corrected after it was uncovered.[2]
- An internal investigation by NASA concluded that multiple programming errors caused the loss of the Mars Global Surveyor orbiter spacecraft in January 2008. Fortunately, by the time of the investigation, the orbiter had lasted four times longer than expected and was successful in mapping the surface of Mars and studying its atmosphere.[3]

- Computer problems are a frequent cause for airline flight cancellations and delays. For example, a computer glitch at All Nippon Airways Company grounded or delayed hundreds of domestic flights in Japan.[4] United Air Lines, Inc. had to cancel 24 domestic flights and delay another 250 flights by more than 90 minutes when its systems for dispatching flights failed.[5]
- A computer error knocked out ATM and wire and ACH transfer services provided by Wells Fargo for two days.[6]

PREVENTING COMPUTER-RELATED WASTE AND MISTAKES

To remain profitable in a competitive environment, organizations must use all resources wisely. Preventing computer-related waste and mistakes should therefore be a goal. Today, nearly all organizations use some type of Computer Based Information System (CBIS). To employ IS resources efficiently and effectively, employees and managers alike should strive to minimize waste and mistakes. Preventing waste and mistakes involves (1) establishing, (2) implementing, (3) monitoring, and (4) reviewing effective policies and procedures.

Establishing Policies and Procedures

The first step to prevent computer-related waste is to establish policies and procedures regarding efficient acquisition, use, and disposal of systems and devices. Computers permeate organizations today, and it is critical for organizations to ensure that systems are used to their full potential. As a result, most companies have implemented stringent policies on the acquisition of computer systems and equipment, including requiring a formal justification statement before computer equipment is purchased, definition of standard computing platforms (operating system, type of computer chip, minimum amount of RAM, etc.), and the use of preferred vendors for all acquisitions.

Prevention of computer-related mistakes begins by identifying the most common types of errors, of which there are surprisingly few. Types of computer-related mistakes include the following:

- Data-entry or data-capture errors
- Errors in computer programs
- Errors in handling files, including formatting a disk by mistake, copying an old file over a newer one, and deleting a file by mistake
- Mishandling of computer output
- Inadequate planning for and control of equipment malfunctions
- Inadequate planning for and control of environmental difficulties (electrical problems, humidity problems, etc.)
- Installing computing capacity inadequate for the level of activity on corporate Web sites
- Failure to provide access to the most current information by not adding new Web links and not deleting old links

To control and prevent potential problems caused by computer-related mistakes, companies have developed policies and procedures that cover the acquisition and use of computers, with a goal of avoiding waste and mistakes. Training programs for individuals and workgroups as well as manuals and documents covering the use and maintenance computer systems also help prevent problems. Other preventive measures include approval of certain systems and applications before they are implemented and used to ensure compatibility and cost-effectiveness, and a requirement that documentation and descriptions of certain applications be filed or submitted to a central office, including all cell formulas for spreadsheets and a description of all data elements and relationships in a database system. Such standardization can ease access and use for all personnel.

Many companies have established strong policies to prevent employees from wasting time using computers inappropriately at work. A survey of 304 U.S. companies determined that over one-fourth of bosses have fired employees for inappropriate use of e-mail and one-third have fired workers for wasting valuable time on the Internet.[7] Three workers were terminated and another 26 workers for the Collier County, Florida government were given unpaid suspensions for inappropriate use of county computers.[8]

After companies have planned and developed policies and procedures, they must consider how best to implement them.

Implementing Policies and Procedures

Implementing policies and procedures to minimize waste and mistakes varies according to the business conducted. Most companies develop such policies and procedures with advice from the firm's internal auditing group or its external auditing firm. The policies often focus on the implementation of source data automation and the use of data editing to ensure data accuracy and completeness, and the assignment of clear responsibility for data accuracy within each information system. Some useful policies to minimize waste and mistakes include the following:

- Changes to critical tables, HTML, and URLs should be tightly controlled, with all changes authorized by responsible owners and documented.
- A user manual should be available covering operating procedures and documenting the management and control of the application.
- Each system report should indicate its general content in its title and specify the time period covered.
- The system should have controls to prevent invalid and unreasonable data entry.
- Controls should exist to ensure that data input, HTML, and URLs are valid, applicable, and posted in the right time frame.
- Users should implement proper procedures to ensure correct input data.

Training is another key aspect of implementation. Many users are not properly trained in using applications, and their mistakes can be very costly. When business intelligence tools were first installed at the Maryland Department of Transportation, inexperienced users began executing queries unrelated to their jobs. The large number of queries and report requests led to system performance problems, with excessive run times and slow response time to queries. The department implemented new policies for running requests and provided training to show users why it was important to access only the data required to do their work. System performance was improved through the implementation of these policies.[9]

Because more and more people use computers in their daily work, it is important that they understand how to use them. Training is often the key to acceptance and implementation of policies and procedures. Because of the importance of maintaining accurate data and of people understanding their responsibilities, companies converting to ERP and e-commerce systems invest weeks of training for key users of the system's various modules.

Monitoring Policies and Procedures

To ensure that users throughout an organization are following established procedures, the next step is to monitor routine practices and take corrective action if necessary. By understanding what is happening in day-to-day activities, organizations can make adjustments or develop new procedures. Many organizations implement internal audits to measure actual results against established goals, such as percentage of end-user reports produced on time, percentage of data-input errors detected, number of input transactions entered per eight-hour shift, and so on.

The Société Générale scandal in France is a classic example of an individual employee circumventing internal policies and procedures. A low-level trader on the arbitrage desk at the French bank created a series of fraudulent and unauthorized investment transactions that built a $72 billion position in European stock index futures.[10] Eventually the house of cards collapsed causing the bank to lose over $7 billion—even though a compliance officer at the bank had been alerted months in advance not once, but twice that something unusual was going on.[11]

Reviewing Policies and Procedures

The final step is to review existing policies and procedures and determine whether they are adequate. During review, people should ask the following questions:

- Do current policies cover existing practices adequately? Were any problems or opportunities uncovered during monitoring?
- Does the organization plan any new activities in the future? If so, does it need new policies or procedures addressing who will handle them and what must be done?
- Are contingencies and disasters covered?

This review and planning allows companies to take a proactive approach to problem solving, which can enhance a company's performance, such as by increasing productivity and improving customer service. During such a review, companies are alerted to upcoming changes in information systems that could have a profound effect on many business activities.

Tokyo Electron, a global supplier of semiconductor production equipment, provides an excellent example of a firm thoroughly reviewing its policies and procedures. As a U.S. subsidiary of Tokyo Electron of Japan, Tokyo Electron U.S. Holdings was required to comply with the Sarbanes-Oxley Act. When Japan's Financial Instruments and Exchange Law, that country's equivalent of the Sarbanes-Oxley Act, went into effect, the firm used it as a motivation to re-examine its entire set of policies regarding user access to data and applications, financial control, and protection of intellectual property.[12]

Information systems professionals and users still need to be aware of the misuse of resources throughout an organization. Preventing errors and mistakes is one way to do so. Another is implementing in-house security measures and legal protections to detect and prevent a dangerous type of misuse: computer crime.

COMPUTER CRIME

Even good IS policies might not be able to predict or prevent computer crime. A computer's ability to process millions of pieces of data in less than one second can help a thief steal data worth millions of dollars. Compared with the physical dangers of robbing a bank or retail store with a gun, a computer criminal with the right equipment and know-how can steal large amounts of money from the privacy of a home. The following is a sample of recent computer crimes:

- Criminals illegally obtained information about the bank accounts of an undetermined number of Citibank customers. They created counterfeit ATM cards encoded with the stolen information to make some 9,000 fraudulent ATM withdrawals totaling millions of dollars. Avivah Litan, Gartner vice president, stated: "Criminals have found ways to basically bypass many of the controls banks have in place. So ATM and debit card fraud is expected to rise. In our surveys, banks themselves expect the rate of fraud to double over the next two years."[13]
- A Chilean hacker gathered personal data about 6 million people from various Chilean government sites including names, addresses, phone numbers, ID numbers, and e-mail addresses and posted them to a blog site for all to see. The hacker's motivation was to protest his country's weak data security.[14]
- A hacker is alleged to have broken into the computers holding the financial results of IMS Health to learn of the firm's disappointing results for the quarter prior to their public announcement. Taking advantage of this knowledge, the hacker purchased over $41,000 in sell options, figuring the stock would go down when results were announced. The investment resulted in profits of nearly $300,000.[15]
- A 15-year-old Pennsylvania student broke into an educational network and saved on a flash drive the names, addresses, and Social Security numbers of some 55,000 people. The student was arrested and charged with four offenses of unlawful duplication and theft.[16]

- When customers initially link their brokerage accounts to their bank account to allow the transfer of funds, firms such as E*Trade and Schwab.com use a test procedure to make micro-deposits of a few cents to a few dollars to the bank account to ensure that the account numbers and routing information are correct. A hacker took advantage of a backdoor to this procedure by opening tens of thousands of banking accounts with the brokerages and linked them to fraudulent brokerage accounts to collect the micro-deposits. The hacker stole more than $50,000 over six months.[17]

Although no one really knows how pervasive cybercrime is, according to the 2007 FBI Internet Crime Report, 206,844 complaints of crime were perpetrated over the Internet during 2007 with a dollar value of $240 million in losses.[18] Unfortunately, this represents a small fraction of total computer related crimes as many crimes go unreported because companies don't want the bad press or don't believe that law enforcement could help. Such publicity makes the job even tougher for law enforcement. Most companies that have been electronically attacked won't talk to the press. A big concern is loss of public trust and image—not to mention the fear of encouraging copycat hackers.

The Computer Security Institute, with the participation of the San Francisco Federal Bureau of Investigation (FBI) Computer Intrusion Squad, conducts an annual survey of computer crime and security. The aim of the survey is to raise awareness of security, as well as to determine the scope of computer crime in the United States. The following are a few of the highlights of the 2007 Computer Crime and Security Survey[19] based on responses from 494 companies and government agencies that are members of the Computer Security Institute:

- Financial fraud, followed by virus attacks, is the leading cause of financial loss from computer incidents.
- For the respondents, the average annual loss from computer incidents was $350,424.
- A full 46 percent of the respondents said they had suffered a security incident, though only 29 percent of the respondents reported computer intrusions to law enforcement. (Surprisingly, 10 percent responded that they did not know if they had been subjected to an incident.)

The tenth annual *InformationWeek* Global Information Security survey reveals that the number one tactical security problem for U.S. companies in 2007 was creating and enhancing user awareness of security policies.[20]

Today, computer criminals are a new breed—bolder and more creative than ever. With the increased use of the Internet, computer crime is now global. It's not just on U.S. shores that law enforcement has to battle cybercriminals. Regardless of its nonviolent image, computer crime is different only because a computer is used. It is still a crime. Part of what makes computer crime so unique and difficult to combat is its dual nature—the computer can be both the tool used to commit a crime and the object of that crime.

THE COMPUTER AS A TOOL TO COMMIT CRIME

A computer can be used as a tool to gain access to valuable information and as the means to steal thousands or millions of dollars. It is, perhaps, a question of motivation—many people who commit computer-related crime claim they do it for the challenge, not for the money. Credit card fraud—whereby a criminal illegally gains access to another's line of credit with stolen credit card numbers—is a major concern for today's banks and financial institutions. In general, criminals need two capabilities to commit most computer crimes. First, the criminal needs to know how to gain access to the computer system. Sometimes, obtaining access requires knowledge of an identification number and a password. Second, the criminal must know how to manipulate the system to produce the desired result. Frequently, a critical computer password has been talked out of a person, a practice called **social engineering**. Or, the attackers simply go through the garbage—**dumpster diving**—for important pieces of information that can help crack the computers or convince someone at the company to give

social engineering
Using social skills to get computer users to provide information to access an information system or its data.

dumpster diving
Going through the trash cans of an organization to find secret or confidential information, including information needed to access an information system or its data.

them more access. In addition, over 2,000 Web sites offer the digital tools—for free—that will let people snoop, crash computers, hijack control of a machine, or retrieve a copy of every keystroke.

Although all the details have not been revealed, it appears that an outsider used social engineering skills to convince an MTV employee to download malicious software onto a corporate computer. Through this ruse, attackers gained access to over 5,000 employees' personal data including names, Social Security numbers, birth dates, and salary data.[21]

Also, with today's sophisticated desktop publishing programs and high-quality printers, crimes involving counterfeit money, bank checks, traveler's checks, and stock and bond certificates are on the rise. As a result, the U.S. Treasury redesigned and printed new currency that is much more difficult to counterfeit.

Cyberterrorism

Cyberterrorism has been a concern for countries and companies around the globe. The U.S. government considered the potential threat of cyberterrorism serious enough that it established the National Infrastructure Protection Center in February 1998. This function was transferred to the Homeland Security Department's Information Analysis and Infrastructure Protection Directorate to serve as a focal point for threat assessment, warning, investigation, and response for threats or attacks against the country's critical infrastructure, which provides telecommunications, energy, banking and finance, water systems, government operations, and emergency services. Successful cyberattacks against the facilities that provide these services could cause widespread and massive disruptions to the normal function of American society. International Multilateral Partnership Against Cyber Terrorism (IMPACT) is a global public and privately supported initiative against cyberterrorism.[22]

cyberterrorist
Someone who intimidates or coerces a government or organization to advance his political or social objectives by launching computer-based attacks against computers, networks, and the information stored on them.

A **cyberterrorist** is someone who intimidates or coerces a government or organization to advance his political or social objectives by launching computer-based attacks against computers, networks, and the information stored on them. Fortunately, only relatively few cases of cyberterrorism have been documented, including the following:

- The small Baltic nation of Estonia was subjected to a cyberterrorism attack for three weeks in 2007 that disabled government and corporate networks. The attack followed deadly riots by the nation's ethnic Russian minority in response to the relocation of a Soviet war memorial. Moscow has denied any involvement.[23]
- Pro-China cyberterrorists launched a brief denial-of-service attack on the CNN Web site, which they believe has been overly critical of China, to protest the news network's coverage of Tibet. The attack was cancelled after less than 30 minutes, but the group threatened to launch another attack in the near future.[24]

Identity Theft

identify theft
A crime in which an imposter obtains key pieces of personal identification information, such as Social Security or driver's license numbers, to impersonate someone else.

Identity theft is a crime in which an imposter obtains key pieces of personal identification information, such as Social Security or driver's license numbers, to impersonate someone else. The information is then used to obtain credit, merchandise, and/or services in the name of the victim or to provide the thief with false credentials. In 2007, 8.4 million adults in the United States were victims of identity fraud, according to Javelin Strategy & Research, which compiles a widely accepted survey.[25] The perpetrators of these crimes employ such an extensive range of methods that investigating them is difficult.

In some cases, the identity thief uses personal information to open new credit accounts, establish cellular phone service, or open a new checking account to obtain blank checks. In other cases, the identity thief uses personal information to gain access to the person's existing accounts. Typically, the thief changes the mailing address on an account and runs up a huge bill before the person whose identity has been stolen realizes there is a problem. The Internet has made it easier for an identity thief to use the stolen information because transactions can be made without any personal interaction.

Another popular method to get information is "shoulder surfing"—the identity thief simply stands next to someone at a public office, such as the Bureau of Motor Vehicles, and watches as the person fills out personal information on a form.

Consumers can help protect themselves by regularly checking their credit reports with major credit bureaus, following up with creditors if their bills do not arrive on time, not revealing any personal information in response to unsolicited e-mail or phone calls (especially Social Security numbers and credit card account numbers), and shredding bills and other documents that contain sensitive information.

The U.S. Congress passed the Identity Theft and Assumption Deterrence Act of 1998 to fight identity theft. Under this act, the Federal Trade Commission (FTC) is assigned responsibility to help victims restore their credit and erase the impact of the imposter. It also makes identity theft a federal felony punishable by a prison term ranging from 3 to 25 years.

Internet Gambling

Many people enjoy Internet gambling as a recreational and leisure activity. Baccarat, bingo, blackjack, pachinko, poker, roulette, and sports betting are all readily available online. The size of the online gambling market is not known, but one estimate is that $10–20 billion is wagered on online poker alone each year.[26] Although Internet gambling is legal in more than 70 countries, the legality of these online activities is far from clear in the U.S.

- The Interstate Wire Act of 1961 has been interpreted by the Department of Justice as banning all Internet gambling. However, various courts have interpreted the Act as covering only sporting events and exempting casino games such as blackjack and poker.[27]
- The Unlawful Internet Gambling Enforcement Act of 2006 (UIGEA) made it illegal to transfer funds from banks or financial institutions to online gambling sites. However, it failed to clarify the issue of the legality of gambling online. The Act simply states that some gambling is unlawful under state or federal law without specifying details other than excluding several specific gambling activities from its purview.
- Various individual states have passed laws regulating Internet gambling. These laws regulate making bets online, taking bets online, and transferring money between the bettor and an online casino anywhere in the world.

CBSSports.com and Facebook were investigated briefly by the FBI for collaborating to make it easier for Facebook users to fill out brackets for the NCAA 2008 Basketball Tournament. Leslie Anne Wade, senior vice president at CBS stated: "These are new issues that are going to require new thought processes and new answers. [CBS will] look at it."[28]

The revenues generated by Internet gambling represent a major untapped source of income for the state and federal governments. A study prepared by PriceWaterhouseCoopers estimates that the taxation of Internet gambling would yield somewhere between $8.7 billion and $42.8 billion in additional federal revenues during its first ten years.[29]

THE COMPUTER AS THE OBJECT OF CRIME

A computer can also be the object of the crime, rather than the tool for committing it. Tens of millions of dollars of computer time and resources are stolen every year. Each time system access is illegally obtained, data or computer equipment is stolen or destroyed, or software is illegally copied, the computer becomes the object of crime. These crimes fall into several categories: illegal access and use, data alteration and destruction, information and equipment theft, software and Internet piracy, computer-related scams, and international computer crime.

Illegal Access and Use

Crimes involving illegal system access and use of computer services are a concern to both government and business. Since the outset of information technology, computers have been plagued by criminal hackers. Originally, a **hacker** was a person who enjoys computer technology and spends time learning and using computer systems. A **criminal hacker**, also called a **cracker**, is a computer-savvy person who attempts to gain unauthorized or illegal access to

hacker
A person who enjoys computer technology and spends time learning and using computer systems.

criminal hacker (cracker)
A computer-savvy person who attempts to gain unauthorized or illegal access to computer systems to steal passwords, corrupt files and programs, or even transfer money.

script bunny
A cracker with little technical savvy who downloads programs called scripts, which automate the job of breaking into computers.

insider
An employee, disgruntled or otherwise, working solo or in concert with outsiders to compromise corporate systems.

computer systems to steal passwords, corrupt files and programs, or even transfer money. In many cases, criminal hackers are people who are looking for excitement—the challenge of beating the system. Today, many people use the term hacker and cracker interchangeably. **Script bunnies** admire crackers, but have little technical savvy. They are crackers who download programs called *scripts* that automate the job of breaking into computers. **Insiders** are employees, disgruntled or otherwise, working solo or in concert with outsiders to compromise corporate systems. The biggest threat for many companies is their own employees who hack into their computers, not external hackers. Insiders have extra knowledge that makes them especially dangerous—they know logon IDs, passwords, and company procedures that help them evade detection.

Some criminals have started phony VoIP phone companies and sold subscriptions for services to unsuspecting customers. Instead of establishing their own network, the criminals hack into the computers that route calls over the networks of legitimate VoIP providers and use this network to carry its customers' calls. One criminal obtained more than $1 million for more than 10 million minutes of VoIP service stolen from a legitimate VoIP service provider.[30]

Catching and convicting criminal hackers remains a difficult task. The method behind these crimes is often hard to determine. Even if the method behind the crime is known, tracking down the criminals can take a lot of time. It took years for the FBI to arrest one criminal hacker for the alleged theft of almost 20,000 credit card numbers that had been sent over the Internet.

Data and information are valuable corporate assets. The intentional use of illegal and destructive programs to alter or destroy data is as much a crime as destroying tangible goods. The most common of these programs are viruses and worms, which are software programs that, when loaded into a computer system, will destroy, interrupt, or cause errors in processing. Such programs are also called *malware* and the growth rate for such programs is epidemic. Internet security firm McAfee estimates that 150 to 200 malware programs emerge each day.[31]

virus
A computer program file capable of attaching to disks or other files and replicating itself repeatedly, typically without the user's knowledge or permission.

A **virus** is a computer program file capable of attaching to disks or other files and replicating itself repeatedly, typically without the user's knowledge or permission. Some viruses attach to files, so when the infected file executes, the virus also executes. Other viruses sit in a computer's memory and infect files as the computer opens, modifies, or creates the files. They are often disguised as games or images with clever or attention-grabbing titles such as "Boss, nude." Some viruses display symptoms, and some viruses damage files and computer systems. Computer viruses are written for several operating systems, including Windows, Macintosh, UNIX, and others.

Virus writers can become very aggressive in their attacks. For example, a variant of the GPcode virus encrypts various file types including .doc, .txt, .pdf, .xls, and images, and then demands a ransom payment for the key required to decrypt the files.[32] An increasing problem is the purchase of computer equipment already infected with malware. Best Buy unknowingly sold digital picture frames that were somehow infected with a computer virus during the manufacturing process.[33] Seagate Technology confirmed that many of its 500 GB hard drives left an Asian manufacturing plant infected with malware designed to steal online gaming passwords.

worm
A parasitic computer program that can create copies of itself on the infected computer or send copies to other computers via a network.

Worms are parasitic computer programs that replicate but, unlike viruses, do not infect other computer program files. Worms can create copies on the same computer or can send the copies to other computers via a network. Worms often spread via Internet Relay Chat (IRC).

Trojan horse
A malicious program that disguises itself as a useful application or game and purposefully does something the user does not expect.

A **Trojan horse** program is a malicious program that disguises itself as a useful application or game and purposefully does something the user does not expect. Trojans are not viruses because they do not replicate, but they can be just as destructive. Many people use the term to refer only to nonreplicating malicious programs, thus making a distinction between Trojans and viruses.

Although security is often cited as a strong point of the Mac computer, an increasing number of malware threats against the Mac OS X operating system have been uncovered. For example, an OS X Trojan horse is disguised either as an AppleScript known as ASthtv05 or bundled as an application named AStht_v06. When executed, this Trojan horse enables the attacker to remotely access the user's iSight camera, log keystrokes, retrieve screen shots, and manipulate file sharing settings.[34]

A *logic bomb* is a type of Trojan horse that executes when specific conditions occur. Triggers for logic bombs can include a change in a file by a particular series of keystrokes or at a specific time or date.

A *rootkit* is a set of programs that enable its user to gain administrator level access to a computer or network. Once installed, the attacker can gain full control of the system and even obscure the presence of the rootkit from legitimate system administrators. The Mebroot rootkit infects the master boot record, the first sector of the hard drive that the personal computer views before loading the operating system, making it all but invisible to security software and administrators. In an especially nefarious attack, hackers have created Web pages that when visited by users with certain browsers, release the Mebroot malware to infect the machine, a process known as a *drive-by download*.[35]

A *variant* is a modified version of a virus that is produced by the virus's author or another person who amends the original virus code. If changes are small, most antivirus products will also detect variants. However, if the changes are significant, the variant might go undetected by antivirus software.

The Storm worm is a Trojan horse that infects personal computers running the Microsoft operating systems. It began infecting computers via e-mail messages with a subject line about weather disasters in Europe, hence the name. Over time, and as users became wiser, the subject line of the malicious e-mail has changed several times. The e-mail contains an attachment that if opened loads a "cocktail" of various malware programs onto a personal computer. The result is that the computer is compromised and acts as a "zombie" computer under control of other computers. Such "zombies" are often used to send spam. It is estimated that as many as 40 million personal computers could have been infected by the Storm worm between January 2007 and February 2008.[36]

In some cases, a virus or a worm can completely halt the operation of a computer system or network for days or longer until the problem is found and repaired. In other cases, a virus or a worm can destroy important data and programs. If backups are inadequate, the data and programs might never be fully functional again. The costs include the effort required to identify and neutralize the virus or worm and to restore computer files and data, as well as the value of business lost because of unscheduled computer downtime.

The F-Secure Corporation provides centrally managed security solutions, and its products include antivirus, file encryption, and network security solutions for all major platforms—from desktops to servers and from laptops to handhelds. F-Secure is headquartered in Helsinki, Finland, and provides real-time virus statistics on the most active viruses in the world at its Web site, *www.f-secure.com/virus-info/statistics*.

McAfee Security for Consumers is a division of Network Associates Inc. that delivers retail and online solutions designed to secure, protect, and optimize the computers of consumers and home office users. McAfee's retail desktop products include premier antivirus, security, encryption, and desktop optimization software. McAfee delivers software through an Internet browser to provide these services to users online through its Web site *www.mcafee.com*, one of the largest paid subscription sites on the Internet with over 2 million active paid subscribers. McAfee provides a real-time map of where the latest viruses are infecting computers worldwide at *http://us.mcafee.com/virusInfo/default.asp*. See Figure 9.1. The site also provides software for scanning your computer for viruses and tips on how to remove a virus.[37]

Using Antivirus Programs

As a result of the increasing threat of viruses and worms, most computer users and organizations have installed **antivirus programs** on their computers. Such software runs in the background to protect your computer from dangers lurking on the Internet and other possible sources of infected files. Some antivirus software is even capable of repairing common virus infections automatically, without interrupting your work. The latest virus definitions are downloaded automatically when you connect to the Internet, ensuring that your PC's protection is current. To safeguard your PC and prevent it from spreading viruses to your friends and coworkers, some antivirus software scans and cleans both incoming and

antivirus program
Software that runs in the background to protect your computer from dangers lurking on the Internet and other possible sources of infected files.

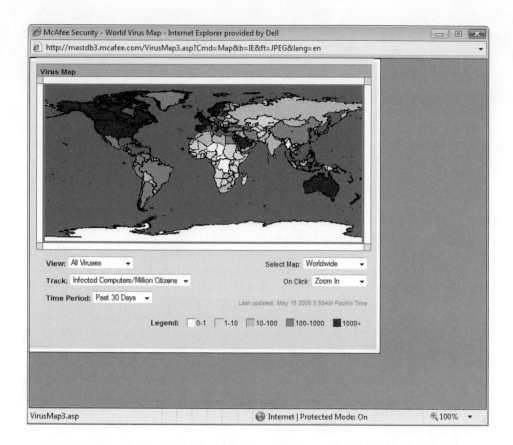

outgoing e-mail messages. Some of the most highly rated antivirus software for 2007 includes BitDefender, Kaspersky, ESET Nod32, AVG AntiVirus, F Secure Anti Virus, Trend Micro, McAfee VirusScan, Norton AntiVirus, and CA Antivirus. This software can be purchased for $20 to $66.

Many e-mail services and ISP providers offer free antivirus protection. For example, AOL and MWEB (one of South Africa's leading ISPs) offer free antivirus software from McAfee.

Disk defragmentation reorganizes the contents of the disk to store the pieces of each file close together and contiguously. It also creates larger regions of free space. Tests have shown that antivirus scans run significantly faster on computers with regularly defragmented files and free space, reducing the time to do a complete scan by 18 to 58 minutes.[38] Consider running disk defragmentation software on a regular basis.

Proper use of antivirus software requires the following steps:

1. **Install antivirus software and run it often.** Many of these programs automatically check for viruses each time you boot up your computer or insert a disk or CD, and some even monitor all e-mail and file transmissions and copying operations.

2. **Update antivirus software often.** New viruses are created all the time, and antivirus software suppliers are constantly updating their software to detect and take action against these new viruses.

3. **Scan all removable media, including CDs, before copying or running programs from them.** Hiding on disks or CDs, viruses often move between systems. If you carry document or program files on removable media between computers at school or work and your home system, always scan them.

4. **Install software only from a sealed package or secure Web site of a known software company.** Even software publishers can unknowingly distribute viruses on their program disks or software downloads. Most scan their own systems, but viruses might still remain.

5. **Follow careful downloading practices.** If you download software from the Internet or a bulletin board, check your computer for viruses immediately after completing the transmission.

6. **If you detect a virus, take immediate action.** Early detection often allows you to remove a virus before it does any serious damage.

Despite careful precautions, viruses can still cause problems. They can elude virus-scanning software by lurking almost anywhere in a system. Future antivirus programs might incorporate "nature-based models" that check for unusual or unfamiliar computer code. The advantage of this type of antivirus program is the ability to detect new viruses that are not part of an antivirus database.

Hoax, or false, viruses are another problem. Criminal hackers sometimes warn the public of a new and devastating virus that doesn't exist to create fear. Companies sometimes spend hundreds of hours warning employees and taking preventive action against a nonexistent virus. Security specialists recommend that IS personnel establish a formal paranoia policy to thwart virus panic among gullible end users. Such policies should emphasize that before users forward an e-mail alert to colleagues and higher-ups, they should send it to the help desk or the security team. The corporate intranet can be used to explain the difference between real viruses and fakes, and it can provide links to Web sites to set the record straight.

Be aware that virus writers also use known hoaxes to their advantage. For example, AOL4FREE began as a hoax virus warning. Then, a hacker distributed a destructive Trojan attached to the original hoax virus warning. Always remain vigilant and never open a suspicious attachment.

Spyware

Spyware is software installed on a personal computer to intercept or take partial control over the user's interaction with the computer without knowledge or permission of the user. Some forms of spyware secretly log keystrokes so that logon IDs and passwords may be captured. Other forms of spyware record information about the user's Internet surfing habits and sites that have been visited. Still other forms of spyware change personal computer settings so that the user experiences slow connection speeds or is redirected to different Web pages than those expected. The number of personal computers infected with spyware has become epidemic and users need to install anti-spyware software. The top four rated anti-spyware software for 2008 includes Spy Sweeper, CounterSpy, Spyware Doctor, and SuperAntiSpyware, which cost in the range of $17 to $30.[39]

DirectRevenue was a major distributor of adware, a form of spyware that monitors the viewing habits of Internet users to display targeted pop-up ads. The company received more than $80 million in ad revenue from its clients before it eventually ceased operations.[40] The company offered consumers free screensavers, games, and utility software but failed to disclose that downloading this software would load adware as well. Once it was installed, it was nearly

spyware
Software that is installed on a personal computer to intercept or take partial control over the user's interaction with the computer without knowledge or permission of the user.

impossible to identify, locate, and remove the adware. DirectRevenue agreed to a settlement with the FTC that barred future downloads of their adware without informed consent on the part of consumers. The firm was also fined $1.5 million.[41]

Information and Equipment Theft

Data and information are assets or goods that can also be stolen. People who illegally access systems often do so to steal data and information. To obtain illegal access, criminal hackers require identification numbers and passwords. Some criminals try different identification numbers and passwords until they find ones that work. Using password sniffers is another approach. A **password sniffer** is a small program hidden in a network or a computer system that records identification numbers and passwords. In a few days, a password sniffer can record hundreds or thousands of identification numbers and passwords. Using a password sniffer, a criminal hacker can gain access to computers and networks to steal data and information, invade privacy, plant viruses, and disrupt computer operations.

In addition to theft of data and software, all types of computer systems and equipment have been stolen from offices. Portable computers such as laptops and portable storage devices (and the data and information stored in them) are especially easy for thieves to take. In many cases, the data and information stored in these systems are more valuable than the equipment and there is a risk that the data can be used in identity theft. In addition, the organization responsible receives a tremendous amount of negative publicity that can cause it to lose existing and potential future customers. Often, the responsible organization offers to pay for credit monitoring services for those people affected in an attempt to restore customer goodwill and avoid law suits.

Perhaps the worst single example in terms of number of people affected by theft of equipment was in May 2006, when the Department of Veterans Affairs announced that a laptop and hard drive containing some 26.5 million personal records of current and former members of the military were stolen.[42] Here are a few more examples of laptops stolen that contained personal information. In most cases, the laptops were left in plain view where others could see them and the data was not encrypted or protected in any manner.

- *February 2007*: The Department of Veterans Affairs admitted that a hard drive containing the identities of almost 50,000 veterans was missing and might have been stolen.[43]
- *May 2007*: The Transportation Security Administration realized that an external disk drive containing names, Social Security numbers, birth dates, bank account and routing data and payroll information of about 100,000 employees who had worked for the agency over three years was missing.[44]
- *July 2007*: A contract worker for VeriSign who worked in the company's Human Resources department had her laptop containing employee data stolen when thieves broke into her parked car overnight.[45]
- *August 2007*: The Connecticut Department of Revenue Services revealed that a laptop containing personally identifiable data about more than 106,000 taxpayers was missing.[46]
- *September 2007*: The Gap, Inc. revealed that a laptop storing personal information on 800,000 job applicants was stolen from a contractor that managed job applicant data for the firm.[47]
- *December 2007*: Laptop computers were stolen from the Davidson County, Tennessee election office containing personal information for more than 337,000 registered voters.[48]
- *January 2008*: Horizon Blue Cross Blue Shield of New Jersey revealed that an employee laptop containing personal information of 300,000 clients was stolen.[49]

Many companies are putting into place tough measures to protect the data on laptops amid the epidemic of thefts. These policies include the following elements:

- Clear guidelines on what kind of data (and how much of it) can be stored on vulnerable laptops. In many cases, private data or company confidential data may not be downloaded to laptops that leave the office.

password sniffer
A small program hidden in a network or a computer system that records identification numbers and passwords.

- Requiring that data stored on laptops be encrypted and doing spot checks to ensure that this policy is followed.
- Requiring that all laptops be secured using a lock and chain device so that they cannot be easily removed from an office area.
- Providing training to employees and contractors on the need for safe handling of laptops and their data. For example, laptops should never be left in a position where they can be viewed by the public, such as on the front seat of an automobile.

In addition to the theft or loss of laptop computers, the U.S. Customs and Border Protection has increased the level of performing random inspection of electronic media. For years, U.S. agents have been taking and searching traveler's laptops, digital cameras, cell phones, PDAs, and other electronic devices. In some reported cases, the electronic devices of travelers entering the United States have been taken for two or more weeks to be inspected for evidence of child pornography or criminal or terrorist activity.[50]

To fight computer crime, many companies use devices that disable the disk drive or lock the computer to the desk.

(Source: Courtesy of Kensington Technology Group.)

Safe Disposal of Personal Computers

Many companies donate personal computers they no longer need to schools, churches, or other organizations. Some sell them at a deep discount to their employees or put them up for sale on Internet auction sites such as eBay. However, care must be taken to ensure that all traces of any personal or company confidential data is completely removed. Simply deleting files and emptying the Recycle Bin does not make it impossible for determined individuals to view the data. Be sure to use disk-wiping software utilities that overwrite all sectors of your disk drive making all data unrecoverable. For example, Darik's Boot and Nuke (DBAN) is free and can be downloaded from the SourceForge Web site.

Patent and Copyright Violations

Works of the mind, such as art, books, films, formulas, inventions, music, and processes that are distinct and "owned" or created by a single person or group are called intellectual property. Copyright law protects authored works such as art, books, film, and music. Patent laws protect processes, machines, objects made by humans or machines, compositions of matter, and new uses of these items.

Each time you use a word processing program or access software on a network, you are taking advantage of someone else's intellectual property. Like books and movies—other intellectual properties—software is protected by copyright laws. Often, people who would never think of plagiarizing another author's written work have no qualms about using and copying software programs they have not paid for. Such illegal duplicators are called *pirates*; the act of unauthorized copying or distribution of copyrighted software is called **software piracy**.

Software piracy often involves the copying, downloading, sharing, selling, or installing of multiple copies onto personal or work computers. When you purchase software, you are purchasing a license to use it; you do not own the actual software. The license states how many times you can install the software. If you make more copies of the software than the license permits, you are pirating.[51]

software piracy
The act of unauthorized copying or distribution of copyrighted software

The Software and Information Industry Alliance (SIIA) was the original software antipiracy organization, formed and financed by many of the large software publishers. Microsoft financed the formation of a second antipiracy organization, the Business Software Alliance (BSA). The BSA, through intense publicity, has become the more prominent organization. Other software companies, including Apple, Adobe, Hewlett-Packard, and IBM, now contribute to the BSA. The BSA estimates that the software industry lost over $48 billion in 2007 in revenue to worldwide software piracy. "Worldwide, for every two dollars of software purchased legitimately, one dollar was obtained illegally."[52]

Penalties for software piracy can be severe. If the copyright owner brings a civil action against someone, the owner can seek to stop the person from using its software immediately and can also request monetary damages. The copyright owner can then choose between compensation for actual damages—which includes the amount it has lost because of the person's infringement, as well as any profits attributable to the infringement—and statutory damages, which can be as much as $150,000 for each program copied. In addition, the government can prosecute software pirates in criminal court for copyright infringement. If convicted, they could be fined up to $250,000 or sentenced to jail for up to five years, or both.[53]

The Web site *www.MDofPC.com* was caught selling unlicensed copies of software from Adobe, McAfee, Microsoft, and Symantec. Visitors to the site could pay online and then be given access to the software for downloading when their payment cleared. The BSA initiated an investigation that led eventually lead to a $36,000 fine against the operators of the site.[54] The Acorn Engineering Company agreed to pay $250,000 to BSA to settle claims that it had unlicensed copies of software from Adobe, Autodesk, and Microsoft on its computers.[55]

Another major issue in regards to copyright infringement is the downloading of music that is copyright protected. Estimates vary widely as to how much music piracy is costing the recording industry. An estimate from the Institute for Policy Innovation (an economic public policy organization) is that the recording industry loses about $5.3 billion and retailers lose about $1.0 billion for a total direct loss of $6.3 billion. In addition, the U.S. government loses about $422 million in tax revenue.[56]

Operation Copycat is an ongoing undercover investigation into Warez groups, which are online organizations engaged in the illegal uploading, copying, and distribution of copyrighted works such as music, movies, games, and software, often even before they are released to the public. The investigation is led by the Computer Hacking and Intellectual Property (CHIP) Unit of the United States Attorney's Office and the FBI. Operation Copycat has resulted in 40 convictions over a period of three years from July 2005 to July 2008. Those convicted have typically been sentenced to over a year in prison, required to pay fines in excess of $200,000 and had to forfeit all computer and other equipment used in committing the offenses.[57]

The Motion Picture Association of America estimates that it loses over $18 billion per year from movie theft. It won a $100 million judgment against TorrentSpy for offering thousands of copyright-protected movies and TV shows.[58]

Patent infringement is also a major problem for computer software and hardware manufacturers. It occurs when someone makes unauthorized use of another's patent. If a court determines that a patent infringement is intentional, it can award up to three times the amount of damages claimed by the patent holder. It is not unusual to see patent infringement awards in excess of $10 million.

To obtain a patent or to determine if a patent exists in an area a company seeks to exploit requires a lengthy (typically longer than 25 months) search by the U.S. Patent Office. Indeed, the patent process is so controversial that there is a broad consensus among manufacturing firms, the financial community, consumer and public interest groups and government leaders demanding patent reform. Here are just a few examples of numerous recent lawsuits involving patent infringement.

- Personal computer manufacturers Acer, Apple, Dell, and Hewlett-Packard were sued for allegedly violating four patents related to a wireless communications privacy method and system held by Saxon Innovations.[59]

- Telecommunications network equipment maker Tellabs filed a patent infringement lawsuit against Fujitsu in regards to technology associated with optical and multiplexing systems and equipment.[60]
- Red Hat settled patent infringement claims over business process software with Firestar Software and DataTern.[61]

Computer-Related Scams

People have lost hundreds of thousands of dollars on real estate, travel, stock, and other business scams. Today, many of these scams are being perpetrated with computers. Using the Internet, scam artists offer get-rich-quick schemes involving bogus real estate deals, tout "free" vacations with huge hidden costs, commit bank fraud, offer fake telephone lotteries, sell worthless penny stocks, and promote illegal tax-avoidance schemes.

Over the past few years, credit card customers of various banks have been targeted by scam artists trying to get personal information needed to use their credit cards. The scam works by sending customers an e-mail including a link that seems to direct users to their bank's Web site. At the site, they are greeted with a pop-up box asking them for their full debit card numbers, their personal identification numbers, and their credit card expiration dates. The problem is that the Web site customers are directed to is a fake site operated by someone trying to gain access to their private information. As discussed previously, this form of scam is called *phishing*. According to the IT research firm Gartner, more than an estimated 124 million people in the U.S. received phishing e-mails and some 3.6 million of them lost a total of $3.2 billion during 2007.[62] One common phishing scam involves e-mails claiming to be from eBay's security team and warning recipients that they have a security issue to resolve. The e-mail includes a link urging the recipient to take action. Clicking the link takes the user to a page requesting personal information that if provided, compromises the victim's identity.[63]

The following is a list of tips to help you avoid becoming a scam victim:

- Don't agree to anything in a high-pressure meeting or seminar. Insist on having time to think it over and to discuss things with your spouse, partner, or attorney. If a company won't give you the time you need to check out an offer and think things over, you don't want to do business with them. A good deal now will be a good deal tomorrow; the only reason for rushing you is if the company has something to hide.
- Don't judge a company based on appearances. Flashy Web sites can be created and published in a matter of days. After a few weeks of taking money, a site can vanish without a trace in just a few minutes. You might find that the perfect money-making opportunity offered on a Web site was a money maker for the crook and a money loser for you.
- Avoid any plan that pays commissions simply for recruiting additional distributors. Your primary source of income should be your own product sales. If the earnings are not made primarily by sales of goods or services to consumers or sales by distributors under you, you might be dealing with an illegal pyramid.
- Beware of shills, people paid by a company to lie about how much they've earned and how easy the plan was to operate. Check with an independent source to make sure that you aren't having the wool pulled over your eyes.
- Beware of a company's claim that it can set you up in a profitable home-based business but that you must first pay up front to attend a seminar and buy expensive materials. Frequently, seminars are high-pressure sales pitches, and the material is so general that it is worthless.
- If you are interested in starting a home-based business, get a complete description of the work involved before you send any money. You might find that what you are asked to do after you pay is far different from what was stated in the ad. You should never have to pay for a job description or for needed materials.
- Get in writing the refund, buy-back, and cancellation policies of any company you deal with. Do not depend on oral promises.
- Do your homework. Check with your state attorney general and the National Fraud Information Center before getting involved, especially when the claims about a product or potential earnings seem too good to be true.

If you need advice about an Internet or online solicitation, or if you want to report a possible scam, use the Online Reporting Form or Online Question & Suggestion Form features on the Web site for the National Fraud Information Center at *http://fraud.org*, or call the NFIC hotline at 1-800-876-7060.

International Computer Crime

Computer crime is also an international issue, and it becomes more complex when it crosses borders. As already mentioned, the software industry loses about $11 to $12 billion in revenue to software piracy annually, with about $9 billion of that occurring outside the United States.[39]

With the increase in electronic cash and funds transfer, some are concerned that terrorists, international drug dealers, and other criminals are using information systems to launder illegally obtained funds. Computer Associates International developed software called CleverPath for Global Compliance for customers in the finance, banking, and insurance industries to eliminate money laundering and fraud. Companies that are required to comply with legislation such as the USA Patriot Act and Sarbanes-Oxley Act might lack the resources and processes to do so. The software automates manual tracking and auditing processes that are required by regulatory agencies and helps companies handle frequently changing reporting regulations. The application can drill into a company's transactions and detect transaction patterns that suggest fraud or other illegal activities based on built-in business rules and predictive analysis. Suspected fraud cases are identified and passed on to the appropriate personnel for action to thwart criminals and help companies avoid paying fines.

PREVENTING COMPUTER-RELATED CRIME

Because of increased computer use today, greater emphasis is placed on the prevention and detection of computer crime. Although all states have passed computer crime legislation, some believe that these laws are not effective because companies do not always actively detect and pursue computer crime, security is inadequate, and convicted criminals are not severely punished. However, all over the United States, private users, companies, employees, and public officials are making individual and group efforts to curb computer crime, and recent efforts have met with some success.

Crime Prevention by State and Federal Agencies

State and federal agencies have begun aggressive attacks on computer criminals, including criminal hackers of all ages. In 1986, Congress enacted the Computer Fraud and Abuse Act, which mandates punishment based on the victim's dollar loss.

For at least five years after the September 11, 2001 terrorist attacks, the U.S. Treasury Department and CIA executed a program called the Terrorist Finance Tracking Program that relied on data in international money transfers from the Society for Worldwide Interbank Financial Telecommunications. The goal of the program was to track and combat terrorist financing. The program was credited with helping to capture at least two terrorists; however, revelation of the secret program's existence stirred up controversy and rendered the program ineffective.

The Department of Defense also supports the Computer Emergency Response Team (CERT), which responds to network security breaches and monitors systems for emerging threats. Law enforcement agencies are also increasing their efforts to stop criminal hackers, and many states are now passing new, comprehensive bills to help eliminate computer crimes. A complete listing of computer-related legislation by state can be found at *www.onlinesecurity.com/forum/article46.php*. Recent court cases and police reports involving computer crime show that lawmakers are ready to introduce newer and tougher computer crime legislation.

International Cyber Espionage

In 2007 and 2008, businesses and government agencies in many countries experienced a spike in targeted attacks originating outside their borders, many from China. Analysis of the attacks lead security experts to believe that many governments are involved in cyber espionage—that is, the use of the Internet to spy on other governments. Not only is the Internet being leveraged for international espionage, but it is also being used for economic espionage. Economic espionage refers to the use of the Internet by nation-states to steal corporate information in an effort to gain economic advantages in multinational deals.

The SANS Institute, a leading information security research group, ranked cyber espionage number three on its list of the Top Ten Cyber Menaces for 2008. Number one was Web site attacks that exploit Web browser vulnerabilities to install malware on PCs, and number two was botnets. Consider the following examples of recent cyber espionage and economic espionage.

The U.K. government has accused the Chinese of hacking into the computer systems of "some of its leading companies." The Director-General of intelligence agency MI5 sent letters to 300 chief executives and security chiefs of financial institutions warning of a sharp rise in instances of electronic espionage. The organization believes that at least 20 foreign intelligence services are engaged in cyber espionage against "U.K. interests." The organization is most concerned about Russia and China.

One report describes how Chinese hackers infected the Rolls-Royce corporate network with a Trojan horse that sent secret corporate information from the network to a remote server. Shell Oil Company discovered a Chinese cyber spy ring in Houston, Texas, working to steal confidential pricing information from servers at its operations in Africa.

Attacks and hacks against the Pentagon's computer system, the Oak Ridge National Lab, and Los Alamos National Lab, where U.S. nuclear weapons technology is developed, have all been traced to China. Information about top scientists was stolen from the Oak Ridge Lab. Germany, France, and New Zealand have reported similar attacks originating in China.

One of the favorite tools for cyber and economic espionage is a rootkit that works at a low level in the computer system, intercepting messages between the operating system and security software. The rootkit avoids detection, while sending secure data out a backdoor of the network to the hacker's server over the Internet—often located in China, and sometimes to servers registered to the Chinese government. These rootkits can make their way into private networks by tricking employees into visiting Web sites or opening attachments containing the rootkit. The trick typically involves detailed and custom-designed phishing e-mail messages that use social-engineered knowledge to persuade the recipient that the e-mail is legitimate.

Although it would be easy to jump to the conclusion that the Chinese government is behind all of these attacks, experts are quick to point out that it is difficult to pinpoint the origin of an attack. The Internet makes it possible for hackers to launch attacks from any server in the world. If an attack originates in China and is engineered by a Chinese citizen, it still cannot be determined if that person is working for the government. The Chinese government vehemently denies any part in cyber espionage. Still, most governments hold the Chinese government accountable for not cracking down on hackers if not actually sponsoring them. It is estimated that 30 percent of malicious software is created in China. The next largest distributor of malware is Russia and Eastern Europe.

Even if the Chinese government is actively involved in cyber espionage, as many governments are accusing, it is far from alone. A report developed by security firm McAfee states that "120 countries are developing ways to use the Internet as a weapon to target financial markets, government computer systems, and utilities." A number of experts are calling this the "cyber cold war."

Government agencies and businesses that may be targeted by cyber espionage and economic espionage are advised to use data-leak prevention products and database-monitoring tools. These tools lock down data and prevent copies from leaving the network. Some companies have gone as far as maintaining two separate networks, one for secure data and the other for Internet communications. This prevents malware from secretly funneling data out network backdoors to hackers.

Discussion Questions

1. How do cyber espionage and economic espionage differ?
2. What tricks are used by hackers to infiltrate systems and gain access to private information?

Critical Thinking Questions

1. What are the dangers if the cyber cold war turns into an actual cyber war?
2. Why are countries and businesses concerned about cyber espionage that originates in China and Russia?

Sources: Messmer, Ellen, "Cyber espionage seen as growing threat to business, government," *Network World*, January 17, 2008, *www.networkworld.com/news/2008/011708-cyberespionage.html*; Kirk, Jeremy, "Shell, Rolls-Royce reportedly hacked by Chinese spies," *Computerworld*, December 3, 2007, *www.computerworld.com/action/article.do?command=viewArticleBasic&articleId=9050538&intsrc=news_list*; Dahdah, Howard, "UK government accuses Chinese of IT espionage," *Computerworld*, December 3, 2007, *www.computerworld.com/action/article.do?command=viewArticleBasic&articleId=9050499&intsrc=news_list*; Reimer, Jeremy, "Chinese government at the center of five cyber attack claims," *Ars Technica*, September 14, 2007, *http://arstechnica.com/news.ars/post/20070914-chinese-government-at-the-center-of-five-cyber-attack-claims.html*; Griffiths, Peter, "World faces 'cyber cold war' threat," Reuters, November 29, 2007, *www.reuters.com/article/technologyNews/idUSL2932083320071129?feedType=RSS&feedName=technologyNews*.

Crime Prevention by Corporations

Companies are also taking crime-fighting efforts seriously. Many businesses have designed procedures and specialized hardware and software to protect their corporate data and systems. Specialized hardware and software, such as encryption devices, can be used to encode data and information to help prevent unauthorized use. Encryption is the process of converting an original electronic message into a form that can be understood only by the intended recipients. A key is a variable value that is applied using an algorithm to a string or block of unencrypted text to produce encrypted text or to decrypt encrypted text. Encryption methods rely on the limitations of computing power for their effectiveness—if breaking a code requires too much computing power, even the most determined code crackers will not be successful. The length of the key used to encode and decode messages determines the strength of the encryption algorithm.

As employees move from one position to another at a company, they can build up access to multiple systems if inadequate security procedures fail to revoke access privileges. It is clearly not appropriate for people who have changed positions and responsibilities to still have access to systems they no longer use. To avoid this problem, many organizations create role-based system access lists so that only people filling a particular role (e.g., invoice approver) can access a specific system.

Fingerprint authentication devices provide security in the PC environment by using fingerprint recognition instead of passwords. Laptop computers from Lenovo, Toshiba, and others have built-in fingerprint readers used to log on and gain access to the computer system and its data. The JetFlash 210 Fingerprint USB Flash Drive requires users to swipe their fingerprints and match them to one of up to 10 trusted users to access the data. The data on the flash drive can also be encrypted for further protection.[64]

Fingerprint authentication devices provide security in the PC environment by using fingerprint recognition instead of passwords.

(Source: Permission granted by Pay By Touch.)

Crime-fighting procedures usually require additional controls on the information system. Before designing and implementing controls, organizations must consider the types of computer-related crime that might occur, the consequences of these crimes, and the cost and complexity of needed controls. In most cases, organizations conclude that the trade-off between crime and the additional cost and complexity weighs in favor of better system controls. Having knowledge of some of the methods used to commit crime is also helpful in preventing, detecting, and developing systems resistant to computer crime (see Table 9.1). Some companies actually hire former criminals to thwart other criminals.

Although the number of potential computer crimes appears to be limitless, the actual methods used to commit crime are limited. The following list provides a set of useful guidelines to protect your computer from criminal hackers.

- Install strong user authentication and encryption capabilities on your firewall.
- Install the latest security patches, which are often available at the vendor's Internet site.
- Disable guest accounts and null user accounts that let intruders access the network without a password.

Methods	Examples
Add, delete, or change inputs to the computer system.	Delete records of absences from class in a student's school records.
Modify or develop computer programs that commit the crime.	Change a bank's program for calculating interest to make it deposit rounded amounts in the criminal's account.
Alter or modify the data files used by the computer system.	Change a student's grade from C to A.
Operate the computer system in such a way as to commit computer crime.	Access a restricted government computer system.
Divert or misuse valid output from the computer system.	Steal discarded printouts of customer records from a company trash bin.
Steal computer resources, including hardware, software, and time on computer equipment.	Make illegal copies of a software program without paying for its use.
Offer worthless products for sale over the Internet.	Send e-mail requesting money for worthless hair growth product.
Blackmail executives to prevent release of harmful information.	Eavesdrop on organization's wireless network to capture competitive data or scandalous information.
Blackmail company to prevent loss of computer-based information.	Plant logic bomb and send letter threatening to set it off unless paid considerable sum.

Table 9.1

Common Methods Used to Commit Computer Crimes

- Do not provide overfriendly logon procedures for remote users (e.g., an organization that used the word *welcome* on their initial logon screen found they had difficulty prosecuting a criminal hacker).
- Restrict physical access to the server and configure it so that breaking into one server won't compromise the whole network.
- Give each application (e-mail, File Transfer Protocol, and domain name server) its own dedicated server.
- Turn audit trails on.
- Consider installing caller ID.
- Install a corporate firewall between your corporate network and the Internet.
- Install antivirus software on all computers and regularly download vendor updates.
- Conduct regular IS security audits.
- Verify and exercise frequent data backups for critical data.

Using Intrusion Detection Software

An **intrusion detection system (IDS)** monitors system and network resources and notifies network security personnel when it senses a possible intrusion. Examples of suspicious activities include repeated failed logon attempts, attempts to download a program to a server, and access to a system at unusual hours. Such activities generate alarms that are captured on log files. Intrusion detection systems send an alarm, often by e-mail or pager, to network security personnel when they detect an apparent attack. Unfortunately, many IDSs frequently provide false alarms that result in wasted effort. If the attack is real, network security personnel must make a decision about what to do to resist the attack. Any delay in response increases the probability of damage from a criminal hacker attack. Use of an IDS provides another layer of protection in the event that an intruder gets past the outer security layers—passwords, security procedures, and corporate firewall.

A firm called Internet Security Systems (ISS) manages security for other organizations through its Managed Protection Services. The company's IDSs are designed to recognize 30 of the most-critical threats, including worms that go after Microsoft software and those that exploit Apache Web servers and other programs. When an attack is detected, the service

intrusion detection system (IDS)

Software that monitors system and network resources and notifies network security personnel when it senses a possible intrusion.

automatically blocks it without requiring human intervention. Taking the manual intervention step out of the process enables a faster response and minimizes damage from a criminal hacker. To encourage customers to adopt its service, ISS guaranteed up to $50,000 in cash if the prevention service failed.

Security Dashboard

Many organizations employ **security dashboard** software to provide a comprehensive display on a single computer screen of all the vital data related to an organization's security defenses including threats, exposures, policy compliance, and incident alerts. The goal is to reduce the effort required for monitoring and to identify threats earlier. Data comes from a variety of sources including firewalls, applications, servers, and other software and hardware devices. See Figure 9.2.

security dashboard

Software that provides a comprehensive display on a single computer screen of all the vital data related to an organization's security defenses including threats, exposures, policy compliance and incident alerts.

Figure 9.2

The Computer Network Defence Internet Operational Picture

The Computer Network Defence Internet Operational Picture, a security dashboard designed for the United Kingdom government and military networks, displays near real-time information on new and emerging cyber threats.

Associated Newspapers publishes six of the United Kingdom's largest newspapers that deliver timely information to some six million daily subscribers. Its journalists work in many countries and time zones. The organization implemented a security dashboard to cut the potential of interruption to its news cycle and raise the security protection of its news stories. As Mark Callaby, IT Security Officer, states: "We have a diverse IT infrastructure, which makes it difficult to track the current status of system patches and identify potential vulnerabilities. We needed to improve our ability to detect spyware, as well as establish a centralized view of our infrastructure and its security status."[65]

Using Managed Security Service Providers (MSSPs)

Keeping up with computer criminals—and with new regulations—can be daunting for organizations. Criminal hackers are constantly poking and prodding, trying to breach the security defenses of companies. Also, such recent legislation as HIPAA, Sarbanes-Oxley, and the USA Patriot Act requires businesses to prove that they are securing their data. For most small and midsized organizations, the level of in-house network security expertise needed to protect their business operations can be quite costly to acquire and maintain. As a result, many are outsourcing their network security operations to managed security service providers (MSSPs) such as Counterpane, Guardent, Internet Security Services, Riptech, and Symantec.

MSSPs monitor, manage, and maintain network security for both hardware and software. These companies provide a valuable service for IS departments drowning in reams of alerts and false alarms coming from virtual private networks (VPNs); antivirus, firewall, and intrusion detection systems; and other security monitoring systems. In addition, some provide vulnerability scanning and Web blocking/filtering capabilities.

Filtering and Classifying Internet Content

To help parents control what their children see on the Internet, some companies provide *filtering software* to help screen Internet content. Many of these screening programs also prevent children from sending personal information over e-mail or through chat groups. This stops children from broadcasting their name, address, phone number, or other personal information over the Internet. The two approaches used are filtering, which blocks certain Web sites, and rating, which places a rating on Web sites. According to the 2004 Internet Filter Review, the five top-rated filtering software packages are, in order: ContentProtect, Cybersitter, Net Nanny, CyberPatrol, and FilterPack.

Business organizations also implement filtering software to prevent employees from visiting nonwork-related Web sites, particularly those related to gambling or those containing pornographic or other offensive material. Before implementing any sort of Web site blocking, the users must be informed about the company's policies and why they exist. It is best if the organization's Internet users, management, and IS organization work together to define the policy to be implemented. The policy should be clear on the repercussions to employees who attempt to circumvent the blocking measures.

The Internet Content Rating Association (ICRA) is a nonprofit organization whose members include Internet industry leaders such as America Online, Bell South, British Telecom, IBM, Microsoft, UUNet, and Verizon. Its specific goals are to protect children from potentially harmful material, while also safeguarding free speech on the Internet. Using the ICRA rating system, Web authors fill out an online questionnaire describing the content of their site—what is and isn't present. The broad topics covered include chat capabilities, the language used, nudity and sexual content, violence depicted, and other areas such as alcohol, drugs, gambling, and suicide. Based on the authors' responses, ICRA then generates a content label (a short piece of computer code) that the authors add to their site. Internet users (and parents) can then set their browser to allow or disallow access to Web sites based on the objective rating information declared in the content label and their own subjective preferences. Reliance on Web site authors to do their own rating has its weaknesses, though. Web site authors can lie when completing the ICRA questionnaire so that their site receives a content label that doesn't accurately reflect the site's content. In addition, many hate group and sexually explicit sites don't have an ICRA rating, so they will not be blocked unless a browser is set to block all unrated sites. Also, this option would block out so many acceptable sites that it could make Web surfing useless. For these reasons, at this time, site labeling is at best a complement to other filtering techniques.

The U.S. Congress has made several attempts to limit children's exposure to online pornography including the Communications Decency Act (enacted 1996) and the Child Online Protection Act (enacted 1998). Within two years of their being enacted, the U.S. Supreme Court found that both these acts violated the First Amendment (freedom of speech) and ruled them to be unconstitutional. The Children's Internet Protection Act (CIPA) was signed into law in 2000 and later upheld by the Supreme Court in 2003. Under CIPA, schools and libraries subject to CIPA do not receive the discounts offered by the "E-Rate" program unless they certify that they have certain Internet safety measures in place to block or filter "visual depictions that are obscene, child pornography, or are harmful to minors."[66] (The E-Rate program provides many schools and libraries support to purchase Internet access and computers).

The Yorba Linda Library Commission applied for and began receiving discounts on Internet access through the federally funded E-Rate program in July 2007. The estimated savings exceeds $10,000. As a result, the library plans to install filter software on its 28 computers for the safety of computer users under the age of 18.[67]

ContentProtect is a filtering software program that helps block unwanted Internet content from children and young adults.

(Source: Courtesy of ContentWatch Inc.)

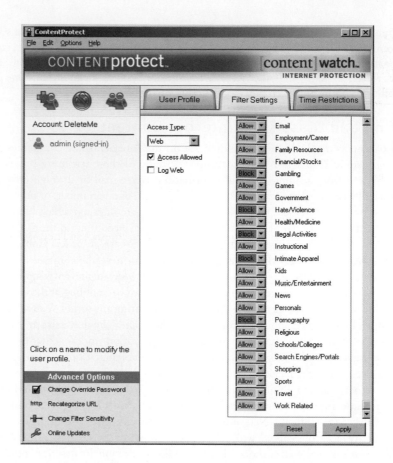

Internet Libel Concerns

With the increased popularity of networks and the Internet, libel becomes an important legal issue. A publisher, such as a newspaper, can be sued for libel, which involves publishing an intentionally false written statement that is damaging to a person's reputation. Generally, a bookstore cannot be held liable for statements made in newspapers or other publications it sells. Online services, such as CompuServe and America Online, might exercise some control over who puts information on their service but might not have direct control over the content of what is published by others on their service. So, can online services be sued for libel for content that someone else publishes on their service? Do online services more closely resemble a newspaper or a bookstore? This legal issue has not been completely resolved, but some court cases have been decided. The *Cubby, Inc. v. CompuServe* case ruled that CompuServe was similar to a bookstore and not liable for content put on its service by others. In this case, the judge stated, "While CompuServe can decline to carry a given publication altogether, in reality, after it does decide to carry a given publication, it will have little or no editorial control over that publication's content." This case set a legal precedent that has been applied in similar, subsequent cases. Companies should be aware that publishing Internet content to the world can subject them to different countries' laws in the same way that exporting physical products does.

Geolocation tools match the user's IP address with outside information to determine the actual geographic location of the online user where the customer's computer signal enters the Internet. This enables someone to identify the user's actual location within approximately 50 miles. Internet publishers can now limit the reach of their published speech to avoid potential legal risks. Use of such technology is also dividing the global Internet into separate content regions, with readers in Brazil, Japan, and the United States all receiving variations of the same information from the same publisher.

Individuals, too, must be careful what they post on the Internet to avoid libel charges. In many cases, disgruntled former employees are being sued by their former employers for material posted on the Internet.

Preventing Crime on the Internet

Internet security can include firewalls and many methods to secure financial transactions. A firewall can include both hardware and software that act as a barrier between an organization's information system and the outside world. Some systems have been developed to safeguard financial transactions on the Internet.

To help prevent crime on the Internet, the following steps can be taken:

1. Develop effective Internet usage and security policies for all employees.
2. Use a stand-alone firewall (hardware and software) with network monitoring capabilities.
3. Deploy intrusion detection systems, monitor them, and follow up on their alarms.
4. Monitor managers and employees to make sure that they are using the Internet for business purposes.
5. Use Internet security specialists to perform audits of all Internet and network activities.

Even with these precautions, computers and networks can never be completely protected against crime. One of the biggest threats is from employees. Although firewalls provide good perimeter control to prevent crime from the outside, procedures and protection measures are needed to protect against computer crime by employees. Passwords, identification numbers, and tighter control of employees and managers also help prevent Internet-related crime.

PRIVACY ISSUES

Another important social issue in information systems involves privacy. In 1890, U.S. Supreme Court Justice Louis Brandeis stated that the "right to be left alone" is one of the most "comprehensive of rights and the most valued by civilized man." Basically, the issue of privacy deals with this right to be left alone or to be withdrawn from public view. With information systems, privacy deals with the collection and use or misuse of data. Data is constantly being collected and stored on each of us. This data is often distributed over easily accessed networks and without our knowledge or consent. Concerns of privacy regarding this data must be addressed. For example, the U.S. Department of Heath and Human Services has received over 26,000 complaints of medical privacy breaches since new privacy rules went into effect in 2003.[68]

With today's computers, the right to privacy is an especially challenging problem. More data and information are produced and used today than ever before. When someone is born, takes certain high school exams, starts a job, enrolls in a college course, applies for a driver's license, purchases a car, serves in the military, gets married, buys insurance, gets a library card, applies for a charge card or loan, buys a house, or merely purchases certain products, data is collected and stored somewhere in computer databases. A difficult question to answer is, "Who owns this information and knowledge?" If a public or private organization spends time and resources to obtain data on you, does the organization own the data, and can it use the data in any way it desires? Government legislation answers these questions to some extent for federal agencies, but the questions remain unanswered for private organizations.

Privacy and the Federal Government

The U.S. federal government is the largest collector of data in the U.S. Over 4 billion records exist on citizens, collected by about 100 federal agencies, ranging from the Bureau of Alcohol, Tobacco, and Firearms to the Veterans Administration. Other data collectors include state and local governments and commercial and nonprofit organizations of all types and sizes. The government must be on guard at all times to safeguard this data. For example, two workers were fired at the State Department when electronic monitoring detected unauthorized accessing of the personal passport information of three 2008 presidential candidates.[69]

The European Union has a data-protection directive that requires firms transporting data across national boundaries to have certain privacy procedures in place. This directive affects virtually any company doing business in Europe, and it is driving much of the attention being given to privacy in the United States.

Privacy at Work

The right to privacy at work is also an important issue. Currently, the rights of workers who want their privacy and the interests of companies that demand to know more about their employees are in conflict. A recent poll uncovered that 78 percent of companies monitor their employees while at work in one form or another.[70] According to another recent survey, nearly one-third of companies have fired an employee for violating corporate e-mail policies.[71] Statistics such as these have raised employee concerns. For example, workers might find that they are being closely monitored via computer technology. These computer-monitoring systems tie directly into workstations; specialized computer programs can track every keystroke made by a user. This type of system can determine what workers are doing while at the keyboard. The system also knows when the worker is not using the keyboard or computer system. These systems can estimate what people are doing and how many breaks they are taking. Needless to say, many workers consider this close supervision very dehumanizing.

E-Mail Privacy

E-mail also raises some interesting issues about work privacy. Federal law permits employers to monitor e-mail sent and received by employees. Furthermore, e-mail messages that have been erased from hard disks can be retrieved and used in lawsuits because the laws of discovery demand that companies produce all relevant business documents. On the other hand, the use of e-mail among public officials might violate "open meeting" laws. These laws, which apply to many local, state, and federal agencies, prevent public officials from meeting in private about matters that affect the state or local area.

E-mail has changed how workers and managers communicate in the same building or around the world. E-mail, however, can be monitored and intercepted. As with other services—such as cellular phones—the convenience of e-mail must be balanced with the potential of privacy invasion.

(Source: © Gary Conner/Photo Edit.)

Instant Messaging Privacy

Using instant messaging (IM) to send and receive messages, files, and images introduces the same privacy issues associated with e-mail. As with e-mail, federal law permits employers to monitor instant messages sent and received by employees. Do not send personal or private IMs at work. Other significant privacy issues depend on the instant messaging client that you use. For example, at one time AOL and ICQ stated in their privacy policy that "You waive any right to privacy" and that they may use your instant messages in any way they see fit. Here are a few other tips:

- Choose a nonrevealing, nongender-specific, unprovocative IM screen name (Sweet Sixteen, 2hot4u, UCLAMBA all fail this test).
- Don't send messages you would be embarrassed to have your colleagues or significant other read.
- Do not open files or click links in messages from people you do not know.
- Never send sensitive personal data such as credit card numbers, bank account numbers, or passwords via IM.

Privacy and Personal Sensing Devices

RFID tags, essentially microchips with antenna, are embedded in many of the products we buy such as medicine containers, clothing, computer printers, car keys, library books, and tires. RFID tags generate radio transmissions that if appropriate measures are not taken, can lead to potential privacy concerns. Once these tags are associated with the individual who purchased the item, there is the potential to track individuals by the unique identifier associated with the RFID chip.

Several states have reacted to the potential for abuse of RFID tags by going so far as passing legislation prohibiting the implantation of RFID chips under people's skin without their approval.[72]

Privacy and the Internet

Some people assume that there is no privacy on the Internet and that you use it at your own risk. Others believe that companies with Web sites should have strict privacy procedures and be accountable for privacy invasion. Regardless of your view, the potential for privacy invasion on the Internet is huge. People wanting to invade your privacy could be anyone from criminal hackers to marketing companies to corporate bosses. Your personal and professional information can be seized on the Internet without your knowledge or consent. E-mail is a prime target, as discussed previously. Sending an e-mail message is like having an open conversation in a large room—people can listen to your messages. When you visit a Web site on the Internet, information about you and your computer can be captured. When this information is combined with other information, companies can know what you read, what products you buy, and what your interests are.

Most people who buy products on the Web say it's very important for a site to have a policy explaining how personal information is used, and the policy statement must make people feel comfortable and be extremely clear about what information is collected and what will and will not be done with it. However, many Web sites still do not prominently display their privacy policy or implement practices completely consistent with that policy. The real issue that Internet users need to be concerned with is—what do content providers want with their personal information? If a site requests that you provide your name and address, you have every right to know why and what will be done with it. If you buy something and provide a shipping address, will it be sold to other retailers? Will your e-mail address be sold on a list of active Internet shoppers? And if so, you should realize that it's no different than the lists compiled from the orders you place with catalog retailers. You have the right to be taken off any mailing list.

A potential solution to some consumer privacy concerns is the screening technology called the **Platform for Privacy Preferences (P3P)** being proposed to shield users from sites that don't provide the level of privacy protection they desire. Instead of forcing users to find and read through the privacy policy for each site they visit, P3P software in a computer's browser will download the privacy policy from each site, scan it, and notify the user if the policy does not match his preferences. (Of course, unethical marketers can post a privacy policy that does not accurately reflect the manner in which the data is treated.) The World Wide Web Consortium, an international industry group whose members include Apple, Commerce One, Ericsson, and Microsoft, is supporting the development of P3P. Version 1.1 of the P3P was released in February 2006 and can be found at *www.w3.org/TR/2006/WD-P3P11-20060210/Overview.html*.

The Children's Online Privacy Protection Act (COPPA) was passed by Congress in October 1998. This act was directed at Web sites catering to children, requiring them to post comprehensive privacy policies on their sites and to obtain parental consent before they collect

Platform for Privacy Preferences (P3P)
A screening technology that shields users from Web sites that don't provide the level of privacy protection they desire.

any personal information from children under 13 years of age. Web site operators who violate the rule could be liable for civil penalties of up to $11,000 per violation.[73] The Act has made an impact in the design and operations of Web sites that cater to children. For example, Lions Gate Entertainment, the operator of the *www.thebratzfilm.com* Web site, had to modify its site after the Council of Better Business Bureaus determined the site failed to meet the COPPA requirements. The Web site requested personally identifiable information to register for the Bratz Newsletter and register for a chance to win a trip to the premiere of *The Bratz Movie* without first obtaining verifiable parental consent.[74]

A social network service employs the Web and software to connect people for whatever purpose. There are thousands of such networks, which have become popular among teenagers. Some of the more popular social networking Web sites include Bebo, Classmates.com, Facebook, Hi5, Imbee, MySpace, Namesdatabase.com, Tagged, and XuQa. Most of these Web sites allow one to easily create a user profile that provides personal details, photos, even videos that can be viewed by other visitors to the Web site. Some of the Web sites have age restrictions or require that a parent register their preteen by providing a credit card to validate the parent's identity. Teens can provide information about where they live, go to school, their favorite music, and interests in hopes of meeting new friends. Unfortunately, they can also meet ill-intentioned strangers at these sites. Many documented encounters involve adults masquerading as teens attempting to meet young people for illicit purposes. Parents are advised to discuss potential dangers, check their children's profiles, and monitor their activities at such Web sites.

Fairness in Information Use

Selling information to other companies can be so lucrative that many companies will continue to store and sell the data they collect on customers, employees, and others. When is this information storage and use fair and reasonable to the people whose data is stored and sold? Do people have a right to know about data stored about them and to decide what data is stored and used? As shown in Table 9.2, these questions can be broken down into four issues that should be addressed: knowledge, control, notice, and consent.

In the past few decades, significant laws have been passed regarding a person's right to privacy. Others relate to business privacy rights and the fair use of data and information.

Fairness Issues	Database Storage	Database Usage
The right to know	Knowledge	Notice
The ability to decide	Control	Consent

Knowledge. Should people know what data is stored about them? In some cases, people are informed that information about them is stored in a corporate database. In others, they do not know that their personal information is stored in corporate databases.

Control. Should people be able to correct errors in corporate database systems? This is possible with most organizations, although it can be difficult in some cases.

Notice. Should an organization that uses personal data for a purpose other than the original purpose notify individuals in advance? Most companies don't do this.

Consent. If information on people is to be used for other purposes, should these people be asked to give their consent before data on them is used? Many companies do not give people the ability to decide if information on them will be sold or used for other purposes.

Table 9.2

The Right to Know and the Ability to Decide Federal Privacy Laws and Regulations

The Privacy Act of 1974

The major piece of legislation on privacy is the Privacy Act of 1974 (PA74). PA74 applies only to certain federal agencies. The act, which is about 15 pages long, is straightforward and easy to understand. The purpose of this act is to provide certain safeguards for people against an invasion of personal privacy by requiring federal agencies (except as otherwise provided by law) to do the following:

- Permit people to determine what records pertaining to them are collected, maintained, used, or disseminated by such agencies
- Permit people to prevent records pertaining to them from being used or made available for another purpose without their consent
- Permit people to gain access to information pertaining to them in federal agency records, to have a copy of all or any portion thereof, and to correct or amend such records
- Ensure that they collect, maintain, use, or disseminate any record of identifiable personal information in a manner that ensures that such action is for a necessary and lawful purpose, that the information is current and accurate for its intended use, and that adequate safeguards are provided to prevent misuse of such information
- Permit exemptions from this act only in cases of an important public need for such exemption, as determined by specific law-making authority
- Be subject to civil suit for any damages that occur as a result of willful or intentional action that violates anyone's rights under this act

PA74, which applies to all federal agencies except the CIA and law enforcement agencies, also established a Privacy Study Commission to study existing databases and to recommend rules and legislation for consideration by Congress. PA74 also requires training for all federal employees who interact with a "system of records" under the act. Most of the training is conducted by the Civil Service Commission and the Department of Defense. Another interesting aspect of PA74 concerns the use of Social Security numbers—federal, state, and local governments and agencies cannot discriminate against people for not disclosing or reporting their Social Security number.

Gramm-Leach-Bliley Act

This act was passed in 1999 and required all financial institutions to protect and secure customers' nonpublic data from unauthorized access or use. Under terms of this act, it was assumed that all customers approve of the financial institutions' collecting and storing their personal information. The institutions were required to contact their customers and inform them of this fact. Customers were required to write separate letters to each of their individual financial institutions and state in writing that they wanted to opt out of the data collection and storage process. Most people were overwhelmed with the mass mailings they received from their financial institutions and simply discarded them without ever understanding their importance.

USA Patriot Act

As discussed previously, the 2001 Uniting and Strengthening America by Providing Appropriate Tools Required to Intercept and Obstruct Terrorism Act (USA Patriot Act) was passed in response to the September 11 terrorism acts. Proponents argue that it gives necessary new powers to both domestic law enforcement and international intelligence agencies. Critics argue that the law removes many of the checks and balances that previously allowed the courts to ensure that law enforcement agencies did not abuse their powers. For example, under this act, Internet service providers and telephone companies must turn over customer information, including numbers called, without a court order if the FBI claims that the records are relevant to a terrorism investigation. Also, the company is forbidden to disclose that the FBI is conducting an investigation.

Other Federal Privacy Laws

In addition to PA74, other pieces of federal legislation relate to privacy. A federal law that was passed in 1992 bans unsolicited fax advertisements. This law was upheld in a 1995 ruling by the Ninth U.S. Circuit Court of Appeals, which concluded that the law is a reasonable way to prevent the shifting of advertising costs to customers. Table 9.3 lists additional laws related to privacy.

Table 9.3

Federal Privacy Laws and Their Provisions

Law	Provisions
Fair Credit Reporting Act of 1970 (FCRA)	Regulates operations of credit-reporting bureaus, including how they collect, store, and use credit information
Tax Reform Act of 1976	Restricts collection and use of certain information by the Internal Revenue Service
Electronic Funds Transfer Act of 1979	Outlines the responsibilities of companies that use electronic funds transfer systems, including consumer rights and liability for bank debit cards
Right to Financial Privacy Act of 1978	Restricts government access to certain records held by financial institutions
Freedom of Information Act of 1970	Guarantees access for individuals to personal data collected about them and about government activities in federal agency files
Education Privacy Act	Restricts collection and use of data by federally funded educational institutions, including specifications for the type of data collected, access by parents and students to the data, and limitations on disclosure
Computer Matching and Privacy Act of 1988	Regulates cross-references between federal agencies' computer files (e.g., to verify eligibility for federal programs)
Video Privacy Act of 1988	Prevents retail stores from disclosing video rental records without a court order
Telephone Consumer Protection Act of 1991	Limits telemarketers' practices
Cable Act of 1992	Regulates companies and organizations that provide wireless communications services, including cellular phones
Computer Abuse Amendments Act of 1994	Prohibits transmissions of harmful computer programs and code, including viruses
Gramm-Leach-Bliley Act of 1999	Requires all financial institutions to protect and secure customers' nonpublic data from unauthorized access or use
USA Patriot Act of 2001	Requires Internet service providers and telephone companies to turn over customer information, including numbers called, without a court order, if the FBI claims that the records are relevant to a terrorism investigation

Controlling Privacy in Finland's Largest Information System

Arek Oy, Ltd develops information systems and provides system services to pension insurance providers in Finland. The government of Finland has created laws to ensure that anyone earning a paycheck in Finland receives a pension upon retirement.

Finnish employers are required to maintain records on every employee, including the employee's name, national ID number, date of birth, work history, and other private information, along with an account of every paycheck issued to the employee. Employers share that information with one of many pension insurance companies. Arek Oy was created by the Finnish Centre for Pensions (ETK) and the country's authorized pension insurance providers to develop and manage the information systems that collect, store, and deliver employee information to the pension insurance industry.

Arek Oy was established in 2004 to perform an important task. The mission of the new company was to develop the largest information system used in Finland. The company had 30 months to complete the task, which may seem generous, until you consider the size of the system. The goal of the pension insurance information system was to manage employment records of every person that works in Finland. If Arek Oy could not provide a flawless system by the deadline, they would put workers' pensions at risk, acquire hefty fines from the government, and ruin their own reputation, which would most likely mean the end of Arek Oy.

What made the systems development especially challenging was that Arek Oy had to apply many privacy rules and regulations as defined by the Finnish government. Today's privacy-sensitive culture makes database development and maintenance a time and resource-consuming affair for businesses and governments around the globe.

In general, sensitive employee data must be hidden from the eyes of all but approved parties. The systems engineers for Arek Oy were not allowed to see the data stored in the pension database. Special data privacy solutions were employed to mask personal identification information in database records—a practice called "de-identification." Arek Oy set up a safe sandbox for development that provided realistic, fictionalized data for developers to use when testing the systems. These types of systems are referred to as test-data management systems; they promote information privacy by allowing database developers to create reliable systems without accessing the actual private data that the system will manage.

Government privacy regulations, although important to customers and citizens, are particularly burdensome to businesses and information system developers. To assist developers in complying with privacy laws, database management systems provided by major information systems companies such as IBM have compliance embedded in their systems. Arek Oy reduced its stress and responsibility by adopting such a system to use for its pension insurance information system.

As you might guess, Arek Oy was successful in meeting its deadline for Finland's largest information system. It has deployed a database management system that includes a safe sandbox for test-data management that meets the high privacy standards of the Finnish government. The many pension insurance companies that work with the system can develop database applications using the secure and private environment that Arek Oy has provided.

Considering the time and effort that Arek Oy invested in complying with government privacy regulations, it's clear why many companies not governed by regulations are hesitant to commit resources to privacy practices. In most cases it isn't a matter of not caring, but of providing the best quality system for the least amount of money. The Arek Oy case provides a good example of the benefits and costs of government regulations.

Discussion Questions

1. What challenges did Arek Oy face in the Finnish pension systems development project?
2. What techniques did the company use to meet project requirements and government regulations?

Critical Thinking Questions

1. Besides government regulations, what other pressure might persuade a business to employ strict privacy practices?
2. What are the risks involved for a company that takes shortcuts and allows systems developers to see private data?

Sources: IBM Staff, "Arek Oy deploys IBM Optim to deliver the largest information management system in Finland," IBM Case Studies, May 30, 2008, *www-01.ibm.com/software/success/cssdb.nsf/CS/LWIS-7F5QWZ? OpenDocument&Site=default&cty=en_us*; Arek Oy Web site, *www.arek.fi*, accessed August 2, 2008.

Corporate Privacy Policies

Even though privacy laws for private organizations are not very restrictive, most organizations are very sensitive to privacy issues and fairness. They realize that invasions of privacy can hurt their business, turn away customers, and dramatically reduce revenues and profits. Consider a major international credit card company. If the company sold confidential financial information on millions of customers to other companies, the results could be disastrous. In a matter of days, the firm's business and revenues could be reduced dramatically. Therefore, most organizations maintain privacy policies, even though they are not required by law. Some companies even have a privacy bill of rights that specifies how the privacy of employees, clients, and customers will be protected. Corporate privacy policies should address a customer's knowledge, control, notice, and consent over the storage and use of information. They can also cover who has access to private data and when it can be used.

Multinational companies face an extremely difficult challenge in implementing data-collection and dissemination processes and policies because of the multitude of differing country or regional statutes. For example, Australia requires companies to destroy customer data (including backup files) or make it anonymous after it's no longer needed. Firms that transfer customer and personnel data out of Europe must comply with European privacy laws that allow customers and employees to access data about them and let them determine how that information can be used.

A few examples of corporate privacy policies are shown in Table 9.4.

Table 9.4

Corporate Privacy Policies

Company	URL
Starwood Hotels & Resorts	www.starwoodhotels.com/corporate/privacy_policy.html
United Parcel Service	www.ups.com/content/corp/privacy_policy.html
Visa	www.corporate.visa.com/ut/privacy.jsp
Walt Disney Internet Group	http://disney.go.com/corporate/privacy/pp_wdig.html

A good database design practice is to assign a single unique identifier to each customer— so that each has a single record describing all relationships with the company across all its business units. That way, the organization can apply customer privacy preferences consistently throughout all databases. Failure to do so can expose the organization to legal risks— aside from upsetting customers who opted out of some collection practices. Again, the 1999 Gramm-Leach-Bliley Financial Services Modernization Act required all financial service institutions to communicate their data privacy rules and honor customer preferences.

Individual Efforts to Protect Privacy

Although numerous state and federal laws deal with privacy, the laws do not completely protect individual privacy. In addition, not all companies have privacy policies. As a result, many people are taking steps to increase their own privacy protection. Some of the steps that you can take to protect personal privacy include the following:

- **Find out what is stored about you in existing databases.** Call the major credit bureaus to get a copy of your credit report. You are entitled to a free credit report every 12 months (see *freecreditreport.com*). You can also obtain a free report if you have been denied credit in the last 60 days. The major companies are Equifax (800-685-1111, *www.equifax.com*), TransUnion (800-916-8800, *www.transunion.com*), and Experian (888-397-3742, *www.experian.com*). You can also submit a Freedom of Information Act request to a federal agency that you suspect might have information stored on you.
- **Be careful when you share information about yourself.** Don't share information unless it is absolutely necessary. Every time you give information about yourself through an 800, 888, or 900 call, your privacy is at risk. Be vigilant in insisting that your doctor, bank, or financial institution not share information about you with others without your written consent.

- **Be proactive to protect your privacy.** You can get an unlisted phone number and ask the phone company to block caller ID systems from reading your phone number. If you change your address, don't fill out a change-of-address form with the U.S. Postal Service; you can notify the people and companies that you want to have your new address. Destroy copies of your charge card bills and shred monthly statements before disposing of them in the garbage. Be careful about sending personal e-mail messages over a corporate e-mail system. You can also get help in avoiding junk mail and telemarketing calls by visiting the Direct Marketing Association Web site at *www.the-dma.org*. Go to the Web site and look under Consumer Help-Remove Name from Lists.
- **When purchasing anything from a Web site, make sure that you safeguard your credit card numbers, passwords, and personal information.** Do not do business with a site unless you know that it handles credit card information securely. (Look for a seal of approval from organizations such as the Better Business Bureau Online or TRUSTe. When you open the Web page where you enter credit card information or other personal data, make sure that the Web address begins with *https* and check to see if a locked padlock icon appears in the Address bar or status bar). Do not provide personal information without reviewing the site's data privacy policy. Many credit card companies issue single-use credit card numbers on request. Charges appear on your usual bill, but the number is destroyed after a single use, eliminating the risk of stolen credit card numbers.

THE WORK ENVIRONMENT

The use of computer-based information systems has changed the makeup of the workforce. Jobs that require IS literacy have increased, and many less-skilled positions have been eliminated. Corporate programs, such as reengineering and continuous improvement, bring with them the concern that, as business processes are restructured and information systems are integrated within them, the people involved in these processes will be removed.

However, the growing field of computer technology and information systems has opened up numerous avenues to professionals and nonprofessionals of all backgrounds. Enhanced telecommunications has been the impetus for new types of business and has created global markets in industries once limited to domestic markets. Even the simplest tasks have been aided by computers, making cash registers faster, smoothing order processing, and allowing people with disabilities to participate more actively in the workforce. As computers and other IS components drop in cost and become easier to use, more workers will benefit from the increased productivity and efficiency provided by computers. Yet, despite these increases in productivity and efficiency, information systems can raise other concerns.

Health Concerns

Organizations can increase employee effectiveness by paying attention to the health concerns in today's work environment. For some people, working with computers can cause occupational stress. Anxieties about job insecurity, loss of control, incompetence, and demotion are just a few of the fears workers might experience. In some cases, the stress can become so severe that workers might sabotage computer systems and equipment. Monitoring employee stress can alert companies to potential problems. Training and counseling can often help the employee and deter problems.

Heavy computer use can affect one's physical health as well. A job that requires sitting at a desk and using a computer for many hours a day qualifies as a sedentary job. Such work can double the risk of seated immobility thromboembolism (SIT), the formation of blood clots in the legs or lungs. People leading a sedentary lifestyle are also likely to experience an undesirable weight gain which can lead to increased fatigue and greater risk of type 2 diabetes, heart problems, and other serious ailments.

Other work-related health hazards involve emissions from improperly maintained and used equipment. Some studies show that poorly maintained laser printers can release ozone into the air; others dispute the claim. Numerous studies on the impact of emissions from display screens have also resulted in conflicting theories. Although some medical authorities believe that long-term exposure can cause cancer, studies are not conclusive at this time. In any case, many organizations are developing conservative and cautious policies.

Most computer manufacturers publish technical information on radiation emissions from their CRT monitors, and many companies pay close attention to this information. San Francisco was one of the first cities to propose a video display terminal (VDT) bill. The bill requires companies with 15 or more employees who spend at least four hours a day working with computer screens to give 15-minute breaks every two hours. In addition, adjustable chairs and workstations are required if employees request them.

In addition to the possible health risks from radio-frequency exposure, cell phone use has raised a safety issue—an increased risk of traffic accidents as vehicle operators become distracted by talking on their cell phones (or operating their laptop computers, car navigation systems, or other computer devices) while driving. As a result, some states have made it illegal to operate a cell phone while driving.

Carpal tunnel syndrome (CTS) is an aggravation of the pathway for the nerves that travel through the wrist (carpal tunnel). CTS involves wrist pain, a feeling of tingling and numbness, and difficulty grasping and holding objects. In the late 1990s, many worker compensation claims were filed by people whose job required them to work at a keyboard many hours a day. However, a 2001 study by the Mayo Clinic found that heavy computer users (up to seven hours per day) had the same rate of carpal tunnel as the general population. It appears that CTS is caused by factors other than the repetitive motion of typing on a keyboard.[75]

Avoiding Health and Environmental Problems

Many computer-related health problems are caused by a poorly designed work environment. The computer screen can be hard to read, with glare and poor contrast. Desks and chairs can also be uncomfortable. Keyboards and computer screens might be fixed in place or difficult to move. The hazardous activities associated with these unfavorable conditions are collectively referred to as *work stressors*. Although these problems might not be of major concern to casual users of computer systems, continued stressors such as repetitive motion, awkward posture, and eye strain can cause more serious and long-term injuries. If nothing else, these problems can severely limit productivity and performance.

Research has shown that developing certain ergonomically correct habits can reduce the risk of adverse health effects when using a computer.

(Source: Courtesy of Balt, Inc.)

ergonomics
The science of designing machines, products, and systems to maximize the safety, comfort, and efficiency of the people who use them.

The science of designing machines, products, and systems to maximize the safety, comfort, and efficiency of the people who use them, called **ergonomics**, has suggested some approaches to reducing these health problems. The slope of the keyboard, the positioning and design of display screens, and the placement and design of computer tables and chairs have been carefully studied. Flexibility is a major component of ergonomics and an important feature of computer devices. People come in many sizes, have differing preferences, and require different positioning of equipment for best results. Some people, for example, want to place the keyboard in their laps; others prefer it on a solid table. Because of these individual differences, computer designers are attempting to develop systems that provide a great deal of flexibility. In fact, the revolutionary design of Apple's iMac computer came about through concerns for users' comfort. After using basically the same keyboard design for over a decade, Microsoft introduced a new split keyboard called the Natural Ergonomic Keyboard 4000. The keyboard provides improved ergonomic features such as improved angles that reduce motion and how much you must stretch your fingers when you type. The design of the keyboard also provides more convenient wrist and arm postures which make typing more convenient for users.

Computer users who work at their machines for more than an hour per day should consider using LCD screens, which are much easier on your eyes than CRT screens. If you stare at a CRT screen all day long, your eye muscles can become fatigued from the screen flicker and bright backlighting of the monitor. LCD screens provide a much better viewing experience for your eyes by virtually eliminating flicker while still being bright without harsh incandescence. Also, remember to blink! We tend to focus hard on the screen and blink much less than normal. The result is red, dry, itchy eyes. A few drops of artificial tears and changing focus away from the screen periodically to rest the eyes has been found to help.

In addition to steps taken by hardware manufacturing companies, computer users must also take action to reduce RSI and develop a better work environment. For example, when working at a workstation, the top of the monitor should be at or just below eye level. Your wrists and hands should be in line with your forearms, with your elbows close to your body and supported. Your lower back needs to be well supported. Your feet should be flat on the floor. Take an occasional break to get away from the keyboard and screen. Stand up and stretch while at your workplace. Do not ignore pain or discomfort. Many workers ignore early signs of RSI, and as a result, the problem becomes much worse and more difficult to treat.

It is estimated that nearly 2 billion personal computers have been sold worldwide. This creates a tremendous disposal problem because personal computers and monitors contain lead, mercury, cadmium, and other metals defined as hazardous according to federal laws that govern their disposal. Congress is considering placing an "e-fee" that would be paid like a sales tax on personal computers, computer monitors, TVs, and some other electronic devices to cover the cost of their safe disposal. The annual cost could be in the neighborhood of $300 million. In the meantime, most personal computer manufacturers have implemented recycling programs and many are trying to redesign their products to reduce material that cannot be easily recycled. Many firms also specialize in the recycling of old personal computers. Unfortunately, some recycling programs ultimately send electronics waste to developing nations in Africa and Asia where it is disposed in environmentally unfriendly ways.[76]

ETHICAL ISSUES IN INFORMATION SYSTEMS

As you've seen throughout this book in the "Ethical and Societal Issues" boxes, ethical issues deal with what is generally considered right or wrong. As we have seen, laws do not provide a complete guide to ethical behavior. Just because an activity is defined as legal does not mean that it is ethical. As a result, practitioners in many professions subscribe to a **code of ethics** that states the principles and core values that are essential to their work and, therefore, govern their behavior. The code can become a reference point for weighing what is legal and what is ethical. For example, doctors adhere to varying versions of the 2000-year-old Hippocratic Oath, which medical schools offer as an affirmation to their graduating classes.

Some IS professionals believe that their field offers many opportunities for unethical behavior. They also believe that unethical behavior can be reduced by top-level managers developing, discussing, and enforcing codes of ethics. Various IS-related organizations and associations promote ethically responsible use of information systems and have developed useful codes of ethics. The Association for Computing Machinery (ACM) is the oldest computing society, founded in 1947, and boasts more than 80,000 members in more than 100 countries. The ACM has a code of ethics and professional conduct that includes eight general

code of ethics
A code that states the principles and core values that are essential to a set of people and, therefore, govern their behavior.

moral imperatives that can be used to help guide the actions of IS professionals. These guidelines can also be used for those who employ or hire IS professionals to monitor and guide their work. These imperatives are outlined in the following list:

As an ACM member I will ...

1. Contribute to society and human well-being.
2. Avoid harm to others.
3. Be honest and trustworthy.
4. Be fair and take action not to discriminate.
5. Honor property rights including copyrights and patents.
6. Give proper credit for intellectual property.
7. Respect the privacy of others.
8. Honor confidentiality.

(Source: ACM Code of Ethics and Professional Conduct, *http://www.acm.org/constitution/ code.html* accessed August 10, 2008)

The mishandling of the social issues discussed in this chapter—including waste and mistakes, crime, privacy, health, and ethics—can devastate an organization. The prevention of these problems and recovery from them are important aspects of managing information and information systems as critical corporate assets. Increasingly, organizations are recognizing that people are the most important component of a computer-based information system and that long-term competitive advantage can be found in a well-trained, motivated, and knowledgeable workforce.

SUMMARY

Principle

Policies and procedures must be established to avoid computer waste and mistakes.

Computer waste is the inappropriate use of computer technology and resources in both the public and private sectors. Computer mistakes relate to errors, failures, and other problems that result in output that is incorrect and without value. Waste and mistakes occur in government agencies as well as corporations. At the corporate level, computer waste and mistakes impose unnecessarily high costs for an information system and drag down profits. Waste often results from poor integration of IS components, leading to duplication of efforts and overcapacity. Inefficient procedures also waste IS resources, as do thoughtless disposal of useful resources and misuse of computer time for games and personal processing jobs. Inappropriate processing instructions, inaccurate data entry, mishandling of IS output, and poor systems design all cause computer mistakes.

A less dramatic, yet still relevant, example of waste is the amount of company time and money employees can waste playing computer games, sending unimportant e-mail, or accessing the Internet. Junk e-mail, also called *spam*, and junk faxes also cause waste.

Preventing waste and mistakes involves establishing, implementing, monitoring, and reviewing effective policies and procedures. Careful programming practices, thorough testing, flexible network interconnections, and rigorous backup procedures can help an information system prevent and recover from many kinds of mistakes. Companies should develop manuals and training programs to avoid waste and mistakes. Company policies should specify criteria for new resource purchases and user-developed processing tools to help guard against waste and mistakes. Spam filters that block unwanted mail should be installed.

Principle

Computer crime is a serious and rapidly growing area of concern requiring management attention.

Some crimes use computers as tools (e.g., to manipulate records, counterfeit money and documents, commit fraud via telecommunications links, and make unauthorized electronic transfers of money). Identity theft is a crime in which an imposter obtains key pieces of personal identification information to impersonate someone else. The information is then used to obtain credit, merchandise, and services in the name of the victim, or to provide the thief with false credentials.

A cyberterrorist is someone who intimidates or coerces a government or organization to advance his political or social objectives by launching computer-based attacks against computers, networks, and the information stored on them. A criminal hacker, also called a *cracker*, is a computer-savvy person who attempts to gain unauthorized or illegal access to computer systems to steal passwords, corrupt files and programs, and even transfer money. Script bunnies are crackers with little technical savvy. Insiders are employees, disgruntled or otherwise, working solo or in concert with outsiders to compromise corporate systems. The greatest fear of many organizations is the potential harm that can be done by insiders who know system logon IDs, passwords, and company procedures.

Computer crimes target computer systems and include illegal access to computer systems by criminal hackers, alteration and destruction of data and programs by viruses (system, application, and document), and simple theft of computer resources. A virus is a program that attaches itself to other programs. A worm functions as an independent program, replicating its own program files until it destroys other systems and programs or interrupts the operation of computer systems and networks. Malware is a general term for software that is harmful or destructive. A Trojan horse program is a malicious program that disguises itself as a useful application and purposefully does something the user does not expect. A logic bomb is designed to "explode" or execute at a specified time and date. A variant is a modified version of a virus that is produced by the virus's author or another person by amending the original virus code. A password sniffer is a small program hidden in a network or computer system that records identification numbers and passwords. Spyware is software installed on a personal computer to intercept or take partial control over the user's interactions with the computer without knowledge or permission of the user.

Identity theft is a crime in which an imposter steals personal identification information to obtain credit, merchandise, or services in the name of the victim. Although Internet gambling is popular, its legality is questionable within the U.S.

Because of increased computer use, greater emphasis is placed on the prevention and detection of computer crime. Antivirus software is used to detect the presence of viruses, worms, and logic bombs. Use of an intrusion detection system (IDS) provides another layer of protection in the event that an intruder gets past the outer security layers—passwords, security procedures, and corporate firewall. It monitors system and network resources and notifies network security personnel when it senses a possible intrusion. Many small and midsized organizations are outsourcing their network security operations to managed security service providers (MSSPs), which monitor, manage, and maintain network security hardware and software.

Software piracy might represent the most common computer crime. It is estimated that the software industry loses nearly $48 billion in revenue each year to software piracy. Computer scams have cost people and companies thousands of dollars. Computer crime is also an international issue.

Many organizations and people help prevent computer crime, among them state and federal agencies, corporations, and people. Security measures, such as using passwords, identification numbers, and data encryption, help to guard against illegal computer access, especially when supported by effective control procedures. Encryption enables users of an unsecured public network such as the Internet to securely and privately exchange data through the use of a public and a private cryptographic key pair that is obtained and shared through a trusted authority. The use of biometrics, involving the measurement of a person's unique characteristics, such as fingerprints, irises, and retinal images, is another way to protect important data and information systems. Virus-scanning software identifies and removes damaging computer programs. Law enforcement agencies armed with new legal tools enacted by Congress now actively pursue computer criminals.

Although most companies use data files for legitimate, justifiable purposes, opportunities for invasion of privacy abound. Privacy issues are a concern with government agencies, e-mail use, corporations, and the Internet. The Children's Internet Protection Act was enacted to protect minors using the Internet. The Privacy Act of 1974, with the support of other federal laws, establishes straightforward and easily understandable requirements for data collection, use, and distribution by federal agencies; federal law also serves as a nationwide moral guideline for privacy rights and activities by private organizations. The USA Patriot Act, passed only five weeks after the September 11 terrorist attacks, requires Internet service providers and telephone companies to turn over customer information, including numbers called, without a court order, if the FBI claims that the records are relevant to a terrorism investigation. Also, the company is forbidden to disclose that the FBI is conducting an investigation. Only time will tell how this act will be applied in the future. The Gramm-Leach-Bliley Act requires all financial institutions to protect and secure customers' nonpublic data from unauthorized access or use. Under terms of this act, it is assumed that all customers approve of the financial institutions collecting and storing their personal information.

A business should develop a clear and thorough policy about privacy rights for customers, including database access. That policy should also address the rights of employees, including electronic monitoring systems and e-mail. Fairness in information use for privacy rights emphasizes knowledge, control, notice, and consent for people profiled in databases. People should know about the data that is stored about them and be able to correct errors in corporate database systems. If information on people is to be used for other purposes, they should be asked to give their consent beforehand. Each person has the right to know and the ability to decide. Platform for Privacy Preferences (P3P) is a screening technology that shields users from Web sites that don't provide the level of privacy protection they desire.

Principle

Jobs, equipment, and working conditions must be designed to avoid negative health effects.

Computers have changed the makeup of the workforce and even eliminated some jobs, but they have also expanded and enriched employment opportunities in many ways. Jobs that involve heavy use of computers contribute to a sedentary lifestyle, which increases the risk of health problems. Some critics blame computer systems for emissions of ozone and electromagnetic radiation. Use of cell phones while driving has been linked to increased car accidents.

The study of designing and positioning computer equipment, called *ergonomics*, has suggested some approaches to reducing these health problems. Ergonomic design principles help to reduce harmful effects and increase the efficiency of an information system. The slope of the keyboard, the positioning and design of display screens, and the placement and design of computer tables and chairs are essential for good health. RSI prevention includes keeping good posture, not ignoring pain or problems, performing stretching and strengthening exercises, and seeking proper treatment. Although they can cause negative health consequences, information systems can also be used to provide a wealth of information on health topics through the Internet and other sources.

Principle

Practitioners in many professions subscribe to a code of ethics that states the principles and core values that are essential to their work.

Ethics determine generally accepted and discouraged activities within a company and society at large. Ethical computer users define acceptable practices more strictly than just refraining from committing crimes; they also consider the effects of their IS activities, including Internet usage, on other people and organizations. The Association for Computing Machinery developed guidelines and a code of ethics. Many IS professionals join computer-related associations and agree to abide by detailed ethical codes.

CHAPTER 9: SELF-ASSESSMENT TEST

Policies and procedures must be established to avoid computer waste and mistakes.

1. Business managers and end users must work with IS professionals to implement and follow proper IS usage policies to ensure effective use of company resources. True or False?

2. Computer-related waste and mistakes are major causes of computer problems, contributing to unnecessarily high ____ and lost ____.

3. Unwanted e-mail is often referred to as _____.

Computer crime is a serious and rapidly growing area of concern requiring management attention.

4. According to the 2007 FBI Internet Crime Report, the dollar amount of Internet crime reported exceeded $250 million. True or False?

5. _____ is using one's skills to get computer users to provide you with information to access an information system or its data.

6. The vast majority of organizations conduct some form of computer security audit. True or False?

7. _____ is a crime in which an imposter obtains key pieces of personal identification information, such as Social Security or driver's license numbers, to impersonate someone else.

8. Internet gambling in the U.S. is completely legal. True or False?

9. A logic bomb is a type of Trojan horse that executes when specific conditions occur. True or False?

10. Malware capable of spreading itself from one computer to another is called a _____.
    a. logic bomb
    b. Trojan horse
    c. virus
    d. worm

11. A(n) _____ is a modified version of a virus that is produced by the virus's author or another person amending the original virus code.

12. The Business Software Alliance estimates that the software industry lost over $48 billion in 2007 due to worldwide software piracy. True or False?

13. Phishing is a computer scam that seems to direct users to a bank's Web site but actually captures key personal information about its victims. True or False?

Jobs, equipment, and working conditions must be designed to avoid negative health effects.

14. CTS, or _____, is the aggravation of the pathway of nerves that travel through the wrist.

Practitioners in many professions subscribe to a code of ethics that states the principles and core values that are essential to their work.

15. Just because an activity is defined as legal, it does not mean that it is ethical. True or False?

CHAPTER 9: SELF-ASSESSMENT TEST ANSWERS

(1) True (2) costs, profits (3) spam (4) False (5) Social engineering (6) True (7) Identity theft (8) False (9) True (10) d (11) variant (12) True (13) True (14) carpal tunnel syndrome (15) True

REVIEW QUESTIONS

1. What is a spam filter? How does such a program work? What is the issue with image-based spam?
2. How can antivirus software reduce computer waste?
3. Outline a four-step process to prevent computer waste and mistakes.
4. Why are all computer crimes not reported to law enforcement agencies?
5. What is a virus? What is a worm? How are they different?
6. What is a variant? What dangers are associated with such malware?
7. What is phishing? What actions can you take to reduce the likelihood that you will be a victim of this crime?
8. Outline measures you should take to protect yourself against viruses and worms.
9. What does intrusion detection software do? What are some of the issues with the use of this software?
10. Identify at least five tips to follow to avoid becoming a victim of a computer scam.

11. What is the difference between a patent and a copyright? What copyright issues come into play when downloading software or music from a Web site?
12. What is the difference between the Children's Online Privacy Protection Act and the Children's Internet Protection Act?
13. What is ergonomics? How can it be applied to office workers?
14. What specific actions can you take to avoid spyware?
15. What is a code of ethics? Give an example.

DISCUSSION QUESTIONS

1. Imagine that your friend regularly downloads copies of newly released, full-length motion pictures for free from the Internet and makes copies for others for a small fee. Do you think that this is ethical? Is it legal? Would you express any concerns with him?
2. Outline an approach, including specific techniques (e.g., dumpster diving, phishing, social engineering) that you could employ to gain personal data about the members of your class.
3. Your 12-year-old niece shows you a dozen or so photos of herself and a brief biography including address and cell phone number she plans to post on MySpace. What advice might you offer her about posting personal information and photos?
4. Imagine that you are a hacker and have developed a Trojan horse program. What tactics might you use to get unsuspecting victims to load the program onto their computer?
5. Discuss the importance of educating employees in preventing computer waste and computer crime. Imagine that you are given the assignment of developing a computer education program for your employer. What topics would you cover in the course?

6. Briefly discuss the potential for cyberterrorism to cause a major disruption in our daily life. What are some likely targets of a cyberterrorist? What sort of action could a cyberterrorist take against these targets?
7. You are the new head of corporate security for a large Fortune 1000 company and are alarmed at the number of laptop computers your firm's employees lose each month. What actions would you take to cut down on the potential for loss of personal and/or company confidential data?
8. Do you believe that the National Security Agency should be able to collect the telephone call records of U.S. citizens without the use of search warrants? Why or why not?
9. Using information presented in this chapter on federal privacy legislation, identify which federal law regulates the following areas and situations: cross-checking IRS and Social Security files to verify the accuracy of information, customer liability for debit cards, your right to access data contained in federal agency files, the IRS obtaining personal information, the government obtaining financial records, and employers' access to university transcripts.
10. Briefly discuss the difference between acting morally and acting legally. Give an example of acting legally and yet immorally.

PROBLEM-SOLVING EXERCISES

1. Access the Web sites for the Recording Industry Association of America (RIAA), Motion Picture Association of America (MPAA), and Business Software Alliance (BSA) to get estimates of the amount of piracy worldwide for at least three years. Use a graphics package to develop a bar chart to show the amount of music, motion picture, and software piracy over a three-year period. Compare the amount of piracy to the total music, motion picture, and software revenue for the same time period.
2. Using spreadsheet software and appropriate forecasting routines, develop a forecast for the amount of piracy for next year. Document any assumptions you make in developing your forecast.
3. Using your word processing software, write a few brief paragraphs summarizing the trends you see from reviewing the data for the past few years. Then cut and paste the information from Exercise 1 and your forecast from Exercise 2 into your report.

TEAM ACTIVITIES

1. Visit your school's library and interview the librarians about the use of Internet software filters. What level and kinds of complaints are made about the use of filtering software? Who is responsible for updating the list of sites that are deemed "off limits" for minors? What is their opinion about the need for and effectiveness of the software filter?

2. Have each member of your team access ten different Web sites and summarize their findings in terms of the existence of data privacy policy statements: Did the site have such a policy? Was it easy to find? Was it complete and easy to understand? Did you find any sites using the P3P standard or ICRA rating method?

WEB EXERCISES

1. The Computer Emergency Response Team Coordination Center (CERT/CC) is located at the Software Engineering Institute (SEI), a federally funded research and development center at Carnegie Mellon University in Pittsburgh, Pennsylvania. Do research on the center and write a brief report summarizing its activities.

2. Search the Web for a site that provides software to detect and remove spyware. Write a short report for your instructor summarizing your findings.

3. Do research on the Web to discover what role the Business Software Alliance plays in the protection of software. Document some of the tactics it uses to identify and punish organizations that it determines to practice software piracy.

CAREER EXERCISES

1. You are a senior member of a marketing organization for a manufacturer of children's toys. A recommendation has been made to develop a Web site to promote and sell your firm's products as well as learn more about what parents and their children are looking for in new toys. Develop a list of laws and regulations that will affect the design of the Web site. Describe how these will limit the operation of your new Web site.

2. You have just begun a new position in customer relations for a mid-sized bank. Within your first week on the job, several customers have expressed concern about potential theft of customer data from the bank's computer databases and identity theft. Who would you talk with to develop a satisfactory response to address your customers' concerns? What key points would you need to verify with bank employees?

CASE STUDIES

Case One
IT Consumerization and Web 2.0 Security Challenges

In recent years, the direction of investment in information technologies has shifted. The shift is in reaction to the fact that in 2004, independent consumers passed business and government in their consumption of digital electronics devices. More digital devices, such as notebooks, cell phones, and media players, are being designed for consumers rather than businesses. New and popular technologies are now being introduced into the workplace by employees rather than

systems analysts. This is a trend that some refer to as IT consumerization. Unfortunately, consumer devices and systems are introducing a host of new systems vulnerabilities.

A big concern regarding IT consumerization is the free flow of communications and data sharing. Today's Web 2.0 technologies make it all too easy for employees to share information that they shouldn't. A study in the United Kingdom revealed that three-quarters of U.K. businesses have banned the use of instant messaging services such as AIM, Windows Live Messenger, and Yahoo Messenger. The primary concern is the loss of sensitive business information. Even though the

IM services could prove useful for business communications, most businesses are concerned about security rather than interested in innovative communication.

Consider the Apple iPhone. Some businesses that have supported RIM's Blackberry smartphone are feeling pressure from their employees to support the iPhone as well. Systems security experts are hesitant to comply due to concerns over information privacy. For example, the iPhone 3G does not include data encryption native to the device. If the phone is lost or stolen, private corporate information is vulnerable. Systems analysts are stuck trying to serve both a demanding workforce and corporate security needs.

CTO Gary Hodge at U.S. Bank is concerned about Web 2.0 applications. "We always said outside the corporation was untrusted and inside the corporation was trusted territory. Web 2.0 has changed all that. We've had to expose the internal workings of the corporation. There's a whole rash of new devices coming out to enable people to compute when they want to, with the iPhones and smartphones." Hodge worries that smartphone manufacturers haven't paid enough attention to security. CTOs and CIOs are feeling as though they are losing control of their systems and data.

Dmitri Alperovitch, principal research scientist for Secure Computing, is also concerned about security and Web 2.0. The concern stems from the browser becoming a computing platform itself. Although businesses have learned to protect traditional operating systems, they have little power when the browser is acting like an operating system. Web 2.0 sites and social networking sites allow anyone to create applications and post files and content. This increases the risks of transmitting malware and revealing corporate secrets. Gary Dobbins, director of information security at the University of Notre Dame, has simple and effective advice for information security: "Never trust the browser."

In banking, minor lapses in security can have devastating results. Bank CIOs see Web 2.0 as expanding their security perimeter. Web 2.0 gives them a much larger area to watch. Because of this, many banks are taking a hard line. For example, U.S. Bank only allows employees to access business-related content on their PCs. The bank restricts the use of any type of portable storage including USB drives and CDs. Every electronic transmission that leaves the bank is monitored.

For Gary Hodge, investing in information security at U.S. Bank isn't a matter of ROI, but rather a survival necessity. "We protect money. It's new for us to have to protect vast amounts of information," Hodge said. "We spend millions of dollars on security but it doesn't generate any new revenue. I haven't been able to show anybody a return on investment. It comes down to can we secure the organization at the right risk and the right cost. You can't spend all the money. You have to figure out what level of risk you're willing to tolerate."

Discussion Questions

1. What are the differences in information security needs for a bank versus a retail store?
2. Why are IT consumerization and Web 2.0 challenging business information security?

Critical Thinking Questions

1. Do you think that over time consumer devices may become as secure as banking systems? Why or why not?
2. Do you think the "hard line" taken by U.S. Bank in regards to information security policies is justified? Why or why not? Would you be willing to work in that environment?

Sources: Stokes, Jon, "Analysis: IT consumerization and the future of work," *Ars Technica*, July 6, 2008, *http://arstechnica.com/news.ars/post/20080706-analysis-it-consumerization-and-the-future-of-work.html*; Skinner, Carrie-Ann, "U.K businesses ban IM over security concerns," *Computerworld*, July 15, 2008, *www.computerworld.com/action/article.do?command= viewArticleBasic&articleId=9110 159*; Brodkin, Jon, "U.S. Bank suffers Web 2.0 security headaches," *Network World*, April 30, 2008, *www.networkworld.com/ news/2008/043008-interop-bank-web-2-security.html*; Hamblen, Matt, "iPhone 3G, business must wait," *Macworld UK*, June 16, 2008, *www.macworld.co.uk/ipod-itunes/news/index.cfm?newsid=21659*.

Case Two
San Francisco WAN Held Captive

Sometimes in protecting a network, the ones to watch are within the organization. That's the lesson learned by the City of San Francisco. The city's network administrator for its multimillion dollar wide area network (WAN) seized control of the network and denied other system administrators access for ten days while jailed.

The network administrator, who had been experiencing conflicts with his supervisor, created a super password that effectively locked out all administrators but himself to the network's switches and routers. When he refused to reveal the password, he was arrested and held on a $5 million bond. The network that he held captive connects various city offices around San Francisco and supports 60 percent of the municipal government's information traffic. During the system administrator's incarceration, the city network continued functioning without incident.

The system administrator's lawyer argued the defendant felt that none of the people who requested the password were qualified to have it. The defendant claimed his supervisor was undermining his work. The defendant wanted to uncover the problems in the city's Department of Telecommunication Information Services (DTIS). His intent was to "expose the utter mismanagement, negligence, and corruption at DTIS, which if left unchecked, will in fact place the City of San Francisco in danger," his motion read. It is assumed that drastic budget cuts that resulted in losing 200 of 350 employees at DTIS were behind the stress that ultimately drove the administrator to extreme measures.

The network administrator finally revealed the super password to the network when after ten days in prison, San Francisco mayor Gavin Newsom visited him. The two had a lengthy private discussion that concluded with the mayor receiving the password, saving the city the hundreds of thousands of dollars it would have cost to sequentially reset hundreds of switches and routers around the city.

This case points to several important lessons for businesses to observe regarding system administration. Rick Cook of *Computerworld* suggests that perhaps policies used by nuclear power plants, NASA, and the military might have prevented San Francisco from losing control of its network. Nuclear power plants deny access to systems at the slightest sign of suspicious activity. In San Francisco's case, by the time the suspicious activity was noticed, it was too late. The system administrator obviously did not have proper oversight and supervision. If the city used a system that logged administrator activities and assigned security officers to review them regularly, the damage could have been prevented.

In the military, two people are required to take simultaneous actions to launch nuclear missiles. Similar requirements could be implemented with important system actions such as managing switches and routers.

A most important preventive step is called identity management and access control (IM/AC). Identity management requires usernames and passwords, which most networks do effectively. Access control, however, is often undermanaged in important networks. Access controls prevent users from accessing systems and commands for which they do not have authority.

Through a combination of close supervision, duplication of responsibilities, and identity management and access control, the San Francisco WAN kidnapping might have been avoided. Unfortunately, security measures come at some cost. Obviously, with budget cutbacks, the city could not afford the level of security needed for such an important network. As global economies become strained and economies increasingly depend on the stability of secure information systems, San Francisco's dilemma could be played out at a much grander scale unless security for information systems becomes as important as for nuclear power plants and missiles.

Discussion Questions

1. What was the cause of the problems for San Francisco's WAN?
2. How might these problems have been prevented?

Critical Thinking Questions

1. Should information system security be considered as important as security at a nuclear facility, as suggested in this article? Why or why not?
2. Did this system administrator's actions create the effect that he obviously intended? Were his actions justified and ethical?

Sources: Cook, Rick, "Opinion: How to protect your network from rogue IT employees," *Computerworld*, July 21, 2008, www.computerworld.com/action/article.do?command=viewArticleBasic&taxonomyNam e=security&articleId=9110385&taxonomyId=17&intsrc=kc_feat; McMillan, Robert, "San Francisco IT admin locks up city network," *Computerworld*, July 21, 2008, www.computerworld.com/action/article.do?command=viewArticleBasic&articleId=322438; McMillan, Robert, "San Francisco's mayor gets back keys to the network," *Computerworld*, July 23, 2008, www.computerworld.com/action/article.do?command=viewArticleBasic&taxonomyNam e=security&articleId=9110520&taxonomyId=17&intsrc=kc_top.

Questions for Web Case

See the Web site for this book to read about the Whitmann Price Consulting case for this chapter. Following are questions concerning this Web case.

Whitmann Price Consulting: Security, Privacy, and Ethical Considerations

Discussion Questions

1. Why do you think extending access to the Whitmann Price network beyond the business's walls dramatically elevated the risk to information security?
2. What was the primary tool used to minimize that risk, and how does it work?

Critical Thinking Questions

1. Why does information security usually come at the cost of user convenience?
2. List the security policies put in place for the AMCI system and the rational that you think is behind them.

NOTES

Sources for the opening vignette: Qualys staff, "eBay, Inc. - Securing the World's Online Marketplace with QualysGuard," Qualys Case Study, May 10, 2008, *www.bitpipe.com/detail/RES/1210427689_288.html*; Gross, Grant, "Oregon man sentenced to four years for piracy, ID theft," *Computerworld*, July 24, 2008, *www.computerworld.com/action/article.do?command=viewArticleBasic&articleId=9110621&source=rss_news10*; Goodin, Dan, "Notorious eBay hacker arrested in Romania," *The Register*, April 18, 2008, *www.theregister.co.uk/2008/04/18/vladuz_arrested*; Sullivan, Bob, "How far has Vladuz hacked into eBay?," MSNBC Red Tape Chronicles, March 2, 2008, *http://redtape.msnbc.com/2007/03/how_far_has_vla.html*; FBI staff, "2007 IC3 Annual Report," FBI, 2008, *www.ic3.gov/media/annualreport/2007-IC3Report.pdf*; About eBay Web Page, *http://news.ebay.com/about.cfm*, accessed July 26, 2008.

1   "Spam Filter Review 2008," Top Ten Reviews, *http://spam-filter-review.toptenreviews.co*, accessed June 23, 2008.
2   Westbrook, Jesse, "SEC Asks About Ratings Errors on Structured Products," *Bloomberg.com*, June 26, 2008.
3   Songini, Marc L., "Computer Glitch Led to Mars Global Surveyor's Demise," *Computerworld*, April 27, 2007.
4   Williams, Martyn, "Computer Glitch Hits Hundreds of Japan Flights," *Computerworld*, May 28, 2007.

5   McMillan, Robert and Mullins, Robert, "United Flights Grounded by Computer Glitch," *Computerworld*, June 21, 2007.

6   Colliver, Victoria and Muscat, Sabine, "Wells Fargo ATM, Other Glitches Last Longer Than First Reported," *San Francisco Chronicle*, August 22, 2007.

7   Barak, Sylvie, "Getting Fired for Using the Internet Becomes Commonplace," *The Inquirer*, March 3, 2008.

8   Stackel, I.M., "Another Collier Employee Fired for Inappropriate E-mails," *NaplesNews.com*, July 2, 2008.

9   Havenstein, Heather, "IT Officials Are Clearing BI Hurdles to Expand Systems," *Computerworld*, May 28, 2007.

10   Gumbel, Peter, "4 Things I Learned from Societe Generale," *CNN Money*, February 1, 2008.

11   Schwartz, Nelson D. and Bennhold, Katrin, "Societe Generale Scandal: 'A Suspicion That This Was Inevitable,'" *International Herald Tribune*, February 5, 2008.

12   Vance, Jeff, "Using Policy and Compliance Tools to Reduce Insider Threats," at *www.cioupdate.com*, accessed June 27, 2008.

13   Miller, Chuck, "ATM Hackers Net Millions Using Stolen Information," *Secure Computing*, June 20, 2008.

14   Barak, Sylvie, "Hacker Exposes Six Million Chilean's Data to Make a Point," *Secure Computing*, May 13, 2008.

15   Thomson, Iain, "Ukrainian Hacker May Get to Keep Profits," *Secure Computing*, February 19, 2008.

16   Thurston, Richard, "U.S. School Network Falls Victim to Child Hacker," *Secure Computing*, May 27, 2008.

17   Carr, Jim, "Californian Indicted in US $50,000 Scam of E*Trade, Schwab.com," *Secure Computing*, May 30, 2008.

18   Regan, Keith, "Web Crime Spikes in 2007, Losses Near $240 M," *Electronic Commerce Times*, April 4, 2008.

19   "CSI Survey 2007," *GoCSI.com*, accessed June 27, 2008.

20   Greenemeier, Larry, "The Threat Within: Employees Pose the Biggest Security Risk," *InformationWeek*, July 16, 2007.

21   Kaplan, Dan, "MTV Breach Impacts 5,000 Employees, Successful Social-Engineering Blamed," *Secure Computing*, March 11, 2008.

22   Ko, Carol, "Malaysia to Build Centre to Study Cyberterrorism," *IT World Canada*, June 12, 2008.

23   Associated Press, "Stung by Cyber Warfare, Estonia, NATO Allies to Sign Deal on Cyber Defense Center," *International Herald Tribune*, May 14, 2008.

24   McMillan, Robert, "CNN Cyberattack Called Off," *CIO*, April 19, 2008.

25   "2007 Identity Fraud Survey Report: Identity Fraud is Dropping, Continued Vigilance Necessary," Javelin Strategy & Research, February 2007.

26   Varrone, Carl, "What Nobody Else is Saying About Online Poker," Dog Ear Publishing, 2007, p. 126.

27   Chiang, Jennifer W., "Don't Bet On It: How Complying with Federal Internet Gambling Law is Not Enough," *Shidler Journal for Law, Commerce + Technology*, June 6, 2007.

28   Havenstein, Heather, "Report: FBI Looks into Facebook March Madness Betting Pools," *Computerworld*, March 17, 2008.

29   James, Clement, "US House Committee Votes on Web Gambling Ban," *vnunet.com*, June 23, 2008.

30   Sowa, Tom, "A Hacker's Wrong Turn," *SpokesmanReview.com*, August 20, 2007.

31   "Newly Discovered Malware," McAfee Web site *http://vil.nai.com/vil/newly_discovered_viruses.aspx*, accessed July 4, 2008.

32   Hulme, George, "New Ransom-Ware Virus Resurfaces," *Information Week*, June 9, 2008.

33   Keizer, Gregg, "Best Buy Sold Infected Digital Picture Frames," *Computerworld*, January 2, 2008.

34   Nichols, Shaun, "Twin Trojans Attack Macs," *Secure Computing*, June 23, 2008.

35   Kirk, Jeremy, "'Mebroot' Proves to be a Tough Rootkit to Crack," *Computerworld*, March 4, 2008.

36   Leyden, John, "Dodgy Drug Sales Underpin Storm Worm," *The Register*, June 12, 2008.

37   "About McAfee," McAfee Web site, *http://us.mcafee.com/root/aboutUs.asp*, accessed on June 28, 2008.

38   "Antivirus Software and Disk Defragmentation," *Tech Republic*, June 23, 2008.

39   "Anti-Spyware Software Reviews for 2008," Top Ten Reviews, *http://anti-spyware-review.toptenreviews.com*, accessed July 1, 2008.

40   Lemos, Robert, "Spyware Purveyor DirectRevenue Closes Down," *SecurityFocus*, October 25, 2007.

41   "DirectRevenue LLC Settles FTC Charges," Federal Trade Commission Web site, February 16, 2007, *www.tfc.gov/opa/2007/02*, accessed July 1, 2008.

42   Keizer, Gregg, "VA Loses Another Hard Drive, Vet Data At Risk," *Information Week*, February 5, 2007.

43   Ibid.

44   Kaplan, Dan, "Lost: Hard Drive with Information of 100,000 Employees in US," *Secure Computing*, May 8, 2007.

45   Gaudin, Sharon, "VeriSign Worker Fired After Laptop, Employee Info Are Stolen," *Information Week*, August 7, 2007.

46   Vijayan, Jaikumar, "Another Day, Another Laptop Theft: Now, It's Connecticut's Revenue Agency," *Computerworld*, August 31, 2007.

47   Fonseca, Brian, "Personal Data on 800,000 Gap Job Applicants Exposed in Laptop Theft," *Computerworld*, September 26, 2007.

48   Claburn, Thomas, "Record Number of Data Breaches Reported in 2007," *Information Week*, December 31, 2007.

49   McGee, Marianne Kolbasuk, "Laptop Stolen with Personal Data on 300,000 Health Insurance Clients," *Information Week*, January 30, 2008.

50   Puzzanghera, Jim, "Laptop Seizures at Customs Raise Outcry," *Los Angeles Times*, June 26, 2008.

51   "What is Software Piracy?" Business Software Alliance Web site, *www.bsa.org/country.aspx*, accessed July 3, 2008.

52   "Worldwide Software Piracy Rate Holds Steady at 35%; Global Losses up 15%," Business Software Alliance, *www.bsa.org/country*, accessed July 3, 2008.

53   US Code: Title 17 Copyrights, Cornell University Law School, *www.law.cornell.edu/uscode/17*, accessed July 3, 2008.

54   "Judgment Entered Against PA Company for Widespread Unauthorized Distribution of Software," Business Software Alliance Web site, April 24, 2008, *www.bsa.org/country*, accessed July 3, 2008.

55   "Los Angeles-Area Engineering Company Pays $250,000 to The Business Software Alliance," April 14, 2008, *www.bsa.org/country*, accessed July 3, 2008.

56   Jones, K.C., "Music Piracy Costs U.S. Economy $12.5 Billion, Report Reveals," *InformationWeek*, August 22, 2007.

57   "Two Site Operators Receive Prison Terms for Criminal Copyright Infringement," United States Department of Justice Web site, May 14, 2008 *www.usdoj.gov/usaso/can/press*, accessed July 3, 2008.

58   "Jones, K.C., "TorrentSpy Ordered to Pay $100 Million in Piracy Case," *InformationWeek*, May 8, 2008.

59   Claburn, Thomas, "Acer, Apple, Dell and HP Sued for Patent Infringement," *Information Week*, June 30, 2008.

60   "Tellabs Sues Fujitsu Claiming Patent Infringement," *PC World*, June 12, 2008.

61   Kanaracus, Chris, "Red Hat Settles Patent Suits with Firestar, DataTern," *PC World*, June 11, 2008.

62   Rogers, John, "Gartner: US $3.2 Billion Lost to Phishing Attacks in One Year," *SC Magazine*, December 19, 2007.

63   Carr, Jim, "Phishing Scam Uses AOL Address to Target eBay Users," *SC Magazine*, November 13, 2007.

64   "Jet Flash 210 Fingerprint USB Flash Drive," Transcend Web site at *www.transcendusa.com/Products/ModDetail.asp?ModNo=108&SpNo=2&LangNo=0*, accessed July 4, 2008.

65   "Customer Success Stories: Associated Newspapers," accessed at Computer Associates Web site, *http://ca.com/us/success/Collateral.aspx?CID=147880*, accessed June 28, 2008.

66   Children's Internet Protections Act, Pub. L. 106-552, *http://ifea.net/cipa.html*, accessed June 22, 2008.

67 Welch, Erin, "Child-Proofing Internet Access," *The Orange County Register*, January 29, 2008.

68 Houser, Mark, "UPMC Admits Privacy Violation," *Pittsburgh Tribune-Review*, April 13, 2007.

69 Jones, K.C., "Obama, Clinton, McCain Passport Breaches Expose Human, Not Tech Weakness," *InformationWeek*, March 21, 2008.

70 "3/4 of Companies Monitor Employee Web Browsing," *Yahoo! Tech*, May 9, 2007.

71 Singel, Ryan, "Nearly Ten Percent of Companies Have Fired Bloggers, Survey Claims," *Wired*, July 19, 2007.

72 Lewan, Todd, "Microchips Everywhere: A Future Vision," *The Seattle Times*, January 29, 2008.

73 "Children's Online Privacy Protection Act – 15 USC 6501 – 6506," *www.softforyou.com/add/COPPA.pdf*, accessed June 23, 2008.

74 Bean, Linda, "CARU Reviews Site Operated by Lions Gate Entertainment," *CARU News*, October 24, 2007.

75 Simon, Ellen, "How Are Your Wrists Feeling?" *Cincinnati Enquirer*, May 28, 2008, p. E1

76 Gross, Grant, "E-Waste Recycling Faces Challenges, Critics Say," *Computerworld*, April 20, 2008.

accounting MIS An information system that provides aggregate information on accounts payable, accounts receivable, payroll, and many other applications.

ad hoc DSS A DSS concerned with situations or decisions that come up only a few times during the life of the organization.

antivirus program Software that runs in the background to protect your computer from dangers lurking on the Internet and other possible sources of infected files.

application program interface (API) Interface that allows applications to make use of the operating system.

application service provider (ASP) A company that provides software, support, and the computer hardware on which to run the software from the user's facilities.

application software The programs that help users solve particular computing problems.

arithmetic/logic unit (ALU) Part of the CPU that performs mathematical calculations and makes logical comparisons.

ARPANET A project started by the U.S. Department of Defense (DoD) in 1969 as both an experiment in reliable networking and a means to link DoD and military research contractors, including many universities doing military-funded research.

artificial intelligence (AI) The ability of computers to mimic or duplicate the functions of the human brain.

artificial intelligence systems People, procedures, hardware, software, data, and knowledge needed to develop computer systems and machines that demonstrate the characteristics of intelligence.

asking directly An approach to gather data that asks users, stakeholders, and other managers about what they want and expect from the new or modified system.

attribute A characteristic of an entity.

auditing Analyzing the financial condition of an organization and determining whether financial statements and reports produced by the financial MIS are accurate.

backbone One of the Internet's high-speed, long-distance communications links.

backward chaining The process of starting with conclusions and working backward to the supporting facts.

batch processing system A form of data processing where business transactions are accumulated over a period of time and prepared for processing as a single unit or batch.

best practices The most efficient and effective ways to complete a business process.

Bluetooth A wireless communications specification that describes how cell phones, computers, faxes, personal digital assistants, printers, and other electronic devices can be interconnected over distances of 10–30 feet at a rate of about 2 Mbps.

bot A software tool that searches the Web for information such as products and prices.

brainstorming A decision-making approach that often consists of members offering ideas "off the top of their heads."

broadband communications A telecommunications system in which a very high rate of data exchange is possible.

business intelligence The process of gathering enough of the right information in a timely manner and usable form and analyzing it to have a positive impact on business strategy, tactics, or operations.

business-to-business (B2B) e-commerce A subset of e-commerce where all the participants are organizations.

business-to-consumer (B2C) e-commerce A form of e-commerce in which customers deal directly with an organization and avoid intermediaries.

byte (B) Eight bits that together represent a single character of data.

Cascading Style Sheet (CSS) A file or portion of an HTML file that defines the visual appearance of content in a Web page.

central processing unit (CPU) Part of the computer that consists of three associated elements: the arithmetic/logic unit, the control unit, and the register areas.

centralized processing Processing alternative in which processing occurs at a single location or facility.

certificate authority (CA) A trusted third-party organization or company that issues digital certificates.

certification A process for testing skills and knowledge, which results in a statement by the certifying authority that states an individual is capable of performing a particular kind of job.

channel bandwidth The rate at which data is exchanged over a communication channel, usually measured in bits per second (bps).

character A basic building block of information, consisting of uppercase letters, lowercase letters, numeric digits, or special symbols.

chat room A facility that enables two or more people to engage in interactive "conversations" over the Internet.

chief knowledge officer (CKO) A top-level executive who helps the organization use a KMS to create, store, and use knowledge to achieve organizational goals.

choice stage The third stage of decision making, which requires selecting a course of action.

client/server An architecture in which multiple computer platforms are dedicated to special functions such as database management, printing, communications, and program execution.

clock speed A series of electronic pulses produced at a predetermined rate that affects machine cycle time.

cloud computing Using a giant cluster of computers to serve as a host to run applications that require high-performance computing.

code of ethics A code that states the principles and core values that are essential to a set of people and, therefore, govern their behavior.

command-based user interface A user interface that requires you to give text commands to the computer to perform basic activities.

communications protocol A set of rules that govern the exchange of information over a communications channel.

compact disc read-only memory (CD-ROM) A common form of optical disc on which data, once it has been recorded, cannot be modified.

competitive advantage A significant and (ideally) long-term benefit to a company over its competition.

competitive intelligence One aspect of business intelligence limited to information about competitors and the ways that knowledge affects strategy, tactics, and operations.

computer network The communications media, devices, and software needed to connect two or more computer systems and/or devices.

computer programs Sequences of instructions for the computer.

computer-aided software engineering (CASE) Tools that automate many of the tasks required in a systems development effort and encourage adherence to the SDLC.

computer-assisted manufacturing (CAM) A system that directly controls manufacturing equipment.

computer-based information system (CBIS) A single set of hardware, software, databases, telecommunications, people, and procedures that are configured to collect, manipulate, store, and process data into information.

computer-integrated manufacturing (CIM) Using computers to link the components of the production process into an effective system.

concurrency control A method of dealing with a situation in which two or more people need to access the same record in a database at the same time.

consumer-to-consumer (C2C) e-commerce A subset of e-commerce that involves consumers selling directly to other consumers.

content streaming A method for transferring multimedia files over the Internet so that the data stream of voice and pictures plays more or less continuously without a break, or very few of them; enables users to browse large files in real time.

control unit Part of the CPU that sequentially accesses program instructions, decodes them, and coordinates the flow of data in and out of the ALU, the registers, primary storage, and even secondary storage and various output devices.

cost center A division within a company that does not directly generate revenue.

counterintelligence The steps an organization takes to protect information sought by "hostile" intelligence gatherers.

criminal hacker (cracker) A computer-savvy person who attempts to gain unauthorized or illegal access to computer systems to steal passwords, corrupt files and programs, or even transfer money.

culture A set of major understandings and assumptions shared by a group.

customer relationship management (CRM) system A system that helps a company manage all aspects of customer encounters, including marketing and advertising, sales, customer service after the sale, and programs to retain loyal customers.

cybermall A single Web site that offers many products and services at one Internet location.

cyberterrorist Someone who intimidates or coerces a government or organization to advance his political or social objectives by launching computer-based attacks against computers, networks, and the information stored on them.

data Raw facts, such as an employee number, number of hours worked in a week, inventory part numbers, or sales orders.

data administrator A nontechnical position responsible for defining and implementing consistent principles for a variety of data issues.

data analysis The manipulation of collected data so that the development team members who are participating in systems analysis can use the data.

data collection Capturing and gathering all data necessary to complete the processing of transactions.

data correction The process of reentering data that was not typed or scanned properly.

data definition language (DDL) A collection of instructions and commands used to define and describe data and relationships in a specific database.

data dictionary A detailed description of all the data used in the database.

data editing The process of checking data for validity and completeness.

data item The specific value of an attribute.

data manipulation The process of performing calculations and other data transformations related to business transactions.

data manipulation language (DML) The commands that are used to manipulate the data in a database.

data mart A subset of a data warehouse.

data mining An information-analysis tool that involves the automated discovery of patterns and relationships in a data warehouse.

data model A diagram of data entities and their relationships.

data preparation, or data conversion Ensuring all files and databases are ready to be used with new computer software and systems.

data storage The process of updating one or more databases with new transactions.

data store Representation of a storage location for data.

data warehouse A database that collects business information from many sources in the enterprise, covering all aspects of the company's processes, products, and customers.

database An organized collection of facts and information.

database administrator (DBA) A skilled IS professional who directs all activities related to an organization's database.

database approach to data management An approach whereby a pool of related data is shared by multiple application programs.

database management system (DBMS) A group of programs that manipulate the database and provide an interface between the database and the user of the database and other application programs.

data-flow diagram (DFD) A model of objects, associations, and activities that describes how data can flow between and around various objects.

data-flow line Arrows that show the direction of data element movement.

decentralized processing Processing alternative in which processing devices are placed at remote locations.

decision room A room that supports decision making, with the decision makers in the same building, combining face-to-face verbal interaction with technology to make the meeting more effective and efficient.

decision support system (DSS) An organized collection of people, procedures, software, databases, and devices used to support problem-specific decision making.

decision-making phase The first part of problem solving, including three stages: intelligence, design, and choice.

delphi approach A decision-making approach in which group decision makers are geographically dispersed; this approach encourages diversity among group members and fosters creativity and original thinking in decision making.

demand report A report developed to give certain information at someone's request.

design report The primary result of systems design, reflecting the decisions made and preparing the way for systems implementation.

design stage The second stage of decision making, in which alternative solutions to the problem are developed.

desktop computer A relatively small, inexpensive single-user computer that is highly versatile.

dialogue manager A user interface that allows decision makers to easily access and manipulate the DSS and to use common business terms and phrases.

digital audio player A device that can store, organize, and play digital music files.

digital camera Input device used with a PC to record and store images and video in digital form.

digital certificate An attachment to an e-mail message or data embedded in a Web site that verifies the identity of a sender or Web site.

digital subscriber line (DSL) A telecommunications service that delivers high-speed Internet access to homes and small businesses over the existing phone lines of the local telephone network.

digital video disc (DVD) A storage medium used to store digital video or computer data.

direct access Retrieval method in which data can be retrieved without the need to read and discard other data.

direct access storage device (DASD) Device used for direct access of secondary storage data.

direct conversion (also called *plunge* or *direct cutover*) Stopping the old system and starting the new system on a given date.

direct observation Watching the existing system in action by one or more members of the analysis team.

disaster recovery plan (DRP) A formal plan describing the actions that must be taken to restore computer operations and services in the event of a disaster.

distributed database A database in which the data can be spread across several smaller databases connected via telecommunications devices.

distributed processing Processing alternative in which computers are placed at remote locations but are connected to each other via a network.

document production The process of generating output records and reports.

documentation Text that describes the program functions to help the user operate the computer system.

domain The allowable values for data attributes. Also, the area of knowledge addressed by an expert system.

domain expert The person or group who has the expertise or knowledge the expert system is trying to capture.

drill-down report A report providing increasingly detailed data about a situation.

dumpster diving Going through the trash cans of an organization to find secret or confidential information, including information needed to access an information system or its data.

dynamic Web pages Web pages containing variable information that are built to respond to a specific Web visitor's request.

economic feasibility The determination of whether the project makes financial sense and whether predicted benefits offset the cost and time needed to obtain them.

economic order quantity (EOQ) The quantity that should be reordered to minimize total inventory costs.

e-Government The use of information and communications technology to simplify the sharing of information, speed formerly paper-based processes, and improve the relationship between citizen and government.

electronic bill presentment A method of billing whereby a vendor posts an image of your statement on the Internet and alerts you by e-mail that your bill has arrived.

electronic business (e-business) Using information systems and the Internet to perform all business-related tasks and functions.

electronic cash An amount of money that is computerized, stored, and used as cash for e-commerce transactions.

electronic commerce (e-commerce) Conducting business activities (e.g., distribution, buying, selling, marketing, and servicing of products or services) electronically over computer networks such as the Internet, extranets, and corporate networks.

electronic exchange An electronic forum where manufacturers, suppliers, and competitors buy and sell goods, trade market information, and run back-office operations.

electronic retailing (e-tailing) The direct sale from business to consumer through electronic storefronts, typically designed around an electronic catalog and shopping cart model.

end-user systems development A systems development project in which business managers and users assume the primary effort.

enterprise data modeling Data modeling done at the level of the entire enterprise.

enterprise resource planning (ERP) system A set of integrated programs capable of managing a company's vital business operations for an entire multisite, global organization.

enterprise system A system central to the organization that ensures information can be shared across all business functions and all levels of management to support the running and managing of a business.

entity A generalized class of people, places, or things for which data is collected, stored, and maintained.

entity symbol Representation of either a source or destination of a data element.

entity-relationship (ER) diagrams Data models that use basic graphical symbols to show the organization of and relationships between data.

environmental design Also called *green design*, it involves systems development efforts that slash power consumption, require less physical space, and result in systems that can be disposed in a way that doesn't negatively affect the environment.

ergonomics The science of designing machines, products, and systems to maximize the safety, comfort, and efficiency of the people who use them.

exception report A report automatically produced when a situation is unusual or requires management action.

executive support system (ESS) Specialized DSS that includes all hardware, software, data, procedures, and people used to assist senior-level executives within the organization.

expert system Hardware and software that stores knowledge and makes inferences, similar to a human expert.

explanation facility Component of an expert system that allows a user or decision maker to understand how the expert system arrived at certain conclusions or results.

Extensible Markup Language (XML) The markup language for Web documents containing structured information, including words, pictures, and other elements.

external auditing Auditing performed by an outside group.

extranet A network based on Web technologies that links selected resources of a company's intranet with its customers, suppliers, or other business partners.

feasibility analysis Assessment of the technical, economic, legal, operational, and schedule feasibility of a project.

feedback Output that is used to make changes to input or processing activities.

field Typically a name, number, or combination of characters that describes an aspect of a business object or activity.

file A collection of related records.

File Transfer Protocol (FTP) A protocol that describes a file transfer process between a host and a remote computer and allows users to copy files from one computer to another.

financial MIS An information system that provides financial information not only for executives but also for a broader set of people who need to make better decisions on a daily basis.

five-forces model A widely accepted model that identifies five key factors that can lead to attainment of competitive advantage, including (1) the rivalry among existing competitors, (2) the threat of new entrants, (3) the threat of substitute products and services, (4) the bargaining power of buyers, and (5) the bargaining power of suppliers.

flash memory A silicon computer chip that, unlike RAM, is nonvolatile and keeps its memory when the power is shut off.

flexible manufacturing system (FMS) An approach that allows manufacturing facilities to rapidly and efficiently change from making one product to making another.

forecasting Predicting future events to avoid problems.

forward chaining The process of starting with the facts and working forward to the conclusions.

game theory The use of information systems to develop competitive strategies for people, organizations, or even countries.

genetic algorithm An approach to solving large, complex problems in which a number of related operations or models change and evolve until the best one emerges.

geographic information system (GIS) A computer system capable of assembling, storing, manipulating, and displaying geographic information, that is, data identified according to its location.

graphical user interface (GUI) An interface that uses icons and menus displayed on screen to send commands to the computer system.

grid computing The use of a collection of computers, often owned by multiple individuals or organizations, to work in a coordinated manner to solve a common problem.

group consensus approach A decision-making approach that forces members in the group to reach a unanimous decision.

group support system (GSS) Software application that consists of most elements in a DSS, plus software to provide effective support in group decision making; also called *group decision support system* or *computerized collaborative work system*.

hacker A person who enjoys computer technology and spends time learning and using computer systems.

handheld computer A single-user computer that provides ease of portability because of its small size.

hardware Any machinery (most of which uses digital circuits) that assists in the input, processing, storage, and output activities of an information system.

heuristics Commonly accepted guidelines or procedures that usually find a good solution.

hierarchy of data Bits, characters, fields, records, files, and databases.

highly structured problems Problems that are straightforward and require known facts and relationships.

HTML tags Codes that let the Web browser know how to format text—as a heading, as a list, or as body text—and whether images, sound, and other elements should be inserted.

human resource MIS An information system that is concerned with activities related to employees and potential employees of an organization, also called a personnel MIS.

Hypertext Markup Language (HTML) The standard page description language for Web pages.

identify theft A crime in which an imposter obtains key pieces of personal identification information, such as Social Security or driver's license numbers, to impersonate someone else.

IF-THEN statements Rules that suggest certain conclusions.

implementation stage A stage of problem solving in which a solution is put into effect.

inference engine Part of the expert system that seeks information and relationships from the knowledge base and provides answers, predictions, and suggestions similar to the way a human expert would.

informatics A specialized system that combines traditional disciplines, such as science and medicine, with computer systems and technology.

information A collection of facts organized in such a way that they have additional value beyond the value of the facts themselves.

information center A support function that provides users with assistance, training, application development, documentation, equipment selection and setup, standards, technical assistance, and troubleshooting.

information service unit A miniature IS department.

information system (IS) A set of interrelated components that collect, manipulate, store, and disseminate data and information and provide a feedback mechanism to meet an objective.

information systems planning Translating strategic and organizational goals into systems development initiatives.

input The activity of gathering and capturing raw data.

insider An employee, disgruntled or otherwise, working solo or in concert with outsiders to compromise corporate systems.

installation The process of physically placing the computer equipment on the site and making it operational.

instant messaging A method that allows two or more people to communicate online using the Internet.

institutional DSS A DSS that handles situations or decisions that occur more than once, usually several times per year or more. An institutional DSS is used repeatedly and refined over the years.

intelligence stage The first stage of decision making, in which potential problems or opportunities are identified and defined.

intelligent agent Programs and a knowledge base used to perform a specific task for a person, a process, or another program; also called *intelligent robot* or *bot*.

intelligent behavior The ability to learn from experiences and apply knowledge acquired from experience, handle complex situations, solve problems when important information is missing, determine what is important, react quickly and correctly to a new situation, understand visual images, process and manipulate symbols, be creative and imaginative, and use heuristics.

internal auditing Auditing performed by individuals within the organization.

Internet The world's largest computer network, actually consisting of thousands of interconnected networks, all freely exchanging information.

Internet Protocol (IP) A communication standard that enables traffic to be routed from one network to another as needed.

Internet service provider (ISP) Any company that provides Internet access to people or organizations.

intranet An internal network based on Web technologies that allows people within an organization to exchange information and work on projects.

intrusion detection system (IDS) Software that monitors system and network resources and notifies network security personnel when it senses a possible intrusion.

Java An object-oriented programming language from Sun Microsystems based on C++ that allows small programs (applets) to be embedded within an HTML document.

joining Manipulating data to combine two or more tables.

joint application development (JAD) A process for data collection and requirements analysis in which users, stakeholders, and IS professionals work together to analyze existing systems, propose possible solutions, and define the requirements of a new or modified system.

just-in-time (JIT) inventory A philosophy of inventory management in which inventory and materials are delivered just before they are used in manufacturing a product.

key A field or set of fields in a record that is used to identify the record.

key-indicator report A summary of the previous day's critical activities; typically available at the beginning of each workday.

knowledge The awareness and understanding of a set of information and ways that information can be made useful to support a specific task or reach a decision.

knowledge acquisition facility Part of the expert system that provides convenient and efficient means of capturing and storing all the components of the knowledge base.

knowledge base A component of an expert system that stores all relevant information, data, rules, cases, and relationships used by the expert system.

knowledge engineer A person who has training or experience in the design, development, implementation, and maintenance of an expert system.

knowledge user The person or group who uses and benefits from the expert system.

learning systems A combination of software and hardware that allows the computer to change how it functions or reacts to situations based on feedback it receives.

legal feasibility The determination of whether laws or regulations may prevent or limit a systems development project.

linking Data manipulation that combines two or more tables using common data attributes to form a new table with only the unique data attributes.

local area network (LAN) A network that connects computer systems and devices within a small area like an office, home, or several floors in a building.

logical design A description of the functional requirements of a system.

magnetic disk Common secondary storage medium, with bits represented by magnetized areas.

magnetic tape Secondary storage medium; Mylar film coated with iron oxide with portions of the tape magnetized to represent bits.

mainframe computer Large, powerful computer often shared by hundreds of concurrent users connected to the machine via terminals.

make-or-buy decision The decision regarding whether to obtain the necessary software from internal or external sources.

management information system (MIS) An organized collection of people, procedures, software, databases, and devices that provides routine information to managers and decision makers.

market segmentation The identification of specific markets to target them with advertising messages.

marketing MIS An information system that supports managerial activities in product development, distribution, pricing decisions, and promotional effectiveness.

material requirements planning (MRP) A set of inventory-control techniques that help coordinate thousands of inventory items when the demand for one item is dependent on the demand for another.

metropolitan area network (MAN) A telecommunications network that connects users and their devices in a geographical area that spans a campus or city.

mobile commerce (m-commerce) Transactions conducted anywhere, anytime.

model base Part of a DSS that provides decision makers access to a variety of models and assists them in decision making.

model management software Software that coordinates the use of models in a DSS.

monitoring stage The final stage of the problem-solving process, in which decision makers evaluate the implementation.

MP3 A standard format for compressing a sound sequence into a small file.

multicore microprocessor Microprocessor that combines two or more independent processors into a single computer so they can share the workload and deliver a big boost in processing capacity.

multiprocessing The simultaneous execution of two or more instructions at the same time.

natural language processing Processing that allows the computer to understand and react to statements and commands made in a "natural" language, such as English.

Near Field Communication (NFC) A very short-range wireless connectivity technology designed for cell phones and credit cards.

network operating system (NOS) Systems software that controls the computer systems and devices on a network and allows them to communicate with each other.

network-management software Software that enables a manager on a networked desktop to monitor the use of individual computers and shared hardware (such as printers), scan for viruses, and ensure compliance with software licenses.

networks Computers and equipment that are connected in a building, around the country, or around the world to enable electronic communications.

neural network A computer system that can simulate the functioning of a human brain.

nominal group technique A decision-making approach that encourages feedback from individual group members, and the final decision is made by voting, similar to the way public officials are elected.

nonprogrammed decision A decision that deals with unusual or exceptional situations.

object-oriented database A database that stores both data and its processing instructions.

object-oriented database management system (OODBMS) A group of programs that manipulate an object-oriented database and provide a user interface and connections to other application programs.

object-oriented systems development (OOSD) An approach to systems development that combines the logic of the systems development life cycle with the power of object-oriented modeling and programming.

object-relational database management system (ORDBMS) A DBMS capable of manipulating audio, video, and graphical data.

off-the-shelf software An existing software program that can be purchased.

online analytical processing (OLAP) Software that allows users to explore data from a number of perspectives.

online transaction processing (OLTP) A form of data processing where each transaction is processed immediately, without the delay of accumulating transactions into a batch.

operational feasibility The measure of whether the project can be put into action or operation.

optimization model A process to find the best solution, usually the one that will best help the organization meet its goals.

organization A formal collection of people and other resources established to accomplish a set of goals.

organizational change The responses that are necessary so that for-profit and nonprofit organizations can plan for, implement, and handle change.

organizational culture The major understandings and assumptions for a business, corporation, or other organization.

output Production of useful information, usually in the form of documents and reports.

parallel processing The simultaneous execution of the same task on multiple processors in order to obtain results faster.

parallel start-up Running both the old and new systems for a period of time and comparing the output of the new system closely with the output of the old system; any differences are reconciled. When users are comfortable that the new system is working correctly, the old system is eliminated.

password sniffer A small program hidden in a network or a computer system that records identification numbers and passwords.

perceptive system A system that approximates the way a person sees, hears, and feels objects.

personal area network (PAN) A network that supports the interconnection of information technology within a range of 33 feet or so.

phase-in approach (also called *piecemeal approach***)** Slowly replacing components of the old system with those of the new one. This process is repeated for each application until the new system is running every application and performing as expected; also called a *piecemeal approach*.

physical design The specification of the characteristics of the system components necessary to put the logical design into action.

pilot start-up Running the new system for one group of users rather than all users.

planned data redundancy A way of organizing data in which the logical database design is altered so that certain data entities are combined, summary totals are carried in the data records rather than calculated from elemental data, and some data attributes are repeated in more than one data entity to improve database performance.

Platform for Privacy Preferences (P3P) A screening technology that shields users from Web sites that don't provide the level of privacy protection they desire.

portable computer Computer small enough to be carried easily.

predictive analysis A form of data mining that combines historical data with assumptions about future conditions to predict outcomes of events, such as future product sales or the probability that a customer will default on a loan.

primary key A field or set of fields that uniquely identifies the record.

problem solving A process that goes beyond decision making to include the implementation stage.

procedures The strategies, policies, methods, and rules for using a CBIS.

process A set of logically related tasks performed to achieve a defined outcome.

process symbol Representation of a function that is performed.

processing Converting or transforming data into useful outputs.

productivity A measure of the output achieved divided by the input required.

profit center A department within an organization that focuses on generating profits.

programmed decision A decision made using a rule, procedure, or quantitative method.

programmer A specialist responsible for modifying or developing programs to satisfy user requirements.

programming languages Sets of keywords, symbols, and a system of rules for constructing statements by which humans can communicate instructions to be executed by a computer.

projecting Manipulating data to eliminate columns in a table.

proprietary software One-of-a-kind program developed for a specific application.

quality control A process that ensures that the finished product meets the customers' needs.

questionnaires A method of gathering data where the data sources are spread over a wide geographic area.

Radio Frequency Identification (RFID) A technology that employs a microchip with an antenna that broadcasts its unique identifier and location to receivers.

random access memory (RAM) A form of memory in which instructions or data can be temporarily stored.

rapid application development (RAD) A systems development approach that employs tools, techniques, and methodologies designed to speed application development.

read-only memory (ROM) A nonvolatile form of memory.

record A collection of related data fields.

redundant array of independent/ inexpensive disks (RAID) Method of storing data that generates extra bits of data from existing data, allowing the system to create a "reconstruction map" so that if a hard drive fails, the system can rebuild lost data.

relational model A database model that describes data in which all data elements are placed in two-dimensional tables, called *relations*, which are the logical equivalent of files.

reorder point (ROP) A critical inventory quantity that determines when to order more inventory.

replicated database A database that holds a duplicate set of frequently used data.

request for proposal (RFP) A document that specifies in detail required resources such as hardware and software.

requirements analysis The determination of user, stakeholder, and organizational needs.

return on investment (ROI) One measure of IS value that investigates the additional profits or benefits that are generated as a percentage of the investment in IS technology.

revenue center A division within a company that generates sales or revenues.

rich Internet application Software that has the functionality and complexity of traditional application software, but does not require local installation and runs in a Web browser.

robotics Mechanical or computer devices that perform tasks requiring a high degree of precision or that are tedious or hazardous for humans.

rule A conditional statement that links conditions to actions or outcomes.

satisficing model A model that will find a good—but not necessarily the best—problem solution.

scalability The ability to increase the capability of a computer system to process more transactions in a given period by adding more, or more powerful, processors.

schedule feasibility The determination of whether the project can be completed in a reasonable amount of time.

scheduled report A report produced periodically, or on a schedule, such as daily, weekly, or monthly.

schema A description of the entire database.

script bunny A cracker with little technical savvy who downloads programs called scripts, which automate the job of breaking into computers.

search engine A valuable tool that enables you to find information on the Web by specifying words that are key to a topic of interest, known as keywords.

Secure Sockets Layer (SSL) A communications protocol is used to secure sensitive data during e-commerce.

security dashboard Software that provides a comprehensive display on a single computer screen of all the vital data related to an organization's security defenses including threats, exposures, policy compliance and incident alerts.

selecting Manipulating data to eliminate rows according to certain criteria.

semistructured or unstructured problems More complex problems in which the relationships among the pieces of data are not always clear, the data might be in a variety of formats, and the data is often difficult to manipulate or obtain.

sequential access Retrieval method in which data must be accessed in the order in which it is stored.

sequential access storage device (SASD) Device used to sequentially access secondary storage data.

server A computer designed for a specific task, such as network or Internet applications.

site preparation Preparation of the location of a new system.

smart card A credit card–sized device with an embedded microchip to provide electronic memory and processing capability.

social engineering Using social skills to get computer users to provide information to access an information system or its data.

software The computer programs that govern the operation of the computer.

software as a service (SaaS) A service that allows businesses to subscribe to Web-delivered business application software by paying a monthly service charge or a per-use fee.

software piracy The act of unauthorized copying or distribution of copyrighted software

software suite A collection of single application programs packaged in a bundle.

speech-recognition technology Enables a computer equipped with a source of speech input such as a microphone to interpret human speech as an alternative means of providing data or instructions to the computer.

spyware Software that is installed on a personal computer to intercept or take partial control over the user's interaction with the computer without knowledge or permission of the user.

stakeholders People who, either themselves or through the organization they represent, ultimately benefit from the systems development project.

start-up (also called *cutover*) The process of making the final tested information system fully operational.

static Web pages Web pages that always contain the same information.

steering committee An advisory group consisting of senior management and users from the IS department and other functional areas.

storage area network (SAN) Technology that provides high-speed connections between data-storage devices and computers over a network.

strategic alliance (strategic partnership) An agreement between two or more companies that involves the joint production and distribution of goods and services.

strategic planning Determining long-term objectives by analyzing the strengths and weaknesses of the organization, predicting future trends, and projecting the development of new product lines.

structured interview An interview where the questions are written in advance.

supercomputers The most powerful computer systems, with the fastest processing speeds.

system performance measurement Monitoring the system—the number of errors encountered, the amount of memory required, the amount of processing or CPU time needed, and other problems.

system performance products Software that measures all components of the computer-based information system, including hardware, software, database, telecommunications, and network systems.

systems analysis The systems development phase that determines what the information system must do to solve the problem by studying existing systems and work processes to identify strengths, weaknesses, and opportunities for improvement.

systems analyst A professional who specializes in analyzing and designing business systems.

systems design The systems development phase that defines how the information system will do what it must do to obtain the problem solution.

systems development The activity of creating or modifying business systems.

systems implementation The systems development phase involving the creation or acquiring of various system components detailed in the systems design, assembling them, and placing the new or modified system into operation.

systems investigation The systems development phase during which problems and opportunities are identified and considered in light of the goals of the business.

systems investigation report A summary of the results of the systems investigation and the process of feasibility analysis and recommendation of a course of action.

systems maintenance A stage of systems development that involves checking, changing, and enhancing the system to make it more useful in achieving user and organizational goals.

systems operation Use of a new or modified system.

systems request form A document filled out by someone who wants the IS department to initiate systems investigation.

systems review The final step of systems development, involving the analysis of systems to make sure that they are operating as intended.

systems software The set of programs designed to coordinate the activities and functions of the hardware and various programs throughout the computer system.

technical feasibility Assessment of whether the hardware, software, and other system components can be acquired or developed to solve the problem.

technology acceptance model (TAM) A model that describes the factors that lead to higher levels of acceptance and usage of technology.

technology diffusion A measure of how widely technology is spread throughout the organization.

technology infrastructure All the hardware, software, databases, telecommunications, people, and procedures that are configured to collect, manipulate, store, and process data into information.

technology infusion The extent to which technology is deeply integrated into an area or department.

technology-enabled relationship management Occurs when a firm obtains detailed information about a customer's behavior, preferences, needs, and buying patterns and uses that information to set prices, negotiate terms, tailor promotions, add product features, and otherwise customize its entire relationship with that customer.

telecommunications The electronic transmission of signals for communications, which enables organizations to carry out their processes and tasks through effective computer networks.

telecommunications medium Anything that carries an electronic signal and serves as an interface between a sending device and a receiving device.

thin client A low-cost, centrally managed computer with essential but limited capabilities and no extra drives, such as a CD or DVD drive, or expansion slots.

total cost of ownership (TCO) The measurement of the total cost of owning computer equipment, including desktop computers, networks, and large computers.

traditional approach to data management An approach whereby separate data files are created and stored for each application program.

transaction Any business-related exchange, such as payments to employees, sales to customers, and payments to suppliers.

transaction processing cycle The process of data collection, data editing, data correction, data manipulation, data storage, and document production.

transaction processing system (TPS) An organized collection of people, procedures, software, databases, and devices used to record completed business transactions.

transaction processing system audit A check of a firm's TPS systems to prevent accounting irregularities and/or loss of data privacy.

Transmission Control Protocol (TCP) The widely used Transport-layer protocol that most Internet applications use with IP.

Trojan horse A malicious program that disguises itself as a useful application or game and purposefully does something the user does not expect.

tunneling The process by which VPNs transfer information by encapsulating traffic in IP packets over the Internet.

ultra wideband (UWB) A wireless communications technology that transmits large amounts of digital data over short distances of up to 30 feet using a wide spectrum of frequency bands and very low power.

Uniform Resource Locator (URL) An assigned address on the Internet for each computer.

unstructured interview An interview where the questions are not written in advance.

user acceptance document A formal agreement signed by the user that states that a phase of the installation or the complete system is approved.

user interface Element of the operating system that allows you to access and command the computer system.

user preparation The process of readying managers, decision makers, employees, other users, and stakeholders for new systems.

users People who will interact with the system regularly.

utility programs Programs that help to perform maintenance or correct problems with a computer system.

value chain A series (chain) of activities that includes inbound logistics, warehouse and storage, production, finished product storage, outbound logistics, marketing and sales, and customer service.

virtual private network (VPN) A secure connection between two points on the Internet.

virtual reality The simulation of a real or imagined environment that can be experienced visually in three dimensions.

virtual reality system A system that enables one or more users to move and react in a computer-simulated environment.

virtual workgroups Teams of people located around the world working on common problems.

virus A computer program file capable of attaching to disks or other files and replicating itself repeatedly, typically without the user's knowledge or permission.

vision systems The hardware and software that permit computers to capture, store, and manipulate visual images.

Voice over Internet Protocol (VoIP) A collection of technologies and communications protocols that enables your voice to be converted into packets of data that can be sent over a data network such as the Internet, a WAN or LAN.

Web 2.0 The Web as a computing platform that supports software applications and the sharing of information between users.

Web auction An Internet site that matches buyers and sellers.

Web browser Web client software, like Internet Explorer, Firefox, and Safari, used to view Web pages.

Web log (blog) A Web site that people can create and use to write about their observations, experiences, and feelings on a wide range of topics.

Web page construction software Software that uses Web editors and extensions to produce both static and dynamic Web pages.

Web services Standards and tools that streamline and simplify communication among Web sites for business and personal purposes.

Web site development tools Tools used to develop a Web site, including HTML or visual Web page editor, software development kits, and Web page upload support.

wide area network (WAN) A telecommunications network that ties together large geographic regions.

Wi-Fi A wireless telecommunication technology brand owned by the Wi-Fi Alliance that enables wireless devices to connect to a wireless access pint which has a wired connection to the Internet.

wireless mesh A way to route communications between network nodes (computers or other devices) by allowing for continuous connections and reconfiguration around blocked paths by "hopping" from node to node until a connection can be established.

workgroup application software Software that supports teamwork, whether in one location or around the world.

workstation A more powerful personal computer that is used for technical computing, such as engineering, but still fits on a desktop.

World Wide Web A collection of tens of millions of server computers that work together as one in an Internet service using hyperlink technology to provide information to billions of users.

Worldwide Interoperability for Microwave Access (WiMAX) The common name for a set of IEEE 802.16 wireless metropolitan-area network standards that support different types of communications access.

worm A parasitic computer program that can create copies of itself on the infected computer or send copies to other computers via a network.

Subject

A boldface page number indicates a key term and the location of its definition in the text.

P

P3P (Platform for Privacy Preferences), **407**
PA74 (Privacy Act of 1974), 408–409
packets, data, 169
Palm OS, 88
PANs (personal area networks), **162**
parallel communication, 277
parallel processing, **64**
parallel start-up (in systems implementation), **367**
password sniffers, **394**
patent violations, 395–397
payments
    See also banking, online banking
    using cell phones, 216–217
PayPal, 14–15, 216, 381
payroll systems
    feedback in, 9
    and transaction processing systems (TPSs), 16–17
PDF files, and knowledge repositories, 297
pen input devices, 72
people and information systems, 13
perceptive system, **301**
performance-based information systems, 35–37
permanent storage, 66
personal area networks (PANs), **162**
personal computers
    See also computers
    input devices, 69–70
    safe disposal of, 395
personal information managers (PIMs), 95–96
personal productivity software, 80, 92–94, 103
personal sphere of influence, 80
personnel hiring
    See also careers
    in systems development, 365
phase-in approach (to systems implementation), 367
phishing, 397
phones
    cell. *See* cell phones
    iPhones, 171–172
PHP (Hypertext Preprocessor), 177
physical design
    in data modeling, 118
    in systems design, **357**
piecemeal approach (in systems implementation), 367
pilot start-up (in systems implementation), **367**
PIMs (personal information managers), 95–96
piracy, software, **395**, 417
pixels and display monitors, 72
place shifting, 13
planned data redundancy, **118**
planning
    disaster recovery plans (DRPs), **223–224**
    enterprise resource planning. *See* ERP
    human resource, 267
    IS, 354
Platform for Privacy Preferences (P3P), **407**
plotters as output devices, 74
plug-ins, Web, 174
plunge (in systems implementation), 367
Pocket PCs, 76
podcasting, 13, 182–183

policies
    computer-use, 384–386
    corporate privacy, 412, 418
portable computers, 77
portals, Web, 153, **174–175**
Porter, Michael, 25, 30
Pratt & Whitney's KMS, 295
prediction, and forecasting, 9
predictive analysis, **136**, 144
preliminary evaluation, systems development, 361
presentation graphics programs, 96
preventing
    computer crime, 398–405
    computer-related mistakes, 384–386
    computer-related waste, 383
price comparisons, mobile, 210
primary keys (data management), **114**
printers, 74
privacy
    and federal government, 405–406
    Finland's information system and, 411
    and Internet, 189, 407–408
    issues overview, 418
    legislation regarding, 408–410
    at work, 406–407
Privacy Act of 1974 (PA74), 408–409
problem solving
    with artificial intelligence, 300
    and decision making, 246–251, 283
    described, **247**
    in DSSs, 270–271
procedures
    described, **13**
    preventing computer waste, mistakes, 384–386
process symbol (in DFDs), **354**
processes described, **6**
processing
    described, **8**, 83
    distributed, 139–140
    multicore, **64**
processing devices, 62–63
processors, 10, 84–85
procurement, electronic, 15–16
product configuration software, 213
product development MIS, 264
product differentiation, 32
production and supply chain management, 231
productivity described, **35–36**
productivity paradox, 48
profit, and IS value, 36
profit center, **258**
program trading systems, 9
programmed decisions, **248**
programmers, **338**
programming, extreme (XP), 343, 344
programming languages
    See also specific language
    described, **100**, 104
    for expert systems, 314
    Web, 177
programs
    computer, **79**
    utility, **89**

Company Names

Numbers